Climbing and Walking Robots

Edited by
Behnam Miripour

Climbing and Walking Robots
Edited by Behnam Miripour

Republished by InTech
Janeza Trdine 9, 51000 Rijeka, Croatia

Additional hard copies can be obtained from orders@intechopen.com

Climbing and Walking Robots,
Edited by Behnam Miripour
p. cm.
ISBN 978-953-307-030-8

For EU product safety concerns:
IN TECH d.o.o., Prolaz Marije Krucifikse Kozulić 3, 51000 Rijeka, Croatia,
info@intechopen.com or visit our website at intechopen.com.

Meet the editor

Behnam Miripour Fard received his B.Sc. and Ph.D. degrees in Mechanical Engineering from the University of Guilan, Iran, in 2005 and 2013, respectively. He also received his M. Sc. degree in Mechanical Engineering from Bu-Ali Sina University, Iran, in 2008. From Apr. 2012 to Oct. 2012, he was a Visiting PhD Student at Robotics Institute of Carnegie Mellon University, USA, under supervision of Professor C. G. Atkeson. He is currently an Assistant Professor and the Head of the Robotics Eng. Dept. of the Hamedan University of Technology. Dr. Behnam Miripour Fard is the author/coauthor of over 25 technical publications, proceedings, editorials and books. He has been a Member of Iranian Elite Foundation and Inventors Association since 2007. He is also a reviewer of the Journal of Robotica and International Journal of Engineering. His research interests include; Dynamical modeling of Mechanical systems, Path Planning and Control of Humanoid Biped Robots, Limit Cycle Walking, Disturbance Rejection of Legged Locomotion, Predictive Dynamics, Optimal Control, Hybrid Dynamical Modeling, Receding Horizon control, and Bio-mimetic approaches in robotics.

Preface

Nowadays robotics is one of the most dynamic fields of scientific researches. The shift of robotics researches from manufacturing to services applications is clear. During the last decades interest in studying climbing and walking robots has been increased. This increasing interest has been in many areas that most important ones of them are: mechanics, electronics, medical engineering, cybernetics, controls, and computers. Today's climbing and walking robots are a combination of manipulative, perceptive, communicative, and cognitive abilities and they are capable of performing many tasks in industrial and non-industrial environments. Surveillance, planetary exploration, emergence rescue operations, reconnaissance, petrochemical applications, construction, entertainment, personal services, intervention in severe environments, transportation, medical and etc are some applications from a very diverse application fields of climbing and walking robots. By great progress in this area of robotics it is anticipated that next generation climbing and walking robots will enhance lives and will change the way the human works, thinks and makes decisions.

This book presents the state of the art achievments, recent developments, applications and future challenges of climbing and walking robots. These are presented in 26 chapters by authors throughtot the world. The book serves as a reference especially for the researchers who are interested in mobile robots. It also is useful for industrial engineers and graduate students in advanced study.

Editor

Behnam Miripour

Contents

1

A Survey of Technologies and Applications for Climbing Robots Locomotion and Adhesion

Manuel F. Silva and J. A. Tenreiro Machado
ISEP - Instituto Superior de Engenharia do Porto
Portugal

1. Introduction

The interest in the development of climbing robots has grown rapidly in the last years. Climbing robots are useful devices that can be adopted in a variety of applications, such as maintenance and inspection in the process and construction industries. These systems are mainly adopted in places where direct access by a human operator is very expensive, because of the need for scaffolding, or very dangerous, due to the presence of an hostile environment. The main motivations are to increase the operation efficiency, by eliminating the costly assembly of scaffolding, or to protect human health and safety in hazardous tasks. Several climbing robots have already been developed, and other are under development, for applications ranging from cleaning to inspection of difficult to reach constructions.

A wall climbing robot should not only be light, but also have large payload, so that it may reduce excessive adhesion forces and carry instrumentations during navigation. These machines should be capable of travelling over different types of surfaces, with different inclinations, such as floors, walls, or ceilings, and to walk between such surfaces (Elliot et al. (2006); Sattar et al. (2002)). Furthermore, they should be able of adapting and reconfiguring for various environment conditions and to be self-contained.

Up to now, considerable research was devoted to these machines and various types of experimental models were already proposed (according to Chen et al. (2006), over 200 prototypes aimed at such applications had been developed in the world by the year 2006). However, we have to notice that the application of climbing robots is still limited. Apart from a couple successful industrialized products, most are only prototypes and few of them can be found in common use due to unsatisfactory performance in on-site tests (regarding aspects such as their speed, cost and reliability). Chen et al. (2006) present the main design problems affecting the system performance of climbing robots and also suggest solutions to these problems.

The major two issues in the design of wall climbing robots are their locomotion and adhesion methods.

With respect to the locomotion type, four types are often considered: the crawler, the wheeled, the legged and the propulsion robots. Although the crawler type is able to move relatively faster, it is not adequate to be applied in rough environments. On the other hand, the legged type easily copes with obstacles found in the environment, whereas generally its speed is lower and requires complex control systems.

Regarding the adhesion to the surface, the robots should be able to produce a secure gripping force using a light-weight mechanism. The adhesion method is generally classified into four

groups: suction force, magnetic, gripping to the surface and thrust force type. Nevertheless, recently new methods for assuring the adhesion, based in biological findings, were proposed. The vacuum type principle is light and easy to control though it presents the problem of supplying compressed air. An alternative, with costs in terms of weight, is the adoption of a vacuum pump. The magnetic type principle implies heavy actuators and is used only for ferromagnetic surfaces. The thrust force type robots make use of the forces developed by thrusters to adhere to the surfaces, but are used in very restricted and specific applications.
Bearing these facts in mind, this chapter presents a survey of different applications and technologies adopted for the implementation of climbing robots locomotion and adhesion to surfaces, focusing on the new technologies that are recently being developed to fulfill these objectives. The chapter is organized as follows. Section two presents several applications of climbing robots. Sections three and four present the main locomotion principles, and the main "conventional" technologies for adhering to surfaces, respectively. Section five describes recent biological inspired technologies for robot adhesion to surfaces. Section six introduces several new architectures for climbing robots. Finally, section seven outlines the main conclusions.

2. Climbing Robots Applications

Climbing robots are mainly adopted in places where direct access by a human operator is very expensive, because of the need for scaffolding, or very dangerous, due to the presence of an hostile environment.
In the last decades different applications have been envisioned for these robots, mainly in the areas of cleaning, technical inspection, maintenance or breakdown diagnosis in dangerous environments, or in the outside of tall buildings and human made constructions.
Several climbing robots have already been developed for the following application areas:

- **Inspection:** bridges (Balaguer et al. (2005); Robert T. Pack and Kawamura (1997)), nuclear power plants (Savall et al. (1999); Yan et al. (1999)), pipelines (Park et al. (2003)), wind turbines (Rodriguez et al. (2008)), solar power plants (Azaiz (2008)), for scanning the external and internal surfaces of gas or oil tanks (Longo and Muscato (2004b); Park et al. (2003); Sattar et al. (2002); Yan et al. (1999)), offshore platforms (Balaguer et al. (2005)), and container ships (Mondal et al. (2002));
- **Testing:** performing non-destructive tests in industrial structures (Choi et al. (2000); Kang et al. (2003)), floating production storage oil tanks (Sattar et al. (2008; 2006)), planes (Backes et al. (1997); Chen et al. (2005); Robert T. Pack and Kawamura (1997)) and ships (Armada et al. (2005); Robert T. Pack and Kawamura (1997); Sánchez et al. (2006));
- **Civil construction:** civil construction repair and maintenance (Balaguer et al. (2005));
- **Cleaning:** cleaning operations in sky-scrapers (Derriche and Kouiss (2002); Elkmann et al. (2002); Gao and Kikuchi (2004); Yan et al. (1999); Zhang et al. (2004); Zhu et al. (2003)), for cleaning the walls and ceilings of restaurants, community kitchens and food preparation industrial environments (Cepolina et al. (2004)) and cleaning ship hulls (Fernández et al. (2002));
- **Transport:** for the transport of loads inside buildings (Minor et al. (2000));
- **Security:** for reconnaissance in urban environments (Elliot et al. (2006); Tummala et al. (2002)) and in anti-terrorist activities (Li et al. (2007)).

Finally, their application has also been proposed in the education (Bell and Balkcom (2006); Berns et al. (2005)) and human care (Balaguer et al. (2005)) areas and in the prevention and fire fighting actions (Chen et al. (2006); Nishi (1991)).

3. Principles of Locomotion

In this section are analyzed the characteristics of the four main locomotion technologies implemented in climbing robots, namely the crawler, wheeled, legged and propulsion types.

3.1 Locomotion using Sliding Segments (Crawling)

With respect to the locomotion type, the simpler alternatives often make use of sliding segments, with suction cups (Backes et al. (1997); Cepolina et al. (2004); Choi et al. (2000); Elkmann et al. (2002); Savall et al. (1999); Zhang et al. (2004); Zhu et al. (2003)) or permanent magnets (Yan et al. (1999)) that grab to surfaces, in order to move (Figure 1). The main disadvantage of this solution is the difficulty in crossing cracks and obstacles.

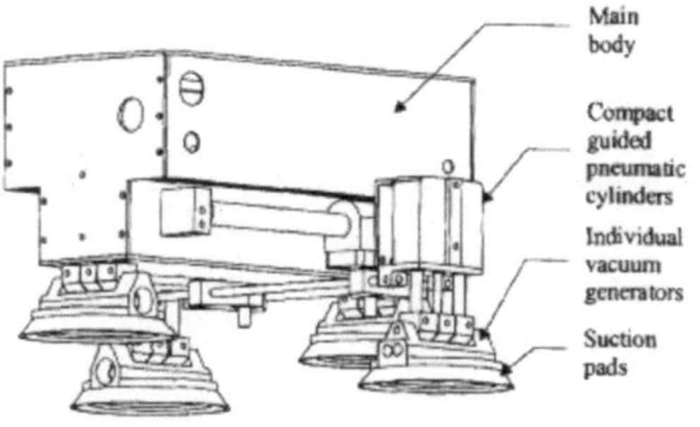

Fig. 1. ROBICEN III climbing robot (Savall et al. (1999))

3.2 Locomotion using Wheels

A second form of locomotion is to adopt wheels (Gao and Kikuchi (2004); Longo and Muscato (2004b); Park et al. (2003); Sánchez et al. (2006); Yan et al. (1999)) (Figure 2). These robots can achieve high velocities. However, some of the wheeled robots that use the suction force for adhesion to the surface, need to maintain an air gap between the surface where they are moving over and the robot base. This technique may create problems either with the loss of pressure, or with the friction with the surface, namely if the air gap is too small, or if some material is used to prevent the air leak (Hirose et al. (1991)).

3.3 Locomotion using Legs

A third form of locomotion consists in the adoption of legs. Legged climbing robots, equipped with suction cups, or magnetic devices on the feet, have the disadvantage of low speed and require complex control systems, but allow the creation of a strong and stable adhesion force to the surface. These machines also have the advantage of easily coping with obstacles or cracks found in the environment (Hirose et al. (1991)). Structures having from two up to eight legs are predominant for the development of these tasks. The adoption of a larger number of limbs supplies redundant support and, frequently, raises the payload capacity and safety. These advantages are achieved at the cost of increased control complexity (regarding leg coordination), size and weight. Therefore, when size and efficiency are critical, a structure with minimum

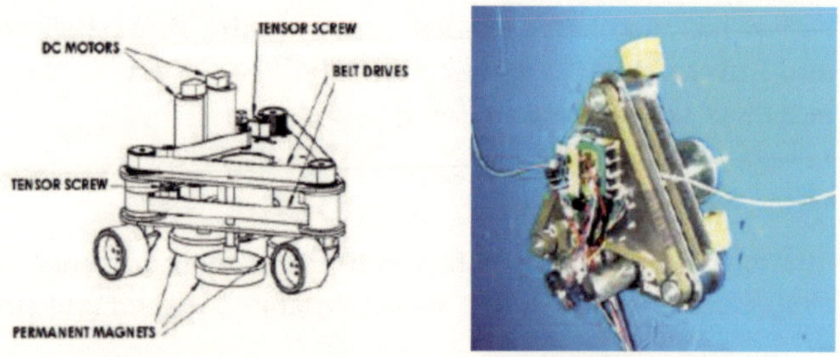

Fig. 2. CAD representation of an wheeled climbing robot (left) and its real aspect (right) (Sánchez et al. (2006))

weight and complexity is more adequate. For these reasons the biped structure is an excellent candidate (Figure 3). Presently there are many biped robots with the ability of climbing over surfaces with different slopes (Armada et al. (2005); Balaguer et al. (2005); Brockmann (2006); Krosuri and Minor (2003); Resino et al. (2006); Robert T. Pack and Kawamura (1997); Shores and Minor (2005); Tummala et al. (2002); Xiao et al. (2003; 2004)).

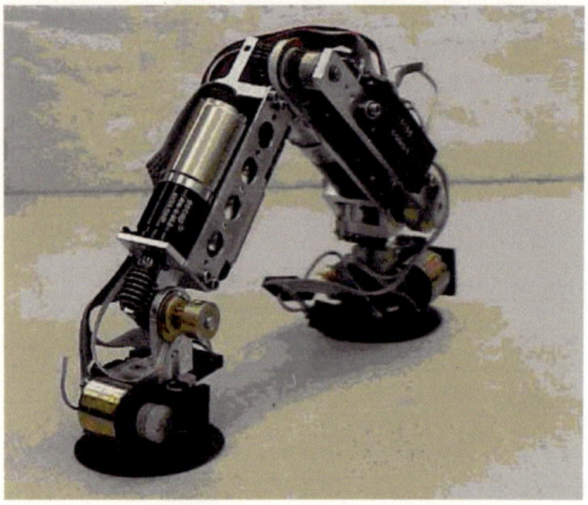

Fig. 3. RAMR1 biped climbing robot (Tummala et al. (2002))

When is needed an increased safety or payload capability are adopted quadrupeds (Armada et al. (2005); Daltorio et al. (2005); Hirose and Arikawa (2000); Hirose et al. (1991); Kang et al. (2003); Kennedy et al. (2006)) (such as MRWALLSPECT III, presented in Figure 4), or robots with a larger number of legs (Armada et al. (2005); Inoue et al. (2006); Li et al. (2007)). The control and leg coordination of these larger robots is, however, more complicated.

3.4 Locomotion through Propulsion

The propulsion type robots make use of the forces developed by propellers to move and to adhere to the surfaces (Nishi (1991)), but are used in very restricted and specific applications. Nishi (1991) developed a climbing robot using the thrust force of propellers to locomote (Figure 5). The contact between the robot and the surface is maintained though a large number of non-actuated wheels. The thrust force is inclined to the wall side to produce the frictional force between the wheels and the surface. Since strong wind is predicted on the wall surfaces

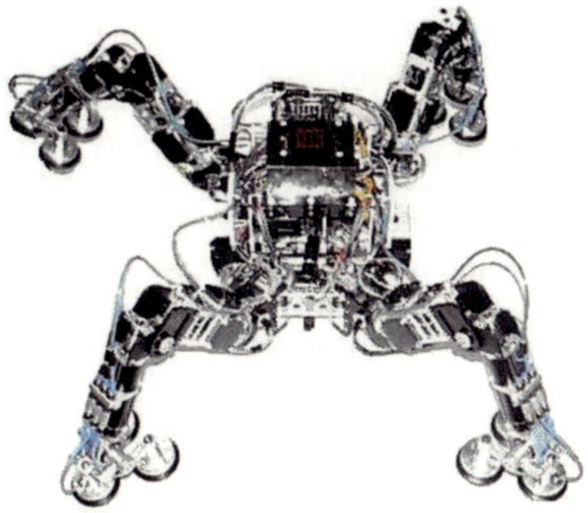

Fig. 4. MRWALLSPECT III quadruped climbing robot (Kang et al. (2003))

of high buildings, the direction of thrust force is controlled to compensate the wind force acting on the robot. A frictional force augmentor is also considered, which is an airfoil to produce the lift force directed to the wall side by the cross wind. Nevertheless, is has been shown that slipping of this robot occurs for abrupt changes in the wind direction or speed.

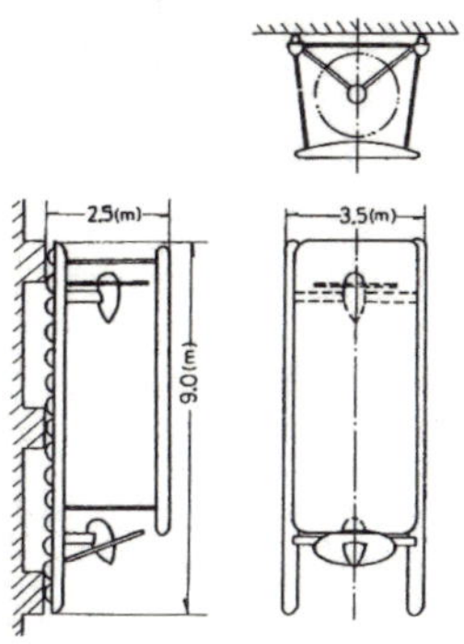

Fig. 5. A conceptual model of a propeller based wall climbing robot (Nishi (1991))

4. Technologies for Adhering to Surfaces

The most important work in developing a climbing robot project is to design a proper adhesion mechanism to ensure that the robot sticks to various wall surfaces reliably without sacrificing mobility (Elliot et al. (2006)).
In this section are reviewed the main aspects of the four adhesion methods usually adopted in climbing robots: suction force, magnetic, gripping to the surface and thrust force type. The next section will review in some depth the new methods for assuring the adhesion, based in biological findings.

4.1 Suction Force

The most frequent approach to guarantee the robot adhesion to a surface is to use the suction force. The vacuum type principle requires light mechanisms and is easy to control. This oper-

ating principle allows climbing over arbitrarily surfaces, made of distinct types of materials, and can be implemented by using different strategies. Usually, more than one vacuum cup is used in each feet in order to prevent loss of pressure (and adhesion force) due to surface curvature or irregularities (Chen et al. (2006); Hirose et al. (1991)). Nevertheless, this type of attachment has some associated drawbacks. The suction adhesion mechanism requires time to develop enough vacuum to generate sufficient adhesion force. This delay may reduce the speed at which the robot can locomote. Another issue associated with suction adhesion is that any gap in the seal can cause the robot to fall. This drawback limits the suction cup adhesion mechanism to relatively smooth, nonporous and non-cracked surfaces. Finally, the suction adhesion mechanism relies on the ambient pressure to stick to a wall and, therefore, is not useful in space applications, because the ambient pressure in space is essentially zero (Menon et al. (2004)). Another problem is the supply of compressed air. The vacuum can be generated through the Venturi Principle (Balaguer et al. (2005); Choi et al. (2000); Elkmann et al. (2002); Savall et al. (1999); Zhang et al. (2004)), or through a vacuum pump, either on-board the robot (Cepolina et al. (2004); Gao and Kikuchi (2004); Kang et al. (2003); Li et al. (2007); Tummala et al. (2002); Yan et al. (1999)), or external to it (Zhu et al. (2003)).
The RAMR1 is an example of a biped climbing robot, adopting suction cups for the adhesion to the surface, being the vacuum generated through an on-board vacuum pump (Figure 3).
When the vacuum is generated through the Venturi Principle, or through vacuum pumps, it makes climbing robots noisy. A solution for this noise problem has been proposed (Li et al. (2007)). Vacuum pumps on-board the robot increase the weight and the costs of a robot, also due to additional vacuum tubes, muffles, valves, and other necessary equipment. This solution causes some level of steady, not negligible, energy consumption. Vacuum pumps external to the robot imply the need for a tether cable, with the inherent problems of the interference of the umbilical cord for the robot with its mobility and dynamics (Chen et al. (2006)). Hence, it is desirable to avoid an active vacuum generation and a separate installation for vacuum transportation.
Bearing these ideas in mind, Brockmann proposed the use of passive suction cups (see Figure 6) because they are low cost, simple and robust and allow a light-weight construction of climbing robots. However, although being a promising approach, in order to construct a proper system, several aspects related to the behavior of passive suction cups have to be better understood (Brockmann (2006)).

Fig. 6. Passive suction cups with (left) and without (right) a strap (Brockmann (2006))

An alternative way to create the adhesion is to adopt air aspiration on a sliding chamber and then to move the robot through wheels (Longo and Muscato (2004a;b)). A variation of this adhesion method is presented by Elliot et al. (2006) and implemented in the City-Climber robot. These researchers designed a device based on the aerodynamic attraction produced by a vacuum rotor package which generates a low pressure zone enclosed by a chamber. The vacuum rotor package consists of a vacuum motor with impeller and exhaust cowling to direct air flow, as shown in Figure 6, left. It is essentially a radial flow device which combines two types of air flow. The high speed rotation of the impeller causes the air to be accelerated toward the outer perimeter of the rotor, away from the center radially. Air is then pulled along the spin axis toward the device creating a low-pressure region, or partial vacuum region if sealed adequately, in front of the device. With the exhaust cowling, the resultant exhaust of air is directed toward the rear of the device, actually helping to increase the adhesion force by thrusting the device forward.

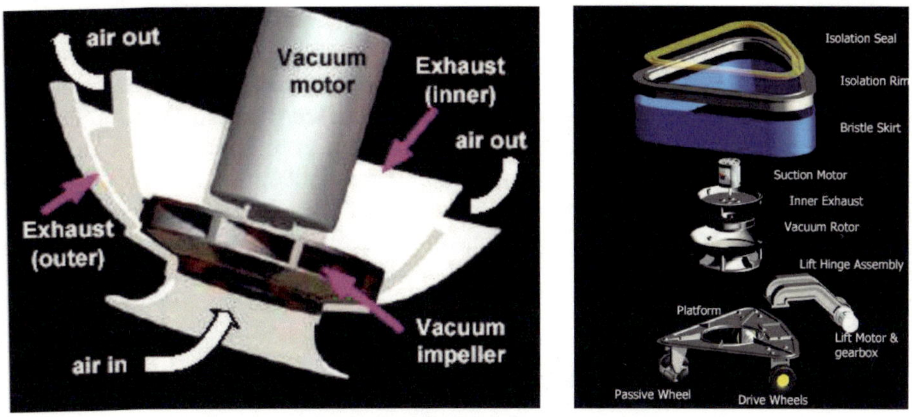

Fig. 7. Vacuum rotor package to generate aerodynamic attraction (left) and exploded view of the City-Climber prototype-II (right) (Elliot et al. (2006))

The experimental test demonstrated that the City-Climber with the module weight of 1 kg (Figure 6, right), can handle 4.0 kg additional payload when moving on brick walls.
Recently, a new technology, named Vortex Regenerative Air Movement (VRAM), was patented (Reinfeld and Illingworth (2002)). This adhesion system adopts vortex to generate high adhesion forces with a low power consumption, and allows the robot to travel on both smooth and rough surfaces. However, the adhesion force generated by the vortex technology is not enough to support large payload (Elliot et al. (2006)) and it is difficult for the robot to make wall-to-wall, and wall-to-ceiling transitions.

4.2 Magnetic Force

The magnetic adhesion is an alternative principle adopted for creating the adhesion force, in specific cases where the surface allows it. Magnetic attachment can be highly desirable due to its inherent reliability. This method is fast, but implies the adoption of heavy actuators. Despite that, magnetic attachment is useful only in specific environments where the surface is ferromagnetic and, therefore, for most applications it represents an unsuitable option (Menon et al. (2004)).
The most frequent solution is the use of electromagnets (Armada et al. (2005); Shores and Minor (2005)). Another possibility is the use of permanent magnets to adhere to the surface, combined with wheels or tracks to move along it (Mondal et al. (2002); Sánchez et al. (2006); Yan et al. (1999)). The main advantages of this last solution are the fact that there is not the need to spend energy for the adhesion process, it will not occur any loss of adhesion in the event of a power failure and permanent magnets are suitable for application in hazardous environments (Berns et al. (2005); Mondal et al. (2002)). A third solution is to use magnetic

wheels that allow to implement the locomotion and the adhesion at the same time (Park et al. (2003)).
The adoption of permanent magnets makes the robot more reliable and safer but there is a drawback: it is more difficult to control the adhesion and release of the robot to the surfaces in which it must work (Yan et al. (1999).

4.3 Gripping to the Surface

The previous adhesion techniques make the robots suitable for moving on flat walls and ceilings. However, it is difficult for them to move on irregular surfaces or surfaces like wire meshes.
In order to surpass this difficulty, some robots climb through man made structures or through natural environments, by gripping themselves to the surface where they are moving. These robots typically exhibit grippers (Balaguer et al. (2005)) (Figure 8), or other special designed gripping systems (Balaguer et al. (2005); Bell and Balkcom (2006); Inoue et al. (2006); Kennedy et al. (2006); Linder et al. (2005)), at the extremity of their limbs.
Examples of this kind of robots, are the ROMA 1 robot (Figure 8), that has two legs with grippers at their ends, for travelling in complex metallic-based environment (Balaguer et al. (2005)).

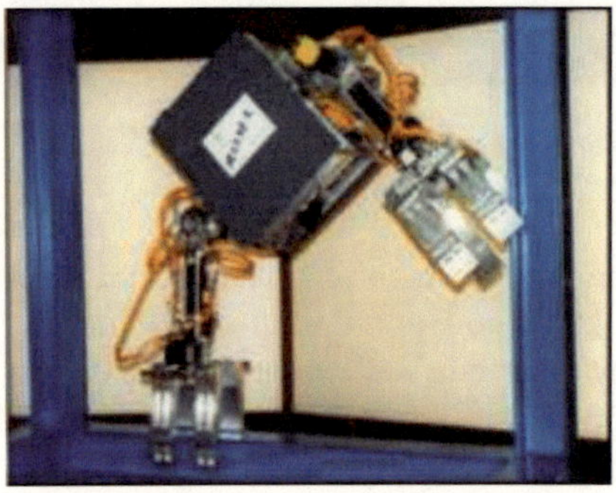

Fig. 8. ROMA1 robot climbing a beam-based structure (Balaguer et al. (2005))

Another example, is the ASIBOT robot (Figure 9), able to move between different points (Docking Stations) of the rooms through an innovative grasping method based on special connectors and a bayonet fitting (Balaguer et al. (2005)).
The Lemur IIb quadruped (Figure 10), intended for free climbing in steep terrain found in space exploration (Kennedy et al. (2006)), climbs over irregular surfaces just like if it was escalating a rock wall.
Finally, the ASTERISK robot (Inoue et al. (2006)) (Figure 11) is equipped with a special mechanism at the extremities of its limbs in order to grab and move on surfaces like wire meshes.
It is also worth mentioning, the toy climbing robot developed at the University of Dartmouth (Figure 12). A major design goal was to keep the project as simple as possible, making feasible for the general public to buy an inexpensive kit for building the robot. Based on these ideas, the robot was built of hobby servo-motors and LEGO pieces, and is capable of climbing a wall of pegs (Bell and Balkcom (2006); Linder et al. (2005)).

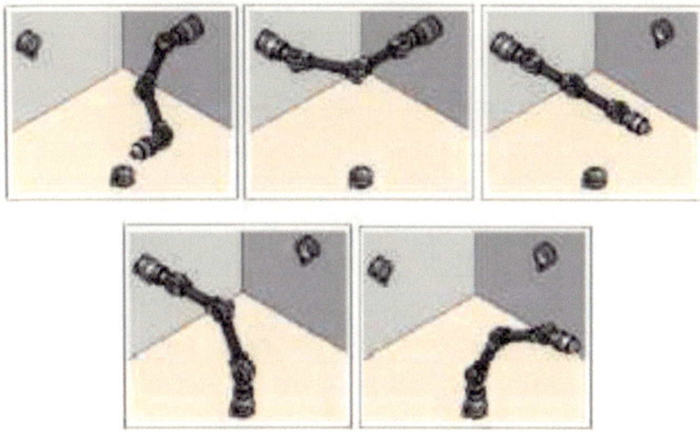

Fig. 9. ASIBOT climbing robot locomotion principle (Balaguer et al. (2005))

Fig. 10. LEMUR IIb robot climbing a test wall (Kennedy et al. (2006))

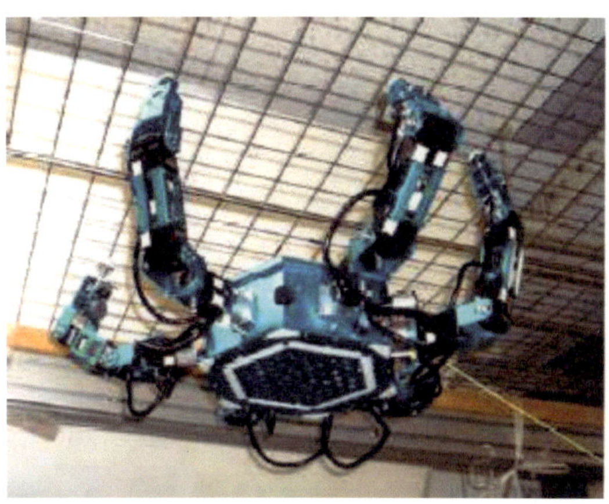

Fig. 11. ASTERISK robot hanging from a grid-like structure (Inoue et al. (2006))

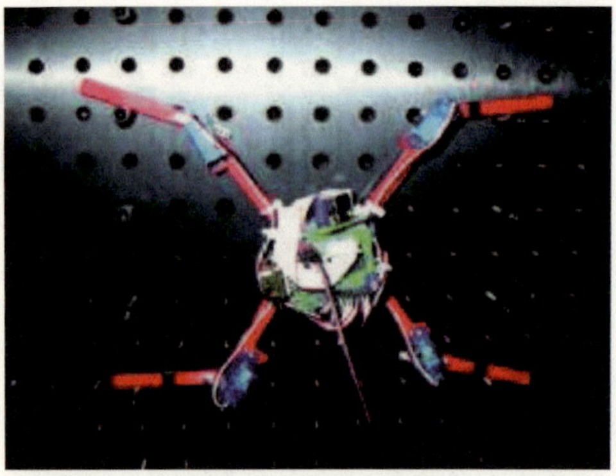

Fig. 12. The robot on the climbing wall (Bell and Balkcom (2006))

4.4 Thrust Force Adhesion

The prototypes that make use of this adhesion principle have been developed for working in submerged applications. These machines mainly allow to perform in-service inspection of the floor and walls of oil, petroleum and chemical storage tanks while submerged in the liquid, thereby saving the cost of emptying, cleaning and manually inspecting the tank (Sattar et al. (2002)).

The RobTank climbing robot, developed by Sattar et al. (2002), can enter oil and chemical storage tanks through 300 mm, or more, diameter openings in their roof, travel on the floor, rotate through any angle within the full 360°, and change surfaces from the floor to the wall and back to the floor. Regarding the locomotion, two servomotors provide the drive for the wheels of the vehicle while one propeller, mounted on top of the vehicle, provides the thrust force for adhesion to the wall. This way, this vehicle is able to climb on all types of surfaces.

Latter, Sattar et al. (2006) developed a climbing robot for Non-Destructive Testing of the internal tank wall and floor surfaces on Floating Production Storage Oil (FPSO) (see Figure 13, left).

This robot is equipped with two independent, speed controlled, thrusters that move the robot in a horizontal plane in the forward and reverse direction or rotate it to face in any direction. After contact with a wall, thrust forces generated by these two thrusters guarantee the adhesion to the wall, while actuated wheels move the robot on the wall. The robot manoeuvres freely on the wall and can be driven down from a wall to the floor of the tank and back on to it (see Figure 13, right).

A variable buoyancy tank was latter developed to change buoyancy around neutral by obtaining volume change (Sattar et al. (2008)). The tank enables the robot to swim to a given depth and to be parked on the floor with negative buoyancy when inspecting the floor.

5. Biological Inspired Adhesion Principles

In spite of all the developments made up to now, the proposed technologies still need to be improved and no definite and stable solution has yet been found. Therefore, developments continue in this research area.

In the last years a considerable inspiration has been gathered from climbing animals (Daltorio et al. (2005); Menon et al. (2004)). Insects, beetles, skinks, anoles, frogs and geckos have been studied for their sticking abilities (Figure 14). Beetles and Tokay geckos adhere to surfaces

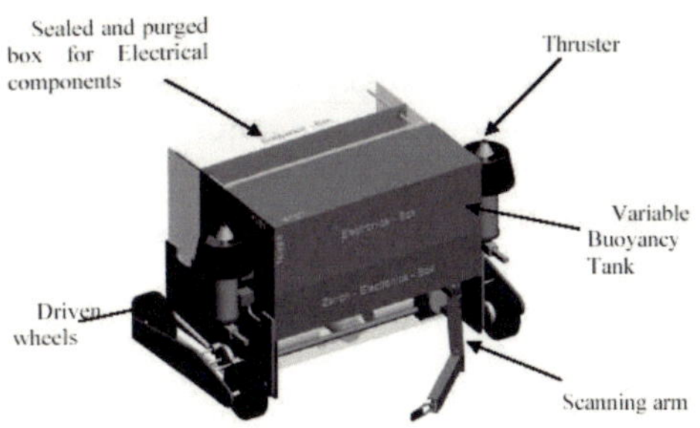

Fig. 13. Conceptual design of FPSO swimming and wall climbing robot (left) and wall-climbing robot climbing a tank glass wall without buoyancy tank (Sattar et al. (2006))

using patches of microscopic hairs that provide a mechanism for dry adhesion by van der Waals forces. Cockroaches climb a wide variety of substrates using their active claws, passive spines, and smooth adhesive pads. Inspired by these animals mechanisms, new methods for assuring the adhesion, based in biological findings, have recently been proposed.

Fig. 14. Gecko foot while climbing a glass surface (Tørrissen (Last Acessed: October 1, 2009))

Using bio-inspired adhesive technology, robots could potentially be developed to traverse a wide variety of surfaces, regardless of the presence of air pressure or the specific material properties of the substrate. Robots using such adhesives might some day be able to climb uneven, wet surfaces.
Bearing these ideas in mind, this section is organized as follows. The next subsection presents some climbing robots that use gecko inspired synthetic dry adhesives. The following subsection introduces climbing robots using micro-structured polymer feet in order to adhere to the surfaces. Lastly, the third subsection, describes some climbing robots using microspines.

5.1 Climbing Robots Using Gecko Inspired Synthetic Dry Adhesives

The ability of Geckos to climb surfaces, whether wet or dry, smooth or rough, has attracted people attention for decades. According to Menon et al. (2004), by means of compliant micro/nano-scale high aspect ratio beta-keratin structures at their feet, geckos manage to adhere to almost any surface with a controlled contact area. It has been shown that adhesion is mainly due to molecular forces such as van der Waals forces. The geckot's ability to stick to surfaces lies in their feet, specifically the very fine hairs on its toes, as can be seen if Figure 15. Those hairs are roughly 5 microns in diameter, and atop each of these micro-fibers sit hundreds of nano-fibers (spatulae) which are 200 nanometers in diameter. There are billions of these tiny fibers which make contact with the surface and create a significant collective surface area of contact. The hairs have physical properties which let them bend and conform to a wide variety of surface roughness, meaning that the adhesion arises from the structure of these hairs themselves. Also, because of their hydrophobic nature, the gecko fibers are self-cleaning.

Fig. 15. Detail of the gecko foot (Wikipedia (Last Acessed: October 1, 2009))

Since dry adhesion is caused by van der Waals forces, surface chemistry is not of great importance. This means that dry adhesion will work on almost any surface.
Dry adhesion is more robust than the suction adhesion mechanism. If the dry adhesion pad encounters a crack or gap, there will still be adhesion on the parts of the pad that have made contact. This behavior allows a robot, using dry adhesion, to climb on a wider variety of surfaces. Also, since dry adhesion does not rely heavily on the surface material or the atmosphere, it is suitable for use in the vacuum of space as well as inside liquid environments
Another benefit of dry adhesion is the speed at which attachment and detachment is possible. The attachment is nearly instantaneous as is the detachment, and they both only depend on the force applied. This leads to almost no delay in the locomotion, thus allowing very fast locomotion speeds. Furthermore, it is not necessary to control the timing of the attachment as critically as with the electromagnetic attachment. There is only the need to exert a pressure against the surface, so the attachment is passive in nature and, therefore, simple to control.
Inspired by these ideas, Menon et al. (2004) presented two alternative methods to replicate the structure of the micro-hairs present at the geckos feet.
The first one is based on the development of a synthetic adhesive. Much like the real gecko material it is expected that, in the future, the synthetic adhesive will be super-hydrophobic and, therefore, will be self-cleaning allowing for long lifetime robots. The nature of the adhesion force is such that no energy is required to maintain attachment after it has been initiated. Therefore, a robot using dry adhesion could hang on a wall indefinitely with no power consumption.

In order to test these synthetic dry fibrillar adhesives inspired in the geckos feet, Menon et al. (2004) developed two different vehicles to show the feasibility of the climbing mechanisms: the first one using legged wheels and the second robot consisting in a tread vehicle with customized tire. The legged-wheeled machine was latter improved by Murphy et al. (2006), giving rise to a small-scale agile wall climbing robot, named Waalbot. The Waalbot is able to navigate over smooth surfaces of any orientation, including vertical and inverted surfaces, taking advantage of adhesive elastomer materials for attachment (Figure 16, left). This robot can climb and steer in any orientation using two actuated legs with rotary motion and two passive revolute joints at each foot. The presented prototype can climb 90^0 slopes at a speed of 6 cm/s and steer to any angle.

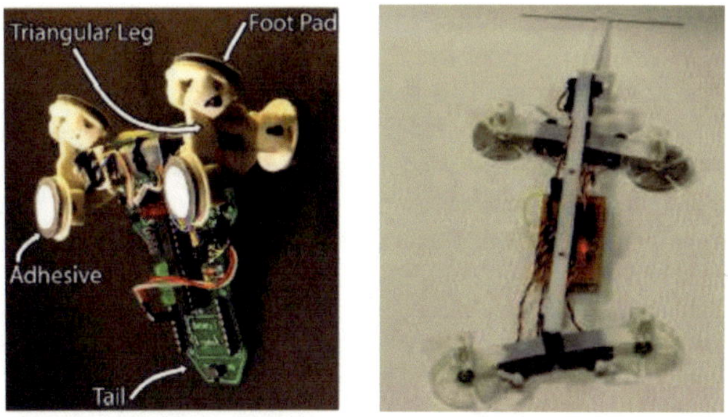

Fig. 16. Photographs of prototype Tri-Foot Waalbot climbing a 90^0 vertical surface (left) (Murphy et al. (2006)) and of Geckobot (right) (Unver et al. (2006))

More recently, Menon and Sitti (2005) developed two other climbing robots with different concepts. The first robot, called the Rigid Gecko Robot (RGR), was designed for operating both in Earth and space environments. Reliability and robustness were the most important requirements for the RGR. It was a relatively large robot actuated by conventional motors. The second robot, called the Compliant Gecko Robot (CGR), was designed using unconventional technologies, allowing its miniaturization up to a few centimeters scale and was designed for terrestrial applications. The CGR prototype had a composite structure and its Gecko mimicking locomotion relies on shape memory alloy wire actuators. Unver et al. (2006) developed the climbing robot Geckobot (Figure 16, right) based on these two. The robot has an overall weight of 100 grams (including the electronic board) and featuring a peeling mechanism for the robot feet, since this aspect is very crucial for climbing robots power-efficient detachment (as seen in geckos). Geckobot can climb up to 85^0 stably on Plexiglas surfaces. However, it was verified that beyond this angle stability diminishes abruptly.
The fibrillar adhesive presented by Menon et al. (2004) is still under development and does not achieve yet as high performances as other soft and dry adhesives. Synthetic gecko adhesive was tested and compared to soft adhesives such as Silly Puttyï£¡ and flat polydimethyl siloxane (PDMS). It was experimentally verified that Silly Puttyï£¡ exerts the highest normal adhesive force and, therefore, it was chosen for testing their robotic application (Menon and Sitti (2005)). For testing the Geckobot it was used the PDMS adhesive (Unver et al. (2006)). Although PDMS is a stable material, it is degraded and contaminated by dust and dirt. Therefore, after some time it looses its adhesive characteristics. This problem would be improved by using micro-patterned PDMS, in order to have self-cleaning characteristics like geckos (Unver

et al. (2006)). For testing the Waalbot, Murphy et al. (2006) equiped the robot feet with polymer adhesive material (Smooth-On Vytaflex 10), which shares many performance characteristics with the envisioned dry adhesive material. As the adhesives used on the feet of the robot gather dust and other contaminants their performance degrades quickly. Therefore, these adhesives are not suitable for dirty outdoor environments, walking across indoor floors, or for long term tasks.

5.2 Climbing Robots Using Micro-structured Polymer Feet

Daltorio et al. (2005) converted Mini-Whegsï£¡ (Figure 17, left), a small robot that uses four wheel-legs for locomotion, to a wall-walking robot with compliant, conventional-adhesive feet (5.4 cm by 8.9 cm, 87 grams). The feet are bonded to contact areas on the ends of the spokes and the flexibility of the feet acts as a hinge between the feet and spokes. The feet contact the substrate, bend as the hub turns, peel off the substrate gradually, and spring back to their initial position for the next contact. These researchers report that the Mini-Whegsï£¡ 7 can climb glass walls and walk on ceilings, and perform transitions between orthogonal surfaces, using standard pressure sensitive adhesives. The main problem with this approach (although some tests were made to find the best foot design and adhesive tape contact area (Daltorio et al. (2007))) is that after some runs, the robot falls with increasing frequency as the tape becomes dirty or damaged.

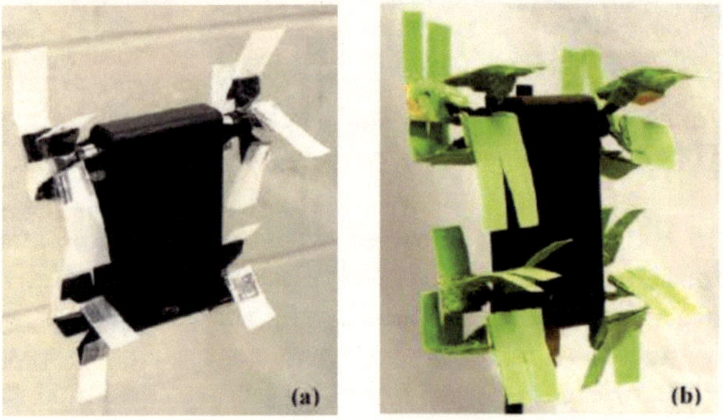

Fig. 17. Mini-Whegsï£¡ 7 on vertical glass (a) with office tape feet and (b) with micro-structured polymer feet and 25 cm long tail (tail not shown) (Daltorio et al. (2006))

Further developments of this robot, reported by Daltorio et al. (2006), lead to the replacement of the feet with a novel, reusable insect-inspired adhesive (Figure 17, right). Two polymer samples were tested: a smooth one and an insect-inspired surface-structured one. The reusable structured polymer adhesive presents less tenacity than the previous adhesive, resulting in an inferior climbing capability. However, after the addition of a tail, changing to off-board power, and widening the feet, the robot was capable of ascending vertical surfaces using the novel adhesive. Comparing with the previous approach, the polymer feet retained their traction/adhesive properties for several hours of testing and could be renewed by washing with soap and water.

While the current robot only walks on clean smooth glass, a practical climbing robot should be able to traverse rougher surfaces as well. This requires adhesives to be resistant to dust and to oils. Additionally, alternative attachment mechanisms, such as insect-like claws or spines, could be added to take advantage of surface roughness.

Based on these ideas, Wei et al. (2006) added claws, spines, and compliant ankles to Mini-Whegs, which allowed the machine to climb on soft or porous surfaces. The new front wheel-legs have each three spokes, with a foot (tarsus) connected at the end of each spoke.

5.3 Climbing Robots Using Microspines

According to what has been described, none of the above approaches is suitable for porous, and typically dusty, building surfaces such as brick, concrete, stucco or stone.

Inspired by the mechanisms observed in some climbing insects and spiders, Asbeck et al. (2006) developed a technology that enables robots to scale flat, hard vertical surfaces, including concrete, brick, stucco and masonry, without using suction or adhesives. The scheme employs arrays of miniature spines that catch on surface asperities. Unlike the claws of a cat, small spines do not need to penetrate surfaces. Instead, they exploit small asperities (bumps or pits) on the surface.

According to these authors, as spines become smaller it is possible to ascend smoother surfaces because the density of useable spine/asperity contacts increases rapidly. However, it is needed a large number of spines because each contact sustains only a limited force. Therefore, the key design principles behind climbing with microspines are to ensure that (i) as many spines as possible will independently attach to the asperities, and that (ii) the total load is distributed among the spines as uniformly as possible.

The above principles have been demonstrated in a 0.4 kg climbing robot, named Spinybot, that readily climbs hard surfaces such as concrete, brick, stucco and sandstone walls (Asbeck et al. (2006)). The robot has six limbs, and each one is an under-actuated mechanism powered using a single actuator in combination with passive compliance, which is responsible for engaging and disengaging the spines. A seventh actuator produces a ratcheting motion that alternately advances the legs in each of two tripods up the wall. Each feet of the Spinybot consists of ten planar toe mechanisms with two spines per toe. The mechanisms are created using a rapid prototyping process that permits hard and soft materials to be combined into a single structure. As shown in Figure 18, each toe includes several hard members, connected by soft links, with the spines embedded in the hard plastic. Each toe mechanism can deflect and stretch independently of its neighbors. This maximizes the probability that multiple spines, on each foot, will find asperities where they can "grab" and share the robot load.

6. New Architectures for Climbing Robots

New architectures have also been proposed for climbing robots in order to allow them to surpass different specific problems and applications.

In most cases, large, clumsy gantries are necessary to guarantee access for cleaning staff, or climbers are hired at great cost, to clean the glass of the inner side of atriums and glass roofs. Therefore, this is an application suited to the use of climbing robots. However, the main problem is finding a means to safeguard the robot against falling. Moreover, it is extremely difficult for technical personnel to reach the robot and repair it, in the event of malfunctioning.

With these ideas on mind, Elkmann et al. (2002) proposed a balloon-based robot for cleaning the inner side of atriums and glass roofs (Figure 19). The solution proposed by the researchers for automating this particular task, consists of a two-legged walking mechanism, with suction cups in contact with the glass, being the balloon guided by the walking mechanism along the roof.

This system consists of a cigar-formed, helium-filled balloon with a walking mechanism and a cleaning tool located at the front end of the balloon. At the other end of the balloon are

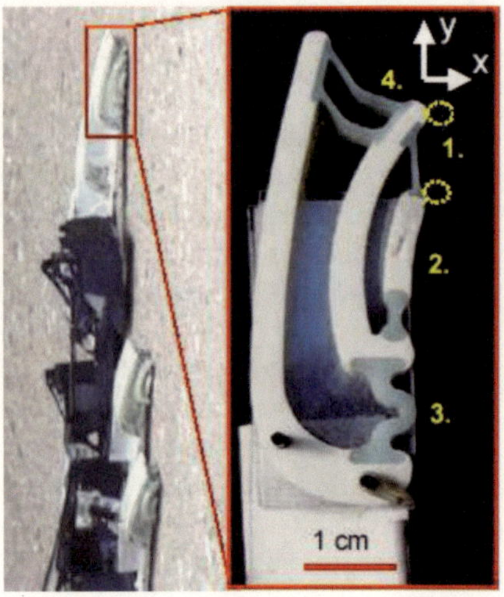

Fig. 18. View of upper section of Spinybot on concrete wall and detailed view of a toe on the foot (Asbeck et al. (2006))

Fig. 19. Balloon-based robot consisting of a walking mechanism and a cleaning tool (Elkmann et al. (2002))

modules like the control box, a water tank, and other systems for weight compensation. The balloon serves to lift the walking mechanism and the cleaning tool up to the glass surface. The robot cannot fall down and, if it is somehow damaged, the robot can be recovered by personnel by simply pulling the cables to the balloon down.
Shores and Minor (2005) presented a morphic bipedal robot with hybrid locomotion, combining the benefits of rolling, walking, and climbing locomotion (Figure 20).

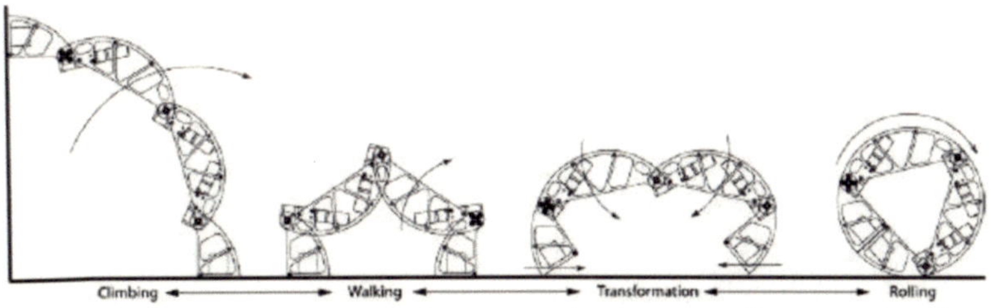

Fig. 20. A hybridized robot transitioning between climbing, walking, and rolling (Shores and Minor (2005))

The design provides these locomotion primitives without the addition of actuators, beyond those required for climbing through the use of a disklike exoskeleton that provides a rolling surface. The feet are equipped with electromagnets that allow the robot to anchor each foot to a ferrous climbing surface. These magnetic feet are centered in the footprint of the exoskeleton to distribute the force of the magnet over a larger area and enable the magnets to support larger moments than they could normally.
Degani et al. (2007) introduced a climbing robot mechanism, which uses dynamic movements to climb between two parallel vertical walls (Figure 21). This robot relies on its own internal dynamic motions to gain height. One benefit of this mechanism is that it allows climbing with only a single actuated degree of freedom.

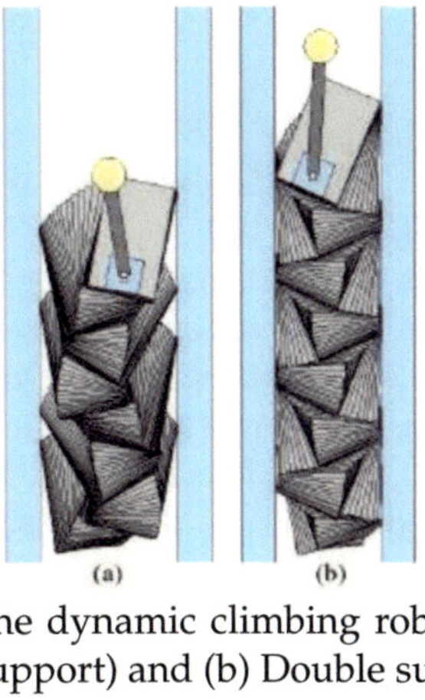

Fig. 21. Two typical motions of the dynamic climbing robot (the main body is traced over time) (a) Purely dynamic (single support) and (b) Double support. (Degani et al. (2007))

7. Conclusions

Considering the severity of many environments where there is the need for human labor, the exploitation of wall-climbing robots has undoubtedly a broad prospect. The main intended applications of these machines ranges from cleaning to inspection of difficult to reach constructions.
Up to now, considerable research was devoted to these machines and over 200 prototypes aimed at such applications had been developed in the world by the year 2006. Nonetheless, the application of climbing robots is still limited. Apart from a couple successful industrialized products, most are only prototypes and few of them can be found in common use due to unsatisfactory performance in on-site tests.
To make wall-climbing robots a popular replacement of manual work, indispensable prerequisites are an high reliability and high efficiency, and, on the other hand, affordable prices. The fulfilment of these requirements is still far, which indicates that there is yet a long way of development and improvement.
Given these considerations, this chapter presented a survey of several climbing robots, adopting different technologies for locomotion and for the adhesion to surfaces. Several possible applications of the presented robots have also been discussed. A special emphasis has been given on the new technologies (mainly biological inspired) that are presently being developed for the robots adhesion to surfaces.

8. References

Armada, M. A., de Santos, P. G., García, E., Prieto, M. and Nabulsi, S. (2005). Design of mobile robots, *Proc. of the 2005 CLAWAR: Introductory Mobile Robotics Workshop*, London, UK, pp. 2890–2895.

Asbeck, A. T., Kim, S., McClung, A., Parness, A. and Cutkosky, M. R. (2006). Climbing walls with microspines, *Proc. of the 2006 IEEE Int. Conf. on Rob. and Aut.*, Orlando, Florida, USA, pp. 4315–4317.

Azaiz, R. (2008). Unifier Ű unified robotic system to service solar power plants, *in* L. Marques, A. de Almeida, M. O. Tokhi and G. S. Virk (eds), *Advances in Mobile Robotics*, World Scientific, pp. 1141–1145.

Backes, P. G., Bar-Cohen, Y. and Joffe, B. (1997). The multifunctional automated crawling system (macs), *Proc. of the 1997 IEEE Int. Conf. on Rob. and Aut.*, Albuquerque, New Mexico, USA, pp. 335–340.

Balaguer, C., Gimenez, A. and Jardon, A. (2005). Climbing robotsŠ mobility for inspection and maintenance of 3d complex environments, *Autonomous Robots* **18**(2): 157–169.

Bell, M. and Balkcom, D. (2006). A toy climbing robot, *Proc. of the 2006 IEEE Int. Conf. on Rob. and Aut.*, Orlando, Florida, USA, pp. 4366–4368.

Berns, K., Braun, T., Hillenbrand, C. and Luksch, T. (2005). Developing climbing robots for education, *in* M. A. Armada and P. G. de Santos (eds), *Climbing and Walking Robots*, Springer, pp. 981–988.

Brockmann, W. (2006). Concept for energy-autarkic, autonomous climbing robots, *in* M. O. Tokhi, G. S. Virk and M. A. Hossain (eds), *Climbing and Walking Robots*, Springer, pp. 107–114.

Cepolina, F., Zoppi, M., Zurlo, G. and Molfino, R. (2004). A robotic cleaning agency, *Proc. of IAS'2004 – 8th Conf. on Intelligent Autonomous Systems*, Amsterdam, The Netherlands, pp. 1153 – 1161.

Chen, H., Sheng, W., Xi, N. and Tan, J. (2005). Motion control of a micro biped robot for nondestructive structure inspection, *Proc. of the 2005 IEEE Int. Conf. on Rob. and Aut.*, Barcelona, Spain, pp. 480–485.

Chen, S., Shang, J., Zhao, Z., Sattar, T. and Bridge, B. (2006). Novel solutions to design problems of industrial climbing robots, *in* M. O. Tokhi, G. S. Virk and M. A. Hossain (eds), *Climbing and Walking Robots*, Springer, pp. 139–146.

Choi, H. R., Ryew, S. M., Kang, T. H., Lee, J. H. and Kim, H. M. (2000). A wall climbing robot with closed link mechanism, *Proc. of the 2000 IEEE/RSJ Int. Conf. on Intelligent Robots and Systems*, pp. 2006 – 2011.

Daltorio, K. A., Gorb, S., Peressadko, A., Horchler, A. D., Ritzmann, R. E. and Quinn, R. D. (2006). A robot that climbs walls using micro-structured polymer feet, *in* M. O. Tokhi, G. S. Virk and M. A. Hossain (eds), *Climbing and Walking Robots*, Springer, pp. 131–138.

Daltorio, K. A., Horchler, A. D., Gorb, S., Ritzmann, R. E. and Quinn, R. D. (2005). A small wall-walking robot with compliant, adhesive feet, *Proc. of the 2005 IEEE/RSJ Int. Conf. on Intel. Robots and Systems*, pp. 4018–4023.

Daltorio, K. A., Wei, T. E., Gorb, S. N., Ritzmann, R. E. and Quinn, R. D. (2007). Passive foot design and contact area analysis for climbing mini-whegs, *Proc. of the 2007 IEEE Int. Conf. on Rob. and Aut.*, Roma, Italy, pp. 1274–1279.

Degani, A., Shapiro, A., Choset, H. and Mason, M. T. (2007). A dynamic single actuator vertical climbing robot, *Proc. of the 2007 IEEE/RSJ Int. Conf. on Intelligent Robots and Systems*, IEEE, San Diego, CA, USA, pp. 2901–2906.

Derriche, O. and Kouiss, K. (2002). A cleaning robot for spherical buildings, *in* P. Bidaud and F. B. Amar (eds), *5th International Conference on Climbing and Walking Robots and the Support Technologies for Mobile Machines*, Professional Engineering Publishing Limited, pp. 993–1001.

Elkmann, N., Felsch, T., Sack, M., Saenz, J. and Hortig, J. (2002). Innovative service robot systems for facade cleaning of difficult-to-access areas, *Proc. of the 2002 IEEE/RSJ Int. Conf. on Intelligent Robots and Systems*, pp. 756 – 762.

Elliot, M., Morris, W. and Xiao, J. (2006). City-climber: A new generation of wall-climbing robots, *Proc. of the 2006 IEEE Int. Conf. on Rob. and Aut.*, Orlando, Florida, USA, pp. 4413–4415.

Fernández, R., Akinfiev, T. and Armada, M. (2002). Control algorithms for an underwater climbing robot, *in* P. Bidaud and F. B. Amar (eds), *5th International Conference on Climbing and Walking Robots and the Support Technologies for Mobile Machines*, Professional Engineering Publishing Limited, pp. 985–992.

Gao, X. and Kikuchi, K. (2004). Study on a kind of wall cleaning robot, *Proc. of the IEEE Int. Conf. on Rob. and Biomimetics 2004*.

Hirose, S. and Arikawa, K. (2000). Coupled and decoupled actuation of robotic mechanisms, *Proc. of the 2000 IEEE Int. Conf. on Rob. and Aut.*, San Francisco, California, USA, pp. 33–39.

Hirose, S., Nagakubo, A. and Toyama, R. (1991). Machine that can walk and climb on floors, walls and ceilings, *Proc. of the Fifth Int. Conf. on Advanced Rob.*, Pisa, Italy, pp. 753–758.

Inoue, K., Tsurutani, T., Takubo, T. and Arai, T. (2006). Omni-directional gait of limb mechanism robot hanging from grid-like structure, *Proc. of the 2006 IEEE/RSJ Int. Conf. on Intel. Robots and Systems*, Beijing, China, pp. 1732–1737.

Kang, T., Kim, H., Son, T. and Choi, H. (2003). Design of quadruped walking and climbing robot, *Proc. of the 2003 IEEE/RSJ Int. Conf. on Intelligent Robots and Systems*, Las Vegas, Nevada, USA, pp. 619 – 624.

Kennedy, B., Okon, A., Aghazarian, H., Badescu, M., Bao, X., Bar-Cohen, Y., Chang, Z., Dabiri, B. E., Garrett, M., Magnone, L. and Sherrit, S. (2006). Lemur iib: a robotic system for steep terrain access, *in* M. O. Tokhi, G. S. Virk and M. A. Hossain (eds), *Climbing and Walking Robots*, Springer, pp. 1077–1084.

Krosuri, S. P. and Minor, M. A. (2003). A multifunctional hybrid hip joint for improved adaptability in miniature climbing robots, *Proc. of the 2003 IEEE Int. Conf. on Rob. and Aut.*, Taipei, Taiwan.

Li, Y., tian Li, M. and ning Sun, L. (2007). Design and passable ability of transitions analysis of a six legged wall-climbing robot, *Proc. of the 2007 IEEE Int. Conf. on Mechatronics and Aut.*, Harbin, China, pp. 800–804.

Linder, S. P., Wei, E. and Clay, A. (2005). Robotic rock climbing using computer vision and force feedback, *Proc. of the 2005 IEEE Int. Conf. on Rob. and Aut.*, Barcelona, Spain, pp. 4696–4701.

Longo, D. and Muscato, G. (2004a). Design of a single sliding suction cup robot for inspection of non porous vertical wall, *Proc. of the* 35th *Int. Symposium on Rob.*, Paris, France, pp. 1153 – 1161.

Longo, D. and Muscato, G. (2004b). A modular approach for the design of the alicia^3 climbing robot for industrial inspection, *Industrial Robot: An International Journal* **31**(2): 148–158.

Menon, C. and Sitti, M. (2005). Biologically inspired adhesion based surface climbing robots, *Proc. of the 2005 IEEE Int. Conf. on Rob. and Aut.*, Barcelona, Spain, pp. 2726–2731.

Menon, C., Murphy, M. and Sitti, M. (2004). Gecko inspired surface climbing robots, *Proc. of the IEEE Int. Conf. on Rob. and Biomimetics 2004.*

Minor, M., Dulimarta, H., Danghi, G., Mukherjee, R., Tummala, R. L. and Aslam, D. (2000). Design, implementation, and evaluation of an under-actuated miniature biped climbing robot, *Proc. of the 2000 IEEE/RSJ Int. Conf. on Intelligent Robots and Systems*, pp. 1999–2005.

Mondal, S., Sattar, T. P. and Bridge, B. (2002). Tofd inspection of v-groove butt welds on the hull of a container ship with a magnetically adhering wall climbing robot, *in* P. Bidaud and F. B. Amar (eds), *5th International Conference on Climbing and Walking Robots and the Support Technologies for Mobile Machines*, Professional Engineering Publishing Limited, pp. 955–961.

Murphy, M. P., Tso, W., Tanzini, M. and Sitti, M. (2006). Waalbot: An agile small-scale wall climbing robot utilizing pressure sensitive adhesives, *Proc. of the 2006 IEEE/RSJ Int. Conf. on Intel. Robots and Systems*, Beijing, China, pp. 3411–3416.

Nishi, A. (1991). A wall climbing robot using propulsive force of propeller, *Proc. of the Fifth Int. Conf. on Advanced Rob.*, Pisa, Italy, pp. 320–325.

Park, S., Jeong, H. D. and Lim, Z. S. (2003). Design of a mobile robot system for automatic integrity evaluation of large size reservoirs and pipelines in industrial fields, *Proc. of the 2003 IEEE/RSJ Int. Conf. on Intelligent Robots and Systems*, Las Vegas, Nevada, USA, pp. 2618–2623.

Reinfeld, D. and Illingworth, L. (2002). *VORTEX GENERATOR – patent:EP1224396*, http://www.freepatentsonline.com/EP1224396A1.html.

Resino, J. C., Jardón, A., Gimenez, A. and Balaguer, C. (2006). Analysis of the direct and inverse kinematics of roma ii robot, *in* M. O. Tokhi, G. S. Virk and M. A. Hossain (eds), *Climbing and Walking Robots*, Springer, pp. 869–874.

Robert T. Pack, J. L. C. J. and Kawamura, K. (1997). A rubbertuator-based structure-climbing inspection robot, *Proc. of the 1997 IEEE Int. Conf. on Rob. and Aut.*, Albuquerque, New Mexico, USA, pp. 1869–1874.

Rodriguez, H. L., Bridge, B. and Sattar, T. P. (2008). Climbing ring robot for inspection of offshore wind turbines, *in* L. Marques, A. de Almeida, M. O. Tokhi and G. S. Virk (eds), *Advances in Mobile Robotics*, World Scientific, pp. 555–562.

Sánchez, J., Vázquez, F. and Paz, E. (2006). Machine vision guidance system for a modular climbing robot used in shipbuilding, *in* M. O. Tokhi, G. S. Virk and M. A. Hossain (eds), *Climbing and Walking Robots*, Springer, pp. 893–900.

Sattar, T. P., Rodriguez, H. E. L. and Shang, J. (2008). Amphibious inspection robot, *in* L. Marques, A. de Almeida, M. O. Tokhi and G. S. Virk (eds), *Advances in Mobile Robotics*, World Scientific, pp. 613–620.

Sattar, T. P., Rodriguez, H. E. L., Shang, J. and Bridge, B. (2006). Automated ndt of floating production storage oil tanks with a swimming and climbing robot, *in* M. O. Tokhi, G. S. Virk and M. A. Hossain (eds), *Climbing and Walking Robots*, Springer, pp. 935–942.

Sattar, T. P., Zhao, Z., Feng, J., Bridge, B., Mondal, S. and Chen, S. (2002). Internal in-service inspection of the floor and walls of oil, petroleum, and chemical storage tanks with a mobile robot, *in* P. Bidaud and F. B. Amar (eds), *5th International Conference on Climbing and Walking Robots and the Support Technologies for Mobile Machines*, Professional Engineering Publishing Limited, pp. 947–954.

Savall, J., Avello, A. and Briones, L. (1999). Two compact robots for remote inspection of hazardous areas in nuclear power p1ants, *Proc. of the 1999 IEEE Int. Conf. on Rob. and Aut.*, Detroit, Michigan, USA, pp. 1993–1998.

Shores, B. E. and Minor, M. A. (2005). Design, kinematic analysis, and quasi-steady control of a morphic rolling disk biped climbing robot, *Proc. of the 2005 IEEE Int. Conf. on Rob. and Aut.*, Barcelona, Spain, pp. 2732–2737.

Tørrissen, B. C. (Last Acessed: October 1, 2009). `http://en.wikipedia.org/wiki/File:Gecko_foot_on_glass.JPG`.

Tummala, R. L., Mukherjee, R., Xi, N., Aslam, D., Dulimarta, H., Xiao, J., Minor, M. and Danghi, G. (2002). Climbing the walls, *IEEE Robotics and Automation Magazine* **9**(4): 10–19.

Unver, O., Uneri, A., Aydemir, A. and Sitti, M. (2006). Geckobot: A gecko inspired climbing robot using elastomer adhesives, *Proc. of the 2006 IEEE Int. Conf. on Rob. and Aut.*, Orlando, Florida, USA, pp. 2329–2335.

Wei, T. E., Daltorio, K. A., Gorb, S. N., Southard, L., Ritzmann, R. E. and Quinn, R. D. (2006). A small climbing robot with compliant ankles and multiple attachment mechanisms, *Proc. of the 9th Int. Conf. on Climbing and Walking Robots*, Brussels, Belgium, pp. 579–585.

Wikipedia (Last Acessed: October 1, 2009). `http://en.wikipedia.org/wiki/File:Gecko_foot_on_glass.JPG`.

Xiao, J., Xi, N., Xiao, J. and Tan, J. (2003). Multi-sensor referenced gait control of a miniature climbing robot, *Proc. of the 2003 IEEE/RSJ Int. Conf. on Intel. Robots and Systems*, Las Vegas, Nevada, USA, pp. 3656 – 3661.

Xiao, J., Xiao, J., Xi, N. and Sheng, W. (2004). Fuzzy system approach for task planning and control of micro wall climbing robots, *Proc. of the 2004 IEEE Int. Conf. on Rob. and Aut.*, pp. 5033 – 5038.

Yan, W., Shuliang, L., Dianguo, X., Yanzheng, Z., Hao, S. and Xueshan, G. (1999). Development and application of wall-climbing robots, *Proc. of the 1999 IEEE Int. Conf. on Rob. and Aut.*, Detroit, Michigan, USA, pp. 1207–1212.

Zhang, H., Zhang, J., Liu, R. and Zong, G. (2004). A novel approach to pneumatic position servo control of a glass wall cleaning robot, *Proc. of the 2004 IEEE/RSJ Int. Conf. on Intel. Robots and Systems*, Sendai, Japan, pp. 467–472.

Zhu, J., Sun, D. and Tso, S.-K. (2003). Application of a service climbing robot with motion planning and visual sensing, *Journal of Robotic Systems* **20**(4): 189Ű–199.

2

Mechanical Synthesis for Easy and Fast Operation in Climbing and Walking Robots

Antonio Gonzalez-Rodriguez, Angel G. Gonzalez-Rodriguez and Rafael Morales
University of Castilla-La Mancha, University of Jaen
Spain

1. Introduction

This chapter deals with the importance of the mechanical design in devices used in mobile robots. A good synthesis of mechanisms will improve the robot's operation. This idea will be explained via two examples.

In the first example, the mechanical design of a staircase climbing wheelchair will be presented. A wheelchair is intended to be a commercial unit, and its control unit must, therefore, be robust, efficient and low-cost.

The second example deals with the mechanical design of an easy-to-operate leg for a mobile robot. This is a research project, but easy operation is fundamental if we are to ensure that the steps that the leg takes are as rapid as possible, which is of great importance in making actual walking robots faster.

2. Design of a new Design for a Staircase Wheelchair

2.1 Review of the current approaches

People with disabilities find that their mobility is improved with the help of powered wheelchairs. However, these chairs are often rendered useless by architectural barriers whose total elimination from the urban landscape is expensive, if not impossible. These barriers appear in many different geometrical shapes, of which staircases are the most difficult obstacle to overcome.

Various designs have been developed to allow a wheelchair climb a stair. One of the first and most common solutions are tracks (Yoneda et al., 2001; Lawn et al., 2001) owing to the simplicity of their control, and their robustness in adapting to different shapes such as spiral staircases. However this solution has important drawbacks: the vehicles that use tracks are uncomfortable and of low efficiency when they work in barrier-free environments; a high friction coefficient between the edge of the step and the track can deteriorate this edge, and the entrance to and exit from the staircase are dangerous and difficult to control.

Another common solution consists of various wheels attached to a rotation link (Lawn & Ishimatzu, 2003). The main problem with this solution is its fixed geometry which cannot be adjusted to the step, and the prototype therefore only works satisfactorily with obstacles which are similar to the step used to define the geometry of the rotating link. Further problems with this solution are that each of the wheels must have their own transmission, which increases the wheelchair's weight, or that the user's resulting trajectory is uncomfortable and difficult to control.

An alternative strategy for the design of the staircase climbing wheelchair will be presented in this paper. This strategy is based on splitting the climbing problem into two sub-problems (Morales et al., 2004; Morales et al., 2006): single step-climbing of each axle, and front and rear axle positioning. Two independent mechanisms have been designed to overcome these sub-problems: the climbing mechanism and the positioning mechanism, respectively.

The final mechanism must be able to successfully negotiate all the staircases designed under international standards. This paper describes the latest prototype, together with the experimental results obtained when the wheelchair climbs different staircases.

2.2 Description and performance of the mechanical system

Fig. 1 shows a CAD model of our proposed design. The prototype can be seen in Fig. 2. The kinematical scheme of the overall system can be seen in Fig. 3, in which the sub-scheme labeled 1 corresponds to the climbing mechanism and the sub-scheme labeled 2 corresponds to the positioning mechanism.

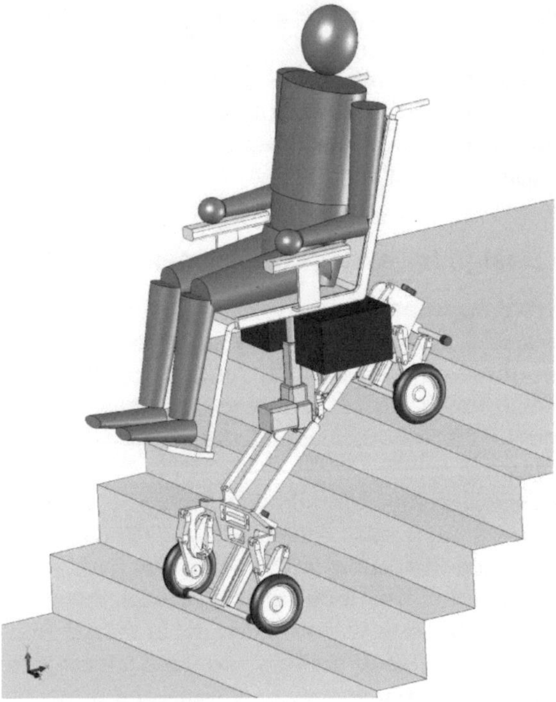

Fig. 1. CAD model of the proposed design

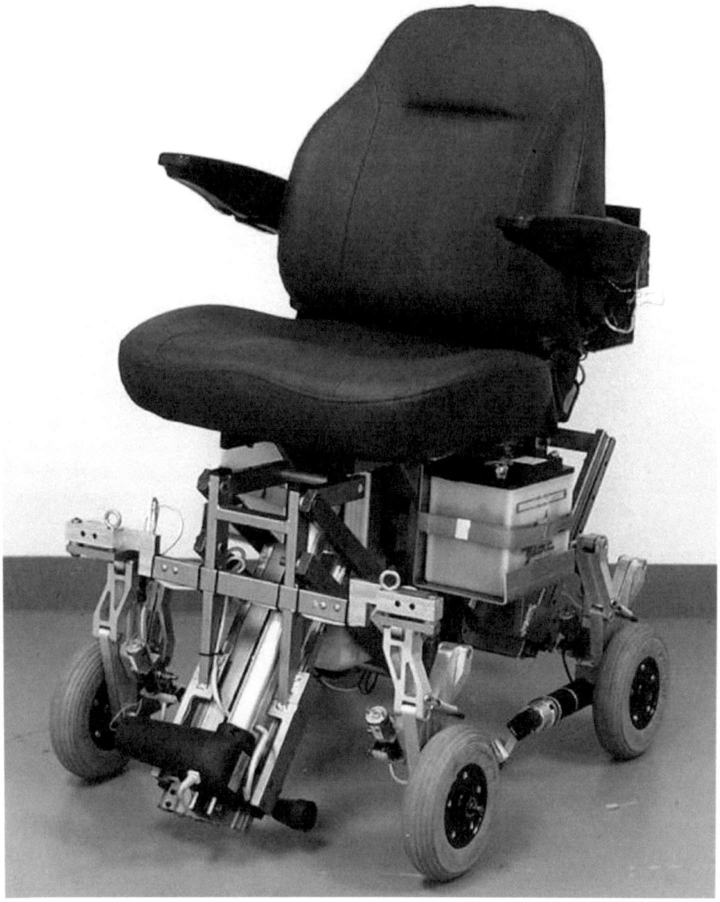

Fig. 2. The staircase climbing mechanism proposed

2.2.1 Climbing mechanism

The climbing mechanism allows an axle climb a single step. There is one climbing mechanism for the rear axle and another for the front axle. These have been designed to adapt to different obstacle geometries, and to guarantee that the system is always in stable equilibrium. This last objective is satisfied by permanently ensuring a wide support polygon with four contact points, two for each axle.

When the climbing mechanism reaches a step, a sliding support (1.5 in Fig. 3) is deployed. A prismatic joint connects this support to the chassis (2.3) at a fixed angle μ.

A new degree of freedom resulting from a four link mechanism (bars 1.1, 1.2, 1.3 and 1.4 in Fig. 3) allows the wheel (1.6) to move backwards to avoid interference from the step. An electromagnetic lock cancels this degree of freedom, e.g. when the system is in a barrier-free environment. The climbing sequence is presented in Fig. 4.

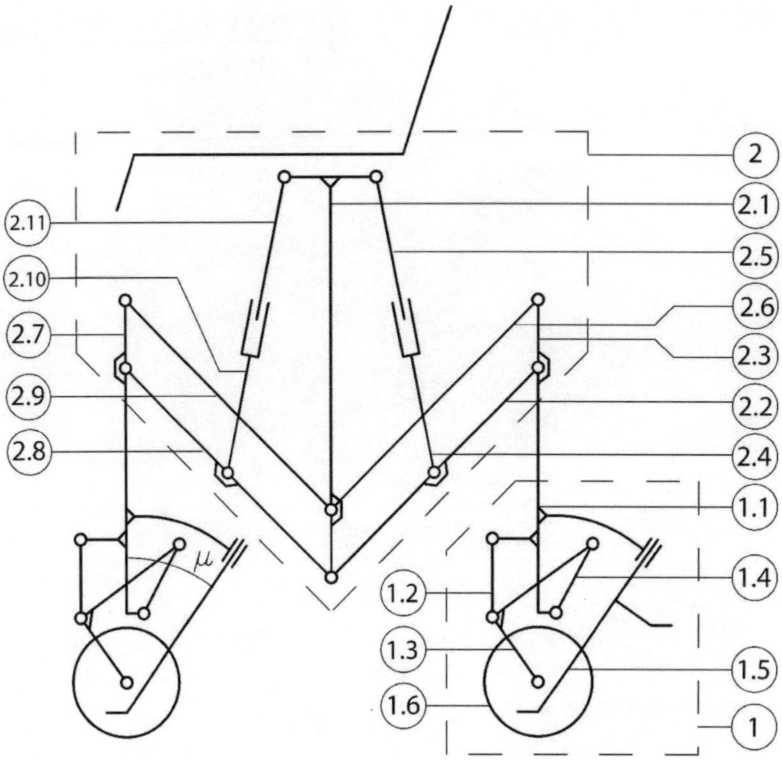

Fig. 3. Scheme of the entire prototype

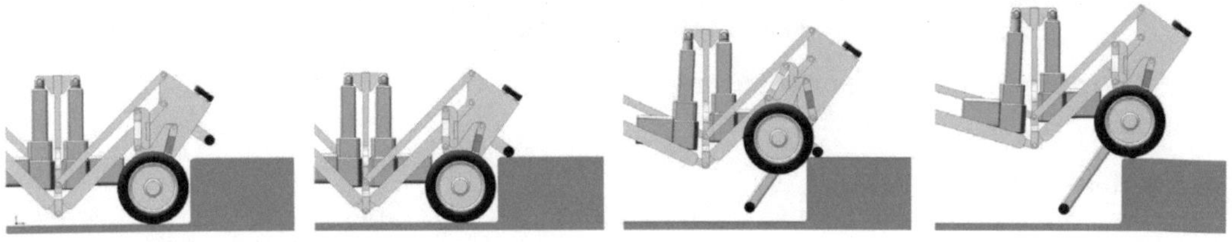

Fig. 4. Climbing sequence for the rear axle

Upon completion of this process, the sliding support is retracted to prepare the system for the following step. The descent process is essentially the same, but the sequence of actions is inverted. In this case, the orientation of the wheelchair follows the normal direction of movement, and hence, the first operating axle is the front one.

One important feature of this system is its high payload capacity, which is of great importance in the carriage of large patients and heavy batteries. The proposed prototype can climb a staircase with a 200kg load (batteries not included). Table 1 shows the weight and weight-payload ratio for other climbing systems. The ratio of the proposed prototype is not achieved with actual tracks or rotating wheel clusters.

Vehicule	Locomotion system	Weight (kg)	Payload/ Weight
Presented vehicle	Hybrid locomotion	72	2.53
XEVIUS (Yoneda et al. 2001)	Single Tracks	65	0.92
IBOT 3000	Wheel cluster	131	0.86
Stair-Climbing Wheelchair with High Single-Step Capability. (Lawn et al 2003)	Wheel cluster	160	0.5
ALDURO (Germann et al. 2005)	Hybrid locomotion	1500	0.32
Stair-Climbing Wheelchair in Nagasaki (Lawn et al. 2001)	Double tracks	250	0.32

Table 1. Weight and Weight-Payload Ratio for Actual Climbing Vehicles

2.2. Positioning mechanism

A closed-loop mechanism has been added to accomplish the positioning task, which is responsible for placing the climbing mechanism in such a way that the stability of the system is ensured. If only one step needs to be climbed then this is the only task accomplished by the positioning mechanism. But if it is necessary for both (rear and front) axles to be coordinated in order to climb a staircase, then the positioning mechanism must also accommodate the wheel base to the stair tread. Besides a time reduction, the coordinated climbing of both axles also facilitates control and increases energy efficiency.

The positioning mechanism is a closed-loop mechanism, and thus has a good performance in terms of rigidity, which consists of three platforms. The central platform (2.1 in Fig. 3) houses the seat and the batteries. The two lateral platforms (2.3 and 2.7) house the climbing mechanisms. The platforms are joined by two parallelograms (2.2, 2.6, 2.8 and 2.9, in gray) that prevent relative rotation between platforms. The system has two degrees of freedom which are driven by two linear actuators (2.4-2.5 and 2.10-2.10). These allow the system to alter both the vertical and the horizontal distance between the wheels, which allows the wheel base to be accommodated to the stair treads. The two degree of freedom system can also alter the height and orientation of the seat.

International standards impose a maximum and minimum width and height for steps. The positioning mechanism has been synthesized to maintain system stability for all the staircases built according to German Standard DIN 18065 (Fig. 5a). There are four extreme positions:

- N: maximum width and height. In this position the wheels are at maximum separation and both parallelograms will be collinear.
- N': minimum width and maximum height. This is the staircase with the maximum slope (dark gray stairs in Fig. 5a).
- N'': minimum width and height.
- N''': maximum width and minimum height. This is the staircase with the minimum slope (light gray stairs in Fig. 5a).

These four points are the corner of a rectangle called an objective rectangle. When one of the wheels is in contact with the upper step, if the positioning mechanism is able to place the other wheel in the four corners of the objective rectangle, then the accommodation process for any staircase is achievable.

The design of the mechanism is an iterative process to synthesize the parallelograms. This process searches for a mechanism which can reach points *N* and *N′* (in this case points *N′′* and N′′′ can be also reached, as is shown by the dashed lines in Fig. 5a).

Fig. 5b shows the vectors used in the synthesis process, where r and s represent the lower bars of both parallelograms when the centre of the wheel is at *N*. When the wheel moves to *N′* these bars are represented by *r′* and *s′*. Vectors *R2* and *R3* belong to the lateral platforms and join the centers of the wheels with the joints of the parallelograms. The point *P* is the common joint of the parallelograms with the central platform.

The first step consists of defining vectors *R2* and *R3* according to the geometrical restrictions of the wheelchair. For example, the vertical component of *R2* must be as short as possible because a large value implies that the seat is too high.

L will be defined as $L = r + s$, therefore $r = cL$, where *c* is a constant. The equation of the vector-pair *r*-*s* can therefore be written as follows (Erdman & Sandor, 1994):

$$cL\left(e^{i\alpha} - 1\right) + (1 - c)L\left(e^{i\beta} - 1\right) = D \tag{1}$$

where *D* joins points *N* and *N′*.

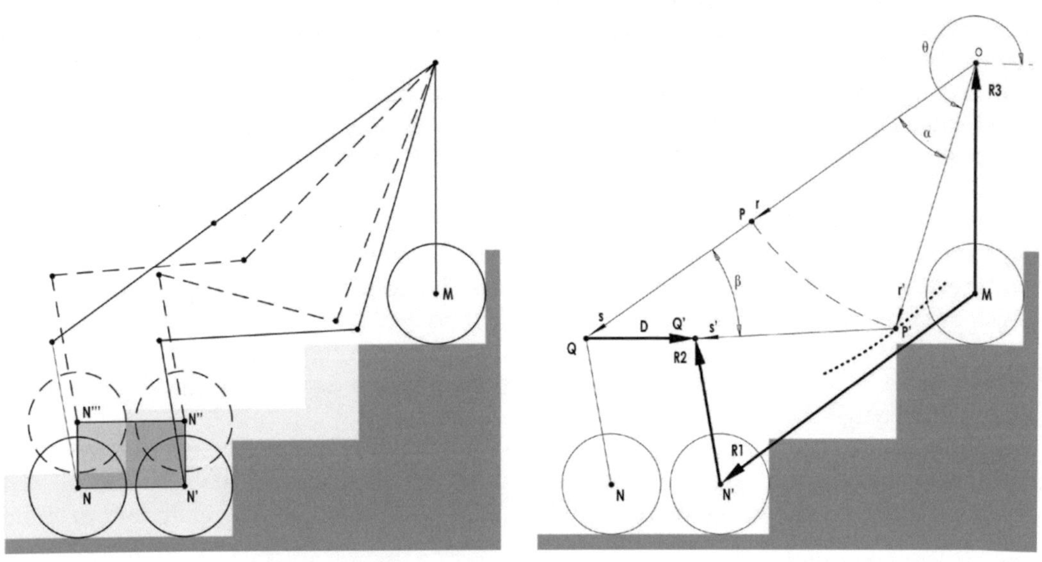

Fig. 5. a) Objective Rectangle and b) vectors used for the dimensioning

In this vectorial equation α, β, and *c* are unknown variables. If β is taken as a parameter, the analytical solution for α can be obtained.

$$\tan\alpha = \frac{\left(V_Y^2 - V_X^2 + 1\right)\sin\beta + 2V_X V_Y \cos\beta - 2V_Y}{\left(V_X^2 - V_Y^2 + 1\right)\cos\beta + 2V_X V_Y \sin\beta - 2V_X} \tag{2}$$

where

$$V = \frac{D}{L} - e^{i\beta} + 1 \tag{3}$$

The geometry of the system can be easily rebuilt when α is known. The dotted line in figure 5b represents the position of *P* for different values of parameter β. The position of *P* allows us to verify the suitability of the mechanism in order to avoid interferences with stairs. If a valid solution has not been found the process returns to the first step, and the initial values for *R2* and *R3* are altered.

The final geometry for Fig. 6 is obtained by repeating the iterative process for the positioning mechanism. The figure also shows the workspace (light gray) and objective rectangle (dark gray). It is worth mentioning that the wheelchair can climb the staircase even when the accommodating process is not carried out. It may thus be reasonable to use a narrower objective rectangle in order to obtain a more compact wheelchair. This rectangle should be chosen in such a way that the most usual staircases are included.

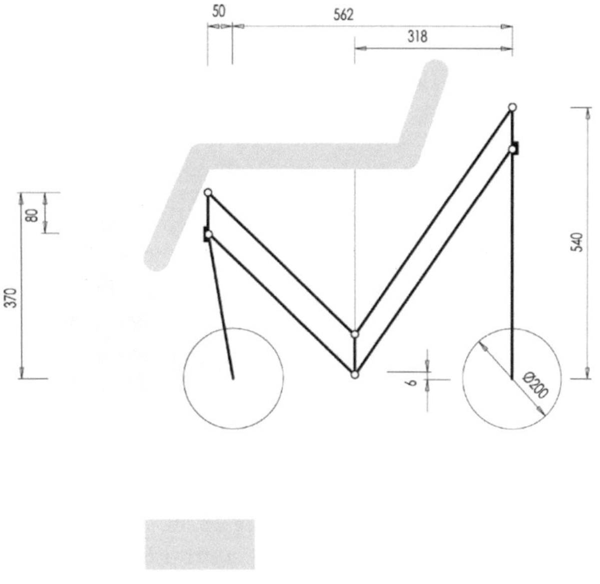

Fig. 6. Workspace, objective rectangle and final geometry

2.3 Experimental results

This section shows the experimental results obtained when the wheelchair climbs a single step of different heights, and when the wheelchair climbs a three-step staircase. The 3D positions of several points of interest have been measured with the Optotrack motion system, which is prepared with several infrared markers to record the trajectories of the platform and wheels, as is shown in Fig. 7.

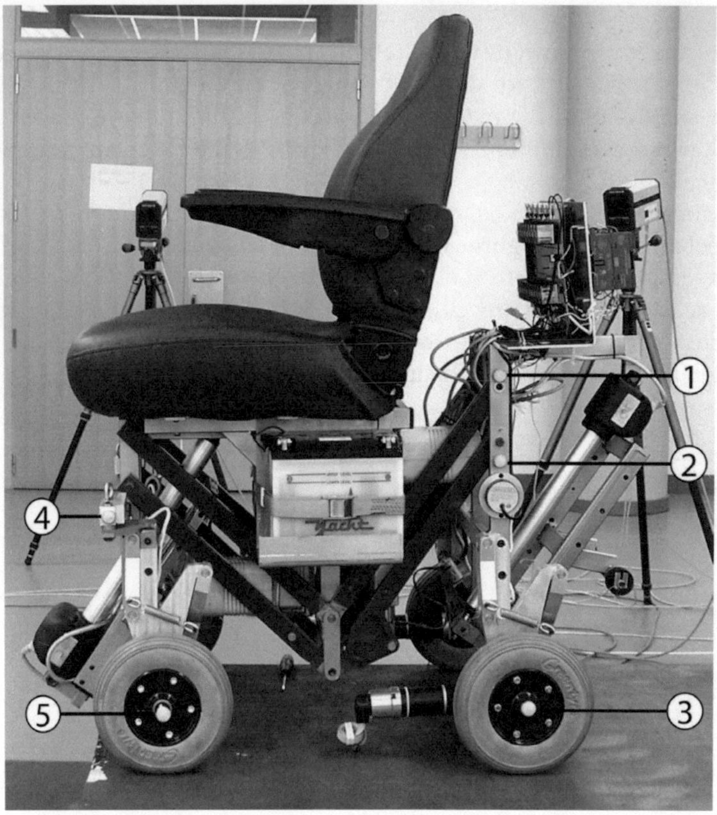

Fig. 7. Position of the markers

In the first experiment, the wheelchair must separately climb single steps of 0.16m, 0.18m and 0.2m, with the aim of studying the horizontality of the seat.

The horizontality is maintained with a bang-bang control that receives the measurement of an inclinometer placed on the rear platform as the fed backward signal. This type of control has been chosen owing to the wide dead band of the linear actuators that make the use of continuous law control unsuitable. This gives rise to performances with slight oscillations due to natural or forced hysteresis in the control (see Fig. 8). Its frequency and amplitude can be reduced at the expense of a higher control effort.

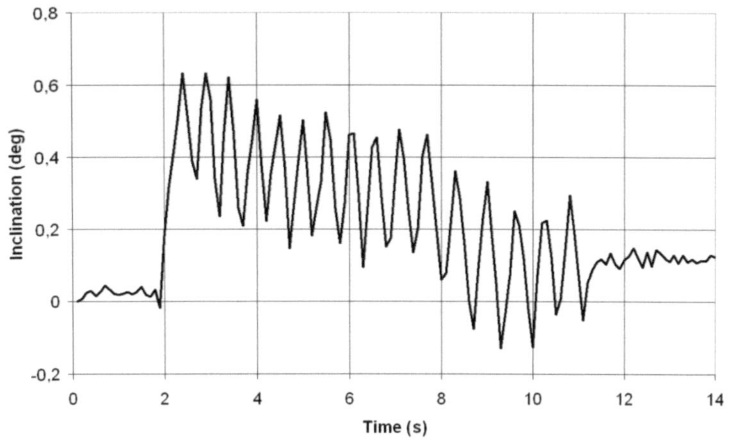

Fig. 8. Inclination of the prototype while climbing a 0.2m height step.

As Fig. 9 shows, markers 1 and 2 follow the trajectory of the sliding support (1.5 of Fig. 3), while marker 3 - the center of the rear wheel - presents a curved trajectory owing to the movement of the four link mechanism that allows the wheel to move backwards and avoid interference from the step.

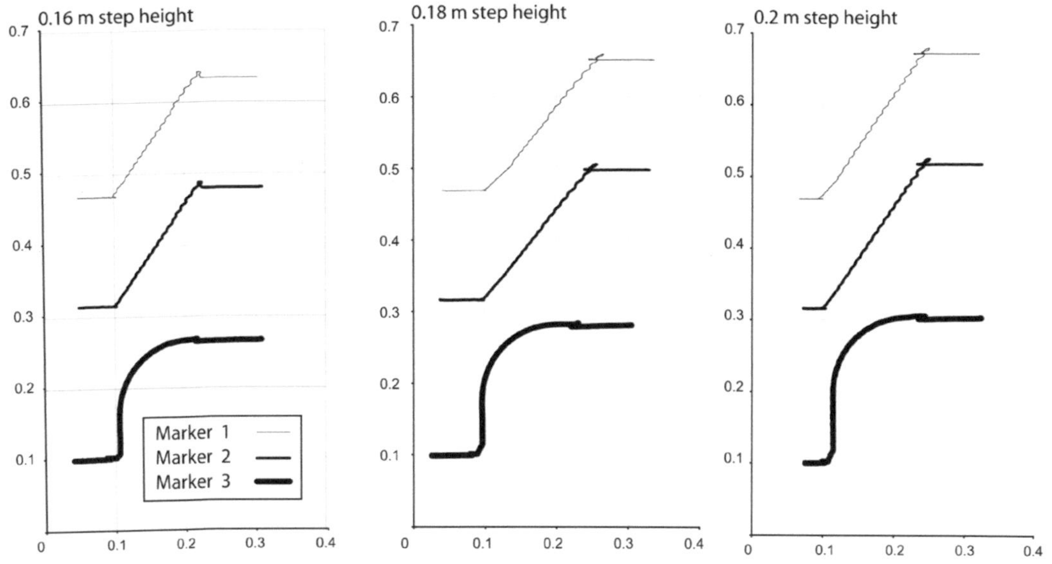

Fig. 9. Climbing of steps with 0.16, 0.18 and 0.2m step height

In the second experiment, the wheelchair climbs a three-step staircase. In order to maintain the center of masses as low as possible, the wheelchair is positioned backwards before accomplishing the climb. Figure 10 shows the trajectories regarding the rear axle markers in thick gray lines (markers 1, 2 and 3) and those of the front axle markers (4 and 5), in thin black lines. As pointed out in Fig. 10, the experiment passes through three stages:

A. Climbing of rear axle while front axle remains on the floor. Segments of the trajectories that belong to this stage are labeled A in Fig. 10. The amplitude and the frequency of the oscillations are wider in this experiment because the hysteresis of the control loop has been increased.
B. Simultaneous climbing of the rear and front axles. The segments of the trajectories that belong to this stage are labeled B in Fig. 10. The accommodation process must be performed in order to climb with both axles at once. In this stage the actuators of both parallelograms remain inactive and, therefore, the oscillations of the platforms are completely eliminated.
C. Climbing of front axle with the rear axle on the upper floor. Segments of the trajectories that belong to this stage are labeled C in Fig. 10.

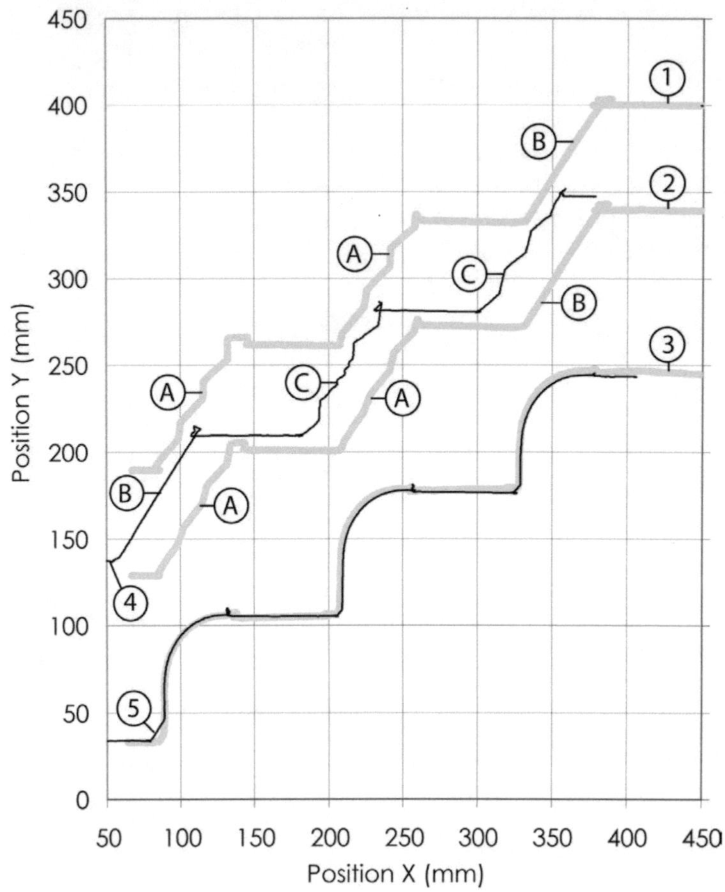

Fig. 10. Trajectories of rear and front platforms while climbing a three step staircase

3. Design of a Fast Controlled Leg for Walking Robots

3.1 Review of the current trend

In the present day, and owing to the power and low price of control units, a disregard is shown for the mechanical structure design of mobile robot legs. The structures currently used are based on serial combinations of joint + link, in which the actuator directly drives the joint. These structures are apparently similar to those in the human body or to those of certain animals.

However, these structures have several drawbacks:

- It is necessary to solve the inverse kinematics in order to establish the particular trajectories of each joint, which take the end of the leg to a determinate trajectory.
- The joint trajectories must be defined point to point, which implies:
  - A control unit for every joint
  - Continuous speed changes during acceleration and braking that drastically increase the energy wasted during the trajectory execution.
- There is a coupling between different joints to perform a determinate movement, which forces the designer to select the actuators in order to satisfy the requirements of force/torque and speed for the more demanding movements.
- The impacts suffered by the end of the leg against the floor or unavoided obstacles are directed towards the actuator shafts, which reduces its life time.

A partial copy of the human or animal structure therefore requires a high response speed from the control units, and a complexity that is not observed in the walking process of humans or animals. The low efficiency of the walking cycles and overdimensioning of the actuators signify that these robots have little autonomy, which is one of their main drawbacks.

In order to overcome this disadvantage, McGeer in (McGeer, 1990 and McGeer, 1990) introduced a new design of low-energy robots based on the concept of passive-dynamic walker, that could walk downhill a slope without consuming energy, only exchanging gravitational energy and kinetic energy, and finally converting them into losses due to friction and collisions.

With the same purpose but with a different approach, a new leg has been designed whose mechanical structure decouples the vertical and horizontal movement. A single control unit is therefore sufficient to set the trajectory of all the legs through the designation of few (4 or 5) points per cycle and leg. Furthermore, the motors are driven at a constant speed or constant acceleration during the majority of this operation, which increases efficiency.

Other advantages of the presented design are:

- It is possible to correctly select dimensioned actuators, with different characteristics, for each kind of movement: faster but with a lower load capacity for horizontal movements, and with a higher force/torque but slower for vertical movements.
- The decoupling of the movements simplifies the introduction of muscles with adaptable compliance (Gonzalez-Rodriguez et al., 2009), although this kind of mechanisms are not ready in present days to be used in active robots, except in the case of pneumatic robots (Grizzle et Poulakakis, 2008).
- A design that does not aim to mimic animal structures allows the actuator to be located at the hip, and far from the directions of reaction impact. Therefore and

respectively, the leg inertia is reduced – in the same way as for industrial robots – and the lifetime and reliability are increased.

It is also possible to include springs, in order to store and recover part of the kinetic energy, and therefore reduce energy losses.

3.2 Mechanical Design to facilitate the control

If a mechanism is intended to operate solely in obstacle-free terrains, the most suitable option is a wheeled vehicle, with higher performance in terms of efficiency, price, controllability, speed and payload.

However, when the terrain has certain characteristics that impede a wheeled robot from circulating, then it will also require some kind of legs, thus necessitating the configuration of a hybrid robot or a walking robot. The robot must also perform in an appropriate manner in terms of speed and autonomy when operating in obstacle-free terrains, which will probably be the most of the time.

This work presents a new design for a robot leg, whose synthesis searches for the simplification of the walking operation control in order to increase the robot's speed and efficiency when operating on a surface without obstacles. The structure of the leg and its control system must simultaneously be able to overcome obstacles (including steps) that are within the workspace of the end of the leg.

In a first stage, the structure has been designed as a mechanism with two degrees of freedom, which allows the end of the leg to move up/down and forwards/backwards. The (sagital) plane within which the movement is performed will be called the *movement plane* and is parallel to the movement direction and to the vertical line.

Secondly, and with the aim of maintaining the robot's balance, a third degree of freedom has been added which permits movement plane rotation around the direction of the robot's movement. The operation of this actuator will not be dealt with in this work, and only the two degrees of freedom acting on the movement plane will be described.

The mechanism has been designed under the restriction that the traction movement, when the end of the leg is in contact with the floor, is performed by only one actuator, the other actuator being inactive. This implies that when in contact with the floor, the action on the traction actuator gives rise to a straight trajectory. To obtain this goal, two four-bar mechanisms have been included (see Fig. 11): the first is formed of segments a, b, c and d, and the second is formed of f, g, h and i. The input bar of the first four-bar mechanism (triangle $d\ f\ e$) is the frame of the second mechanism, and the coupler of the first four-bar mechanism (segment c) is joined to the input bar of the second mechanism.

Horizontal movement (the first DOF) is thus established by acting on the DC motor, and vertical movement (the second DOF) is determined by changing the length of the output bar b of the first four-bar mechanism, which is accomplished by means of the linear actuator.

Five precision points along the traction trajectory (Fig. 11a) have been used for the synthesis process (Erdmann & Sandor, 1994). The relatively low number of precision points allows us to choose the length of certain segments of the mechanism, and those remaining have been obtained by imposing that point P reaches the five precision points. These values are listed in Fig. 11.

This solution yields a straight segment for the trajectory of point P, and the traction trajectory can therefore be accomplished without the use of the unit control to continuously track the movement. The operation of the leg is therefore considerably easier.

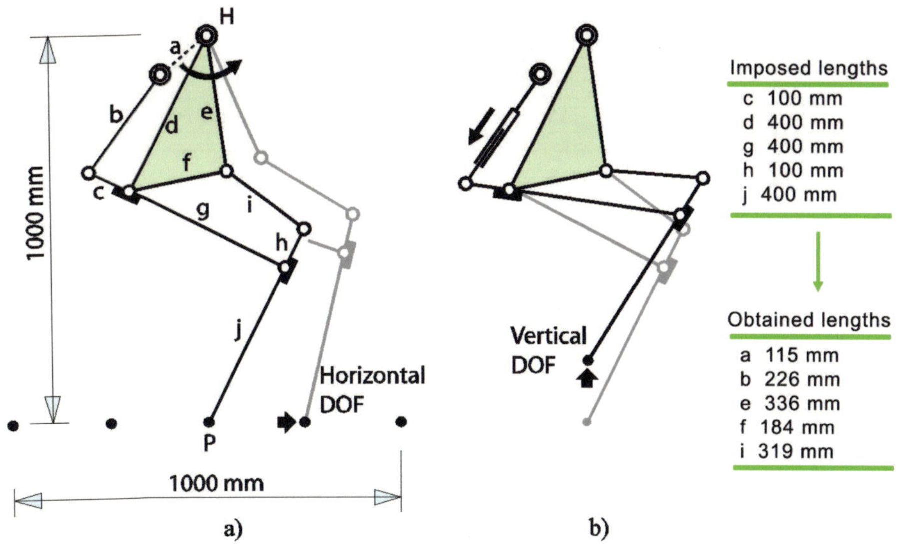

Fig. 11. a) Operation of first DOF of the proposed leg. b) Operation of the second DOF of the proposed leg with the starting and obtained lenghts.

Mechanisms that trace a straight segment have been used in some low cost robots (Ottaviano & Ceccarelli, 2002) with only one DOF, but they are not able to overcome obstacles, and do not therefore show any advantage over wheeled robots. This ability is provided by a second DOF which, in the present design, alters the length of *b* in Fig. 11.

With regard to this DOF, a new condition has been imposed: when the frame of the second four-bar mechanism (which is responsible for vertical movement) is fixed and the point *P* is in the middle point of the horizontal trajectory, the actuation on the second DOF must give rise to a vertical movement of this point *P*. Vertical and horizontal movement can thus be operated quasi-independently, and the length of the remaining segments and the angle of the frame can therefore be obtained.

Fig. 12 shows a scheme that improves the performance of the previous scheme by adding a new four bar mechanism (the segments *k*, *l*, *m* and *n*). The new four bar mechanism has been synthesized as a function generator by using four-point Freudenstein equations as is shown in (Erdman & Sandor, 1994). The points in the function generation have been chosen in order to obtain a constant velocity of point *P* in the central part of the trajectory, specifically within the central 600mm, that is the rated step length of the mechanism. Despite this rated value, the leg is capable of taking shorter or longer steps (up to 1 m).

Far from the conditions in which the synthesis has been made, the trajectories are not straight lines and there is some coupling between vertical and horizontal movement. However, this coupling does not interfere with the good execution of the step in normal operation. A more complicated control of the trajectories is required when it is necessary for the leg to overcome an obstacle.

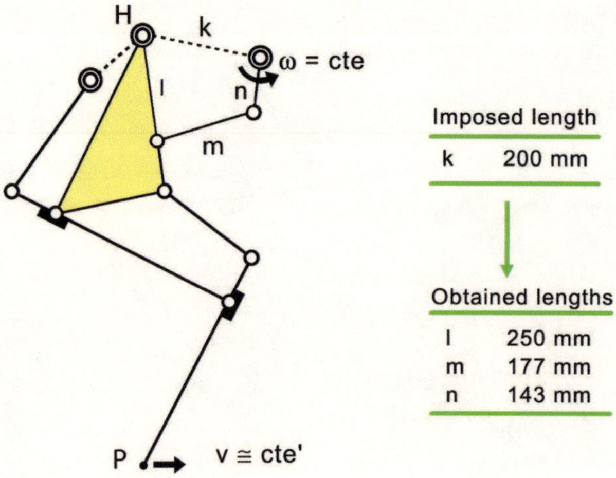

Fig. 12. A constant angular speed in the actuator is translated into a constant linear speed of point P

3.3. Simulation of the leg kinematics

In order to check the motion of the end *P* of the leg, a CAD model of the mechanism has been created, and an application to analyze mechanisms has been exported to ADAMS by following this geometry.

In the first step, simulations were performed to generate the workspace of the mechanism, shown in Fig. 13. The central line of the figure is the trajectory described by the support *P* when it commences a movement from the third precision point when only the linear actuator is being activated. The coordinate origin is at the hip H. As can be seen, the leg is able to overcome obstacles of up to 450 mm.

The validity of the synthesis is proven in Fig. 14, and the deviation of the support point P with regard to a straight line is presented in Fig. 14a) which shows variations lower than 2mm for a step of 0.6m long. This signifies a horizontality error of less than 0.4%, which is difficult to achieve by means of conventional robot legs when they are operated at normal speed.

Additionally, Fig. 14b) shows the linear speed of point P when the DC motor is driven at a constant speed of 150deg/s, and also for a step of 0.6m. For a vehicle speed of 1.5m/s, the speed variations are about 100mm/s, less than 7%. Although these are perfectly acceptable, and are also less than those of other legged robots, these variations between two legs in contact with the ground could give rise to the undesirable sliding of one of the legs.

With this design, since the horizontal movement of the leg can be accomplished by acting on a single actuator, the drive card is easily able to impose a torque control (the same value for both actuators) rather than a speed control to compensate the movements of both legs which, with the help of the inertia, makes the movement smoother and more energetically efficient.

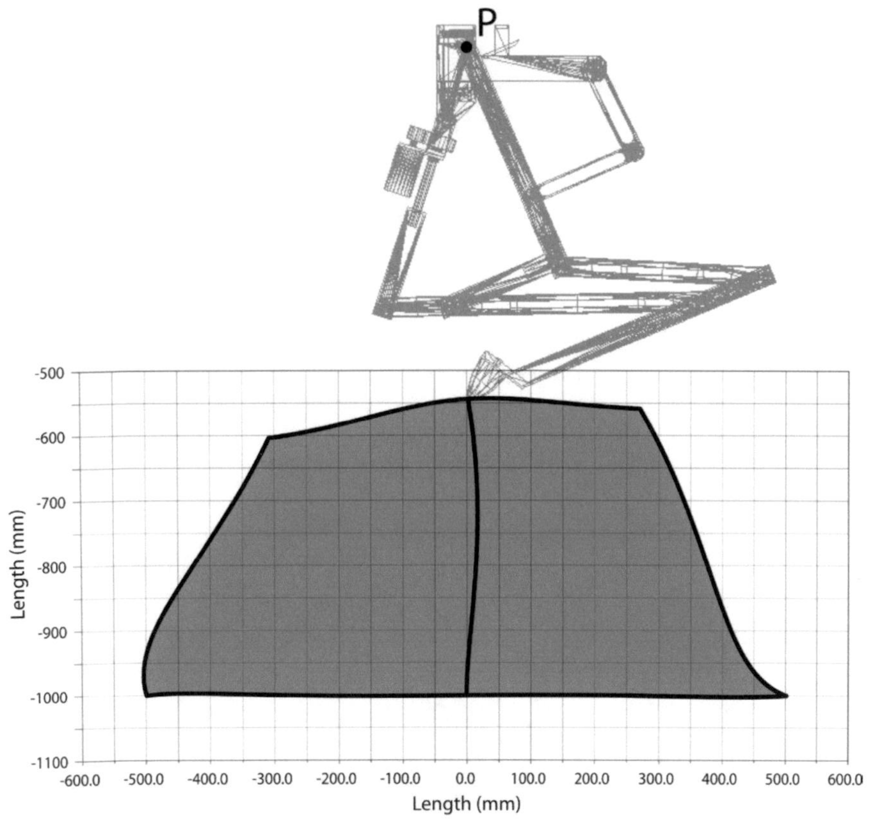

Fig. 13. ADAMS model of the leg and its workspace

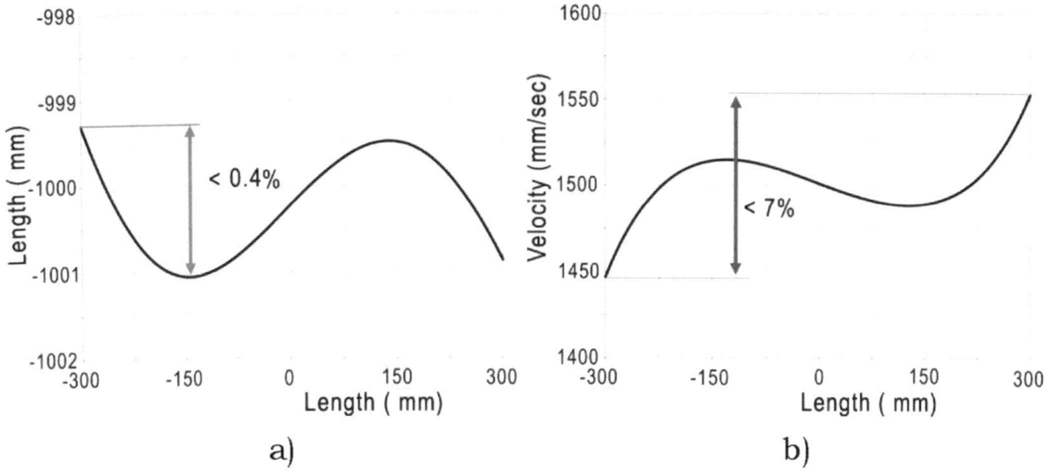

Fig. 14. a) Vertical position and b) horizontal speed of point P when only the horizontal movement motor is actuating

3.4. Simulations of the leg performance in a hybrid locomotion robot

The proposed mechanism was first included in a hybrid motion robot (Fig. 15) in which, as in the model described in (Chevallereau et al. 2003), the problem of lateral stability was initially disregarded. In our case two rear wheels have been added. The robot therefore consists of two articulated legs (those whose structure has been described) and two passive wheels located at the ends of another two articulated legs (those at the rear). These rear legs are fixed for movements on flat terrain, but will be controlled in future in order to overcome obstacles.

The 3D mechanical CAD program SolidWorks was first used to design the model, which was subsequently exported to ADAMS, although the rear legs ending in wheels were simplified as simple wheels, as can be seen in Fig. 15.

The values for stiffness and damping of the rubber support on the ends of the leg and the wheel tires have been experimentally obtained by means of a hydraulic actuator load cell. The obtained values have been incorporated into the ADAMS model in order to accomplish the dynamic simulation of the hybrid motion robot.

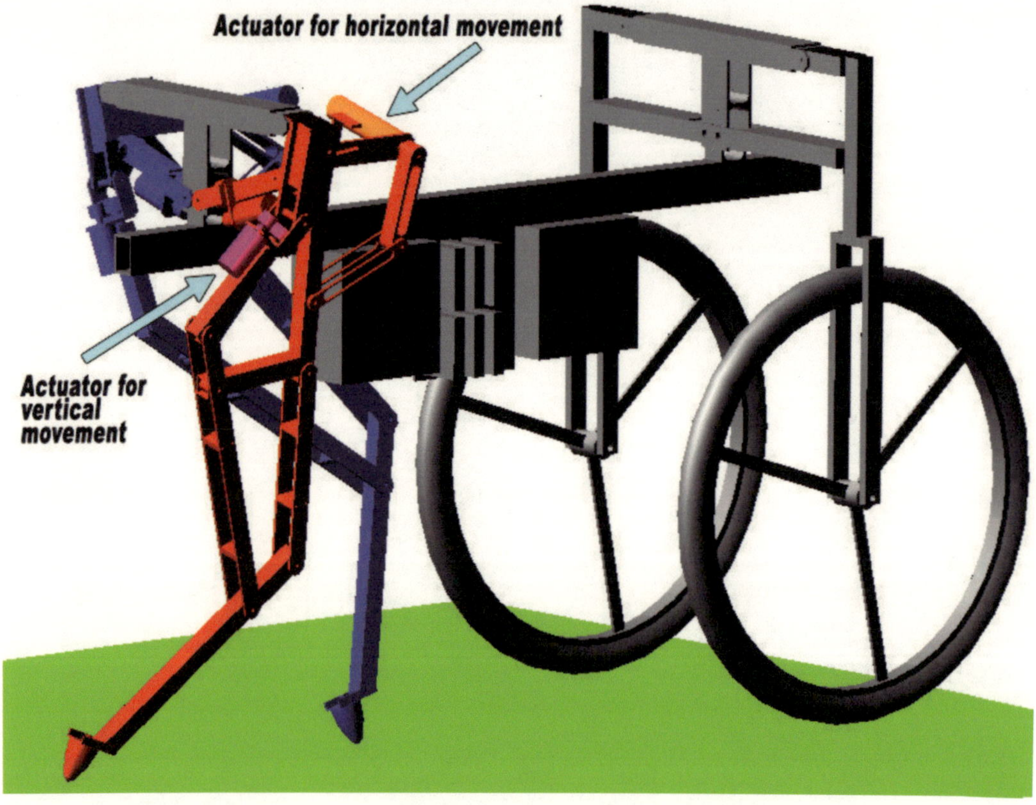

Fig. 15. ADAMS model of the hybrid locomotion robot used to test the proposed leg design

Figures Fig. 16 and Fig. 17 show, respectively, the speed patterns imposed on the linear actuator responsible for the vertical movement and on the DC motors for the horizontal movement. A transient stage can be seen at the beginning of the simulation with regard to the robot acceleration.

As can be seen, both the trajectories and the calculations involved are very simple. The velocity of the chassis center of mass is presented in Fig. 18, and has oscillations of less than 10% of the advance speed.

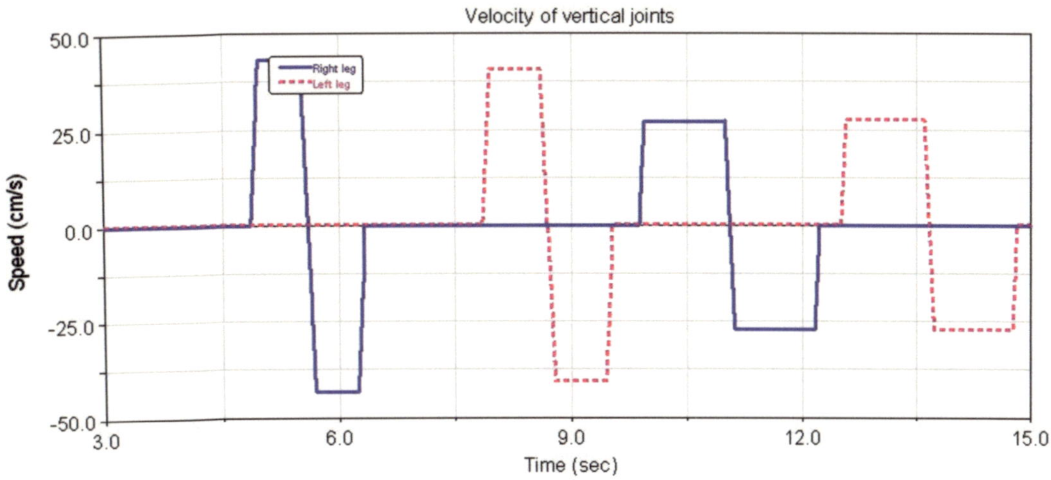

Fig. 16. Speed pattern for the actuator responsible for the vertical movement. The first patterns corresponds to the acceleration transient

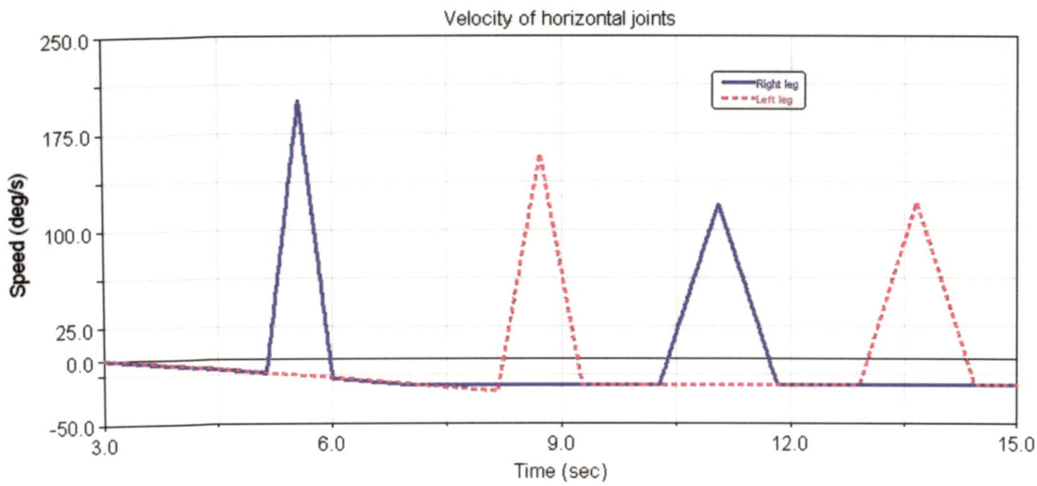

Fig. 17. Speed pattern for the actuator responsible for the horizontal movement

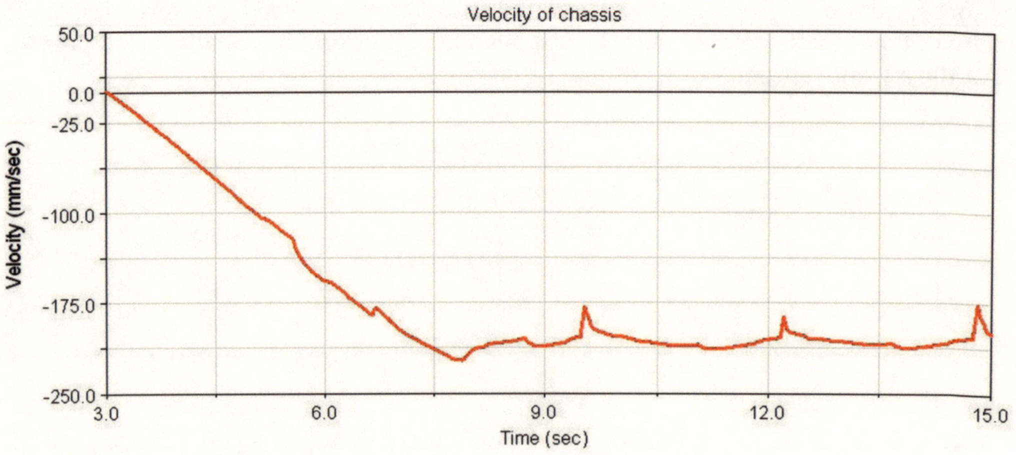

Fig. 18. Velocity of the robot chassis during acceleration and normal operation

Finally, the motor torques required to obtain the aforementioned speed patterns are shown in Fig. 19. The force for the vertical actuator appears as a dashed purple line, while the torque for the horizontal actuator is a solid blue line. The control signals for the actuators have been obtained through a closed loop PI controller, where the output speeds have been fed backward.

In order to perform a more realistic simulation, the control signals have been saturated to the maximum currents that the power electronic control units which will eventually be installed are able to supply.

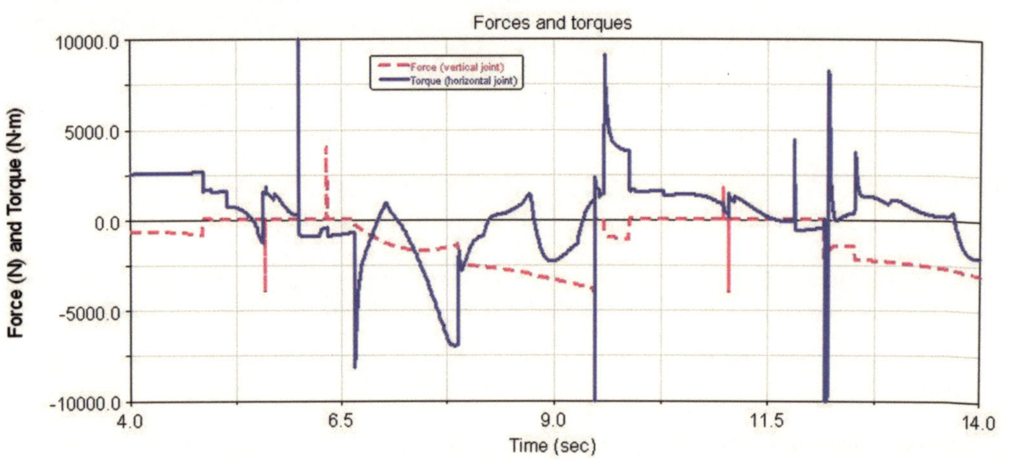

Fig. 19. Force and torque exerted by the actuators responsible for vertical and horizontal movements

4. Conclusions

Two different designs have been presented that contribute new approaches towards existing mechanical schemes.

The first prototype is a stair-climbing wheelchair designed with a new approach, based on splitting the stair-climbing process into two sub-problems: the ascent of each single step, and the positioning of the rear and front axles.
Each sub-problem is solved by using two independent mechanisms which are linked to create the final wheelchair. This new wheelchair has been modeled, built and tested. Its main features are:

a) The ability to climb any staircase built according to international standards.
b) Very high capacity load and weight ratio.
c) Stable equilibrium is guaranteed.

The second prototype consists in a new design for a leg to be used in legged or hybrid robots.
The characteristics pursued and the advantages of the proposed structure are the following:

- The horizontal and vertical movement of the support point are decoupled, at least when in contact with the ground.
- The control task is very simple, with only two vertical movement commands and three horizontal movement commands per step and leg being necessary.
- The low number of movement commands leads to a high speed and high efficiency performance.
- The legged structure allows the robot to overcome obstacles, although in this case the control is not so direct, and the movements are not so fast and efficient, owing to the fact that the trajectory must be imposed not only during the traction but also during the flight stage.
- The motors and actuators can be independently selected for the vertical and horizontal movements, which allows the designer to select the gearboxes in order to achieve a higher speed for the horizontal movement and the load capacity for the vertical movement, without increasing the actuator capacity, and therefore without increasing its volume, height or price.
- A torque control rather than a speed control can be accomplished which facilitates the synchronization between the legs and smoothes the movements.

In the case of both designs, the authors wish to point out the importance of a suitable mechanical scheme, fitted to the problem to be solved, in the global design of a robot. This allows the system to increase its efficiency and performance, discharging the unit control for an unnecessary task.

5. References

Chevallereau, C.; Abba, G.; Aoustin, Y.; Plestan, F.; Westervelt, E. R.; Canudas-De-Wit, C.; Grizzle & J. W. (2003). RABBIT: a testbed for advanced control theory. *Control Systems Magazine, IEEE*. Volume 23, Issue 5, (October 2003) pp 57 - 79, ISSN 0272-1708

Erdman, A. G.; Sandor, G. N. (1994). *Mechanism Design: Analysis and Synthesis,* Prentice-Hall, ISBN 978-0132677820, New Jersey

Germann, D.; Hiller, M. & Schramm, D. (2005). Design and control of the quadruped walking robot ALDURO. *Proceedings of 22nd International Symposium on Automation and Robotics in Construction, ISARC 2005*, pp 1-6. Ferrara, Italy, September 2005.

Gonzalez-Rodriguez, A.; Nava, N. & Gonzalez-Rodriguez, A. G. (2009). Design and validation of a novel actuator with adaptable compliance for application in human-like robotics. *Industrial Robots*. Vol. 36, No. 1, (2009), pp. 84-90, ISSN 0143-991X

Grizzle, J. & Poulakakis, I. (2008) Delft Pneumatic Bipeds. *Control Systems Magazine, IEEE*. Vol.28, no.4, (August 2008), pp.99-101, ISSN 0272-1708

Lawn, M. J.; Sakai, T.; Kuroiwa, M. & Ishimatzu T. (2001). Development and practical application of a stairclimbing wheelchair in Nagasaki. *Journal of Human Friendly Welfare Robotic Systems*. Vol. 2, No.2, (2001) pp 33-39, ISSN 1598-3250

Lawn, M. J. & Ishimatzu, T. (2003). Modeling of a Stair-Climbing Wheelchair Mechanism with High Single-Step Capability. *IEEE Transaction on Neural Systems and Rehabilitation Engineering*, Vol. 11, No. 3, (September 2003) pp 323-332, ISSN 1534-4320

McGeer T., Passive dynamic walking, *International Journal of Robotics Research*, Vol. 9, No. 2, (1990) pp. 62-82, ISSN 0278-3649

McGeer, T. (1990) Passive walking with knees, *Proceedings of IEEE Robotics & Automation Conference*, pp. 1640-1645, ISBN 1050-4729, Cincinnati, Ohio, May 1990

Morales, R.; Gonzalez, A.; Feliu, V. & Pintado, P. (2004). Kinematics of a New Staircase Climbing Wheelchair. *Proceedings of the 7th International Conference on Climbing and Walking Robots and the Support Technologies for Mobile Machines (CLAWAR 2004), pp* 249-264, ISBN 978-3-540-22992-6, September 2004, Springer Berlin

Morales, R.; Feliu, V.; González, A. & Pintado, P. (2006) Kinematic Model of a New Staircase Climbing Wheelchair and its Experimental Validation. *International Journal of Robotic Research*, Vol 25, No. 9, (2006), pp 825-841, ISSN 0278-3649

Ottaviano E. & Ceccarelli M. (2002). Optimal design of CaPaMan (Cassino Parallel Manipulator) with a specified orientation workspace, *Robotica* vol. 20, No.2 (March 2002), pp.159-166, ISSN 0263-5747

Yoneda, K.; Ota, Y. & Hirose, S. (2001). Development of a Hi-Grip Stair Climbing Crawler with Hysteresis Compliant Blocks, *Proceedings of 4th International Conference on Climbing and Walking Robots (CLAWAR 2001)*, pp. 569-576, ISBN 1-86058-365-2, Karlsruhe, Germany, September, 2001, *Professional Engineering Publishing,* Suffolk UK

3

A Wheel-based Stair-climbing Robot with a Hopping Mechanism

Koki Kikuchi, Naoki Bushida, Keisuke Sakaguchi, Yasuhiro Chiba, Hiroshi Otsuka, Yusuke Saito, Masamitsu Hirano and Shunya Kobayashi
Chiba Institute of Technology
Japan

1. Introduction

In this chapter, we introduce a stair-climbing robot developed in our laboratory. This robot consists basically of two body parts connected by springs, and hops as a result of the vibration of a two-degrees-of-freedom (2-DOF) system. The excellent combination between the frequencies of the robotic body vibration and the tread-riser interval of stairs enables a small and simple robot fast stair climbing, soft landing, and energy saving.

In an attempt to give the robot mobility in an environment such as an office building having steps and stairs, various mechanisms have been proposed and developed. Each one of which has different characteristics. For example, wheel-based robots are very simple in terms of both mechanical design and control, and they can travel quickly and stably. But their size tends to be big for climbing stairs, as they cannot surmount a riser higher than their wheel radius. On the other hand, although crawler-type robots can climb over a riser higher than a wheel-based robot, they are slow and noisy. Typical examples of crawler-type robots are TAQT (Hirose et al., 1992) that can carry a human and Kenaf (Yoshida et al., 2007) for rescue operations. Legged robots, especially humanoid ones, are well suited for climbing stairs, but require many DOFs and complex control. Honda's ASIMO (ASIMO OFFICIAL SITE), AIST's HRP (Harada et al., 2006) and Waseda University's legged robot (Sugahara et al., 2007) are good examples. In addition, the hybrids of these types have been proposed and have improved upon mutual demerits. Chari-be (Nakajima et al., 2004), with two wheels and four legs, travels quickly on its wheels over flat terrain, and climbs using its legs in rough terrain such as a step and stairs. A biped-type robot with a wheel at the tip of its legs (Matsumoto et al., 1999) climbs stairs smoothly. RHex (Altendorfer et al., 2001) has six compliant rotary legs and travels speedily not only up and down stairs, but also even uncertain terrain such as a swamp. Moreover, modular robots such as an articulated snake-like robot and special mechanisms for stairs have also been proposed. Yim's snake-like robot (Yim et al., 2001) climbs stairs, transforming its own loop form into a stair shape. These excellent mechanisms have improved the manoeuvrability of the robot for rough terrain, but as most are general-purpose robots for rough terrain, a more specialized mechanism must be developed if we focus solely on stair-climbing ability in an office building.

From this point of view, we have developed a stair-climbing robot specifically for use in an office building (Sakaguchi et al., 2007, Asai et al. 2008 & Kikuchi et al., 2008). Our robot cannot climb stairs with various-height or irregular risers smoothly, but it does climb stairs with a priori determined regular risers rapidly (less than 1.0 s per step), softly (less than the impact at takeoff, at the landing point), and economically. Furthermore, the mechanical design and control are quite simple, and additionally the height of the robot is almost the same as the common stair riser. These features are very important for practical tasks such as monitoring the situation in an office. Here, we introduce the hopping mechanism and property, and show the experimental result of the fast stair climbing and soft-landing.

2. Hopping mechanism and robotic design

The mathematical hopping model consists of two mass points, m_1 and m_2, connected by a spring, as shown in Fig. 1. The lower mass point, m_2, hops if, and only if, the lifting force provided by the spring, $k(z_1\text{-}z_2)$, and the wire, T_w, exceeds the force of gravity on the lower mass, m_2g. The trajectories of the two mass points during hopping evolve based on the reduced mass, the mass ratio between the upper and lower masses, spring constant, k, the friction of the shaft, and stored spring energy. Figure 2 shows a manufactured robot with this hopping mechanism. The robot consists of an upper body part (Body 1) and a lower body part (Body 2) connected by four springs and a wire. Here, the upper body part has a CPU (H8/tiny) for control, a receiver, a position sensitive detector (PSD), a reel with a gear and a motor, a solenoid, and batteries (CPU: 7.4V, motor: 30V for the reel and 22.2V for the travel, and solenoid: 44.4V). The lower body part has two motors for translational travel, four 56-mm diameter wheels, four shafts and two acceleration sensors for the z direction. The robot first stores the spring energy by reeling in the wire. The reel mechanism is then detached by the solenoid, and the robot hops by releasing the stored spring energy. Here, body parts 1 and 2 correspond to the mass points of the two-dimensional mathematical model shown in Fig. 1. The robot is 370 mm tall, 155 mm wide and 140 mm long. Also, the robot has no suspension to act as a damper, as we scrutinize the impact acceleration of body parts.

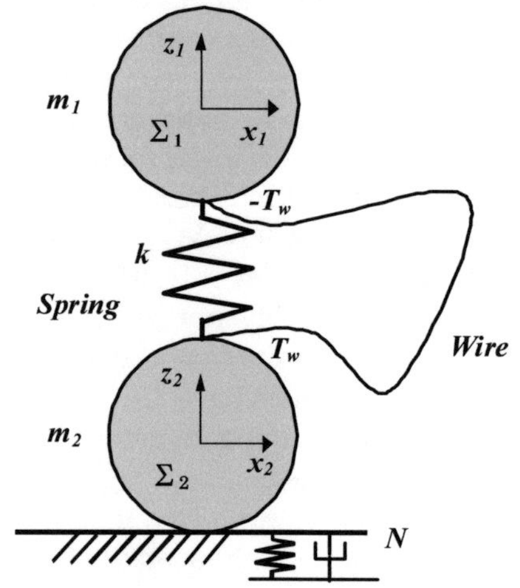

Fig. 1. Two-dimensional mathematical model of hopping mechanism

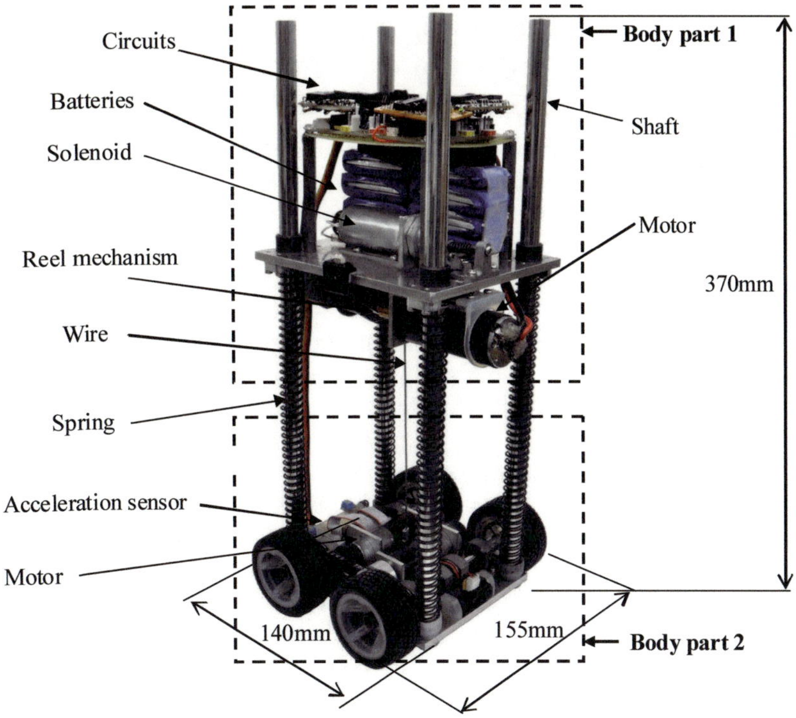

Fig. 2. Wheel-based robot with hopping mechanism

3. Mathematical model for simulation

The robot is modeled simply as a 2-DOF spring-mass system. If the posture of the robot can be neglected, the equations of motion are given by

$$(m_1+m_2)\, x_1'' = f_x \tag{1a}$$

$$m_1 z_1'' + k(z_1 - z_2) = - m_1 g - T_w - \mu_t F_f \tag{1b}$$

$$m_2 z_2'' + k(z_2 - z_1) = - m_2 g + T_w + \mu_t F_f + N$$

where (x_1, z_1) and (x_2, z_2) are coordinates for each body part ($z_1=z_2$ at the natural length of the spring), m_1 and m_2 are the masses of body parts 1 and 2, k is the spring constant, f_x is the motor force for horizontal travel, $\mu_t F_f$ is the friction of the shaft (the magnitude of F_f is determined by the pilot experiment and the sign is determined by the relative vertical velocity between the two masses, $d(z_2-z_1)/dt$), and N is the reaction of the ground. The ground is simply modeled by a spring and a damper, as shown Fig. 1. T_w, the wire tension is a positive value or zero for tensional or relaxant conditions, respectively. Note that as the posture of the robot is neglected for simplification, x_1 is always equal to x_2.

Here, the condition for takeoff of lower mass, z_2, is given by

$$m_2\, g < k(z_1 - z_2) + T_w - \mu_t F_f \tag{2}$$

Assuming that the friction of shaft, $\mu_t F_f$, is neglected ($F_f=0$) and that we do not control the wire tension ($T_w=0$) during hopping, the trajectories of the two masses at the time t after takeoff, i.e., during hopping ($f_x=0$, $N=0$, and $z_2>0$), are as follows,

$$x_1 = x_2 = v_x\, t + D \tag{3a}$$

$$z_1 = h\, M/m_1 \sin(\omega\, t + \varphi) - g\,(t - T)^2/2 + C \tag{3b}$$

$$z_2 = - h\, M/m_2 \sin(\omega\, t + \varphi) - g\,(t - T)^2/2 + C$$

$$\omega = (k/M)^{0.5},\ M=1/(1/m_1 + 1/m_2) \tag{3c}$$

where v_x is the horizontal velocity at takeoff, D is the starting point of takeoff, h, φ, T, and C are constants determined by the initial conditions, ω is the angular frequency, and M is the reduced mass between the two masses. Hence, the horizontal velocity is constant during hopping. The air resistance is neglected. The first terms on the right side of the z equations, Eq. (3b), represent vibration caused by the two body parts. Although two natural frequencies must exist because the system has a 2-DOF configuration, the lower natural frequency is zero in this case. The second terms on the right side of the z equations represent the parabolic motion of the center of mass (COM). Consequently, the hopping motion of the robot after takeoff is represented by the combination of the vibration of the mass points and the parabolic motion of the COM. Hence, the point at which the velocity in the z direction of the vibration of body part 2 and that of the parabolic motion of the COM are canceled out, is the soft-landing point. For later discussion, we define "soft-landing" as a landing in which the vertical velocity of the lower body part is zero ($z_2' \approx 0$), the acceleration is zero ($z_2'' \approx 0$), and the third differential is zero or negative ($z_2''' \leq 0$) at landing height, H (where H is the riser height). Figure 3 shows two typical examples of the trajectories of the mass points, z_1 and z_2, during hopping. The red dashed line depicts the hopping for the higher riser, and the blue solid line is that for the lower riser. The points "A" and "B" are ideal landing points, i.e., mathematical stationary and inflection points. These robots can softly climb stairs with risers of H = 0.1 [m] and H = 0.2 [m], respectively. Note that the closer the z_2''' is to zero, the lower the landing impact for noise, because the lower body part can fly parallel to the tread.

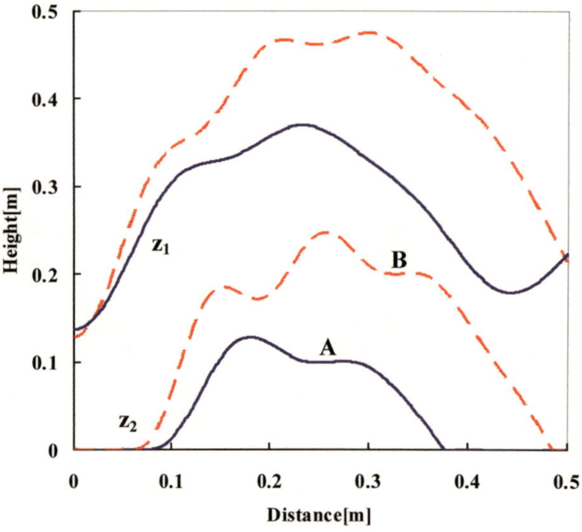

Fig. 3. Two typical examples of trajectories of mass points, z_1 and z_2, during hopping

4. Hopping properties

In this section, we present the characteristics during the hopping motion. The trajectories of the robot change, depending mainly on passive parameters such as reduced mass, the mass ratio between the upper and lower masses, and the spring constant and active parameters such as initial spring contraction, wire tension, and horizontal traveling driving force. Passive parameters are mechanical design parameters and should be designed a priori for the specification of the stairs in an office building. On the other hand, active parameters are control parameters and can be changed in accordance with a local irregular step, etc.

We first show the frequency characteristic, one of the characteristics of the passive parameters, during the hopping motion. Figure 4 shows the relation between the reduced mass and the angular frequency for three spring constants. Here, the lines depict the results obtained by Eq. (3c) and the points are the results performed by 10 hopping experiments for each point. The passive parameters are as follows: The spring constants, k, are 800, 1,200, and 1,600 N/m. In the experiments, reduced masses, M, of 0.4, 0.5, 0.6, 0.7 and 0.8 kg are used. Here, when the reduced mass is 0.6 kg and the mass ratio is 2.0, the masses of the upper and lower body parts mean 1.8 kg and 0.9 kg, respectively, and the total mass is 2.7 kg. This figure implies that the angular frequency can almost be controlled by ω in Eq. (3c), that is, the reduced mass and the spring constant, because, although the experimental values became slightly higher than the simulation results in accordance with the increase in the spring constant, the errors can be estimated simply from the figure and the standard deviations were also very small.

Second, we show the trajectories of body parts 1 and 2 for mass ratios, m_1/m_2, of 0.5 (Case A) and 4.0 (Case B), as two typical examples obtained in numerical simulations (Fig. 5). Here, the reduced mass, M, is constant at 0.7 kg, the spring constant, k, is 2,000 N/m, and the initial contraction of the spring, h, is 0.1 m. The thin lines represent body part 1 and the bold lines are body part 2. The dashed thin lines represent the velocities of body part 1 and the dashed bold lines are the velocity of boby part 2. This figure shows that the amplitudes of body parts 1 and 2 depend on the mass ratio, as shown by the vibration term in Eq. (3b). That is, if the reduced mass is constant, the amplitude of body part 1 is large when the mass of body part 2 is large, and the amplitude of body part 2 is large when the mass of body part 1 is large. In addition, the hopping height of the COM increases with decreasing the mass of body part 2, m_2. Furthermore, the possible points of soft-landing ($z_2'\approx 0$, $z_2''\approx 0$, and $z_2''' \leq 0$) exist in the neighborhood of crests of the vibration of body part 2, as shown in Cases A and B. Note that we cannot choose the highest hopping point, i.e., a vertex, as the landing point, because the robot has a physical body length, L, and cannot land both the front and rear wheels on the stairs simultaneously. Additionally, we also cannot reduce the mass of body part 2 dramatically, as a main drive unit such as a motor is mounted on body part 2.

Next, as active hopping characteristics, Fig. 6 shows examples of the hopping motion for the initial contractions of the spring of h = 0.070, 0.085, and 0.100 [m]. Here, the reduced mass, M, is 0.7 kg, the mass ratio, m_1/m_2, is 2.0, the spring constant, k, is 1,600 N/m and the horizontal velocity, v_x, is 1.2 m/s. From this, we find that with the increase in the initial contraction of the spring, h, the temporal axis of the parabolic motion of the COM, T, is shifted to the right, the hopping height of the COM is increased, but the angular frequency, ω, is unchanged. Also, we can simply find that the spatio-frequency represented by x-z coordinates can be controlled by the horizontal velocity v_x. The slower the horizontal velocity, the higher the spatio-frequency. The correspondence between the spatio-frequencies of the lower body part trajectory and the stair configuration, i.e., the tread-riser interval, enables economic and stable stair climbing by the robot. Moreover, as the wire can pull the body parts mutually, the robot can actively control the possible point of soft landing by the wire tension.

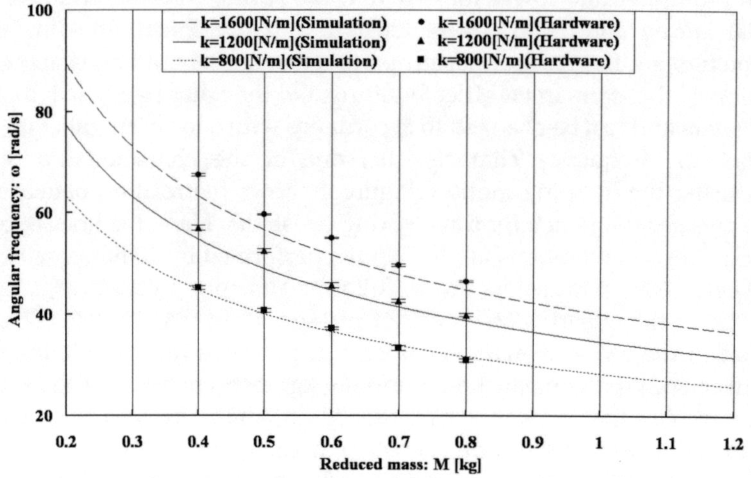

Fig. 4. Relation between reduced mass and angular frequency for different spring constants

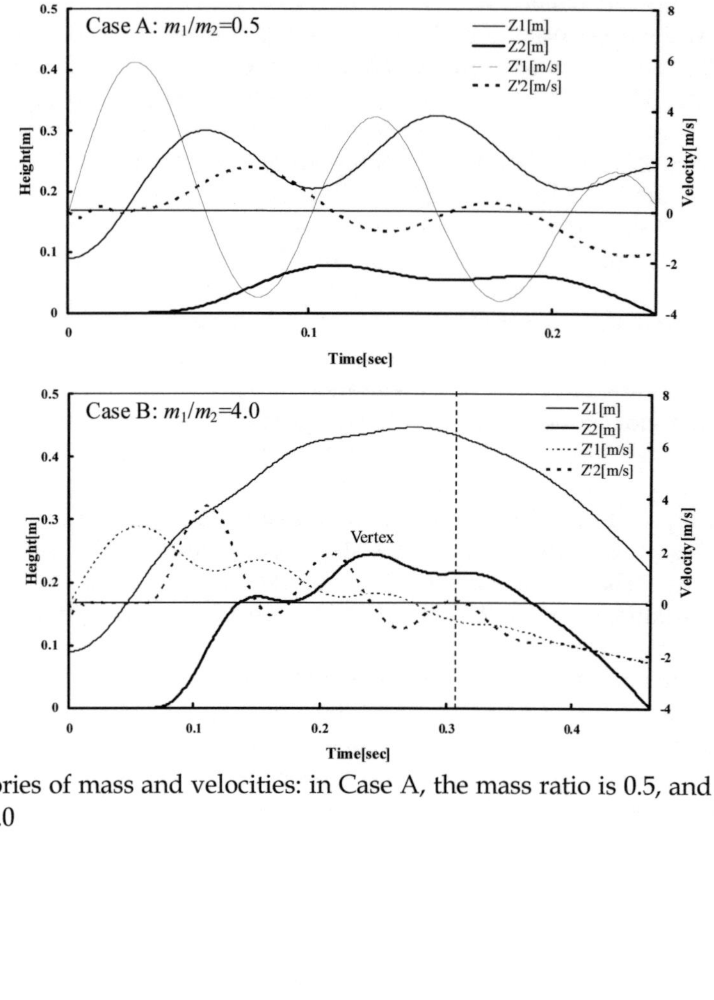

Fig. 5. Trajectories of mass and velocities: in Case A, the mass ratio is 0.5, and in Case B, the mass ratio is 4.0

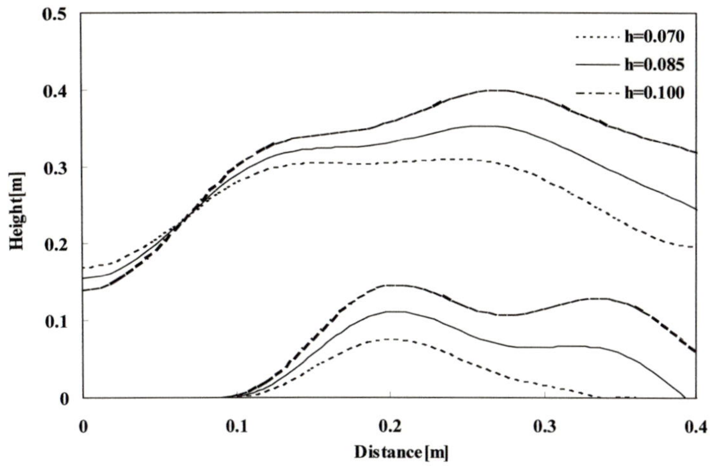

Fig. 6. Trajectories of mass points 1 and 2 during hopping for different initial spring contractions

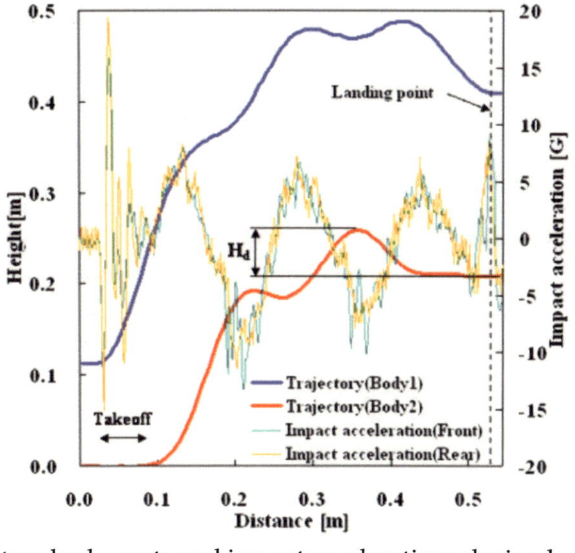

Fig. 7. Trajectories of two body parts and impact accelerations during hopping motion

5. Demonstration of stair climbing

5.1 Climbing up a step

We first demonstrate fast and soft climbing up a riser 0.21 m in height. From the above discussion, we can control the landing point and condition by adjusting the mechanical design and control parameters. Here, we set the parameters as follows: the reduced mass, M, of 0.55 kg, the mass ratio, m_1/m_2, of 1.32, the spring constant, k, of 1,600 N/m, the initial contraction of the spring, h, of 0.13 m, and the horizontal velocity, v_x, of 1.2 m/s. Note that the value of 0.21 m is 7.5 times higher than the wheel radius of the robot and corresponds approximately to the common riser height of stairs. Also, the contraction of the spring, h, is measured by the PSD with a sampling time of 30 ms. Figure 7 shows the trajectories of the robot during the hopping motion and the impact accelerations obtained from the front and rear sensors mounted in the lower body part. The impact acceleration at the moment of landing was approximately 8 G, which was less than the maximum acceleration during takeoff, 19 G, and was close to that experienced during flight, i.e., almost 10 G. As the impact acceleration by free-fall from the maximum hopping height to the step (H_d shown in Fig. 7) was approximately 33 G, the soft-landing of this robot reduced the impact by 76%. However, high impact acceleration at the moment of takeoff was observed, unfortunately. This is not due to impact with the ground, but rather due to the plate deflection of body part 2. The actual acceleration of lower body part at the moment of takeoff was less than 10 G. We need to improve the geometrical moment of inertia of body part 2. Note that we realized fast and soft step climbing by 0.27 m in height.

5.2 Climbing up a flight of stairs

Next, we demonstrate fast and soft stair climbing. The trick in stair climbing is to synchronize the spatio-frequency of the stairs, i.e., the tread-riser intervals, and the body vibration. It is simple and easy in a mathematical model, but it is not in practice. Figure 8 shows three processes for continuous hopping–takeoff, landing, and reeling in–and the tread length required for each hop. The horizontal traveling distance during these three processes, the required tread length, can be quite simply controlled by the horizontal velocity v_x, if the robot length, L, is zero. However, the following constraints exist in practice. First, after takeoff, the front wheels must jump up to the edge of the step (takeoff phase in Fig. 8), next, before landing, the rear wheels must clear the edge of the step (landing phase in Fig. 8), and then the robot must reel in the wire for next hopping (reeling-in phase in Fig. 8). When the horizontal traveling distance during these three phases, $D_T+D_L+D_R$, is equal to or less than the tread length, the robot can climb a flight of stairs. Thus, this is the minimum required tread length, and the shorter, the better. From these constraints, the robot must jump up to the riser height, H, at $t = D_T/v_x$ and the minimum landing phase distance, D_L, is equal to the body length, L. Also, since the reeling-in phase distance, D_R, depends only on the motor torque to reel in the wire, the larger the motor torque, the shorter the reeling-in phase distance. However, as the exceedingly high-power motor makes the upper mass heavy and the wire tension strong, it lifts up the lower body part. Additionally, although the reeling-in phase distance can be shortened by reeling in the wire before landing, the control of the soft landing point becomes difficult as the passive vibration characteristics change. Thus, the wire is reeled in after landing for simplification in this experiment.

Figure 9 shows an example of trajectories of two body parts (blue and red lines) and impact accelerations (green and orange lines) based on the following parameters: the reduced mass, M, of 0.79 kg, the mass ratio, m_1/m_2, of 2.17, the spring constant, k, of 2,000 N/m, the initial spring constriction, h, of 0.12 m, and the horizontal velocity, v_x, of 0.90 m/s for the riser height, H, of 0.20 m. Additionally, Fig. 10 shows stroboscopic pictures of continuous hopping to climb two steps. As shown in Fig. 9, the required tread length was 0.74 m (Note that to avoid clashing into the riser wall, a margin safety of 2.1 was introduced. The required tread length obtained by the numerical simulation was 0.35 m). The impact accelerations were 28 G and 37 G for first and second takeoffs and 10 G and 6 G for first and second landings. The impact acceleration at the moment of landing was less than that during flight.

Also, the stair climbing time per step was 0.77 s. To shorten the required tread length is one of the future tasks as the common tread length is almost 0.4 m.

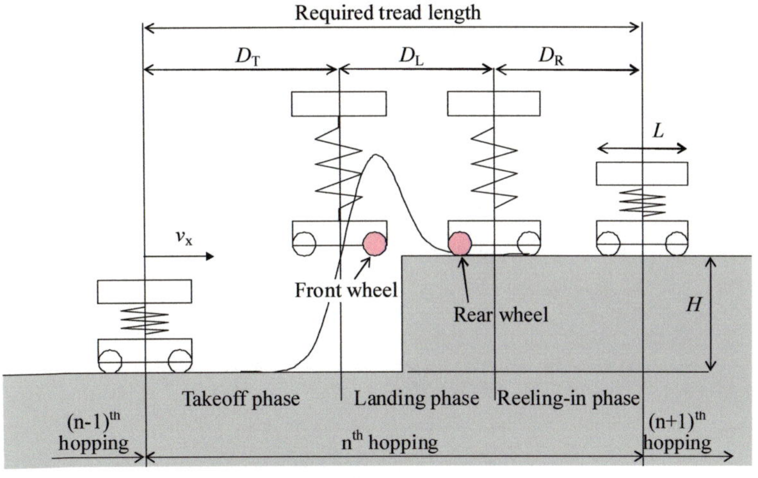

Fig. 8. Hopping processes and required tread length

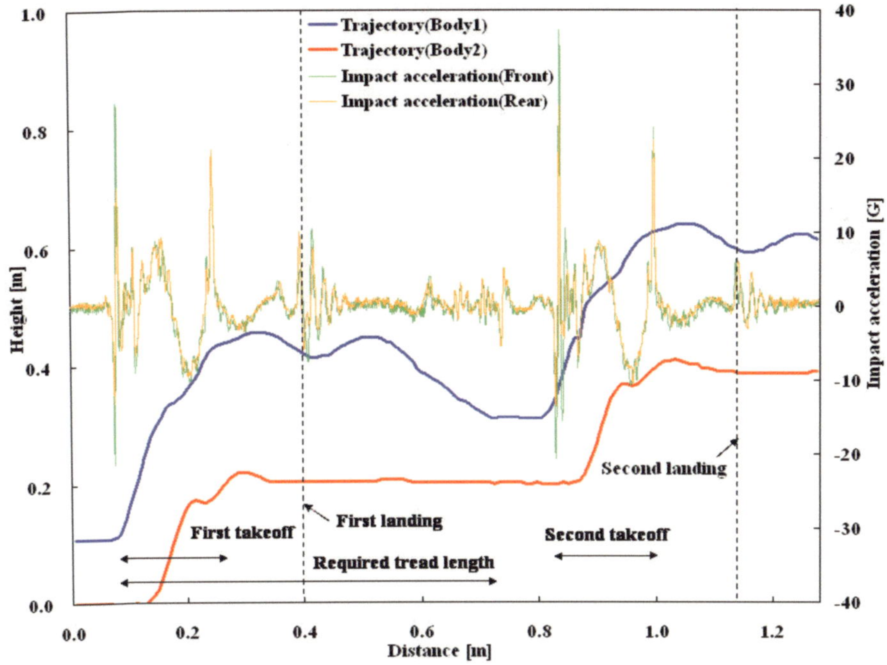

Fig. 9. Trajectories of two body parts and impact acceleration during stair climbing

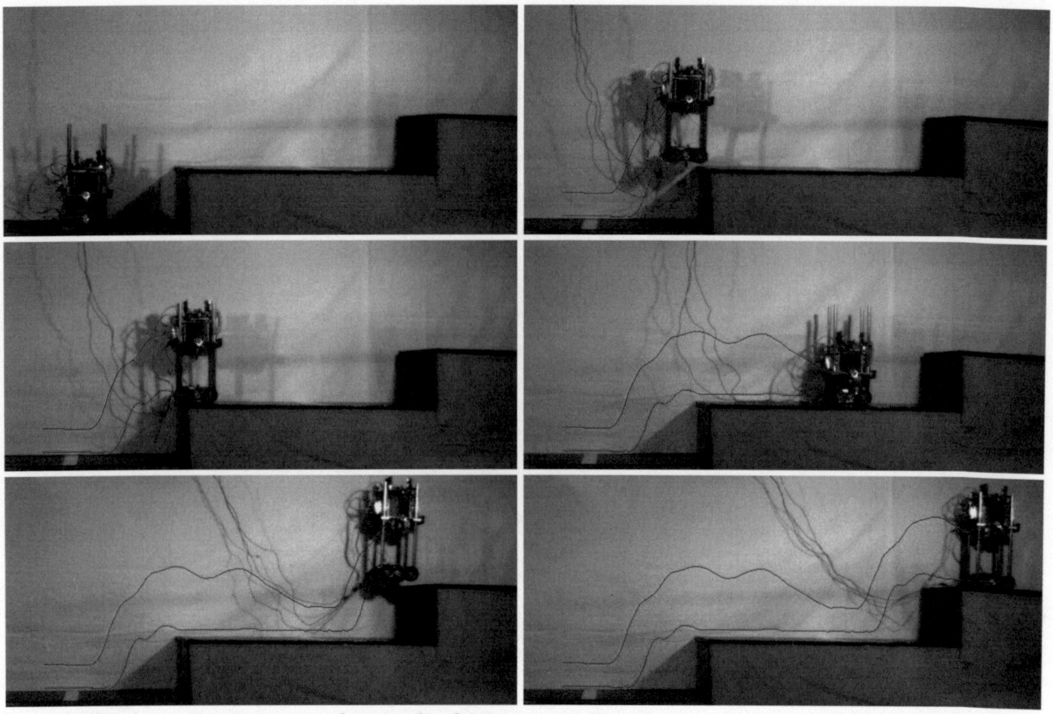

Fig. 10. Stroboscopic images of stair climbing

6. Demonstration of step descending

Finally, we demonstrate fast and soft descending of steps 0.20 m in height. The hopping mechanism is almost the same as for climbing. However, the degree of difficulty is quite different. As mentioned in Section 3, the soft-landing point is the location at which the velocity in the z-direction of vibration of lower body part, $-hM\omega/m_2 \cos(\omega t+\varphi)$, and that of the parabolic motion of the COM, $-g(t-T)$, are canceled out. Here, although the maximum of the former is $hM\omega/m_2$, the latter becomes a large negative value with time, t, because of descending. In climbing, as the robot lands near the top of the parabolic motion, as shown in Fig. 8, and the descending velocity by parabolic motion is low, there are many parameters, $hM\omega/m_2 \cos(\omega t+\varphi)$, which can cancel out the descending velocity. In contrast, in descending, as the robot lands considerably below the top of the parabolic motion, as the dashed line shows in Fig. 11, and the descending velocity is very high, the parameters, $hM\omega/m_2 \cos(\omega t+\varphi)$, which can cancel it out, decrease dramatically. Thus, we use another technique in descending. Hence, the robot does not jump up, but glides from the step horizontally, starts to vibrate by detaching the reel mechanism while descending, and then lands softly, as the solid line shows in Fig. 11. This method requires posture control at takeoff, but decreases the descending velocity by the parabolic motion on landing and makes the required tread length short.

Figure 12 shows the trajectories of two body parts (blue and red lines) and impact accelerations (green and orange lines) during the hopping motion. Here, the parameters are: the reduced mass, M, of 0.74 kg, the mass ratio, m_1/m_2, of 2.04, the spring constant, k, of

1,200 N/m, the initial contraction of the spring, h, of 0.11 m, and the horizontal velocity, v_x, of 1.0 m/s. The impact acceleration at the moment of landing was approximately 14 G, which was close to that experienced during flight, i.e., almost 10 G. As the impact acceleration under free-fall from the riser height to the step was 77 G, the soft-landing of this robot reduced the impact by 82%. Figure 13 shows stroboscopic images of step descending. The posture at takeoff was controlled by a wheelie.

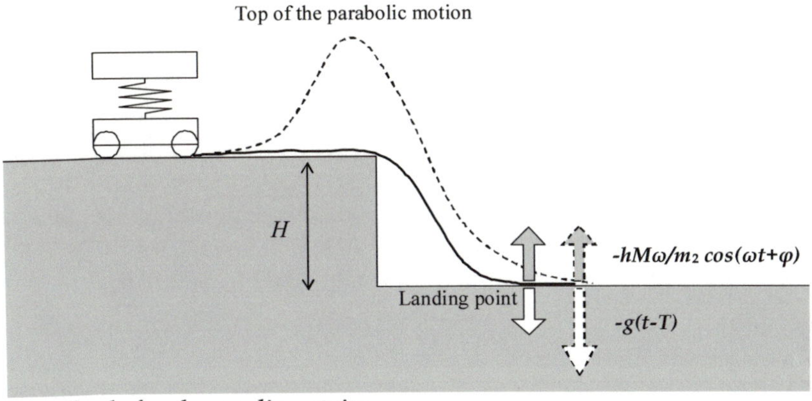

Fig. 11. Two methods for descending stairs

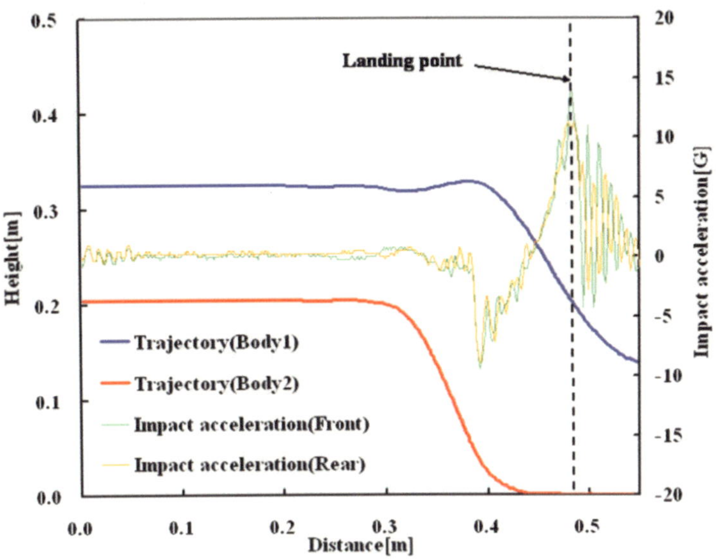

Fig. 12. Trajectories of the two body parts and impact accelerations during hopping motion

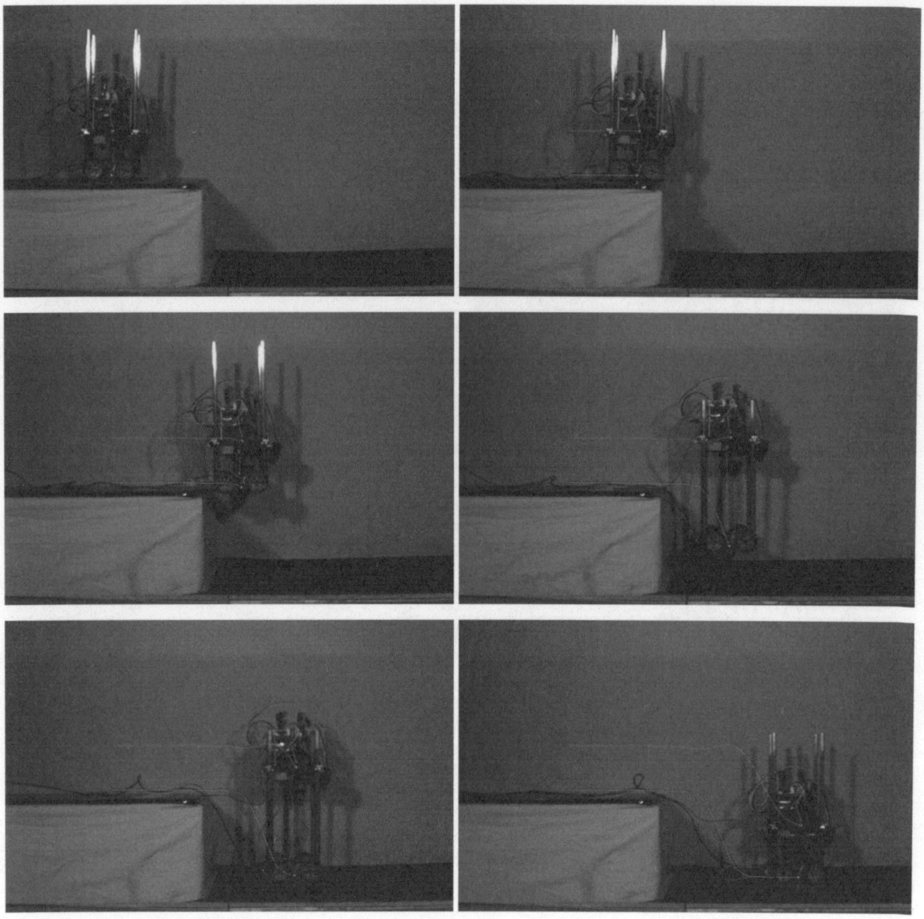

Fig. 13. Stroboscopic images of step descending

7. Conclusion

We introduced a wheel-based stair-climbing robot with a hopping mechanism for stair-climbing. The robot, consisting of two body parts connected by springs, climbed stairs quickly, softly, and economically by using the vibration of a two-degrees-of-freedom system. In the future, we intend to shorten the required tread length by controlling the wire tension and minimizing the body length to realize a practical stair-climbing robot.

8. References

Altendorfer, R.; Moore, E.Z.; Komsuoglu, H.; Buehler, M.; Brown, H.; McMordie, D.; Saranli, U.; Full, R. & Koditschek, D.E. (2001). A Biologically Inspired Hexapod Runner, Autonomous Robots, Vol. 11, (month 2001), pp. 207 - 213

Asai, Y.; Chiba, Y.; Sakaguchi, K.; Sudo, T.; Bushida, N.; Otsuka, H.; Saito, Y. & Kikuchi, K. (2008). Wheel-Based Stair-climbing Robot with Hopping Mechanism - Demonstration of Continuous Stair Climbing Using Vibration-, Journal of Robotics and Mechatronics, Vol. 20, No. 2, Apr. 2008, pp. 221-227

ASIMO OFFICIAL SITE : http://www.honda.co.jp/ASIMO/

Harada, K.; Kajita, S.; Kaneko, K. & Hirukawa, H. (2006). Dynamics and Balance of a Humanoid Robot during Manipulation Tasks, IEEE Transaction on Robotics, 2006, vol. 22, no. 3, pp. 568-575.

Hirose, S.; Sensu, T. & Aoki, S. (1992). The TAQT Carrier: A Practical Terrain-Adaptive Quadru-Track Carrier Robot, Proceedings of IEEE/RSJ International conference on Intelligent Robots and Systems, July 1992, pp. 2068-2073, Tokyo

Kikuchi, K.; Sakaguchi, K.; Sudo, T.; Bushida, N.; Chiba, Y. & Asai, Y. (2008). A study on wheel-based stair-climbing robot with hopping mechanism, MECHANICAL SYSTEMS AND SIGNAL PROCESSING (MSSP), Aug. 2008, Vol. 22, Issue 6, 1316-1326, ELSEVIER

Matsumoto, O.; Kajita, S.; Saigo, M. & Tani, K; (1999). Biped-type leg-wheeled robot, Advanced Robotics, 13(3), Oct. 1999, pp.235-236.

Nakajima, S.; Nakano, E.; & Takahashi, T.; (2007). Motion Control Technique for Practical Use of a Leg-Wheel Robot on Unknown Outdoor Rough Terrains, Proceedings of IEEE/RSJ International conference on Intelligent Robots and Systems, vol.1, (Month 2004), pp. 1353-1358

Sakaguchi, K.; Sudo, S.; Bushida, N.; Chiba, Y.; Asai, Y. & Kikuchi, K. (2007). Wheel-Based Stair-climbing Robot with Hopping Mechanism -Fast Stair-climbing and Soft-landing by Vibration of 2-DOF system-, Journal of Robotics and Mechatronics, Vol. 19, No. 3, Jun. 2007, pp. 258-263

Sugahara, Y.; Carbone, G.; Hashimoto, K.; Ceccarelli, M.; Lim, H. & Takanishi, A. (2007). Experimental Stiffness Measurement of WL-16RII Biped Walking Vehicle during Walking Operation, Journal of Robotics and Mechatronics, Vol. 19, No. 3, Jun. 2007, pp. 272-280

Yim, M. H.; Homans, S. B. & Roufas, K. D. (2001). Climbing with snake-like robots, IFAC Workshop on Mobile Robot Technology, Korea, May 2001, pp. 21-22, Jejudo.

Yoshida, T.; Koyanagi, E.; Tadokoro, E.; Yoshida, K.; Nagatani, K.; Ohno, K.; Tsubouchi, T.; Maeyama, S.; Noda, I.; Takizawa, O. & Hada, Y. (2007). A High Mobility 6-Crawler Mobile Robot "Kenaf", Proceedings of 4th International Workshop on Synthetic Simulation and Robotics to Mitigate Earthquake Disaster (SRMED2007), July, 2007, p. 38, Atlanta

4

Motion Control of a Four-wheel-drive Omnidirectional Wheelchair with High Step Climbing Capability

Masayoshi Wada
Dept. of Mechanical Systems Engineering
Tokyo University of Agriculture and Technology
Japan

1. Introduction

In recent years, aging problem has been arising to be among the most serious social issues world wide, especially in some European and Asian countries, involving Japan. It is reported in Japan that the population of over 65 years old would reach 30,000,000 in 2012 and grow over 30% of total population in 2025[1].

Electric wheelchairs, personal mobiles, scooters are currently commercially available not only for handicapped persons but also for elderly. However, such a rapid grow of aging populations suggest that requirements for electric mobile systems will soon increase dramatically for supporting mobility and activity of elderly people and reducing labor of care-givers.

However, those mobile systems do not have enough functionalities and capabilities for moving around existing environments including step, rough terrain, slopes, gaps, floor irregularities as well as insufficient traction powers and maneuverability in crowded areas. Promotion of barrier-free environments will be required for a large number of users of wheelchairs and other electric mobile systems however, re-constructing of the existing facilities could not be a feasible solution because of the limitations in economy and time.

For overcoming the problem, to improve the mobility of the electric mobile systems to adapt to existing environments could be one solution. For this objective, we propose a new type of wheelchair, four-wheel-drive (4WD) omnidirectional system, with enhanced step climb capability together with high maneuverability. In this chapter, omnidirectional control of a wheelchair with 4WD mechanism would be mainly discussed.

The mobile systems realizing holonomic and omnidirectional motion is one of the important research area in mobile robots. It provide flexibility and high maneuverability to motion planners and human drivers. The holonomic and omnidirectional mobile capability is very convenient for human drivers since they do not have to understand drive mechanisms and its configuration at all. A human only commands the direction and velocity of motion he/her wants to perform since a holonomic and omnidirectional mechanism can start to move in any direction with any configuration of the mechanism such as directions of wheels.

This characteristics is vary suitable for wheelchairs and personal mobiles which is used for daily life for maneuvering crowded area at home.
In the following sections, a new type of omnidirectional system is proposed which realizes the holonomic and omnidirectional capability together with high mobility on irregular terrains or steps.

2. Conventional Omnidirectional Systems For Wheelchairs

A standard wheelchair cannot move sideways. It needs a complex series of movements resembling parallel automobile parking when a wheelchair user wants to move sideways. A lot of omnidirectional drive systems were developed and applied to electric wheelchairs to enhance standard wheelchair maneuverability by enabling them to move sideways without changing the chair orientation. In Fig. 1, an omnidirectional wheelchair with Mechanum wheels [2] uses barrel-shaped rollers on the large wheel's rim inclining the direction of passive rolling 45 degrees from the main wheel shaft and enabling the wheel to slide in the direction of rolling. The standard four-Mechanum-wheel configuration assumes a car-like layout. The inclination of rollers on the Mechanum wheel causes the contact point to vary relative to the main wheel, resulting in energy loss due to conflictions in motion among the four wheels. Because four-point contact is essential, a suspension mechanism is definitely needed to ensure 3-degrees-of-freedom (3DOF) movement. Fig.2 shows an omnidirectinal wheelchair with ball wheel mechanisms developed at MIT [3]. Each ball wheel is driven by an individual motor which provides active traction force in a specific direction while perpendicular to the active direction. With this drive system, the point of contact of a wheel is stable relative to the wheelchair body that enables accurate motion control and smooth movements with no vibration.

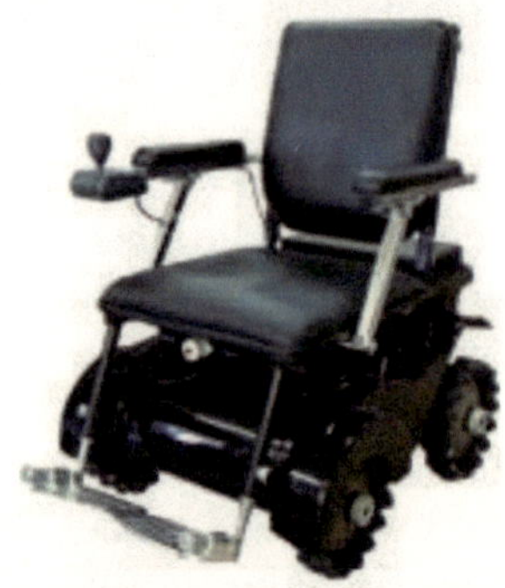

Fig. 1. Omnidirectional wheelchair with Mechanum wheels [2]

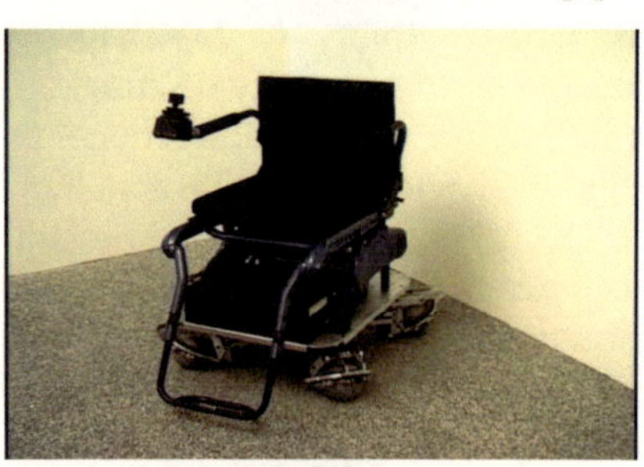

Fig. 2. Ball wheel omnidirectional wheelchair [3]

The other omnidirectional mechanism is VUTON crawler[4] which consists of many cylindrical free rollers. Since VUTON mechanism allows the multiple rollers to touch the ground simultaneously, heavy load can be applied on the platform.

All of the above omnidirectional systems need one motor to drive one wheel mechanism therefore four motors are needed to drive a four-wheeled wheelchair, while a wheelchair has three degrees of freedoms (DOF) on the floor. Thus, it involves 1 DOF redundancy in actuation which causes conflictions in motion among the four wheels.

3. Four-Wheel-Drive (4WD) Mechanism

To give a high mobile capability to a wheelchair, we introduce a four-wheel-drive (4WD) mechanism to our omnidirectional mobile system. At first, the original 4WD design is simply mentioned.

The 4WD drive system was invented in 1989 [5] for enhancing the traction and step climbing capability of the differential drive systems which schematic is illustrated in Fig.3. This 4WD mechanism has recently applied to a product design by a Japanese company [6]. The wheelchair equips four wheels, two omni-wheels in front and two normal tires in rear. A normal wheel and an omni-wheel, mounted on the same side of the chair, are interconnected by a chain or a belt transmission to rotate in unison with a drive motor. A common motor is installed to drive normal and omni wheel pair via synchro-drive transmission on each side of the mechanism. Then two motors provide deferent velocity on each side witch presents differential drive motion of 4WD mechanism. Thus all four wheels on 4WD can provide traction forces. Since the center of rotation shifts backward, when it turns about a steady point on the floor, it requires large space when the wheelchair is controlled in the standard differential drive manner. The offset distance between drive wheels and a center of a chair makes the maneuverability of the wheelchair worse.

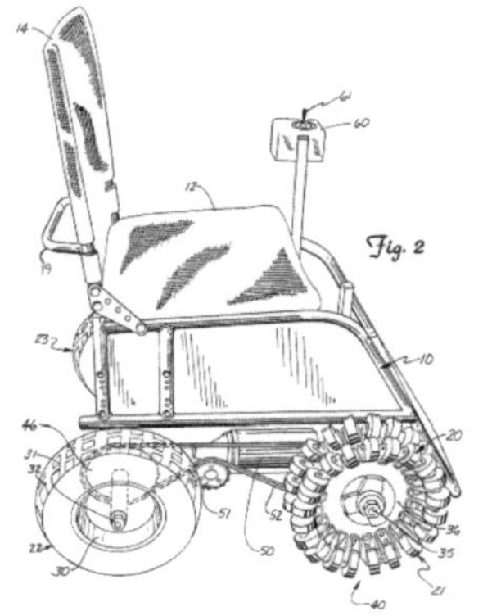

Fig. 3. Original 4WD synchronized transmission

4. Powered-caster Omnidirectional Control

We apply powered-caster control to 4WD mechanism to give an omnidirectional mobile capability to a wheelchair with 4WD. In this section, The powered-caster omnidirectional control for the original single type configuration[7] is breafly mentioned followed by the control of 4WD mechanism in the next section.

4.1 Powered-caster Mechanism

Fig.4 shows a top view of a powered-caster. The original design of the powered-caster is a single wheel type in which normal wheel is off-centered from steering shaft. The wheel shaft and the steering shaft of the powered-caster is driven by independent motors. When only the wheel shaft is rotated by the motor, the caster moves in forward direction which is denoted as $\dot{x}_w$ in Fig.4. When only the steering shaft is rotated by an another motor, the mechanism rotates about the point of contact which is also shown in the figure. By this motion of rotation, the steering shaft moves in lateral in $\dot{y}_w$ at the instant which is tangential of the circle which center is at the point of contact with the radius is s, the caster-offset. These velocity vectors are independently controlled and directing right angle for each other. To generate a velocity V in the direction θ at the center of the steering shaft, the wheel and the steering shaft rotations, ω_W and ω_S, are derived by the following kinematics.

$$\begin{bmatrix} \omega_w \\ \omega_s \end{bmatrix} = \begin{bmatrix} \frac{1}{r}\cos\theta & -\frac{1}{r}\sin\theta \\ \frac{1}{s}\sin\theta & \frac{1}{s}\cos\theta \end{bmatrix} \begin{bmatrix} \dot{x}_w \\ \dot{y}_w \end{bmatrix} \tag{1}$$

where s and r are the caster offset and the wheel radius respectively. Thus shaft rotations are determined by a function of θ, the relative angle between the desired direction and the wheel mechanism.

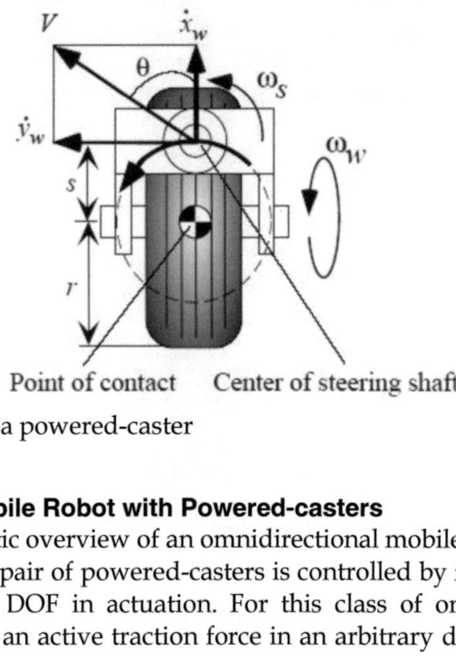

Fig. 4. Velocity control of a powered-caster

4.2 Omnidirectional Mobile Robot with Powered-casters

Figure 5 shows a schematic overview of an omnidirectional mobile robot with two powered-casters. The robot with a pair of powered-casters is controlled by four electric motors which involves one redundant DOF in actuation. For this class of omnidirectional robots, the powered-caster provides an active traction force in an arbitrary direction for propelling the robot. To coordinate the multiple powered-casters, motors on a powered-caster are controlled based on the velocity based robot model.

The inverse kinematics of two-wheeled mobile robot is represented as (2) which represents a relationship between the commanded robot velocity in 3DOF[$\dot{x}_v, \dot{y}_v, \dot{\phi}_v$] and a A wheel velocity [$\dot{x}_a, \dot{y}_a$] and a B wheel velocity[$\dot{x}_b, \dot{y}_b$].

$$\begin{bmatrix} \dot{x}_a \\ \dot{y}_a \\ \dot{x}_b \\ \dot{y}_b \end{bmatrix} = \begin{bmatrix} 1 & 0 & -\frac{W}{2}\cos\phi_v \\ 0 & 1 & -\frac{W}{2}\sin\phi_v \\ 1 & 0 & \frac{W}{2}\cos\phi_v \\ 0 & 1 & \frac{W}{2}\sin\phi_v \end{bmatrix} \begin{bmatrix} \dot{x}_v \\ \dot{y}_v \\ \dot{\phi}_v \end{bmatrix} \quad (2)$$

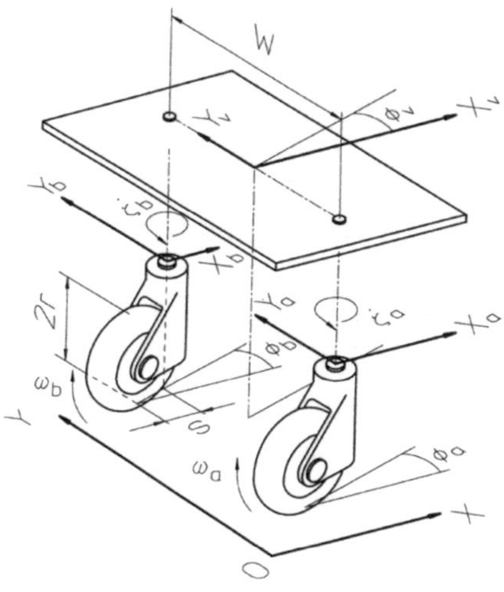

Fig. 5. A two-wheeled omnidirectional robot

5. Omnidirectional Control of 4WD Mobile system

In our project, it is a goal to develop an omnidirectional wheelchair with high mobility and maneuverability in a single design which can be used in multiple environments including outdoor and indoor. To enable a wheelchair to move in any direction instantaneously, omnidirectional control method, called "powered-caster control" which was introduced in previous section, is extended and applied to the 4WD mechanism [7]. Fig.6 shows a schematic of the 4WD omnidirectional wheelchair. The wheelchair has two omniwheels in front and standard pneumatic tires in rear which form 4WD configuration. A pair of an omniwheel and a pneumatic tire mounted on the same side of the wheelchair are connected by belt transmission for rotating unison and driven by a common motor which configuration is completely identical to the original 4WD system shown in Fig.3.
In our design, an additional third motor is mounted on the conventional 4WD platform for rotating a chair about the vertical axis which is also illustrated in Fig.6. Those three motors including two wheel motors and the chair rotation motor enable the wheelchair to realize independent 3DOF omnidirectional motion by a coordinated motion control [8],[9].

To achieve coordinated control for omnidirectional motion of a chair, the powered-caster omnidirectional control for a twin-caster configuration has been applied to the 4WD system. Fig.7 illustrates a schematic top view of a 4WD mechanism. In Fig.7, it is found that rear two drive wheels and center axis form a twin-caster configuration, i.e. parallel two wheels are located on the off-centered position which midpoint is distant from vertical steering axis, which is emphasized by thick lines in the Fig.7 and a vehicle with a twin caster drive mechanism is shown in Fig.8. The powered-caster omnidirectional control enables the caster mechanism to emulate the caster motion by actuating wheel and steering axes.

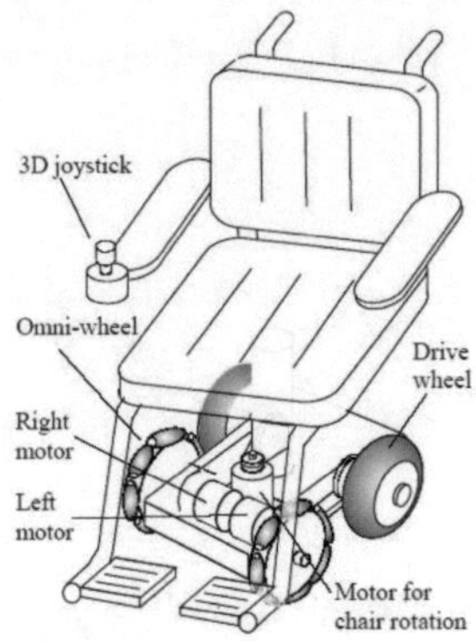

Fig. 6. A 4WD omnidirectional wheelchair

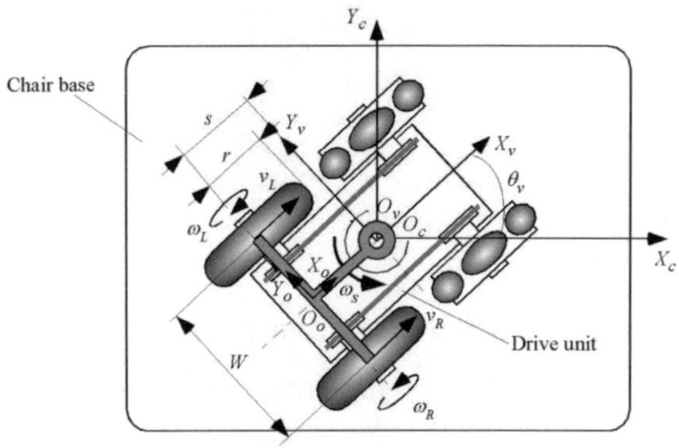

Fig. 7. Omnidirectional vehicle with 4WD mechanism

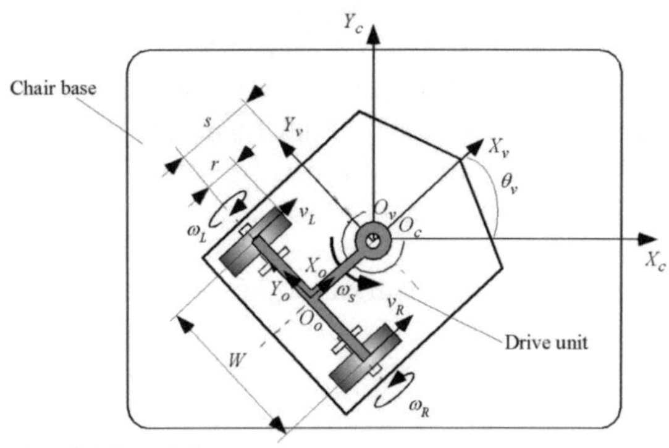

Fig. 8. Omnidirectional vehicle with a twin-caster

The powered-caster-based coordinate control of three motors needs a kinematic model of the 4WD[10]. The kinematic model represents the relationships between the motion of the 4WD and the three motor angular velocities of the drive wheels and the chair rotation axis. First, we consider the fundamental motions of a twin-caster drive (TCD). Figure 9a shows the translational motion of the vehicle in which two wheels rotate in same angular velocity to travel in same direction. In this case, TCD travels also straight forward, therefore TCD velocity and its rotation are represented as follows.

$$\begin{aligned} \dot{x}_v &= \frac{1}{2}(v_R + v_L) = v \\ \dot{\phi}_v &= \frac{1}{W}(v_R - v_L) = 0 \end{aligned} \tag{3}$$

Figure 9b shows another motion in which two wheels are rotated at same angular velocities but in opposite directions resulting in spin of TCD about the midpoint of two wheels. This motion provides only rotation but no translation velocity which is represented as,

$$\begin{aligned} \dot{x}_v &= \frac{1}{2}(v_R + v_L) = 0 \\ \dot{\phi}_v &= \frac{1}{W}(v_R - v_L) = \frac{2v}{W} \end{aligned} \tag{4}$$

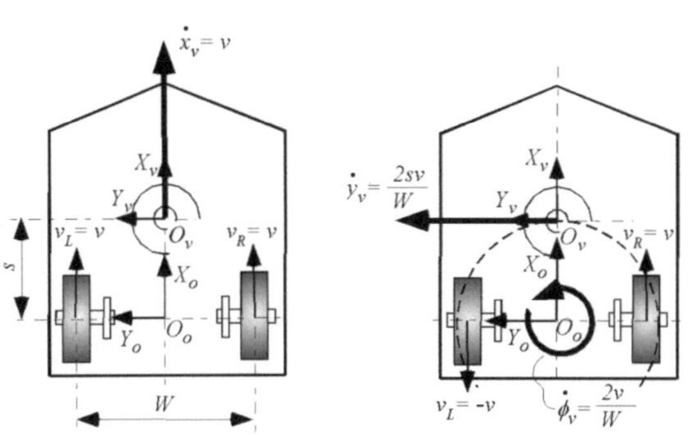

(a): Motion in X-direction (b): Motion in Y-direction
Fig. 9. Omnidirectional control for twin-caster drive (TCD)

Now we focus on the motion of the steering center whose location is identical to the rotation center of a chair. When TCD rotates about the midpoint of two wheels, the center of the steering presents a circular motion whose center is at the midpoint and the radius equals the caster offset, s. At each moment, velocity at the center is directing tangential direction of the circle which directs the lateral direction of TCD at all times. The lateral velocity denoted by $\dot{y}_v$ in Fig. 9b is represented by,

$$\dot{y}_v = s\dot{\phi}_v = \frac{s}{W}(v_R - v_L) = \frac{2sv}{W} \tag{5}$$

The translation velocities $\dot{x}_v$ and $\dot{y}_v$ are directed at right angles to each other. Note here, the rotation of TCD is not independently controlled since the rotation $\dot{\phi}_v$ is determined by creating the lateral velocity $\dot{y}_v$ to satisfy Eq (5). From Eqs. (3)-(5), the relationships between the vehicle translation velocity and wheel velocities are derived as,

$$\begin{bmatrix} \dot{x}_v \\ \dot{y}_v \end{bmatrix} = \begin{bmatrix} 1/2 & 1/2 \\ s/W & -s/W \end{bmatrix} \begin{bmatrix} v_R \\ v_L \end{bmatrix} \tag{6}$$

Thus, translation velocities along the X- and Y-directions are completely determined and independently controlled by wheel velocities. To generate the required velocity vector that directs in an arbitrary direction with arbitrary magnitude, the reference vector is projected in X- and Y-directions of vehicle coordinate system (Fig.10). The velocity component in each direction, $\dot{x}_v$ or $\dot{y}_v$, can be independently achieved by using kinematics of TCD in Eq. (6).

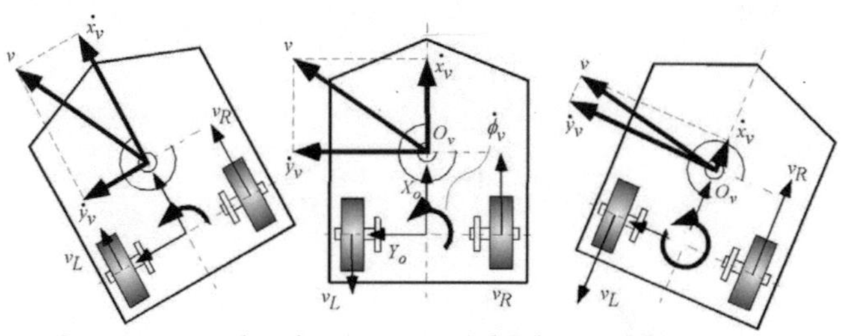

Fig. 10. Projection of a command velocity into vehicle coordinate system depending on the TCD orientation.

When the reference velocity is steady to the ground, the velocity components in X- and Y-directions vary depending on the TCD orientation relative to the ground. Therefore, wheel velocities also vary which results in straight motion of the TCD center (see Fig. 11). TCD shows spontaneous flipping behavior during the motion, which is often found in passive casters installed on legs of chairs and tables, etc. It is said the powered-caster control emulates caster motion by actively actuating the wheel axis.

Fig.11 shows a one of the simulation results in which an omnidirectional control of the twin-caster mechanism are tested. In the simulation, twin-caster mechanism is controlled to track a straight line with a center of a mechanism locating on the line at all times. During the motion, the orientation of the mechanism is rapidly flipped over and orient to the direction of motion. This flip motion is often seen on passive casters which installed on the legs of office chairs and tables. Thus the powered-caster control achieves the emulation of caster motions by coordinated control of multiple actuators.

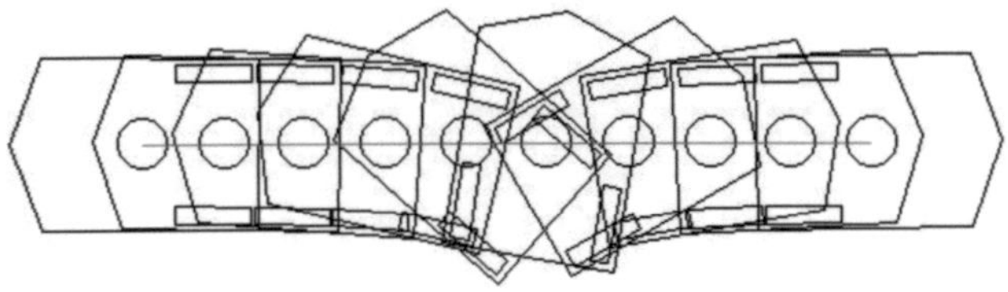

Fig. 11. Omnidirectional control for twin-caster mechanism

Translation in arbitrary direction is achieved by TCD as presented above. However, orientation of TCD can not be controlled independently by the wheel rotations. To control 3DOF motion of a chair, the chair rotation axis must be also coordinated. The velocity command is given based on the chair orientation since a joystick is fixed on the chair. Then the command velocity is translated into TCD coordinate by the relative orientation of the chair to the vehicle, θ_v. as

$$\begin{aligned} \dot{x}_c &= \dot{x}_v \cos\theta_v - \dot{y}_v \sin\theta_v \\ \dot{y}_c &= \dot{x}_v \sin\theta_v + \dot{y}_v \cos\theta_v \\ \dot{\theta}_c &= \dot{\theta}_v + \omega_s \end{aligned} \tag{7}$$

From eqs. (6) and (7), an overall kinematic model of the omnidirectional 4WD wheelchair is represented as follows.

$$\begin{pmatrix} \dot{x}_c \\ \dot{y}_c \\ \dot{\theta}_c \end{pmatrix} = \begin{pmatrix} J_{11} & J_{12} & 0 \\ J_{21} & J_{22} & 0 \\ r/W & -r/W & 1 \end{pmatrix} \begin{pmatrix} \omega_R \\ \omega_L \\ \omega_S \end{pmatrix} \tag{8}$$

where,

$$\begin{aligned} J_{11} &= \frac{r\cos\theta_v}{2} - \frac{rs\sin\theta_v}{W} \\ J_{12} &= \frac{r\cos\theta_v}{2} + \frac{rs\sin\theta_v}{W} \\ J_{21} &= \frac{r\sin\theta_v}{2} + \frac{rs\cos\theta_v}{W} \\ J_{22} &= \frac{r\sin\theta_v}{2} - \frac{rs\cos\theta_v}{W} \end{aligned} \tag{9}$$

Where r,W and s are the wheel radius, tread and caster-offset, respectively. A 3x3 matrix in the right side of eq.(8), called as Jacobian, is a function of orientation of the 4WD unit with relative to the chair base, θ_v. All elements in the Jacobian can always be calculated and a determinant of the Jacobian may not be zero for any θ_v. Therefore there is no singular point on the mechanism and an inverse Jacobian always exits. The three motors are controlled to realize a 3DOF angular velocity commands $\dot{x}_v$, $\dot{y}_v$ and $\dot{\phi}_v$ by independent speed controllers for omnidirectional movements. Thus, holonomic 3DOF motion can be realized by the proposed mechanism. This class of omnidirectional mobility, so called "holonomic mobility", is very effective to realize the high maneuverability of wheelchairs by an easy and simple operation.

6. Prototype Design

6.1 Mechanical Design

Wheelchair specifications for prototype design are shown in Table 1. The wheelbase and tread of the 4WD mechanism are 400mm and 535mm respectively. Those dimensions are determined to satisfy the limitation of the standard wheelchair specification for the dimension, 600mm in width and 700mm in length as shown in the spec. The required step height which can be surmounted by the wheelchair is approx. 100mm for accessing to a train car from a station platform with no assistance. The maximum running speed for continuous drive is 6km/h which is same as conventional standard wheelchairs in Japan.
Fig.12 illustrates a 3D drawing of a prototype designed by 3D CAD. Fig.13 shows an overview of the prototype wheelchair.

Dimension	Width	600mm
	Length	700mm
	Height	450mm
Weight	Total	180kg(human+wheelchair)
	Wheelchair	100kg (including batteries)
Speed	6km/h (Max.)	
Surmountable step	100mm in height	

Table 1. Wheelchair Specifications

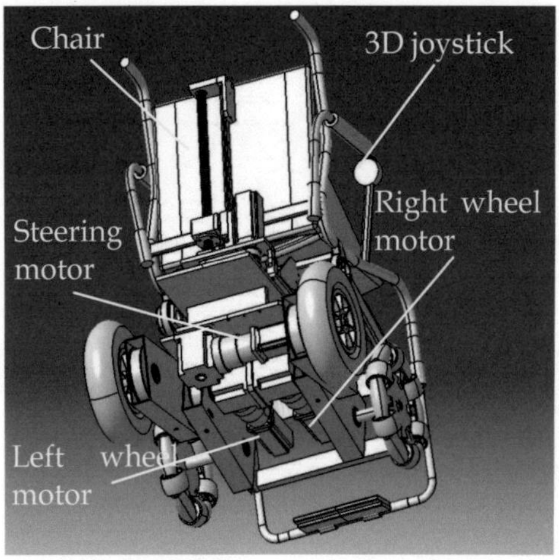

Fig. 12. Prototype bottom view by 3D CAD

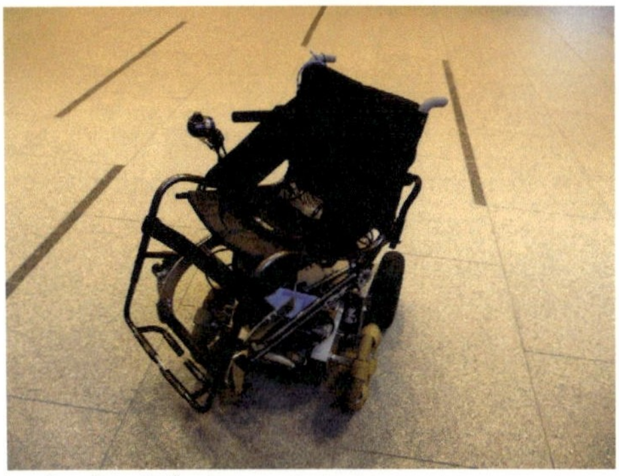

Fig. 13. 4WD omnidirectional wheelchair prototype

6.2 Control system

Figure 14 shows a control system of the wheelchair prototype. Most of equipments including a controller, motors, motor drivers, a battery and sensors are installed on 4WD mechanism side. Electric power is provided by a car battery and distributed to all components on the chair after the inversion to AC 100V. A tablet PC controls three motors to realize an omnidirectional and holonomic motions of a wheelchair based on reference velocity commanded by a 3D joystick. A velocity command is sent to each motor driver via a D/A interface while a encoder pulse is sent back to the PC via a pulse counter interface which form a velocity feedback loop of the axis. An absolute encoder is installed only on the chair rotation shaft which detects relative angle between the 4WD and the chair which needs no initialization process at power on reset. Since a chair have to rotate continuously with no mechanical limit, slip rings are also installed on the chair rotation axis. A USB hub on the chair side enables extension of devices additional to an A/D converter for the 3D joystick.

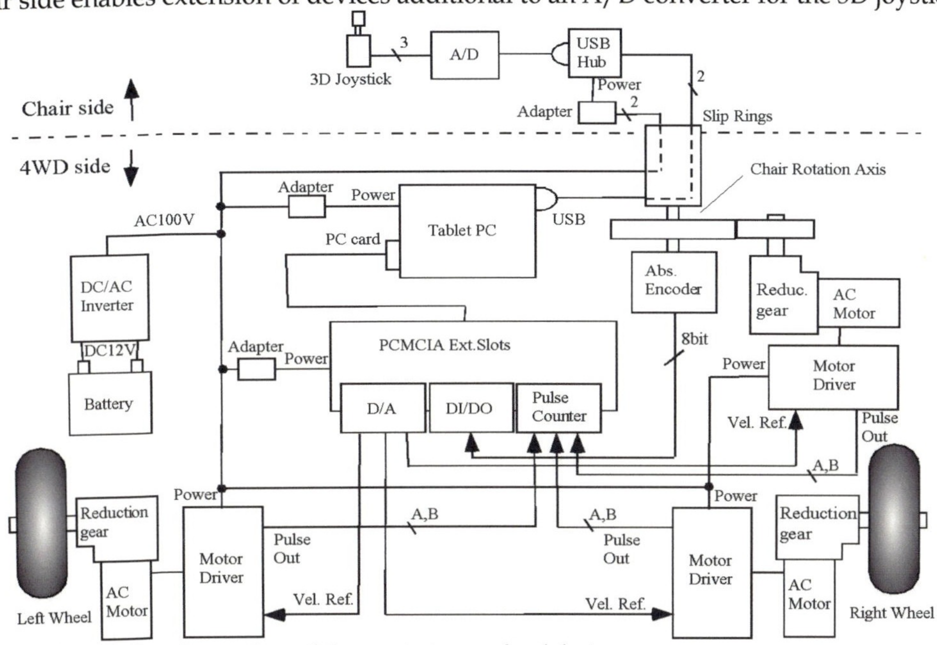

Fig. 14. System configuration of the prototype wheelchair

7. Experiments

7.1 Omnidirectional motion

To verify the omnidirectional mobility of the proposed system, lateral motion was presented by the prototype wheelchair while keeping the chair orientation constant.

A small ball was attached to the chair to measure the location of the center of the steering axis. A stereo camera system (Quick Mag IV from OKK Inc, Japan) detects the ball location in 3D coordinates. The prototype was controlled by a remote PC via a LAN connection through which the remote PC send the velocity command to the wheelchair controller. Thus, a complete linear trajectory in lateral direction was commanded to the prototype. Figure 15 shows a chair motion detected by the stereo camera system. An actual and a reference path are plotted in the figure which closely agree. Figure 16 shows tracking errors between the actual and the reference. The error was suppressed to within 10mm (+5 to -5mm) despite the flipping behavior of the 4WD occurred during the experiment motion.

The camera system also provided real-time video images with over-writing rectangle window(s) and the path of the target(s) (in this experiment, a small ball). Figure 17(a)-(f) shows a series of video images. The prototype wheelchair moved from the right side of the picture frame to the left. The final picture, Fig. 17(f) shows a straight path which was created by the wheelchair movement. During the motion, the flipping behavior of the 4WD is found in Fig. 17(b), (c) and (d), which is also shown in a computer simulation in Fig. 11.

7.2 Variable center of rotation

The proposed system realizes holonomic and omnidirectional motion of a chair by the coordinated control of three motors. The holonomic mobile capability makes it possible to change the location of a center of rotation at any point to fit to customer requests. In a normal setup, a chair rotates about its center when the operator commands a spin turn by twisting the joystick. However, the location of the center is variable in the control program to any point including the out-of-footprint area of the wheelchair. However, usual requests may be to shift it to the back side to simulate a rear drive wheelchair, or to the front side to simulate a front drive wheelchair. To present the flexibility of the proposed system, two patterns of spin turn motion were performed. Figure 18(a) and (b) shows the final image of each test run provided by the stereo camera system. In Fig. 18(a), the wheelchair rotated about the front position which was located approx. 500 mm forward of the center of the chair. In Fig. 18(b), the center was shifted also approx. 500 mm towards the rear. From these results, the center of rotation can be customized to an individual.

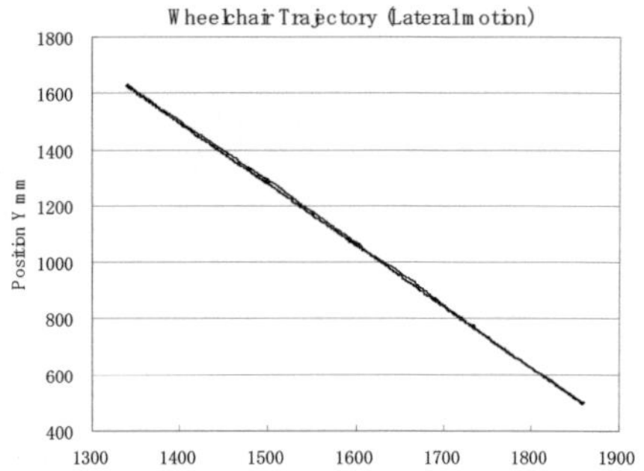

Fig. 15. Reference and actual trajectories of the 4WD wheelchair

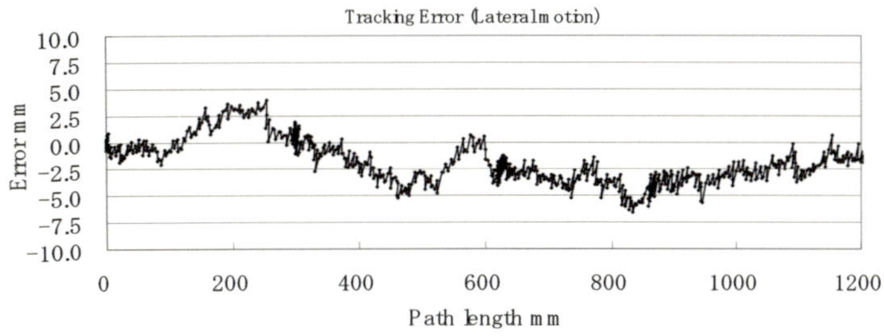

Fig. 16. Tracking error on the experiment using 4WD wheelchair

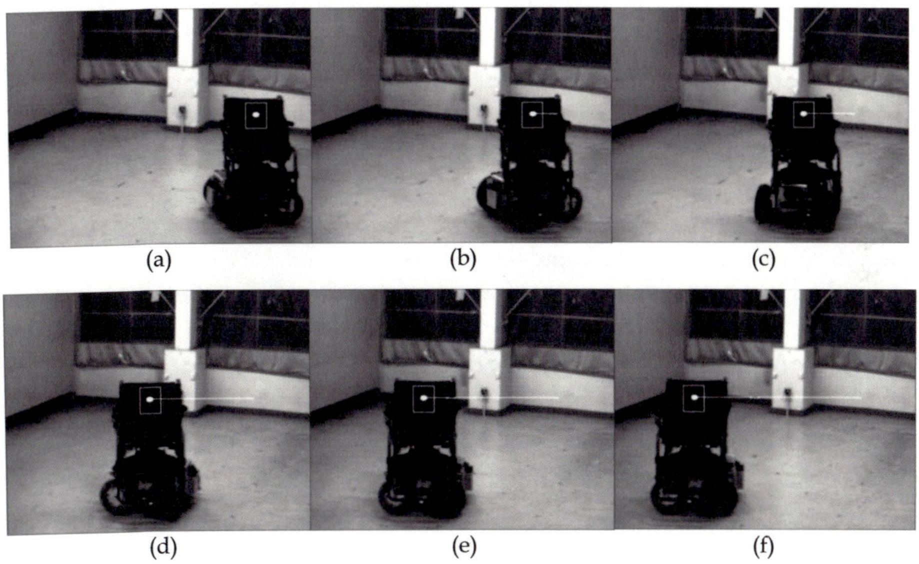

Fig. 17. An example of omnidirectional motion 1: the lateral motion of the wheelchair prototype; it moves in sideways from the right side to the left of the picture frames while maintaining the chair orientation to be stable.

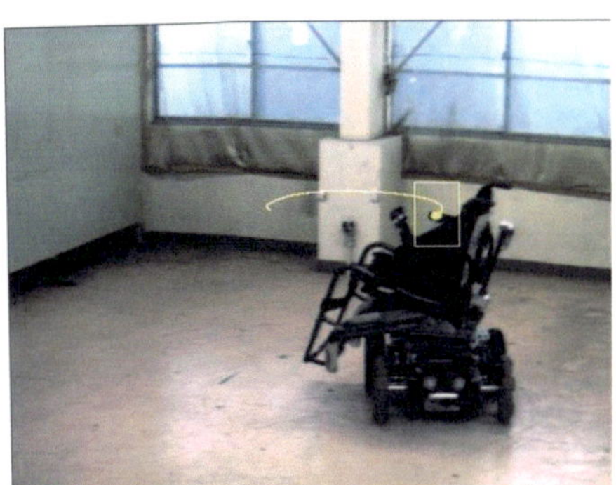

(a) Spin turn about the front position

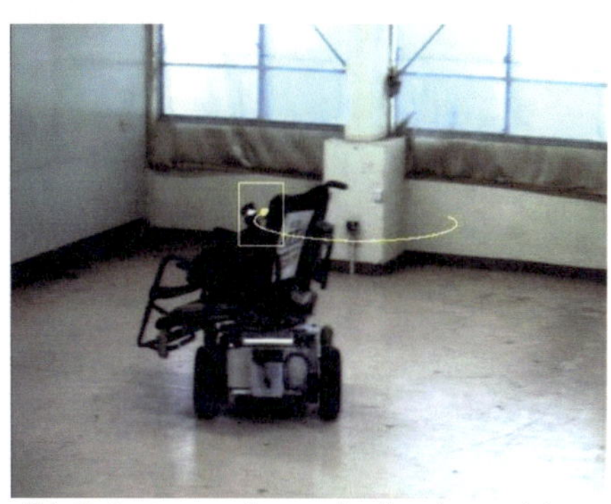

(b) Spin turn about the back position

Fig. 18 Variable center of rotation

7.3 Task example

The holonomic and omnidirectional mobile systems are easy to maneuver because 3D command in X- and Y- directions and rotation are directory commanded and an operator does not have to consider the wheel motions and its configurations.

To demonstrate maneuverability, a task example was performed by the prototype. The task of getting out of a room by pulling the door open is one of difficult tasks for a wheelchair user. A task of pulling the door is relatively more difficult than pushing the door. This task example includes sub tasks such as: 1) approaching to the door to grasp the door knob, 2) moving backward to pull the door open, 3) going through the door and getting out of the room, and 4) driving to another location.

Figure 19(a)-(j) show screen shots of the experiment. In this test, the prototype wheelchair was operated using a 3D joystick by a human operator who was sitting on the chair. Since the wheelchair has high maneuverability, the task was successfully achieved with no collision with the door or a wall. The operator did not see the 4WD mechanism during the task, however, the 4WD mechanism changed its orientation widely when the operator moved sideways to avoid colliding with the door (Fig. 19(d)-(g)).

Figure 20 shows the 3D commands (X, Y and Rotation) to the wheelchair during the task of Fig.19 which shows 3D simultaneous motion to complete the door opening task. Those are found in sub tasks including, 1) approaching the door and 4-1) turning after exiting the room. Individual lateral translations are often found in subtask 3) exiting the room.

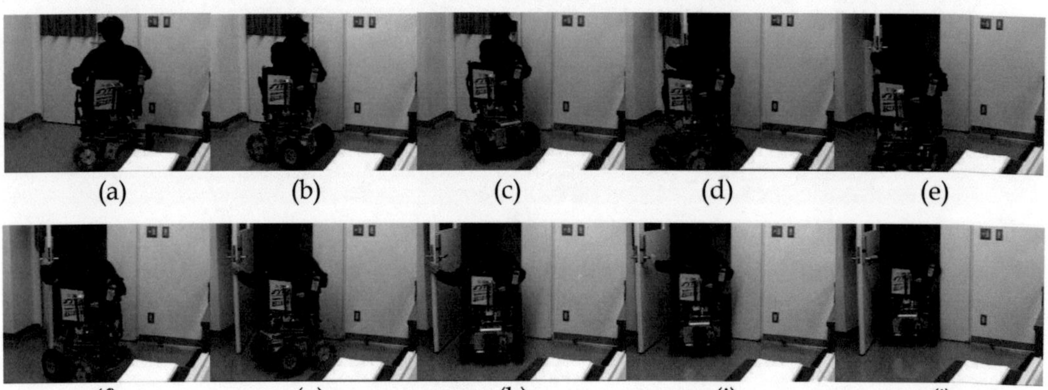

Fig. 19. An example of tasks using a electric wheelchair: Getting out of a room with pulling a door open.

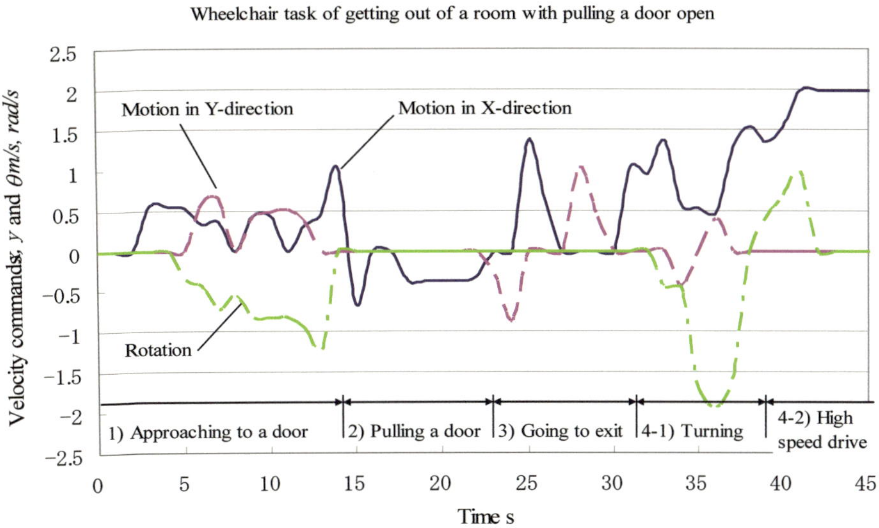

Fig. 20. 3DMotion of wheelchair presenting a door opening task for exiting a room

8. Conclusion

Mechanism and omnidirectional control of a 4WD mechanism for wheelchairs are presented in this chapter. The omnidirectional wheelchair system is proposed for improving maneuverability of standard wheelchairs The 4WD mechanism has high mobility which equips four wheels, two omni-wheels in the front and two normal tires in the rear, and all wheels provide traction even with two motors to drive these wheels. To realize holonomic and omnidirectional motion of a chair by utilizing the 4WD mechanism, the proposed system includes the third motor to rotate the chair at the center of the 4WD mechanism about the vertical axis.

For omnidirectional control of the 4WD mechanism, powered-caster control has been applied. To achieve a coordinated control of three motors, kinematics of the 4WD wheelchair was analyzed and a kinematic model was derived which represents the relationships between 3DOF wheelchair motion and the rotations of three motors. In the powered-caster control, two wheel motors are coordinated to translate the center of the chair in an arbitrary direction while the chair orientation is controlled by the third motor separately.

The omnidirectional motion was verified by a series of experiments using a wheelchair prototype. First, omnidirectional mobility was tested in which the wheelchair made a lateral motion without changing its orientation. Next, one of a applications of the holonomic mobility was performed in which the center of rotation was varied by a control program to customize per user request for simulating the wheelchair drive types, such as front drive, rear drive, or center drive.

To present the high maneuverability of the proposed omnidirectional mobile system, a task example was performed in which an operator maneuvered the wheelchair by 3D joystick to exit a room with pulling the door open. The task was successfully achieved with no collision with the door or walls. During the task, it was found that the simultaneous 3DOF motions, lateral translation are often commanded as well as the forward translation for pursuing the task.

From these experiments, the omnidirectional and holonomic mobile capability are shown to be very effective and useful for maneuvering in crowded areas and achieving complicated tasks.

9. Acknowledgment

This project was supported by the Industrial Technology Research Grant Program in 2006 from the New Energy and Industrial Technology Development Organization (NEDO), Japan, research ID 05A06715a.

10. References

[1] Population Statistics Japan 2006, by the National Institute of Population and Social Security Research, Japan.

[2] All-direction Power-driven Chair "FJ-UEC-600", Fujian Fortune Jet Mechanical & Electrical Technology Co., Ltd.

[3] M. Wada and H. H. Asada (1999) Design and Control of a Variable Footprint Mechanism for Holonomic and Omnidirectional Vehicles and its Application to Wheelchairs, IEEE Trans on Robotics and Automation, Vol.15, No. 6, pp. 978-989.

[4] S. Hirose and S. Amano (1993) The VUTON : High Payload High Efficiency Holonomic Omni-Directional Vehicle, 6th Int. Symp. on Robotics Research.

[5] Jefferey Farnam (1989), Four-wheel Drive Wheel-chair with Compound Wheels, US patent 4,823,900.

[6] Kanto Automobile Corp. "Patrafour" http://www.kanto-aw.co.jp/jp/products/wheelchair/

[7] M. Wada and S. Mori (1996), Holonomic and Omnidirectional Vehicle with Conventional Tires, Proceedings of the 1996 IEEE International Conference on Robotics and Automation (ICRA96), pp. 3671-3676.

[8] M. Wada, A. Takagi and S. Mori (2000), Caster Drive Mechanisms for Holonomic and Omnidirectional Mobile Platforms with no Over Constraint, Proceedings of the 2000 IEEE International Conference on Robotics and Automation (ICRA2000), pp. 1531 -1538.

[9] M. Wada (2007), Omnidirectional and Holonomic Mobile Platform with Four-Wheel-Drive Mechanism for Wheelchairs, JSME Journal of Robotics and Mechatronics, Vol. 19, No. 3, pp. 264-271.

[10] M. Wada (2007), Holonomic and Omnidirectional Wheelchairs with synchronized 4WD Mechanism, Proceedings of the 2007 IEEE/RSJ International Conference on Intelligent Robots and Systems (IROS2007), pp. 1196-1202.

5

Stair Climbing Robots and High-grip Crawler

[1)]Kan Yoneda, [1)]Yusuke Ota and [2)]Shigeo Hirose
[1)]Chiba Institute of Technology, [2)]Tokyo Institute of Technology
Japan

1. Introduction

Stair climbing is one of the most attractive performance of mobile robot for both legged and wheeled. (e.g. Stoeter et al., 2002; Murphy, 2000; Yim et al., 2000; Krishna et al., 1997; Granosik et al., 2005; Liu et al., 2005; Arai et al., 2006; Tanaka et al., 2006; Miyanaka et al., 2007; Tsukagoshi et al.,2005)

Authors have been developing various kinds of stair climbers, considering how to make its climbing ability higher and its mechanical complexity reasonable and practical. The research includes realizing a large step negotiating height, controlling to keep its centre of gravity almost the centre of ground contacting area, higher speed of climbing up, large load tolerance to carry on. Reducing body weight and energy consumption is also the important matter of developing.

In the first part of this chapter, we introduce some solutions to realize stair climbing machines that we developed. Each of them has good performance as in a category of their kind, e.g. various numbers of legged and wheeled shapes. Then, we discuss a development of high-grip crawler, which we think one of the best solutions as the stair climber.

2. Various Stair Climbers

2.1 Biped

We have been developing biped configuration robot named "YANBO" since 1985.

YANBO-1, as shown in Figure 1, was the first developed model (Yoneda, 1987). YANBO-1 consists of five Degrees of Freedom (DOF) which can be considered almost minimum DOF necessity for walking on the level ground. YANBO-1 is confirmed that by walking on horizontal level ground and climb stairs.

YANBO-2 (Ota et al., 2001a; Ota et al., 2002), as shown in Figure 2, and YANBO-3 (Ota et al., 2003), as shown in Figure 3, were second model which could be performed not only mobility but also manipulability to attach two more DOF. Additional DOF is allocated to each ankle joint and can help to generate three ankle joint motions freely. This helps to walk on incline slopes and appropriate manipulate motions. Then total DOF increases to eight, however, the number of actuators are still much less than other biped structure robot, like humanoid (Yoneda & Ota, 2003).

Fig. 1. YANBO-1, Stair climbing motion

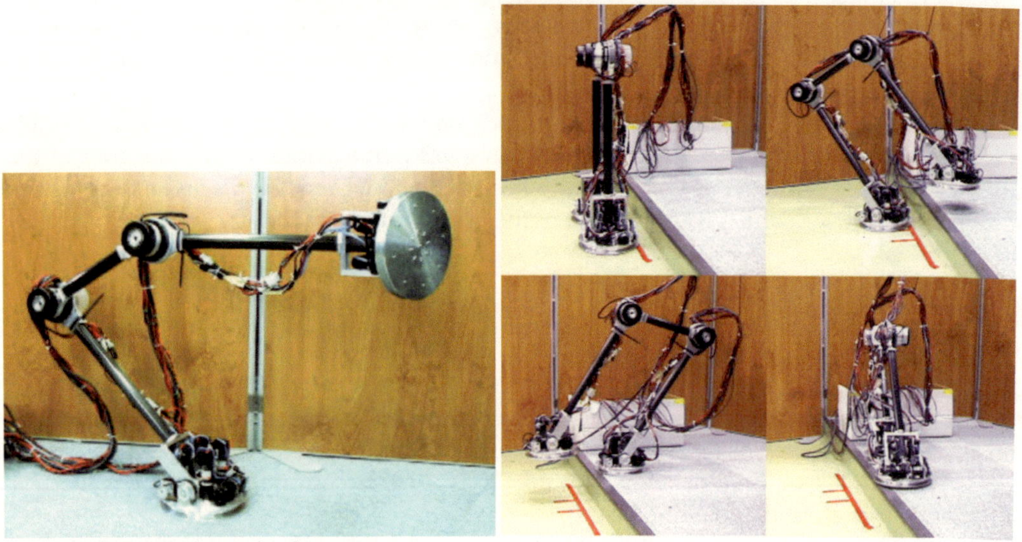

Fig. 2. YANBO-2, Step climbing motion.

2.2 Quadruped

Authors also have been developing various kinds of quadruped walking robot (Hirose et al. 2009). TITAN-VI, as shown in Figure 5, has succeeded to walk on stairs (Hirose et al. 1995; Yoneda et al., 2000). TITAN-VI consists of two separated body segments, where front and rear segment can slide linearly up and down in order to negotiate a large height difference of front and rear landing points on the stairs.

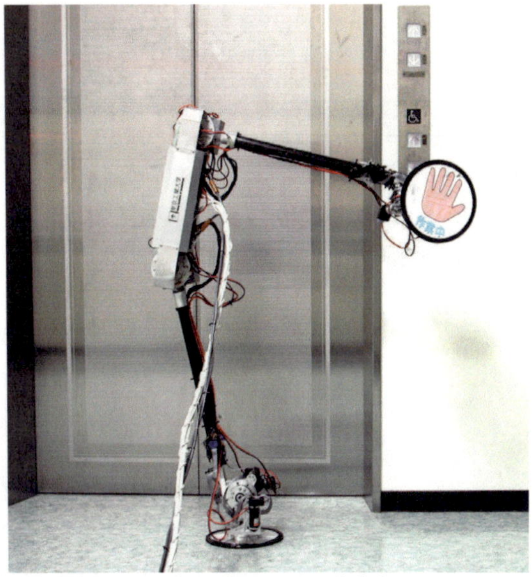

Fig. 3. YANBO-3, which also equipped with manipulability

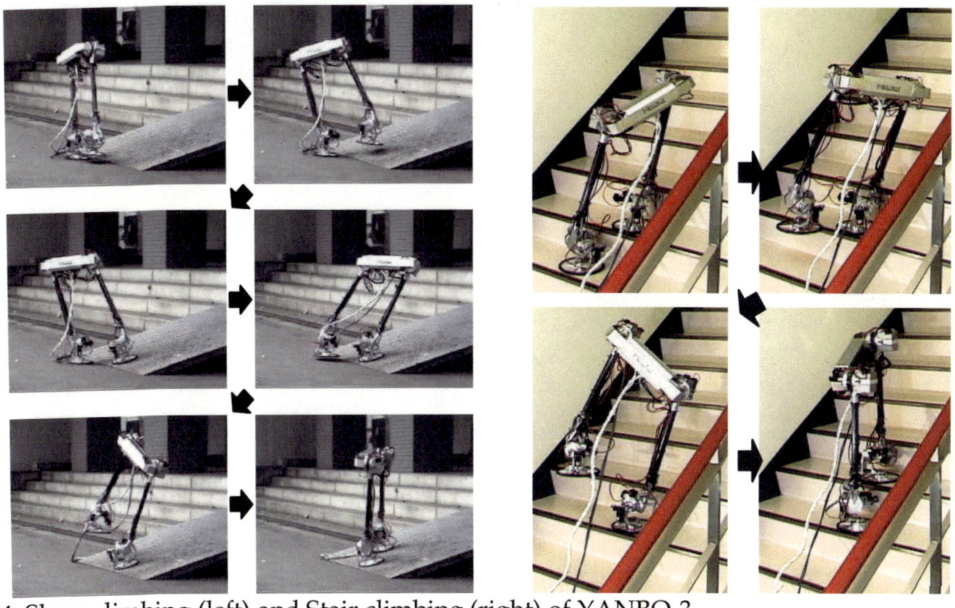

Fig. 4. Slope climbing (left) and Stair climbing (right) of YANBO-3

Authors have also been developed quadruped with another strategy, according to the design concept in which total number of actuators can be used as small as possible. Hyperion, as shown in Figure 6, was quadruped walking robot with minimum actuated-DOF for walking motion. (Yoneda et al., 2001a; Yoneda & Ota, 2003; Yoneda, 2007) Taking lightweight advantages with small number of actuators that is three motors are used in Hypeion-1 and 5 motors are used in Hyperion-2, a wall climbing robot "Hyperion-1SP" and "Hyperion-2SP" was developed. Wall climbing motion and ceiling walking motion of

Hyperion-1SP are shown in Figure 7 and Figure 8, respectively (Yoneda et al., 2001b; Ota et al., 2006). In Hyperion 2SP, transfer motion from ground to wall can be succeeded.

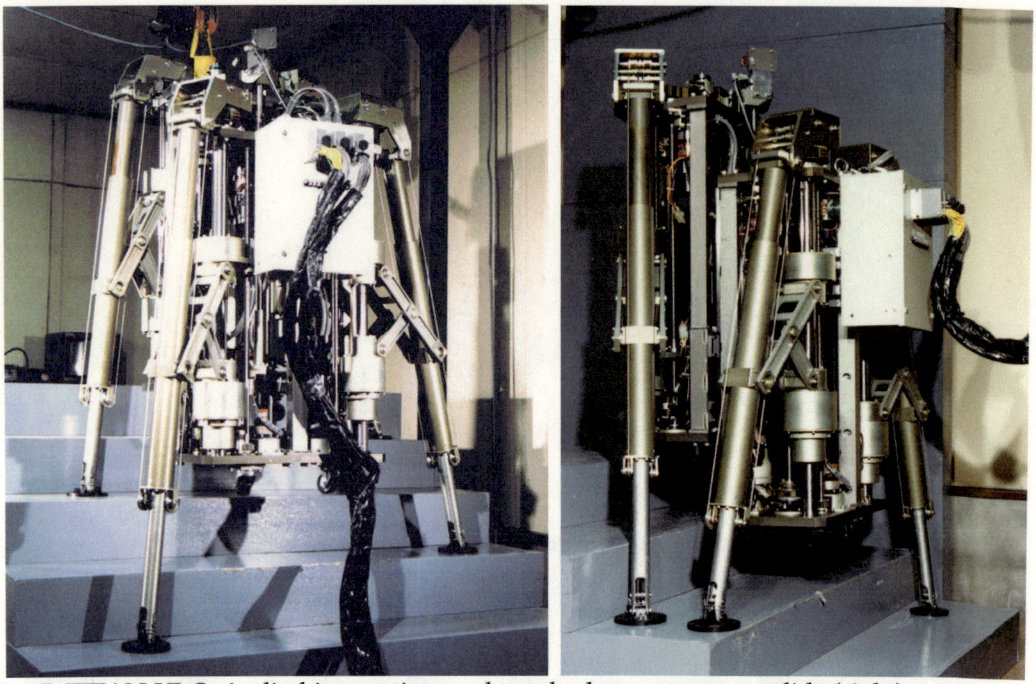

Fig. 5. TITAN-VI, Stair climbing motion, and two body segments can slide (right)

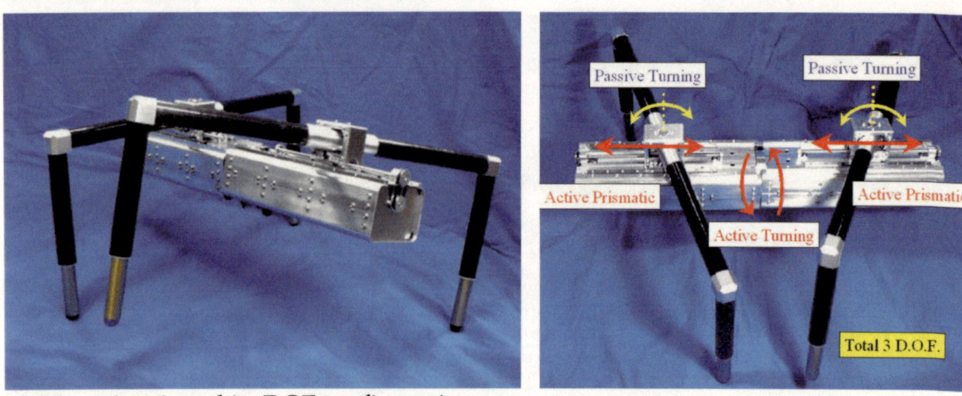

Fig. 6. Hyperion-1, and its DOF configuration

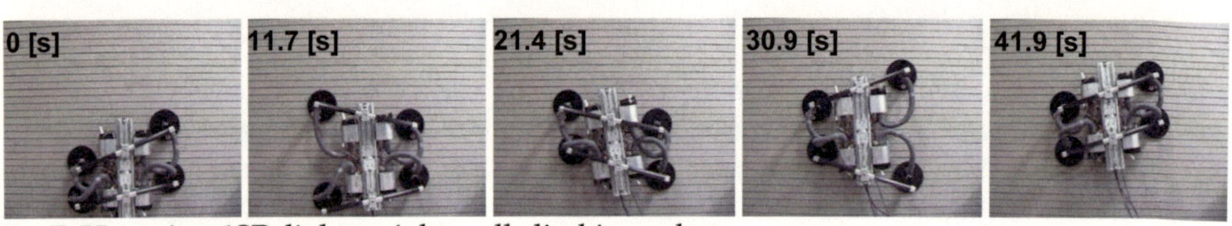

Fig. 7. Hyperion-1SP, light weight wall climbing robot

Fig. 8. Hyperion-1SP can succeed to walk on ceiling using blower and suction feet

2.3 Six-legged

We have been also developing six-legged walking robot. ParaWalker-II (Ota et al., 2001a;Yoneda & Ota, 2003), as shown in Figure 9, composed two frames with three legs each, and each frames are connected by three arms with two actuated joints. Total numbers of actuators are six to generate each frame 6-DOF motion that is necessary and sufficient DOF to lead walking and tasking motion. Moreover, in order to acquire higher step adaptabilities to move stairs, one more actuator was attached for extending a leg and keeping frame balance during stair climbing.

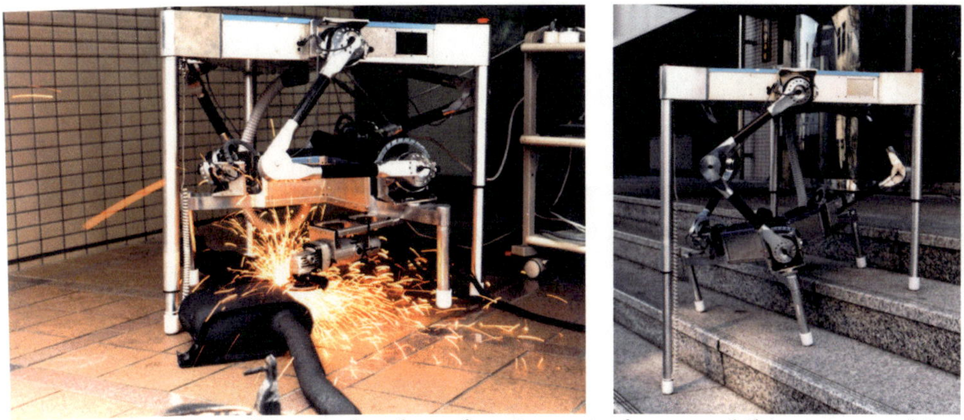

Fig. 9. ParaWalker-II, walking and task performing twin-frame mobile system, and its stair climbing motion to extend one leg

2.4 Leg-wheel Hybrid

Generally, legged-locomotion has very high adaptabilities. However, mobility of wheeled locomotion is much more interesting than legged-locomotion if the moving place is limited to a flat plane. Therefore, much higher mobility can be obtained if both methods are adopted; wheeled locomotion on flat planes and legged-locomotion on uneven terrain. We have developed several leg-wheel hybrid robots.

Hyperion-W has developed one of leg-wheel hybrid robot (Takahashi et al., 2006), whose base body is used the Hyperion-1, which is a quadruped robot with minimum three actuators mentioned above, and actuated wheels are attached to each legs. Hyperion-W can perform a hybrid motion of walking and running on uneven terrain as shown in Figure 10 and show its high mobility.

And we have another maneuver to realize the leg-wheel hybrid locomotion. YANBO-2 and YANBO-3, mentioned above, have 3-DOF ankle joint with eternity rotatable circle shaped sole. Therefore using these characteristics, both YANBO-2 and YANBO-3 can establish leg and wheel hybrid locomotion manoeuvres (Ota et al., 2002; Ota et al., 2003), as shown in Figure 11.

Fig. 10. Hyperion-W, a leg-wheel Hybrid robot

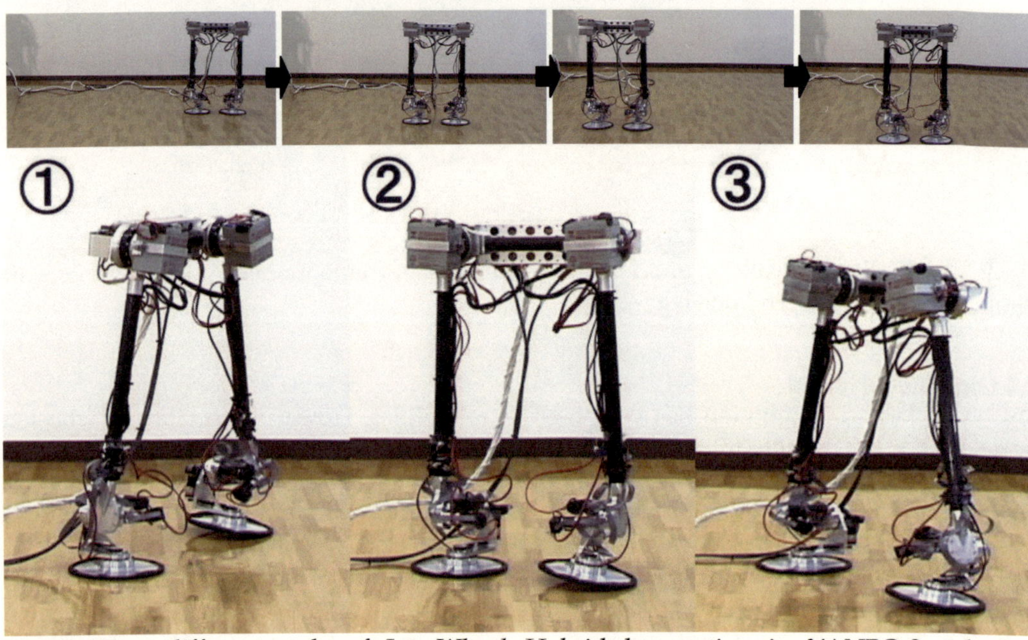

Fig. 11. Two different style of Leg-Wheel Hybrid locomotion in YANBO-3, wheeled locomotion (upper) and wheeled and leg mixture locomotion (lower)

3. Advantages of Crawler type Vehicles in practical use

Through from above mentioned various types of researches, we believe one of the realistic solution which the robots should support and help human tasks in our daily lives is to carry heavy baggage especially by wheeled or crawler type climbers. Because wheeled or crawler type vehicles have much more payload capacity than legged-walking robots have. Therefore, when carrying heavy objects, a cart is useful only on flat ground, and the load must be carried up or down stairs by hand. Conventional approaches to transporting heavy loads on the stairs have yet to be developed. Moreover, humans themselves sometimes require assistance in traversing stairs. Mobile robots require the ability to move with versatility, smoothly and with high efficiency in various environments. Robots with high mobility can easily be used in rescue operations as the robot can move over irregular terrain of collapsed and destroyed buildings.

In our living environment, the most difficult artificial obstacles to move over are stairs. There have been many studies to improve the ability to traverse stairs using legged-type, crawler-type and wheeled-type robots that have special shapes. Among these mobile robots, crawler-type and wheeled-type robots are easier to control and so there are many examples, including crawlers with attached grousers (Hirose et al. 1989; Hirose et al., 1990; Hirose et al., 1992), crawlers for rescue operations (Takayama et al., 2000; Granosik et al., 2005; Liu et al., 2005; Arai et al., 2006; Tanaka et al., 2006; Miyanaka et al., 2007), wheels with coil springs (Hirose et al., 1991), special tires (Uchida et al., 1999; Uchida et al., 2000), and legs that rotate along wheels (Taguchi et al., 1995; Schempf et al., 1999).

In the present research, our goal is to design a practical vehicle to obtain high terrain adaptability and mobility in the human living environment, especially to traverse stairs or steps. To acquire reliable mobility, we developed a new crawler that can obtain a high grip force not by grousers hooking the stairs, but by the deformation of the face material that touches the edges of the stairs. Experimental results revealed that the characteristics of the material that composes the face of the crawler belt affect the grip force. In the present research, a tracked climber vehicle with powder-filled belts carrying heavy loads is proposed and developed, and the efficiency and practical applicability of the proposed tracked climber vehicle are verified.

4. Comparison of a Crawler with Rigid Grousers and with Soft Deformation Belts

A previous crawler was equipped with grousers in order to obtain a certain grip force on stairs. Grousers work very well when the crawler moves over sand or mud, and such crawlers can support heavy loads. However, in the case of traversing stairs or steps, such crawlers have a number of disadvantages, as described below.

1. The intervals of the grousers and the steps do not generally coincide. Thus, the crawler is held by only one grouser on one crawler belt, as shown in Figure 12. Carrying heavy loads with this gripping grouser causes the vehicle to vibrate and may destroy the stair edge. Gripping force is lost easily after the stair edge is destroyed or one of the grousers becomes caught on an obstacle on the stairs. These phenomena reduce the vehicle's stability and safety, and thus these should be avoided.

2. Slippage may occur when the crawler descends the stairs. When ascending the stairs, the crawler belt simply spins until the grouser touches the step. However, when descending the stairs, the crawler moves forward, down the stairs, even when the grousers do not catch a step.
3. A grouser that has caught a step moves backward as the crawler moves forward up the stairs, as shown in Figure 13. The grouser leaves the step when the grouser comes to the end of crawler. In this situation, other grousers may not necessarily be touching the next step. Therefore, the crawler may slip down or spin off the belt until another grouser catches the next step.
4. If it does not climb the stairs in a straight path, the crawler may not obtain sufficient grip force because the grousers, which have a wide structure for easy attachment to the crawler belt, would touch the stair edges at an angle. This would hamper the mobility of the crawler when adjusting the trajectory to the right or to the left when climbing stairs.

These disadvantages can be partially solved by arranging the grousers in shorter intervals. However, as shown in Figure 14, grousers arranged in short intervals do not have large support areas. Furthermore, if the intervals between the grousers become shorter, the ability to climb stairs in non-straight trajectories becomes worse, thus increasing the consequences of the fourth disadvantage.

As shown in Figure 15, deforming the crawler belt adaptively to the stairs to obtain a grip force from all of the steps appears to be an effective method by which to address these disadvantages. Supporting the crawler at several points prevents slippage accidents caused by the lack of stair edges or by an obstacle becoming caught between the stairs and the crawler belt. In addition, by changing the support points when the vehicle is moving also avoids a freely spinning belt. The required grip force at each grip point is far smaller than in the case of only one grip point.

To obtain a grip force from each step as described above, a rubber material with a large friction coefficient can be easily attached to the face of the crawler belt. However, this method is not necessarily practical for versatile application. For example, outdoor stairs with rounded edges, stairs with metallic edges that have a low coefficient of friction, oily stairs in factories or stairs covered by fallen leaves or dust may cause slippage. In such environments, a greater grip force may be obtained by making a ditch on the belt at the edges of the steps so that the crawler belt will match the stair edge shape and the effectiveness of the crawler will not depend on the friction at the face of the belt. Aligning materials with soft deformation characteristics to the face of the crawler belt is considered to be an effective and practical method by which to achieve these characteristics.

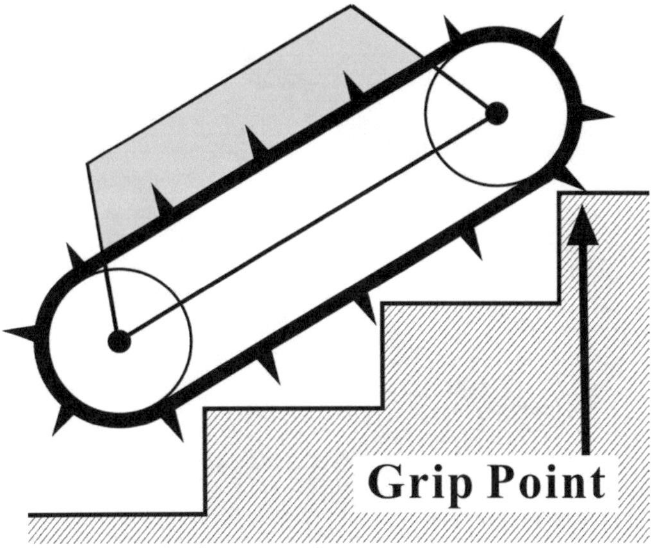

Fig. 12. Grouser-attached stair-climbing crawler

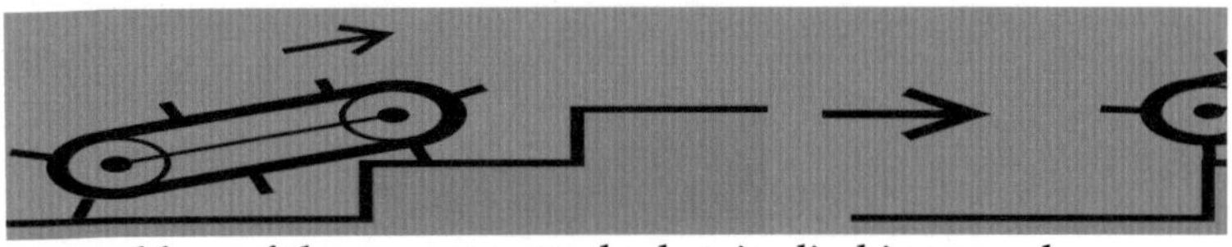

Fig. 13. Slippage problem of the grouser-attached stair-climbing crawler

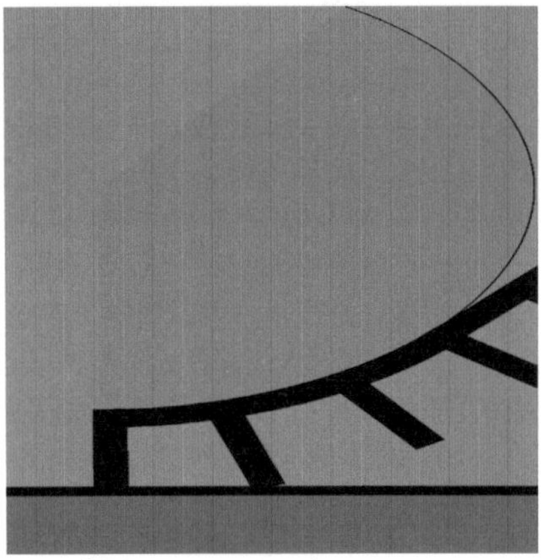

Fig. 14. Problem of short distance between grousers

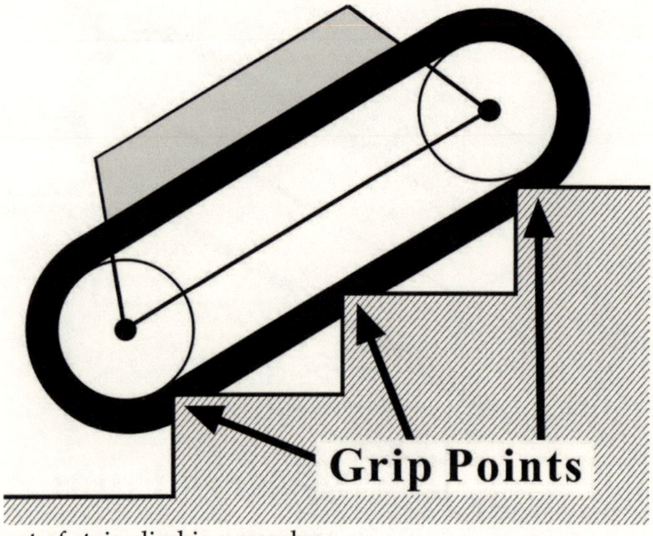

Fig. 15. New concept of stair-climbing crawler

5. Blocks Filled with Powder and Comparison of the Characteristics of Materials

5.1 Blocks Filled with Powder

Usually, rubber or a urethane sponge (which have soft deformation characteristics) are used as the track material, as mentioned earlier. However, as shown in Figure 16, we have developed special blocks that attach to the crawler belt and rely on the deformation characteristics of fluids. Tubes with durability and flexibility are filled with powder and the edges of the tubes are bent for the purpose of attachment to the crawler belt. In the present study, flour is used as the powder. Sand was also found to be an effective powder. A fire hose is used as the tube material. The hose is turned inside out so that the cloth side faces inward and the resinous side faces outward. There is room for improvement in the durability and water-resistance of these materials.

Next, a comparison of the characteristics between the developed blocks filled with powder and the previous soft materials will be performed. Furthermore, the suitability of materials for the crawler belt for a stair-climbing crawler is examined.

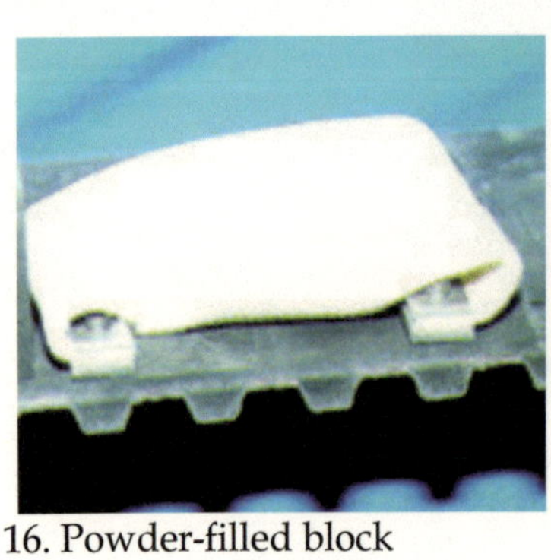

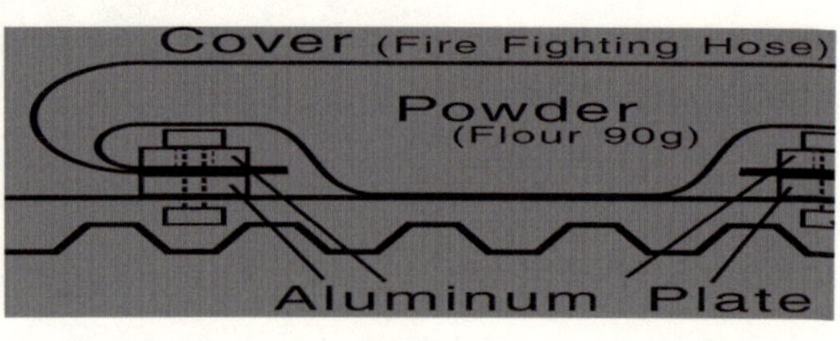

Fig. 16. Powder-filled block

5.2 Friction Characteristics of each Block

For measuring the characteristics of the face material used for the crawler belt of a stair-climbing crawler, the experimental device shown in Figure 17 was prepared. An aluminum block acts as a stair edge and presses against the measured soft material, applying a sideways force. First, the relationship between vertical force and vertical deformation when the experimental edge is pressed was measured. Next, for measuring the grip ability against the stair edge, vertical and horizontal forces were measured when slight slippage occurred due to a horizontal force during vertical loading. The equivalent frictional coefficient for each vertical loading is calculated as:

$$\text{Equivalent Friction coefficient} = \frac{\text{Horizontal Load (Grip Force)}}{\text{Vertical Load}} \tag{1}$$

The equivalent frictional coefficient is measured for cases of increasing vertical load and decreasing vertical load from the maximum load because of the hysteresis characteristics of the materials. The measured materials were the newly developed powder-filled block, a urethane rubber block with approximately the same vertical deformation, a urethane rubber block in the tube used in the newly developed powder-filled block, and the tube itself. The size of these experimental materials is the same as that of the powder-filled block, as shown in Figure 5 (90L × 50W × 30H, 100 g). In order to examine the change in the characteristics with the diameter of the powder, the blocks were filled with aluminum balls of 3 mm in diameter and plastic balls of 6 mm in diameter.

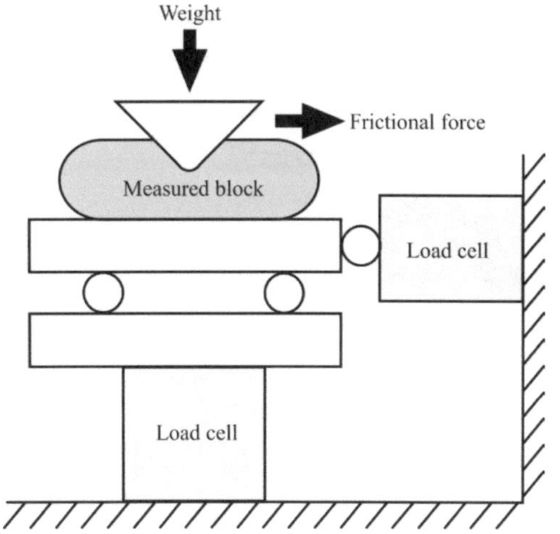

Fig. 17. Experimental system

5.3 Measurement Results of Deformation

First, the results of a comparison of the deformation between the urethane rubber block and the powder-filled block are shown in Figure 18. The same deformation characteristics are observed with an increasing vertical load. However, with a decreasing vertical load, the powder-filled blocks retain their previous deformation, whereas the urethane rubber blocks do not. Next, the results of a comparison of the deformation for different types of powder are shown in Figure 19. This comparison includes the powder-filled block, and the blocks contained 3 mm aluminum balls and 6 mm plastic balls. The results show that the blocks had approximately the same characteristics in each case of increasing and decreasing loads, whereas the maximum deformations differed. Moreover, the results reveal that the blocks have large hysteresis characteristics in common.

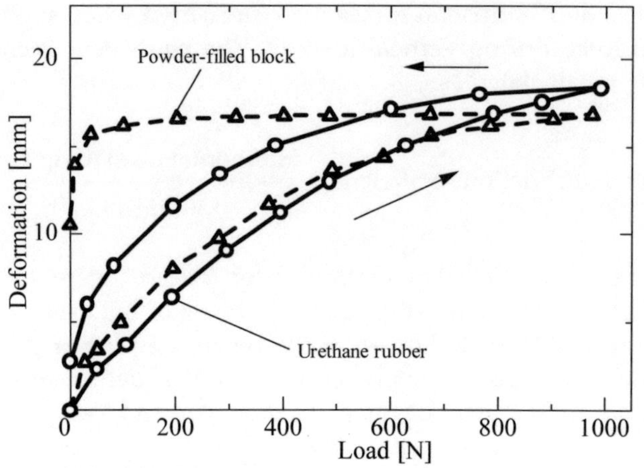

Fig. 18. Characteristics of block deformations

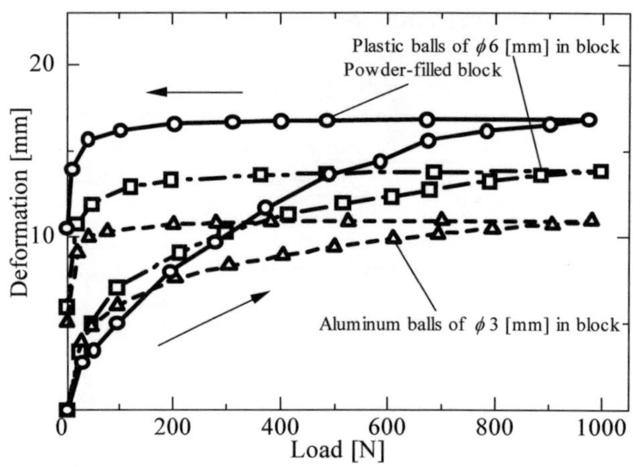

Fig. 19. Comparison of deformation with inner particle size

5.4 Results of Equivalent Frictional Coefficient

Figure 20 shows the results of the measurements of the equivalent frictional coefficient for the four types of blocks: urethane rubber block, the tube itself, urethane rubber in the tube and the powder-filled block. The results show that the equivalent frictional coefficient of the powder-filled blocks becomes much higher than the equivalent frictional coefficients of the other blocks. A very high equivalent frictional coefficient was obtained in the case of a weight reduction. This appears to depend on the hysteresis characteristics of the powder-filled block, because the block maintains its deformation after load reduction. This characteristic benefits the crawler because larger friction forces can be obtained from the middle of the crawler belt where the low-pressure area is located, even while climbing stairs, as shown in Figure 21. The total friction force of the blocks is expressed as the sum of the adhesive friction force, which depends on the face characteristics of the material and the friction force due to deformation that occurs during motion. The adhesive friction force depends only on the facing material, and the friction force due to deformation depends only on the inner materials. For example, friction forces due to deformation are the same between the urethane rubber block and the urethane rubber blocks inside the tube. The difference is the adhesive friction force due to the face material of the tube. Moreover, the friction force due to deformation of the inner powder can be calculated as the total friction force of the powder-filled blocks minus the friction of the tube, which is adhesive friction. Thus, the ratio of adhesive friction to the friction due to deformation for a specific loading can be expressed as shown in Figure 22. Almost all of the friction of the powder-filled blocks is attributed to the deformation. Therefore, it appears that a stable grip force can be always obtained, despite the grounding state of the environment. However, the friction force of the rubber blocks depends on the friction at the surface, and this is not desirable.

This result also shows that the crawler with the powder-filled belt has a relatively smaller friction force on flat surfaces, such as asphalt or concrete. When the crawler moves over a flat surface, the powder-filled blocks deform little because the ground presses equally towards the powder-filled blocks; little energy is lost by rolling resistance which depends on the hysteresis loss. Therefore, the crawler with powder-filled blocks also has better mobility for tasks on flat surfaces such as curving or pivot turning (by relatively small surface friction) and for climbing stairs (by large frictional force due to deformation).

Next, the same experiments were performed in order to compare the effects of the size of particles and materials. The results are shown in Figure 22, which compares the 3 mm diameter aluminum balls with 6 mm plastic balls. The large equivalent frictional coefficient and hysteresis characteristics were approximately the same. Therefore, variations in the inner material and size do not play a very important role in defining the friction force generated by the block. Flour, however, becomes harder and stiff and does not change its form once it has been subjected to loads greater than 2500 N. Thus, the size and the materials used for the inner powder should be decided according to the intended environments and the load carried. Otherwise, the particles can be destroyed and the block will no longer be able to change its form.

After several experiments, the following results were obtained.

1. Sand can generate large friction forces but is heavy.
2. The 3 mm diameter aluminum ball can also can generate large friction forces, but is also heavy (150 g) and very expensive.
3. Plastic balls or rice, which is fragile, cannot maintain their frictional performance because the characteristics of the particles change as they break into smaller particles.
4. The sack should be composed of a non-expandable material.

Based on these considerations, we have developed a stair climber with powder-filled blocks filled with flour.

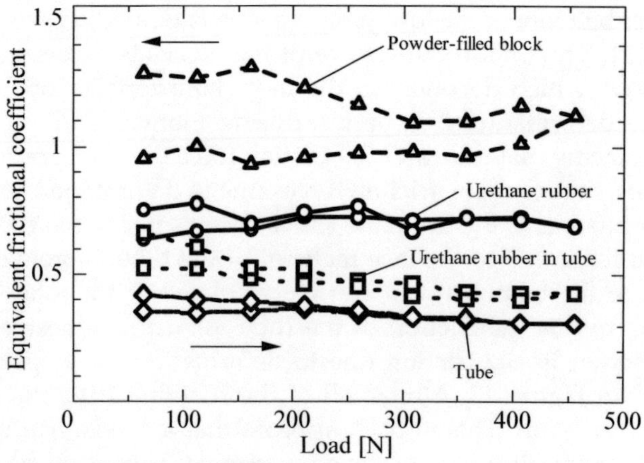

Fig. 20. Characteristics of equivalent coefficient

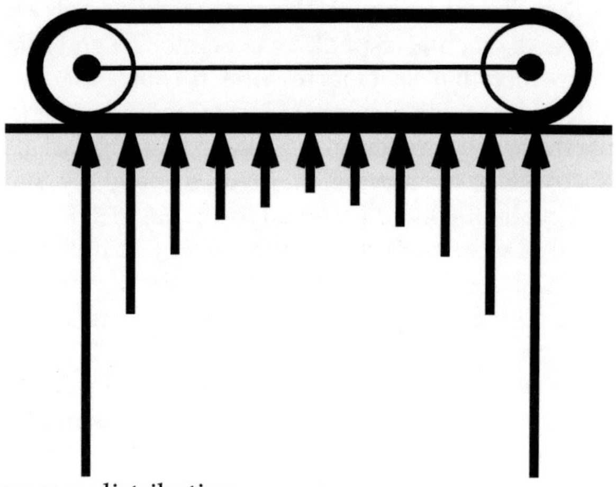

Fig. 21. Grounding pressure distribution

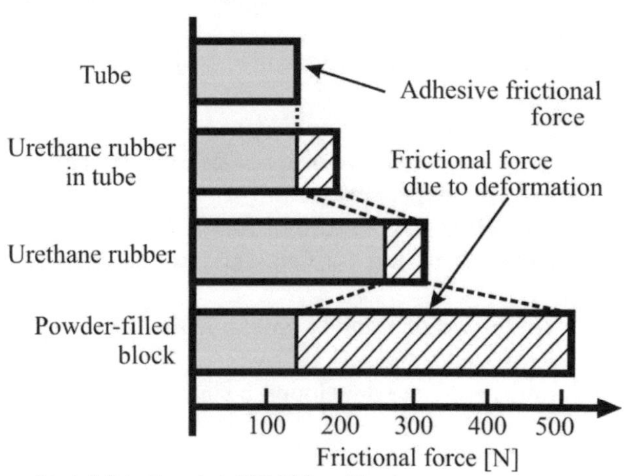

Fig. 22. Comparison of total friction (at 455 N loading)

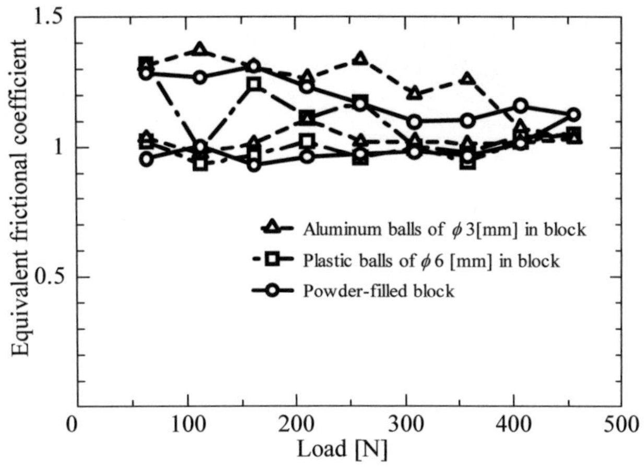

Fig. 23. Comparison of equivalent coefficients of friction with inner particle size

6. Design of Crawler Vehicle

To verify the advantages of using powder-filled blocks when considering stair-climbing safety and reliability, the stair-climbing crawler (Yoneda et al., 2001) as shown in Figure 24 was developed. The climber has a total length of 1180 mm, a width of 830 mm and a weight of 65 kg, including the batteries. This vehicle has a maximum speed of 500 mm s^{-1} and the batteries have a lifespan of 45 min.

To design the deformable powder-filled tracks a total of 112 powder-filled blocks, which were tested from the previous chapter, were attached to each crawler belt (Figure 25). Twenty-eight powder-filled blocks are aligned in two rows per belt. The blocks on the left and right rows are longitudinally shifted by one-half pitch so as to prevent their gaps from coinciding. Thus, the edge of the stair cannot fit within a gap of the block. We can therefore omit the effect of gripping by gaps and check the actual grip performance of powder deformation.

This crawler is also equipped with the belt tension mechanism shown in Figure 26, which was developed to achieve equally distributed grounding pressure. This crawler is also equipped with the active swing idler mechanism shown in Figure 27. This idler is located at the same height as the front and rear main idlers in order to achieve grounding pressure at the middle area of crawler belt, as shown in Figure 28(a). When the crawler approaches the top of the stairs, the swing arm moves and pulls the idler up, bending the crawler belt as shown in Figure 28(b). This motion prevents the sudden change of the posture of the crawler. When the crawler is required to perform pivot turning, the idler is pushed out and the grounding area becomes small, as shown in Figure 28(c). This motion makes pivot turning easier on high-friction surfaces, such as an asphalt road.

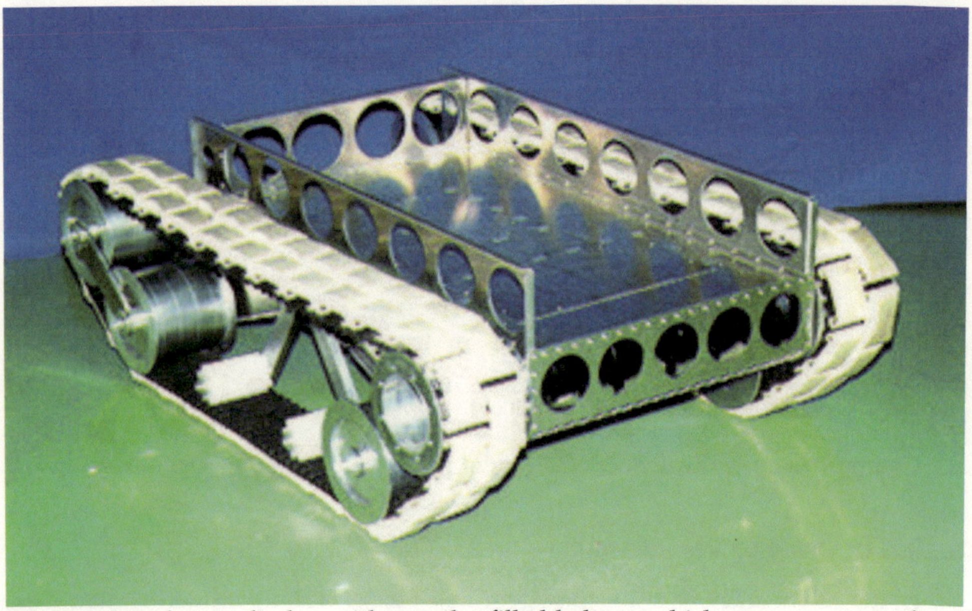

Fig. 24. Developed stair climber with powder-filled belts to which numerous powder-filled blocks are attached

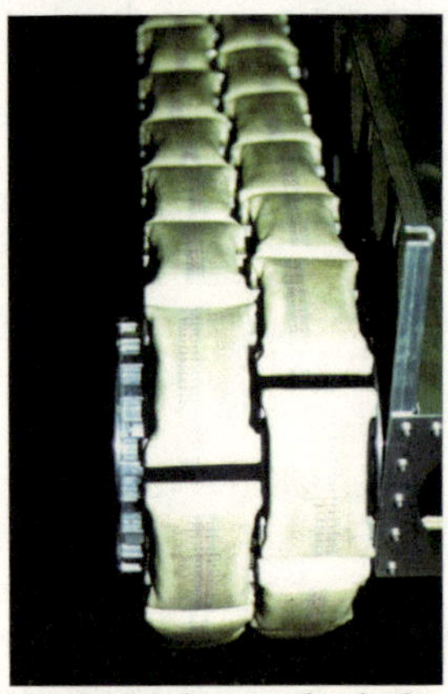

Fig. 25. Alignment of the powder-filled blocks on the belt

Fig. 26. Belt tension mechanism

Fig. 27. Active swing idler mechanism

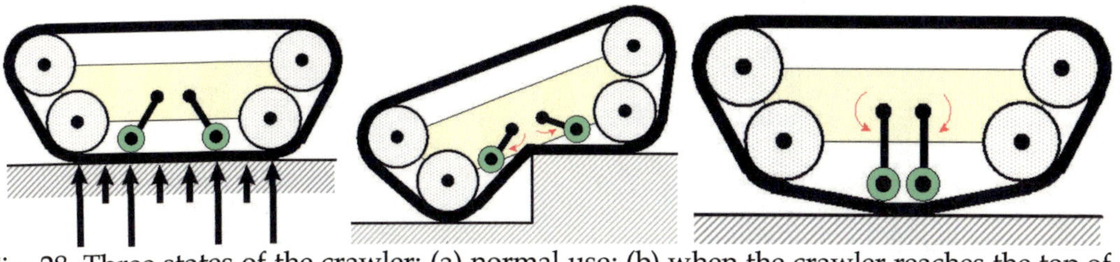

Fig. 28. Three states of the crawler: (a) normal use; (b) when the crawler reaches the top of a stair; and (c) during pivot turning

7. Stair-Climbing Experiment

To verify the abilities of the developed stair-climbing crawler with powder-filled belts, comparison experiments between a crawler with powder-filled belts, a crawler with grouser-attached tracks (Figure 29) and a crawler with urethane rubber blocks (Figure 30) were performed. The stairs used in these experiments have steps of 270 mm in length and 150 mm in height having R2 edges that are sharper than ordinary stairs. All of the crawlers were able to ascend and descend the stairs. In addition the traction forces, which give an indication of the margin of stability and payload, were measured. The results of traction forces are shown in Table 1. It was observed that the developed crawler with powder-filled belts can generate a large traction force that is approximately twice as large as that of the crawler with urethane rubber blocks. The crawler with grouser-attached tracks was able to generate large traction forces when the grousers achieve a good grip on the stair edges. However, as mentioned above, slippage or spinning has been observed when the support point changes. Figure 31 shows the measurement of the pitching angle of the inclination while ascending the stairs. The crawler with grouser-attached tracks generates a larger change in inclination angle than the crawlers with powder-filled belts and urethane rubber blocks.

Furthermore, the crawler with powder-filled belts was able to climb steeper stairs (step length 270 mm, step height 160 mm and edge radius 5 mm), although the crawler with urethane rubber blocks could not ascend because of an insufficient grip force. Moreover, climbing experiments involving the crawlers moving on stairs in non-straight trajectories were performed. Although the crawler with grouser-attached tracks could not ascend the stairs because the grousers could not obtain a sufficient traction from the stair edges, the crawler with powder-filled belts could ascend and descend the stairs stably. In addition, the crawler with powder-filled belts can also adjust its path to the right or to the left stably while ascending and descending stairs. Thus, climbing spiral stairs, which is a difficult task for most conventional stair-climbing vehicles, can be realized. The developed crawler with powder-filled belts can carry the heavy loads, as shown in Figure 32, and the maximum payload capacity is approximately 60 kg when ascending 30 degrees stairs. Furthermore, it was confirmed that the change in the posture becomes smooth at the top of the stairs and easy pivot turning is performed even if the grounding pressure becomes high because of the heavy load on the belt tension mechanism and active swing idler mechanism.

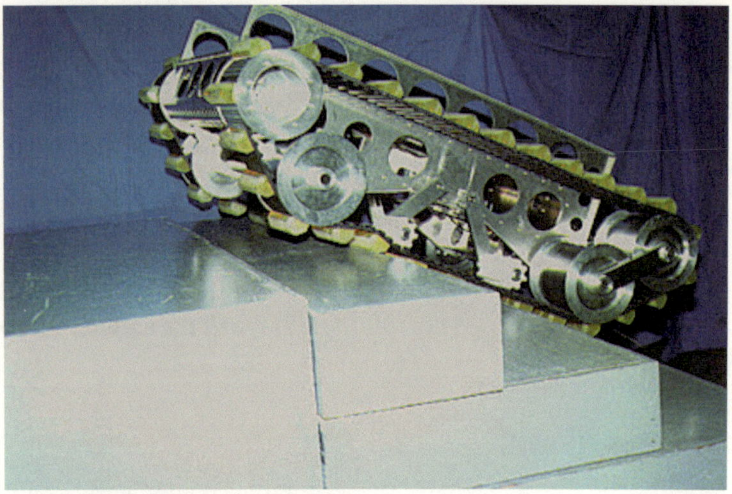

Fig. 29. Crawler with grouser-attached tracks

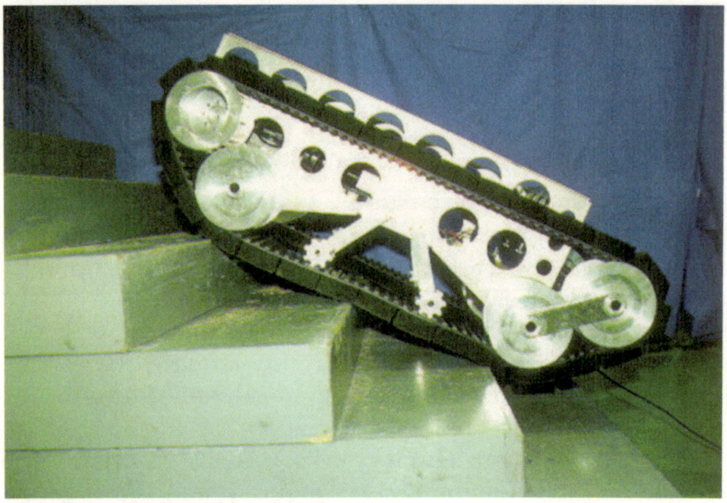

Fig. 30. Crawler with urethane rubber blocks

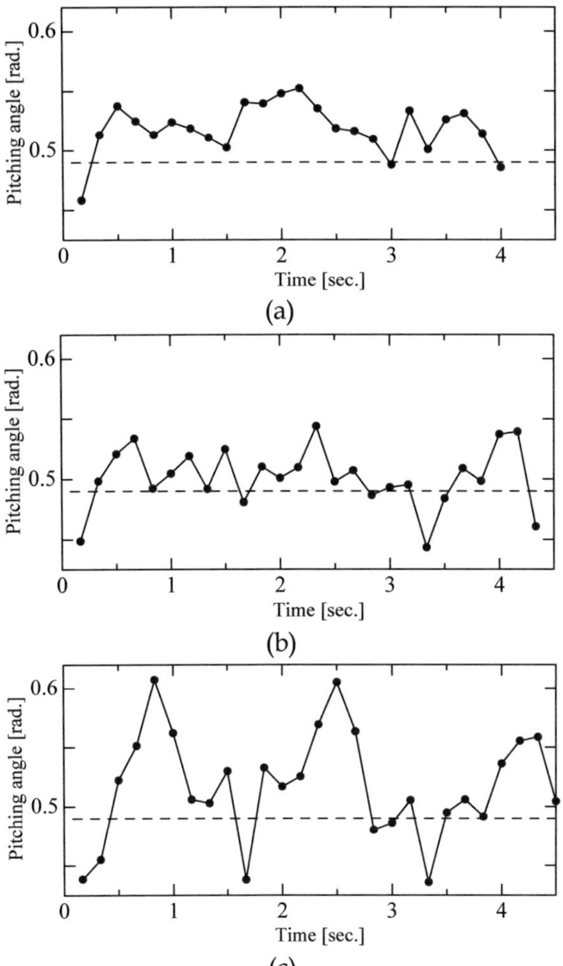

Fig. 31. Pitch angle variation of stair climbing with (a) powder-filled belts; (b) urethane rubber belts; and (c) grouser-attached tracks.

8. Conclusion

We describe a practical stair-climbing crawler and the mechanisms required to obtain sufficient grip force on the stairs. We developed powder-filled belts, which consists of several powder-filled blocks attached to the surface of the crawler belt, and compared the characteristics between the powder-filled blocks and other conventionally used materials. The results reveal that after the powder-filled belts deform to match the stair edge, the belts become harder and are therefore able to keep their shapes. This hysteresis characteristic of the attached powder-filled blocks is due to the fact that the powder flow generates a large equivalent friction coefficient at the middle area of the crawler belt, where there is a lower grounding pressure area after the pressure has been increased once. This has been verified experimentally.

After these experimental verifications, we used this high-grip climber for practical application in helping to carry heavy baggage. We can use the developed climber under several ground conditions with a variety of frictional conditions, such as asphalt, concrete and carpet. Several types of stairs, such as steep stairs (approximately 50 degrees), spiral stairs, narrow stairs, round edged stairs and wet stairs, were also ascended and descended successfully. Under these difficult conditions, the powder-filled belt and composed blocks always deliver sufficient grip force without breaking down. These findings reveal that the newly developed stair-climbing crawler with powder-filled belts has sufficient durability for practical application.

Fig. 32. Ascending stairs while carrying heavy objects

Powder-filled belt	441
Urethane rubber belt	226
Grouser-attached tracks	> 490

Table 1. Results of traction force experiments (N).

9. References

Arai, M.; Tanaka, Y.; Hirose, S.; Tsukui, S. & Kuwahara, H. (2006). Improved Driving Mechanism for Connected Crawler Vehicle "Souryu-IV" for in Rubble Searching Operations. *Proceedings of IEEE International Workshop on Safety Security and Rescue Robotics (SSRR2006)*, pp. TUE-AM 1-1, August 2006, Washington, USA,

Granosik, G.; Hansen, M. & Borenstein, J. (2005). The OmniTread Serpentine Robot for Industrial Inspection and Surveillance, *International Journal of Industrial Robots*, No. 32, Vol.2, 139–148

Hirose, S.; Aoki, S. & Miyake, J. (1989). Terrain Adaptive Tracked Vehicle HELIOS-I, *Proceedings of 4th International Conference on Advanced Robotics*, pp. 676–687

Hirose, S.; Aoki, S. & Miyake, J. (1990). Design and Control of Quadru-Truck Crawler Vehicle HELIOS-II, *Proceedings of 8th RoManSy Symposium*, pp. 1–10.

Hirose, S.; Usa, M.; Ohmori, N.; Aoki, S. & Tsuruzawa, K. (1991). Terrain Adaptive Quadru-Track Vehicle HELIOSIII, *Proceedings of 9th Annual Conference on RSJ*, pp. 305–306 (in Japanese)

Hirose, S.; Sensu, T. & Aoki, S. (1992). The TAQT Carrier:A Pratical Terrain-Adaptive Quadru-Track Carrier Robot, *Proceedings of IEEE/RSJ International Conference on Intelligent Robots and Systems*, pp. 2068–2073

Hirose S.; Yoneda K.; Arai K. & Ibe T. (1995). Design of a quadruped walking vehicle for dynamic walking and stair climbing, *Advanced Robotics*, Vol.9, No.2, 107-124

Hirose S.; Fukuda Y.; Yoneda K.; Nagakubo A.; Tsukagoshi H.; Arikawa K.; Endo G., Doi T. & Hodoshima R. (2009). Quadruped Walking Robots at Tokyo Institute of Technology, *IEEE Robotics and Automation Magazine*, Vol.16, No. 2, 104-114

Krishna M.; Bares J. & Mutschler Ed, (1997). Tethering System Design for Dante II, *Proceedings of IEEE International Conference on Robotics and Automation*, pp.1100-1105

Liu, J.; Wang, Y.; Ma, S. & Li, B. (2005). Analysis of Stairs- Climbing Ability for a Tracked Reconfigurable Modular Robot, *Proceedings of IEEE International Workshop on Safety, Security and Rescue Robotics*, pp. 36–41, Kobe, Japan

Miyanaka, H.; Wada, N.; Kamegawa, T.; Sato, N.; Tsukui, S.; Igarashi, H. & Matsuno, F. (2007). Development of a unit type robot "KOHGA2" with stuck avoidance ability, *Proceedings of 2007 IEEE International Conference on Robotics and Automation*, pp.3877–3882. Roma, Italy

Murphy R. Robin (2000). Biomimetic Search for Urban Search and Rescue, *Proceedings of the IEEE/RSJ Intelligent Robots and Systems*, pp. 2073-2078, Takamatsu Japan, October 2000

Ota, Y.; Yoneda, K.; Ito, F.; Hirose, S. & Inagaki, Y. (2001a). Design and Control of 6-DOF Mechanism for Twin-Frame Mobile Robot, *Autonomous Robots*, Vol.10,No.3, 297-316

Ota, Y.; Yoneda, K.; Muramatsu, Y.; & Hirose S. (2001b). Development of Walking and Task Performing Robot with Bipedal Configuration, *Proceedings of 2001 IEEE/RSJ International Conference on Intelligent Robots and Systems*, pp.247-252, Hawaii USA

Ota Y.; Yoneda K., Tamaki T. & Hirose S. (2002), A Walking and Wheeled Hybrid Locomotion with Twin-Frame Structure Robot, *Proceedings of 2002 IEEE/RSJ International Conference on Intelligent Robots and Systems*, pp.2645-2651, Lausanne Switzerland

Ota Y.; Tamaki T.; Yoneda K. & Hirose S. (2003), Development of Walking Manipulator with Versatile Locomotion, *Proceedings of 2003 IEEE International Conference on Robotics and Automation*, pp.477-483, Taipei Taiwan, September 2003

Ota Y.; Kuga T. & Yoneda K. (2006). Deformation Compensation for Continuous Force control of a Wall Climbing Quadruped with Reduced-DOF, *Proceedings of 2006 IEEE International Conference on Robotics and Automation*, pp.468-474, Florida USA, May 2006

Schempf, H.; Mutschler, E.; Piepgras, C.; Warwick, J.; Chemel, B.; Boehmke, S.; Crowley, W.; Fuchs, R. & Guyot, J. (1999). Pandora: Autonomous Urban Robotic Reconnaissance System, *Proceedings of International Conference on Robotics and Automation*, pp. 2315–2321, Detroit USA, May 1999

Stoeter A. Sascha; Rybski E. Paul; Gini Maria & Papanikolopoulos Nikolao (2002). Autonomous Stair-Hopping with Scout Robots, *Proceedings of the IEEE/RSJ Intelligent Robots and Systems*, pp.721-726, 2002

Taguchi, K. (1995). Enhanced wheel system for step climbing, *Advanced Robotics*, Vol.9, No.2, 137–147

Takahashi M.; Yoneda K. & Hirose S.(2006). Rough Terrain Locomotion of a Leg-Wheel Hybrid Quadruped Robot, *Proceedings of 2006 IEEE International Conference on Robotics and Automation*, pp.1090-1095, Florida USA, May 2006

Takayama T. & Hirose S. (2000). Development of Souryu-I : Connected Crawler Vehicle for Inspection of Narrow and Winding Space, *Proceedings of IEEE International Conference on Industrial Electronics, Control and Instrumentation (IECON-2000)*, pp.143-149

Tanaka, Y.; Arai, M.; Hirose, S. & Tsukui, S. (2006). Development of "Souryu-V" with Mono-Tread-Crawlers and Elastic-Rods Joint. *Proceedings of IEEE International Workshop on Safety Security and Rescue Robotics (SSRR2006)*, pp. TUE-AM 1-1, Washington, USA,

Tsukagoshi H.; Sasaki M.; Kitagawa A. 6 Tanaka T. (2005). Design of a higher jumping rescue robot with the optimized pneumatic drive, *Proceedings of the 2005 IEEE International Conference on Robotics and Automation*, pp.1288-1295

Uchida, Y.; Furuichi, K. & Hirose, S. (1999). Consideration of stair-climbing performance of a six-wheeled off-road vehicle 'HELIOS-V', *Proceedings of 2nd International Conference on Climbing and Walking Robots*, pp. 383–391

Uchida, Y.; Furuichi, K. & Hirose, S. (2000). Evaluation of Wheel Performance on Rough Terrain and Development of HS Wheel. *Journal of Robotics and Mechatronics*, Vol.12, No.5, 586–595

Yim M.; Duff D. & Roufas K. (2000). Modular Reconfigurable Robots, An Approach To Urban Search and Rescue, *Proceedings of 1st International Workshop on Human-friendly Welfare Robotic Systems*, pp.69-76

Yoneda K. (1987). The Development of Biped Walking Robot for HC-plane, *5th Annual conf. of The Robotics Society of Japan*, pp.585-586 (in Japanese)

Yoneda, K.; Iiyama, H. & Hirose, S. (2000). Active Suspension Control of Quadruped Walking Robot, *Machine Intelligence and Robotic Control* (MIROC), Vol.2, No.2, 45-50

Yoneda K.; Ota Y.; Ito F. & Hirose S. (2001a). Quadruped Walking Robot with Reduced Degrees of Freedom, *Journal of Robotics and Mechatronics*, Vol.13, No.2, 190-197

Yoneda K.; Ota Y.; Hirano K. & Hirose S. (2001b). Development of a Light-Weight Wall Climbing Quadruped with Reduced Degrees of Freedom, *Proceedings of 4th International Conference on Climbing and Walking Robots*, pp. 907-912, Karlsruhe Germany

Yoneda K. & Ota Y. (2003) : Non-Bio-Mimetic Walkers, *The International Journal of Robotics Research*, Vol. 22 No.3-4, 241-249

Yoneda K (2007). Light weight Quadruped with Nine Actuators, *Journal of Robotics and Mechatronics*, Vol.19, No.2, 160-165

Yoneda, K.; Ota, Y. & Hirose, S. (2009) Development of a Hi-Grip Stair Climbing Crawler with Hysteresis Compliant Blocks, *The International Journal of Robotics Research*, Vol. 28, No.1, 81-89

A Climbing-Flying Robot for Power Line Inspection

Jaka Katrašnik, Franjo Pernuš and Boštjan Likar
University of Ljubljana
Slovenia

1. Introduction

Our society is becoming increasingly more dependent on reliable electric power supply. Since power outages cause substantial financial losses to the producers, distributors and also to the consumers of electric power, it is in the common interest to minimize failures on power lines. To detect the defects early and to accordingly schedule the maintenance activities, the distribution networks are inspected regularly. Inspection of overhead power lines is usually done manually, either directly on the lines or indirectly from the ground and/or from the helicopters. All these tasks are tedious, expensive, time consuming and dangerous. Consequentially, more and more research has been focused on automating the inspection process by means of mobile robots that would possibly surpass the abovementioned disadvantages. Namely, robot-assisted inspection could be carried out faster, cheaper and more reliable, thus improving the long-term stability and reliability of electric power supply. Most importantly, the safety of the inspection workers could be increased significantly.

In this chapter the requirements for all types of robots for power line inspection and the key research problems and proposed solutions for flying and climbing robots are surveyed. Next, a new so-called climbing-flying robot, which inherits most of the advantages of climbing and flying robots, is proposed. The proposed robot is critically assessed and related to the other inspection robots in terms of design and construction, inspection quality, autonomy and universality. In conclusion, the remaining research challenges in the field of power line inspection that will need to be addressed in the future are outlined.

2. Robot Requirements

2.1 Power Line Features and Faults

Power lines are a dangerous environment. The electric potential differences between the lines are in the order of 100 kV, yielding the electric field in the vicinity of the lines close to 15 kV/cm under normal conditions and even more in the presence of defects. The magnetic field is not small either, due to the currents that are in the order of 1000 A the magnetic field

Based on "New Robot for Power Line Inspection", by Jaka Katrašnik, Franjo Pernuš and Boštjan Likar which appeared in 2008 IEEE Conference on Robotics, Automation and Mechatronics.

on the surface of the conductor reaches values as high as 10 mT. Power lines are also a complex environment, difficult for robots to navigate. The simplest power lines have one conductor per phase, while others may have more. The conductors are hung on insulator strings, which can either be suspension insulators or strain insulators. Besides insulators, there are other obstacles on the conductors, such as dampers, spacers, aircraft warning lights and clamps (Fig. 1).

The faults on the power lines usually occur on conductors and insulator strings (Aggarwal et al., 2000). Aeolian vibrations gradually cause mechanical damage to conductors. Strands brake, the conductor loses its strength and starts overheating. Other important conductor damaging factors are the corona effect and corrosion. Insulator strings are also prone to mechanical damage due to impact, weather and corrosion (Aggarwal et al., 2000). During inspection, it is also necessary to check for vegetation on and beneath power lines, pylon and other power line equipment condition and safety distance between conductors and other objects.

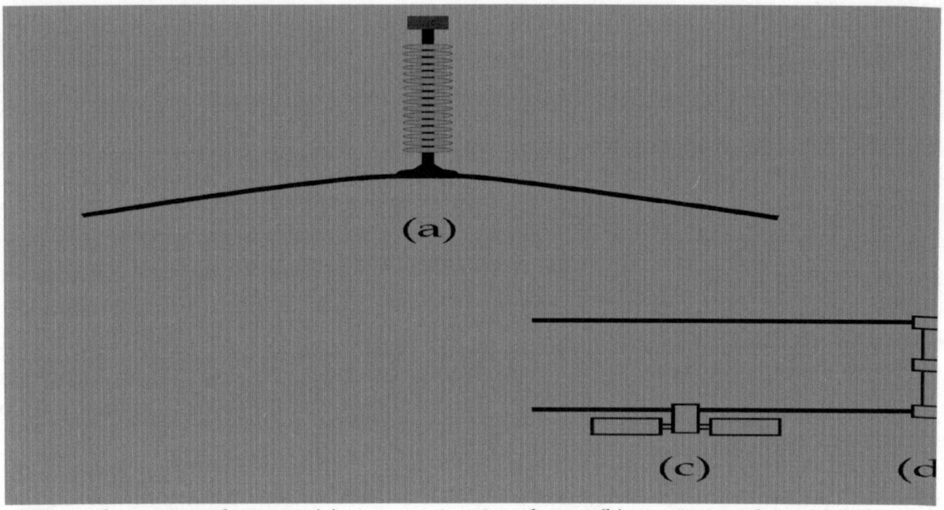

Fig. 1. Obstacles on conductors: (a) suspension insulator, (b) strain insulator, (c) damper, (d) spacer and (e) aircraft warning light.

2.2 Robot Functionality

The design of the robot determines its functionality. In helicopter-assisted inspection the helicopter is flown along the power lines and the camera operator has to track and film the lines with a normal, IR and UV camera. The video footage is then carefully inspected on the ground. This is a very quick method of inspection but tedious for the camera operator and quite inaccurate. That is why the requirements for the automation system are automatic power line tracking, automatic visual inspection and automatic measurement of power line safety distance. Another problem that needs to be solved for these systems to work is also the acquisition of high-quality images, which is very important for visual power line tracking and visual inspection.

Similar problems need to be solved when developing an UAV (Unmanned Aerial Vehicle) for power line inspection. A small helicopter is usually used for the UAV, because it has the ability to hover. The UAV has to be able to autonomously travel along the power lines, find and document faults. It also has to be as energy-independent as possible. The problems associated with this approach are similar to those of helicopter-assisted inspection, but even more demanding. The key issues are position control, automatic power line tracking, obstacle avoidance, communication, image acquisition, automatic fault detection, measuring power line safety distance and power pick-up from the power line.

Another inspection approach, which has been developed for many years, is the climbing robot. The robot travels suspended from the conductor and has to cross obstacles along the power line, which requires complex robotic mechanisms. The robot functionality should include autonomous traveling along the conductor, automatic visual inspection and at least semi-autonomous obstacle crossing. The main problems associated with this approach are thus robotic mechanism design and construction, the conductor grasping system, the driving system, conductor obstacle detection and recognition, the robot control system, communication, visual inspection, power supply and electromagnetic shielding.

3. Automated Helicopter Inspection

One of the first articles on automating helicopter-assisted inspection (Whitworth et al., 2001) addressed some of the problems, specifically, tracking the power line, especially the poles that need careful inspection, and image acquisition stabilization. A tracking algorithm for power line poles was developed and tested on a scaled laboratory test rig. The initial position of the pole would be obtained with DGPS (Differential Global Positioning System). The pole recognition was done on the basis of two vertical lines and the two horizontal lines of the top cross arm. The reported success rate of the pole recognition algorithm was 65-92% on videos recorded at helicopter inspection, but the image processing rate of 2 to 8 images per second was rather small and the recognition did not work well when background was cluttered. The authors concluded that the concept is feasible, although problems with robustness could arise in real environments with complex backgrounds and varying lighting conditions. Visual tracking of the poles with corner detection and matching was investigated in (Golightly and Jones, 2003). For corner detection the zoom invariant CVK (named by the authors: Cooper, Venkatesh, Kitchen) method described in (Cooper et al., 1993) was proposed. The method was found suitable for corner detection at the tops of the power line poles. Because the method detects multiple matches for one physical corner, detected corners have to be aggregated. Corner matching is then done on two consecutive images, using a basic corner matcher. Relatively good stability of the whole system was reported.

For accurate inspection, the quality of images taken from the helicopter has to be as good as possible. Images taken from an on board camera often get blurred, due to constant vibration and translational movement of the helicopter. In (Jones and Earp, 2001) this problem was thoroughly investigated and minimal optical stabilization requirements defined. Small movements of the helicopter can be compensated by mounting the camera on gyro-stabilized gimbals, which lock the sightline to an inertial reference. Translational helicopter movement can only be compensated with visual tracking of the inspected object. It was found that for sufficient inspection detail, the image blur should not be more than 1% and that the stabilized platform must achieve optical stabilization better than 100-200 μr (micro radians).

4. Inspection with an UAV

Inspection with an UAV is an upgrade of automated helicopter inspection so both concepts have some common problems. An evaluation of using an UAV for power line inspection (Jones and Earp, 1996) indicated that this inspection method could be faster than foot patrol and would yield the same or better accuracy than costly helicopter inspection. It was concluded that the system is feasible from a technical point of view. The concept was further investigated in (Jones, 2005). A small electrically driven rotorcraft, which can pick up energy from power lines, was presented. This vehicle would be equipped with gyro-stabilized cameras, navigation and position regulation, a computer for image and other sensor data processing, a communication link and a system for electric power pick up. Power would be obtained from the power line using a pantograph mechanism. The most research was devoted to the development of a vision system for power line tracking and to image quality assurance. Namely, good power line tracking is important for visual position control and navigation, while image quality is of utmost importance for inspection purposes.

4.1 Position Control

Since power lines have to be inspected from a small distance but must under no circumstances get damaged even in strong wind, position control of the UAV is difficult yet very important. Because conductors have to be in the field of view of the camera almost all the time, determining position of the helicopter visually from the images of the conductors seems very attractive (Campoy et al., 2001, Golightly and Jones, 2005, Jones et al., 2006). Position control is thus closely related to automatic tracking of power lines. The helicopter is a very complex, unstable and nonlinear system with cross couplings. In (Campoy et al., 2001) a Linear Quadratic Gaussian (LQG) controller was chosen for roll and pitch control and a PID controller for yaw control. The controllers were implemented on the basis of measured dynamic characteristics of the helicopter. Because only position of the helicopter could be measured, all other required variables were estimated by the Kalman filter. The robustness was tested when the helicopter was in hover by pulling it with a cable. The regulation worked well in the presence of such external disturbances.

A rotorcraft model and a position control system for a power line inspection robot were also presented in (Jones et al., 2006). A mathematical model of a ducted-fan rotorcraft with the center of gravity above the aircraft center was derived and used for the development of control system. The control was achieved by moving a mass, positioned above the center of gravity, left or right. When the mass is moved, the craft tilts in the same direction and accelerates in that direction. The control system is closely linked with visual tracking of power lines and controls the height and lateral position of the craft to the lines. Lateral position and height are both measured with image analysis.

4.2 Automatic Power Line Tracking

Visual tracking of power lines with an UAV is similar to visual tracking with a helicopter. The only major difference is that the UAV can get closer to the lines. The tracking methods are therefore a little different. Jones and Golightly developed a simple tracking algorithm that could track the power line with three lines based on the Hough Transform (Jones et al., 2006). The main purpose of this tracking algorithm was to provide height and lateral displacement of the vehicle to the control system. The method was tested on a scaled model and was proven to be successful even when the background was cluttered. Another method for visual power line tracking (Campoy et al., 2001) utilized a vector-gradient Hough transform for line detection. Only one line was tracked and simultaneously inspected. The position of the helicopter with regard to the line was determined with stereo vision.

4.3 Obstacle Avoidance

Another problem related to robot mobility is obstacle avoidance and path planning. The space around power lines is usually obstacle free; nevertheless, the robot must be able to avoid obstacles on its way, when it is not controlled by a human operator. A computer vision solution to this problem was proposed (Williams et al., 2001). Positions of the obstacles were determined by optical flow. The obtained positions were used in the path planning algorithm based on the distance transform. The algorithms were tested in a laboratory environment using a test rig with a scaled version of a power line. It was established that the principles used were correct but the method was sensitive to the variations in background, lighting and perspective. An important problem was also the computing power because image analysis demands were high and rapid obstacle detection was required.

4.4 Power Supply

An important characteristic of an inspection vehicle is the duration of its power supply. The longer the craft can stay operational the more lines can be inspected. Current battery technology does not permit long durations of flight for small electrically driven helicopters. Power lines are an abundant source of energy but obtaining that power is far from trivial. A concept of a power line power pick-up device was presented in (Jones, 2007). The power would be acquired by touching two lines of different phases with a special pantograph mechanism (Fig. 2). For this concept to work, line tracking and position control algorithms have to be highly reliable.

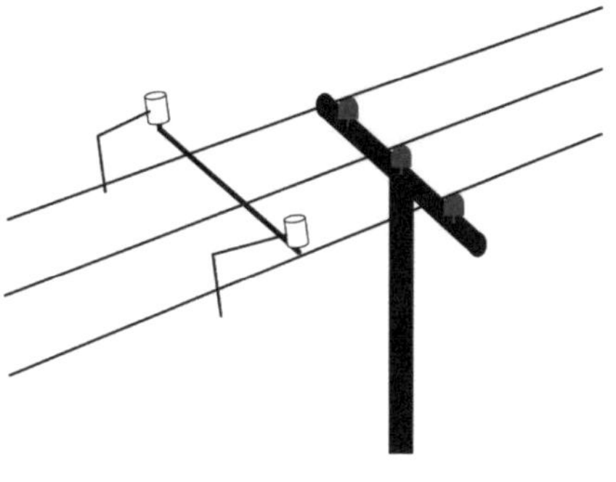

Fig. 2. The proposed pantograph power pick-up mechanism

4.5 Other Problems

A difficult problem that has not been researched thoroughly is automatic power line fault detection. It would be most convenient if the robot would be able to automatically detect faults on-site, so it could re-inspect them more thoroughly. On the other hand, automatic fault detection could be done in the ground station after the inspection, which would be easier to implement but would not provide detailed information about the defects. A big difficulty with fault detection is the quality of images taken from the UAV. Because of the distance from the line and constant movement of the craft, the quality of images is usually poor, which makes automatic fault detection especially demanding.

A big problem with the UAV concept of inspection is that almost every system on the robot (position control, obstacle avoidance, fault detection and power pick up) depends on visual tracking of power lines, which is not very reliable. Although visual power line tracking was successful in the laboratory, the real environment is much more demanding. Contrast between the lines and the background is usually very low. Lighting varies a great deal and depends on unpredictable weather conditions. The UAV is in constant motion and vibrates, so the images acquired by the robot would be of a poor quality and the faults very difficult to detect even for a human. Unintentional detection of other straight lines on the image, such as other power lines or railroad tracks, would also pose a serious problem.

5. Climbing Robots

An alternative approach to power line inspection is by means of a climbing robot, which can climb on the conductor and has to somehow overcome all various obstacles on the power lines. The main advantage of this concept is the inspection accuracy. Namely, close proximity to the line and low vibrations increase the quality of image acquisition. On the other hand, development of a robot mechanism for overcoming obstacles on the line is extremely difficult. The main research problem with climbing robots is therefore the development of a robot mechanism and a control system for obstacle crossing. The proximity of the conductor also brings problems related to electromagnetic shielding. Sensitive electronics and sensors have to be protected from the electric and magnetic fields of the conductor.

5.1 Robot Mechanisms and Obstacle Traversing

One of the first operational robot mechanisms for power line inspection was the robot presented in (Sawada et al., 1991). The robot consisted of a drive, an arc shaped rail, a guide rail manipulator and a balancer with controller. It could travel on slopes of up to 30°. When the robot would come across an obstacle it would unpack its rail and mount it on the conductor on both sides of the obstacle. Then the drive mechanism would release the conductor and travel on the rail to the other side. The robot was able to negotiate towers and other equipment on overhead ground wires. Not having proper shielding and mechanisms for overcoming obstacles, the proposed robot could not travel on phase conductors.

A more complex robot mechanism, presented in (Tang et al., 2004), had two arms (front arm and rear arm) and a body. Each arm had 4 degrees of freedom and a gripper with a running wheel. The body also had a running wheel with a gripper. When overcoming obstacles, the robot would release the conductor with the front arm, elongate it over the obstacle and grasp the conductor on the other side. Then the body would release and the two arms would move it across the obstacle, where it would grip the conductor again. Finally, the rear arm would move across the obstacle. This robot could overcome all standard obstacles on phase conductors of overhead power lines. However, it could not travel on bundled conductors.

The robot configuration in (Xinglong et al., 2006) had two arms and a special gripper combined with a driving wheel. The specialty of this mechanism is that the gripper could always grasp the conductor, when it was in contact with the running wheel. The gripper presses on the conductor from the left and right side of the wheel. The main disadvantage was that the gripper could not handle large torque, which can easily occur when crossing obstacles. For that reason, a special very effective obstacle crossing strategy that also simplifies the design of the robot was presented (Fig. 3). When the robot would detect an obstacle ahead, it would stop, grasp the conductor with the front arm and move its body under the front arm in order to minimize the torque when crossing the obstacle (Fig. 3(a)). Next, the rear arm would lift the running wheel up and the front arm would rotate the robot around its own axis. Finally, the rear arm would lower the wheel on the conductor (Fig. 3(b)). The same process would then be repeated with the arms' roles changed. Because of this obstacle traversing strategy, the robot arms need only two degrees of freedom, the torques in the joints and on the conductor are small and, consequently, the motors do not need to be as powerful and heavy.

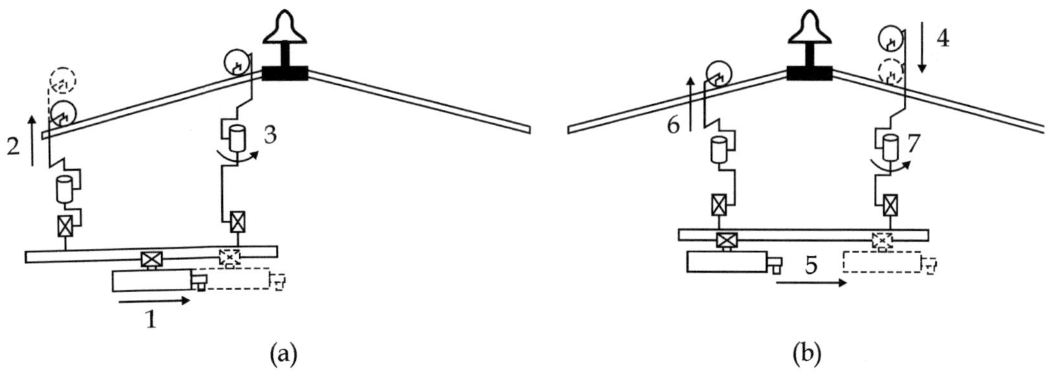

Fig. 3. Obstacle traversing strategy proposed in (Xinglong et al., 2006).

5.2 Robot Control System

The main purpose of the robot control system is to navigate the robot over obstacles on the line. One of the first robot control algorithms for power line inspection was described in (Sawada et al., 1991). A more complex control system, using a distributed expert system that was divided between the robot and the ground station, was described in (Tang et al., 2004). The robot control system would run on an embedded PC/104 based computer, connected to the ground station with a wireless data link and a separate image transmission channel. The robot expert system consisted of an inference engine, knowledge base, static database, external information input module and decision-making module. The inference engine would decide what commands to execute on the basis of sensor information and information in the static database. Sensors would provide information about current position of the robot and the obstacles around it, while the static database would contain data about towers and other obstacles on the line. The robot expert system would plan the path of the robot arms so that the robot would overcome the obstacle successfully. The ground station would be used for monitoring and guiding the robot as well as for detecting faults on the power line from the images sent by the robot. Similar distributed expert system designs were presented in (Ludan et al., 2006).

5.3 Obstacle Detection and Recognition

Obstacle detection is usually done with a proximity sensor, which is simple yet effective but the detection of the obstacle is usually not enough to overcome it. In most cases the type of the obstacle has to be known. In (Zhang et al., 2006) a computer vision method for obstacle recognition and distance measurement was presented. The method determines the obstacle types from the shapes on the image. An ellipse represents a suspension insulator string and two circles left and right of the conductor a strain insulator string. After the obstacle is recognized, its position is also located with a stereo vision. The method was tested on a real power line for which the accuracy of 7 % or better was reported. Another important problem associated with visual obstacle detection and recognition is the elimination of motion blur from the captured images (Fu et al., 2006). Although climbing robot is fixed on the conductor, it also swings under the influence of wind and when traveling along the line.

5.4 Power Supply

Power lines could provide the inspection robot with energy for its operation. Energy from the line could be extracted from the magnetic field of the line. This concept was presented in (Peungsungwal et al., 2001). A magnetic iron core was placed around the conductor. Current induced in the secondary coil around the core was measured at different numbers of windings of the secondary. It was shown that the current reaches its maximum value at a certain number of secondary windings and that the power transferred to the secondary coil increases with the current of the power line.

6. Climbing-Flying Robot

The abovementioned advantageous features of both robot types can be combined. For that reason, we propose a new robot type, i.e. the so-called climbing-flying robot (Fig. 4). The proposed robot would combine a helicopter for flying over the obstacles and a special drive mechanism for traveling on the conductor. During inspection, the robot would travel on the conductor up to an obstacle. Then it would fly off the conductor over the obstacle, land on the other side and continue traveling along the conductor. Traveling on the conductor would be automated, while flying over the obstacles would likely have to be done manually. Some of the problems that would need to be solved are similar to those described in the two previous sections. For instance, power pick-up system, obstacle detection and recognition, and drive mechanism for traveling on the conductor. Besides the advantages that arise from the proposed combination of the two robot types, there are also some specific new problems, which we address in the following.

6.1 Robot design

Design of the climbing-flying robot is much more difficult than design of the flying robot, although not as difficult as design of the climbing robot. When designing the proposed robot one must take into consideration the weight limitations of the helicopter, which are much stricter than for the flying robot. The reason for this is the addition of the drive mechanism and the electromagnetic shielding, which significantly increase the weight of the robot.

Another major problem is also the weight distribution in the robot. In order to achieve a good degree of stability on the power line, the center of gravity of the robot must be below the conductor. This conflicts with the design of the helicopter, where the majority of the weight is placed directly below the rotor to achieve good maneuverability. The parts of the robot must therefore be carefully positioned to achieve the optimal position of the center of gravity. A coarse distribution of robot parts inside the robot is proposed in Fig. 4.

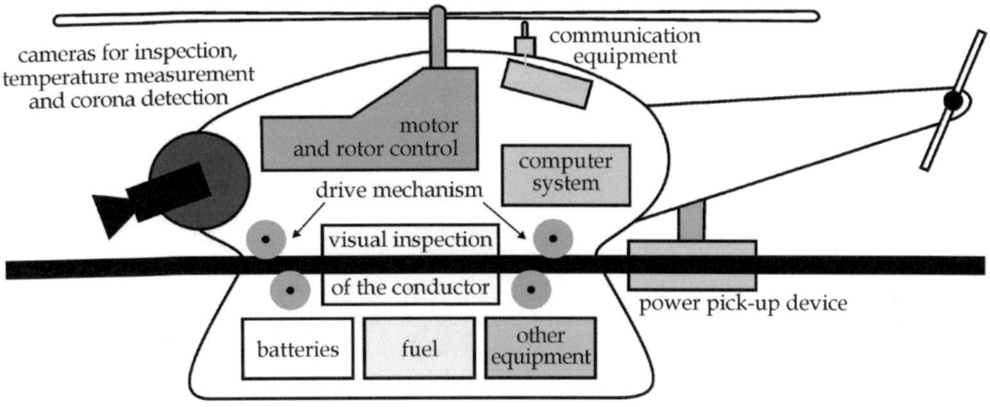

(a)

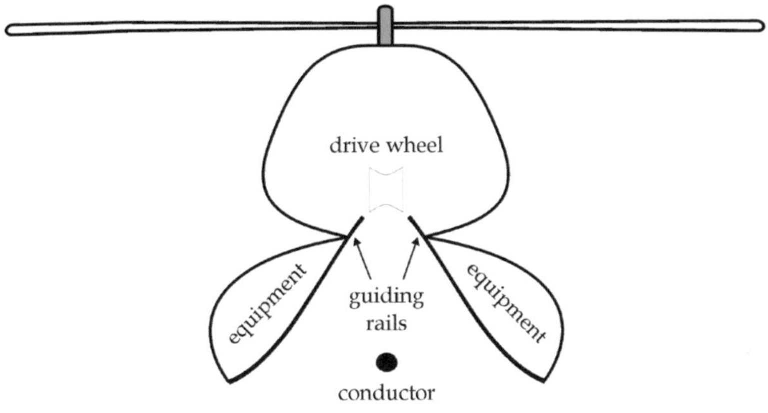

(b)

Fig. 4. The proposed robot: (a) An illustration of the proposed climbing-flying robot and its components; (b) a sketch of the robot from the front, showing the rails for easier landing on conductors and equipment placement for stability.

Setting the weight limitation and distribution problems aside, the most important problems of the climbing-flying robot are the design of individual systems, which are the helicopter, drive mechanism, visual inspection system, power pick-up device and communication system. The design of these systems is discussed in the following subsections.

6.2 Helicopter

When choosing the rotor configuration of the helicopter for the climbing-flying robot we have three choices. The most common is the Sikorsky configuration. Ninety percent of all the helicopters in the world are made in this configuration. It is simple to produce, has good maneuverability and sufficient lift. The tandem rotors configuration has worse maneuverability, but produces more lift as there is no power needed for balancing the main rotor torque. This configuration is also more difficult to make and maintain, as it has more moving parts and a more complex design. The coaxial rotors configuration is also more expensive to build and maintain. However it requires less space, while producing the same amount of lift as the other two configurations. This results in a smaller and more maneuverable helicopter for the same payload limitations. In comparison to the Sikorsky configuration the coaxial configuration has better maneuverability but is more expensive and has more frequent maintenance. For the climbing-flying robot the coaxial configuration is therefore the best choice.

6.3 Drive mechanism

The drive mechanism would consist of the front and the rear drive mechanism. Each of the two drive mechanisms would consist of two wheels (Fig. 5). The upper wheel would be the drive wheel while the lower wheel would provide stability for the robot on the power line. The drive wheel would be connected to an electrical motor with a drive chain, whereas the lower support wheel would run freely. The wheels would be made of aluminum and the conductor contact surfaces of the wheels would be covered with conductive rubber to increase traction, damp vibrations and to keep the robot on the same electric potential as the conductor.

Grasping the conductor would be done with the support wheel. At landing the robot would sit down onto the drive wheels with the help of special rails (Fig. 4(b) and Fig. 5). After the robot would be positioned on the drive wheels the support wheels would be moved into position with servomotors. The contact force with the conductor would be applied with springs. Before takeoff, the support wheels would be retracted and the robot would be free to lift off the conductor.

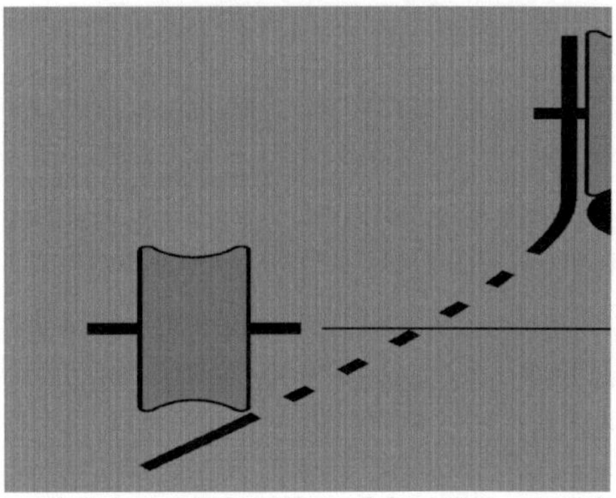

Fig. 5. A part of the proposed drive mechanism. After the robot lands on the conductor the lower wheels grasp the conductor from the sides.

6.4 Visual inspection system

Visual inspection of the conductor would consist of two systems. Visual inspection system at the front of the robot would perform visual inspection of the power line and obstacle detection with a wide angle camera. Visual inspection of the power line would consist of detection of conductor, insulator, supporting tower and other equipment defects. As visual detection of defects on all these different systems would not be very reliable, the conductor would be more accurately inspected with the second visual inspection system, while defects on other equipment would be detected with infrared and ultraviolet cameras, also a part of the front visual inspection system. Infrared cameras would be used to easily detect overheating of any part of the power line equipment. Ultraviolet cameras, on the other hand, would make detection of corona, which is usually a sign of a defect, fairly straightforward.
The second visual inspection system inside the robot would perform a more accurate visual inspection of the conductor. This visual inspection system would consist of three line scan cameras placed around the conductor 120 degrees apart (Fig. 6(b)). The conductor would be illuminated with two LED based lights for each camera (Fig. 6(a)). The lights would be placed on both sides of the cameras. This lighting configuration would provide diffuse illumination of the conductor, which would enable efficient visual defect detection. For triggering the line scan cameras an incremental encoder on one of the wheels of the drive mechanism would be used.

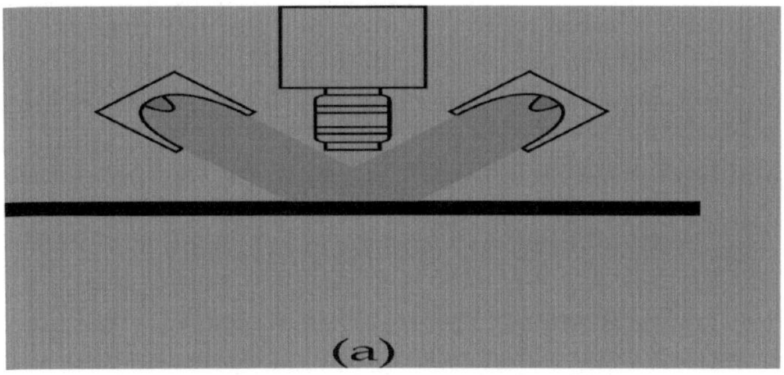

Fig. 6. Conductor visual inspection system. (a) Illumination, camera and conductor configuration for one camera. (b) Camera configuration around the conductor.

6.5 Power pick-up device

The power pick-up device consists of two parts, the toroidal core and the clasping mechanism. The toroidal core (Fig. 7) is made from a ferromagnetic iron core and is split into two halves. On each half is a winding that transforms the energy of the magnetic field in the iron core to electrical energy, which is then further treated with a special converter circuit to obtain a useable voltage to power the systems onboard the robot. The converter must be capable of handling a large range of input voltages as the voltage in the winding changes linearly with the power line current. The clasping mechanism takes care of the closing and opening of the toroidal core after landing and before takeoff (Fig. 8). It is extremely important that the clasping mechanism closes the two halves of the toroidal core as closely together as possible, as even a small slit between the two halves significantly affects the efficiency of the power pick-up device. Its precision is therefore of great importance.

A very important parameter of the power pick-up device is its power to weight ratio. It is crucial that a power pick-up device is as light as possible, as weight is limited on the robot. The power produced by the power pick-up device depends on the power line current and also on the geometry of the toroidal core. An analysis of the power to weight ratio in dependence of the geometry of the power pick-up device has to be performed to determine the feasibility of this device. A preliminary analysis was done in (Katrašnik, 2007). The analysis showed that such a power pick-up device is feasible, as the power to weight ratio of more than 250 W/kg can be achieved for a relatively small 400 A power line current.

Fig. 7. The power pick-up device.

6.6 Communication

In order to guide the robot over the obstacles effectively the operator needs real-time visual feedback about the robot's surroundings, while the guiding data have to be sent with minimal latency. For this reasons, a reliable high bandwidth wireless data link with very low latency is required. Another requirement, which somehow conflicts with the high bandwidth requirement, is long communication distance that should reach at least 5 kilometers for an efficient operation.

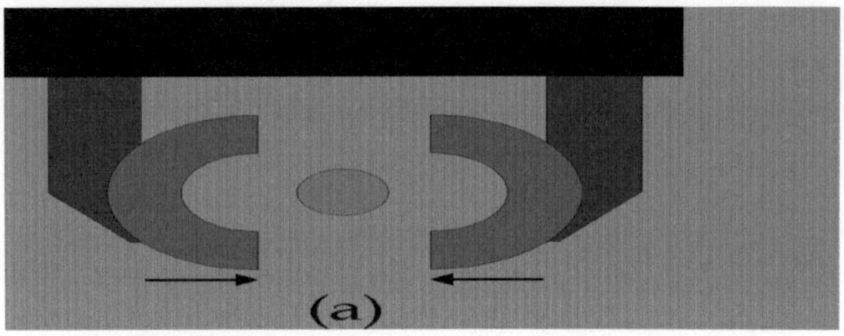

Fig. 8. The power pick-up device with the clasping mechanism. (a) Opened power pick-up device. (b) Closed power pick-up device.

7. Comparison of robot types

Because it is not practically feasible to objectively assess all three robot types, we decided to conduct a subjective scoring of the three inspection concepts according to some important characteristics. Specifically, we selected and weighted four evaluation categories (table I): design and construction requirements (weight 4), inspection quality (weight 3), autonomy at inspection, obstacle avoidance and energy requirements (weight 2), and universality or generality of the inspection principle (weight 1). The three inspection concepts were then ranked in each of the categories, as explained in the following paragraphs, and accordingly assigned the weighted scores (rank × weight).

The climbing robot is definitely the most difficult to design and construct, while the flying robot is the less so. Namely, there are a large number of commercial UAVs available on the market that can well serve the purpose if equipped by the appropriate sensory and computing equipment. The most challenging task for the flying robot is software development. Similar observations can be made for the climbing-flying robot but due to additional components and shielding requirements, this robot would be more difficult to design and construct. The climbing robot ranks last in this the most important category due to a number of reasons. First, complex grippers, drive mechanism and controller are required for effectively crawling on the line. Another problem is the expert system for obstacle detection, recognition and overcoming. Next, the whole body of the robot requires appropriate shielding from the powerful electromagnetic fields.

In terms of inspection quality, the flying robot is the most problematic due lower image quality (vibrations), lower resolution (grater inspection distance) and limited field of view, especially when inspecting the conductors. The latter can be much better inspected by the climbing and flying-climbing robots if using line scan cameras for high-resolution inspection from all sides. Other power line equipment can be efficiently inspected from the conductor by both robots just before crossing the obstacles. However, climbing-flying robot can further inspect the equipment from additional angles when flying over the obstacles and this is why it ranks best in the inspection category.

Autonomy at inspection, obstacle crossing and the terms of energy independence fit into another important category. Developing a flying robot for autonomous inspection, flying and avoiding obstacles is certainly more difficult than making the climbing robot autonomous at inspection and when climbing over the obstacles. In this respect, the climbing-flying robot ranks in between. In terms of energy independence, the climbing and climbing-flying robots can use induction system for power pick-up from the conductor, while the pantograph mechanism proposed for the flying robot is more dangerous, less reliable and more complex.

The last evaluation category deals with the universality of the inspection concept or its flexibility when inspecting different power lines systems. It is very difficult if not impossible to design a robot that would work on all power lines without any modifications. The flying robot is certainly the most flexible in this respect, followed by the climbing-flying robot, which would need adaptations for traveling along different conductors. The less general is certainly the climbing robot as major modifications would be required for adaptation to different conductors and especially to other types of obstacles.

	w	**Climbing**	**Climbing-flying**	**Flying**
Design and construction	4	1 | 4	2 | 8	**3** | 12
Inspection quality	3	2 | 6	**3** | 9	1 | 3
Autonomy	2	**3** | 6	2 | 4	1 | 2
Universality	1	1 | 1	2 | 2	**3** | 3
Total score		17	**23**	20

w = weight, rank | weighted score

Table 1. Robot type comparison

8. Conclusion

The most important decisive factor for choosing the robot type is certainly performance to price ratio. A big portion of the final price is usually the cost of development. The robot performance is a combination of inspection quality, autonomy and universality. We can conclude that the flying robot would likely have the smallest performance to price ratio (table I). Namely, this robot, although universal and easy to design, would offer the lowest inspection quality and low inspection autonomy. The development costs would certainly be very high for the climbing robot but the robot would be much more autonomous and could offer better inspection quality. The latter also holds true for the climbing-flying robot, which would probably not be as autonomous as the climbing robot when crossing obstacles but the cost of development should be much lower.

In conclusion, the power pick-up system and automatic visual inspection are problems that still need to be solved efficiently for the flying robot, while automatic power line tracking and obstacle avoidance need further improvements to become practically feasible. The climbing robot also needs further developments of the power pick-up system, automatic visual inspection, electromagnetic shielding, robot mechanism and the control system. The proposed climbing-flying concept for power line inspection has not been researched yet, offering a number of specific challenges, such as high communication link bandwidth and reliable system for landing on the conductor. Nevertheless, the proposed concept seems feasible from the practical point of view and because good performance to price ration could be obtained.

9. References

Aggarwal, R. K., Johns, A. T., Jayasinghe, J. A. S. B. & Su, W. (2000) An overview of the condition monitoring of overhead lines. *Electric Power Systems Research,* 53, 15.

Campoy, P., Garcia, P. J., Barrientos, A., Cerro, J. D., Aguirre, I., Roa, A., Garcia, R. & Muñoz, J. M. (2001) An Stereoscopic Vision System Guiding an Autonomous Helicopter for Overhead Power Cable Inspection. *Proc., Robot Vision: International Workshop RobVis 2001, Auckland, New Zealand.*

Cooper, J., Venkatesh, S. & Kitchen, L. (1993) Early jump-out corner detectors. *IEEE Transactions on Pattern Analysis and Machine Intelligence,* 15, 823.

Fu, S., Zhang, Y., Zhao, X., Liang, Z., Hou, Z., Zou, A., Tan, M., Ye, W. & Bo, L. (2006) Motion Deblurring for a Power Transmission Line Inspection Robot. *Computational Intelligence.* Berlin, Springer Berlin / Heidelberg.

Golightly, I. & Jones, D. (2003) Corner detection and matching for visual tracking during power line inspection. *Image and Vision Computing,* 21, 827.

Golightly, I. & Jones, D. (2005) Visual control of an unmanned aerial vehicle for power line inspection. *Proc. 12th International Conference on Advanced Robotics (ICAR '05).*

Jones, D. (2005) Power line inspection - a UAV concept. *The IEE Forum on Autonomous Systems 2005 (Ref. No. 2005/11271).*

Jones, D., Golightly, I., Roberts, J. & Usher, K. (2006) Modeling and Control of a Robotic Power Line Inspection Vehicle. *Proc. of IEEE International Conference on Control Applications (CCA '06).*

Jones, D. I. (2007) An experimental power pick-up mechanism for an electrically driven UAV. *Proc. of IEEE International Symposium on Industrial Electronics - ISIE 2007.*

Jones, D. I. & Earp, G. K. (1996) Requirements for aerial inspection of overhead electrical power lines. *Proc. 12th International Conference on Remotely Piloted Vehicles.* Bristol.

Jones, D. I. & Earp, G. K. (2001) Camera sightline pointing requirements for aerial inspection of overhead power lines. *Electric Power Systems Research,* 57, 73.

Katrašnik, J. (2007) Robotic Power Line Inspection. *Faculty of Electrical Engineering.* Ljubljana, University of Ljubljana.

Ludan, W., Hongguang, W., Lijin, F. & Mingyang, Z. (2006) Research on Obstacle-navigation Control of a Mobile Robot for Inspection of the Power Transmission Lines Based on Expert System. *Proceedings of the 8th International Conference on Climbing and Walking Robots and the Support Technologies for Mobile Machines (CLAWAR 2005)* Springer Berlin Heidelberg.

Peungsungwal, S., Pungsiri, B., Chamnongthai, K. & Okuda, M. (2001) Autonomous robot for a power transmission line inspection. *Proc. of the 2001 IEEE International Symposium on Circuits and Systems (ISCAS 2001).*

Sawada, J., Kusumoto, K., Maikawa, Y., Munakata, T. & Ishikawa, Y. (1991) A mobile robot for inspection of power transmission lines. *IEEE Transactions on Power Delivery,* 309.

Tang, L., Wang, H. & Fang, L. (2004) Development of an inspection robot control system for 500KV extra-high voltage power transmission lines. *Proc. of SICE 2004 Annual Conference.*

Whitworth, C. C., Duller, A. W. G., Jones, D. I. & Earp, G. K. (2001) Aerial video inspection of overhead power lines. *Power Engineering Journal [see also Power Engineer],* 15, 25.

Williams, M., Jones, D. I. & Earp, G. K. (2001) Obstacle avoidance during aerial inspection of power lines. *Aircraft Engineering and Aerospace Technology,* 73, 472-479.

Xinglong, Z., Hongguang, W., Lijin, F., Mingyang, Z. & Jiping, Z. (2006) A Novel Running and Gripping Mechanism Design Based on Centroid Adjustment. *Proc. of the 2006 IEEE International Conference on Mechatronics and Automation.*

Zhang, Y., Fu, S., Zhao, X., Liang, Z., Tan, M. & Zhang, Y. (2006) Visual Navigation for a Power Transmission Line Inspection Robot. *Computational Intelligence.* Berlin, Springer Berlin / Heidelberg.

A Fuzzy Control Based Stair-Climbing Service Robot

Ming-Shyan Wang
Department of Electrical Engineering, Southern Taiwan University
Taiwan, R.O.C.

1. Introduction

Recently, Taiwan and many developed countries have been experienced an emergence of growing aging population and decreasing working population and birth rate. The number of home-caring workers in need has surpassed that of available working force. Furthermore, the expense of home-caring elders will substantially increase. The employment of foreign workers is just a temporary solution. It is therefore necessary to develop a future substitute for the human work force. Home-caring robot is an excellent candidate capable of supporting such an aging society. Especially, the elders can control the robots directly to move up-and-down stairs for service.

It is well-known that the most effective style of movement of a robot on a plane field is the wheel type. However, as obstacles and stairs exist, crawler-type and leg-type robots become better candidates for application. The robots of the stick type and the biped type are generally designed to carry elders up and down stairs. The former uses tires, rubber belts, and the handrail to assist the elders while walking and moving up-and-down stairs, and the latter uses two legs and the handrail to assist walking and moving up-and-down stairs (Takahashi et al., 1998). A control bar is attached to the robot waist to assist the aged person by stepping onto the feet of the robot. The robot "Zero Walker-1" (Konuma and Hirose, 2001) focused on ascending and descending stairs. An electric wheelchair named "iBOT" (Independence Technology, 2009), capable of ascending and descending stairs and slopes was released by a Japanese company. In (Sugahara, 2005), a bipedal robot was developed to provide only lower-limbs and a waist, named "WL-16RII", which can walk independently and allow users to build the upper body based on their requirements. This biped locomotor would be applicable as a walking wheelchair or as a walking support machine that is able to walk up and down stairs carrying or assisting an elder. The developed biped locomotor with Stewart Platform Legs successfully achieved walking up and down on stairs for 250 mm continuously and carrying one 60-kg man on it. Additionally, the authors (Nishiwaki et al., 2002) successfully controlled their robot "H6" to walk up stairs for 250 mm by utilizing toe joints; and the robot "HRP-2" from Harada successfully climbed up 280 mm stairs by grasping the stair rail (Harada et al., 2002). A self-standing type eight-wheeled robot in (Takita et al., 2004) is able to climb up and down stairs and is supported by a mechanism with a planetary gear without an inner gear to eliminate the disadvantages of a wheeled system. However, the height of climbing an obstacle generally is the same as the diameter of the robot's wheel. The aforementioned robots generally need tremendous effect on expense and time. Furthermore, it is very difficult to lift an aged person by human force and not very easy to have a large and heavy-weight lift machine in a normal house.

The main difference between a fuzzy logic control (FLC) and the conventional control is that the former is not based on a properly defined model of the system but instead implement the same control "rules" that a skilled expert would operate. FLC has been applied to robot applications, such as mobile robots (Malima et al., 2006; Chen et al., 2004; Song & Wu, 1999), humanoid robots (Wang et al., 2004), and soccer robots (Wang & Tu, 2008). In the chapter, the FLC will steer the robot based on the outputs of DC bus current sensor and an inclinometer. A control board including a digital signal processor (DSP) TMS320F28335 realizes the fuzzy rules.

The chapter is organized in 6 sections including introduction as follows for further discussion. Section 2 describes the robot mechanism design and each component, the ways of climbing up and going down stairs, and the friction during motion. The image processing for the CMOS camera and FPGA and the tracking, capturing, and putting back the target object by the arm based on results of image processing are introduced in section 3. Section 4 states the fundamental theory of fuzzy logic control. Section 5 presents the experimental results of two kinds of stair-motion and the image processing. Finally, section 6 claims the conclusion and future work.

2. Stair-Climbing Robot

The designed stair-climbing robot consists of a main body, roller chains, a front arm, and a rear arm. The lateral-view and vertical-view sketches of the robot by AutoCAD are shown in Figs. 1 and 2. The main body is equipped with two brushless DC motors (BLDCMs) and their drives for locomotion, worm gears for torque magnification, two DC motors to control two arms, and DSP-based board as control center. The chassis size of the main body is 58.5cm ×53cm and each arm is 48cm × 40cm, such that the maximum and minimum lengths of the moving robot will be 154.5cm and 58.5cm, respectively. There are 3 pairs of roller chains in the main body and two arms, respectively. Some polyurethane rubber blocks, each with size of 3cm×2cm×1cm, attached to the roller chains are applied for generating friction with ground and stairs for moving. There are 40 blocks for each arm and 56 for the main body. The distance between any two plastic blocks is properly arranged to fix the stair brink. One DC bus current sensor and one inclinometer provide the information for the robot to steer two motors.

Fig. 3 shows the way of climbing up stairs based on the physical constraint. The front arm will be pushed down to flat top so that the main body is lifted and will be pulled up for the next stair-climbing. The rear arm keeps flat while the robot climbs up. Fig. 4 displays summary of climbing-up motion step by step. While climbing, two forces have to be overcome. One is the force along the inclined plane due to the robot system force of gravity, $mg \cdot \sin\theta$, and the other is the frictional force, $mg \cdot \cos\theta \cdot \mu$, where m is the total mass of the robot system, g is the gravity acceleration, θ is the inclination angle of the stair, and μ is the frictional coefficient. In order to reduce the electrical specifications and volume size of

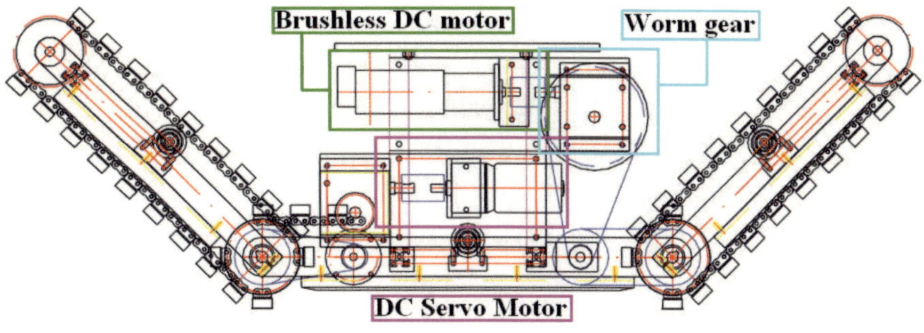

Fig. 1. A lateral view of the robot

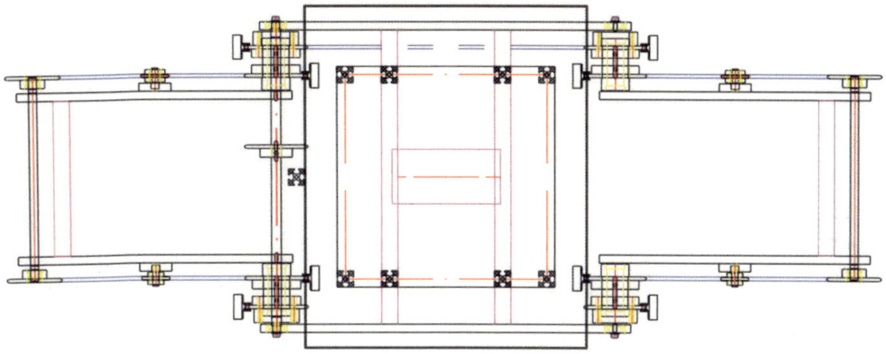
Fig. 2. A vertical view of the robot

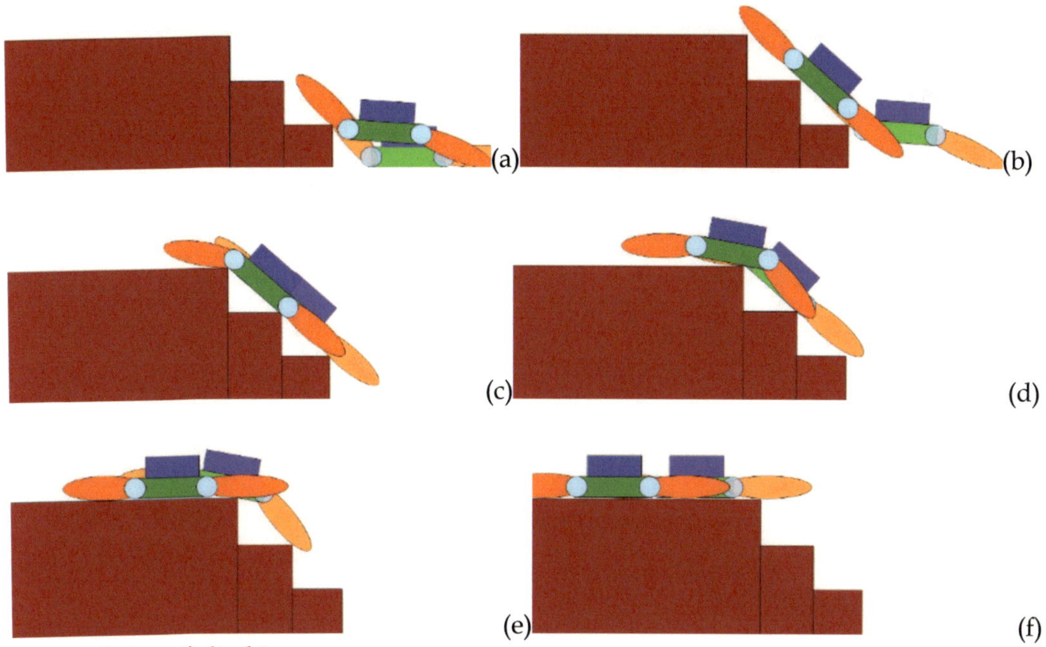

Fig. 3. Motion of climbing up

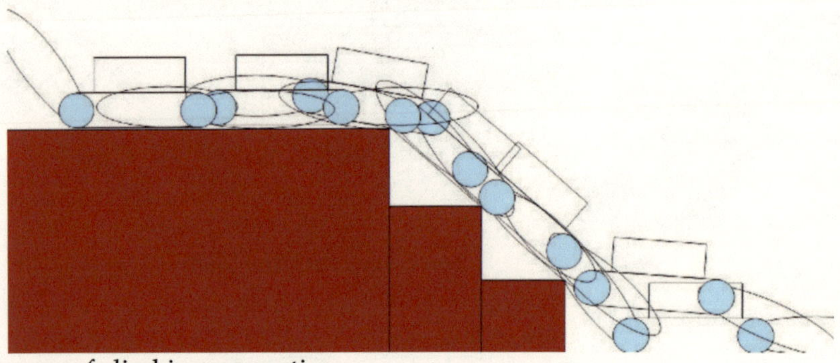

Fig. 4. Summary of climbing-up motion

the motor, gears are considered for torque magnification. The total output torque of the motor, T_e, has to satisfy the following inequality,

$$2T_e \times S_1 \times \eta_1 \times S_2 \times \eta_2 \geq (mg \cdot \sin\theta + mg \cdot \cos\theta \cdot \mu) \times l_m \quad (1)$$

where l_m is the operating radius, $S_1(S_2)$ and $\eta_1(\eta_2)$ are the gear ratio and the efficiency of the first (second) gear, respectively. Consequently, the motor types of low rated input voltage and high rated speed are primary selection.

Similarly, Fig. 5 shows the way of going down stairs and Fig. 6 displays summary of going-down motion step by step. During going down, the output torque from motors can be reduced since it is in the same direction of force of gravity of the robot system.

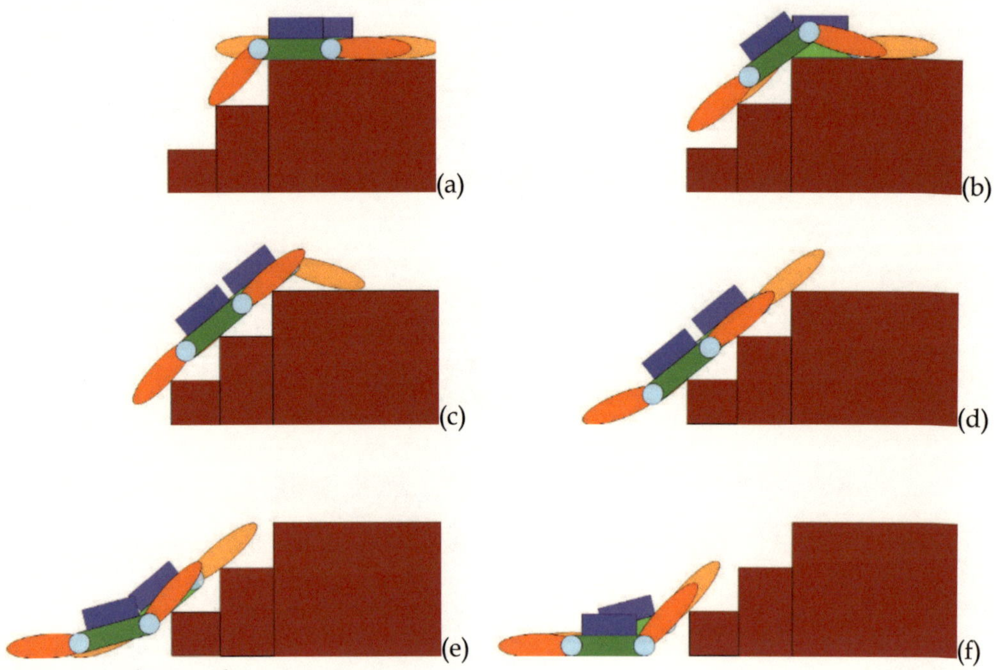

Fig. 5. Motion of going down

Fig. 7 displays the picture of the 45-Kg stair-climbing robot with one 5-kg arm for service. Its operating radius is $l_m = 0.25m$. Worm gears and the charger are then shown in Fig. 8. Since the batteries are prerequisite for the robot, an inbuilt charger is considered for convenience in charging.

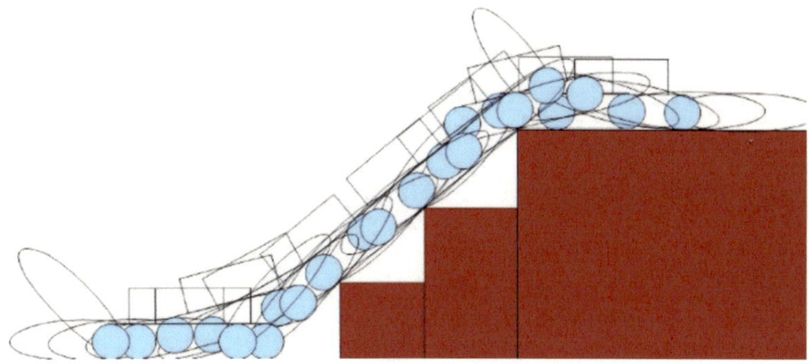

Fig. 6. Summary of going-down motion

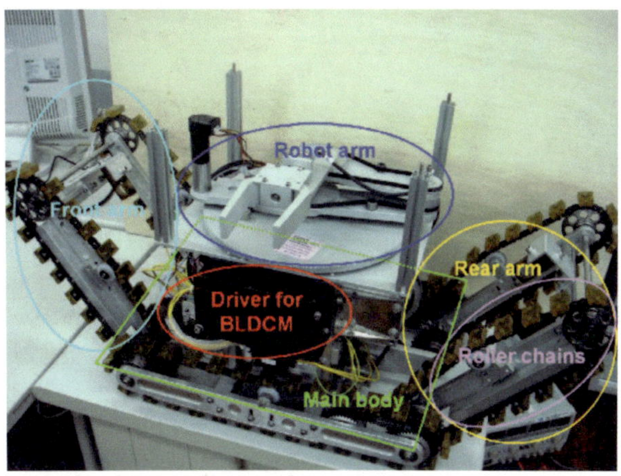

Fig. 7. Picture of the stair-climbing robot with its arm

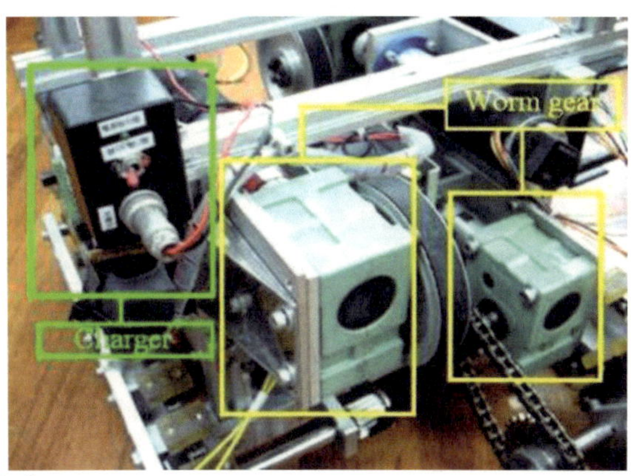

Fig. 8. Worm gear and charger

3. Robot Arm and Image Processing

3.1 Robot arm

The top view of the multi-link arm is shown in Fig. 9. It consists of three couples of gears, three DC motors, four links, and one clamper. Referring to Fig. 9, the first DC motor steers the diving gear S_3 and driven gear S_4 to determine the rotating angle. Gear S_3 links S_4 directly. Due to the limit of four stalls at corners, the span angle is within $(-30^{\circ},30^{\circ})$. The second motor controls the gear couples of $S_5 - S_{6d}$ and $S_{6u} - S_7$ together with belts to stretch the length of the arm. S_{6d} and S_{6u} are mounted in the same shaft and with same number of gears. The lengths of the four links are l_1, l_2, l_2 and l_3, respectively.

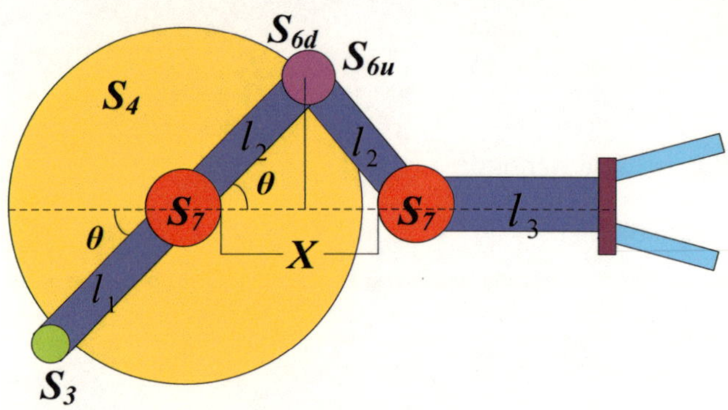

Fig. 9. The multi-link arm

The pixel array of CMOS camera THDB-D5M used in the robot consists of a matrix of 2752 x 2004 pixels addressed by column and row (Terasic, 2008). The address (column 0, row 0) represents the upper-right corner of the entire array. The 2592 x 1944 array in the centre called active region represents the default output image, surrounded by an active boundary region and a border of dark pixels, shown in Fig. 10. The boundary region can be used to avoid edge effects when achieving colour processing the result image of the active region, while the optically black columns and rows can be used to monitor the black level.

Pixels of active region are output in a Bayer pattern format consisting of four "colours", Green1, Green2, Red, and Blue (G1, G2, R, B) to represent three filter colours (Terasic, 2008). The first row output alternates between G1 and R pixels, and the second row output alternates between B and G2 pixels, shown in Fig. 11. The Green1 and Green2 pixels have the same colour filter, but they are treated as separate colours by the data path and analogue signal chain.

The image raw data is sent from D5M to DE2-70 board (Terasic, 2008) where the FPGA on DE2-70 board will handle image processing and convert the data to RGB format to display on the VGA display. As a result, we first capture the image of experiment background to find the ranges of colours of RGB, and then define their location regions for colour discrimination, shown in Fig. 12.

The target object in the experiment is a cola can with the weight of 330 g and red color surface. Referring to Fig. 12, the ranges of RGB intensities locate at (50, 70), (25, 30), and (23, 28), respectively. In order to reduce the effect of light variation, the image in RGB space will be converted into YC_bC_r space (Benkhalil et al., 1998; Hamamoto et al., 2002). In addition, the ranges of RGB from D5M are four times of the general image.

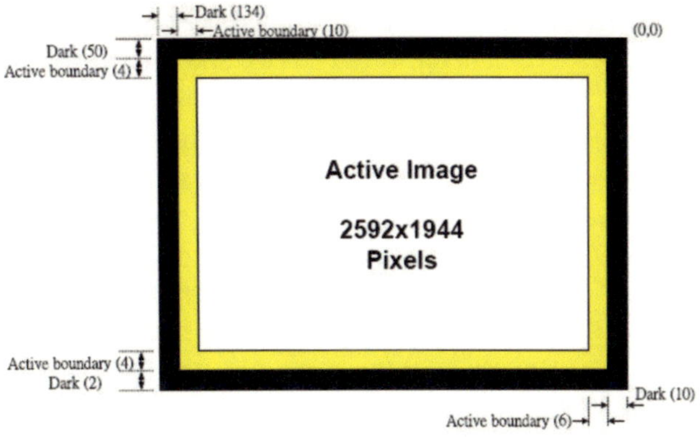

Fig. 10. Pixel array description (Terasic, 2008)

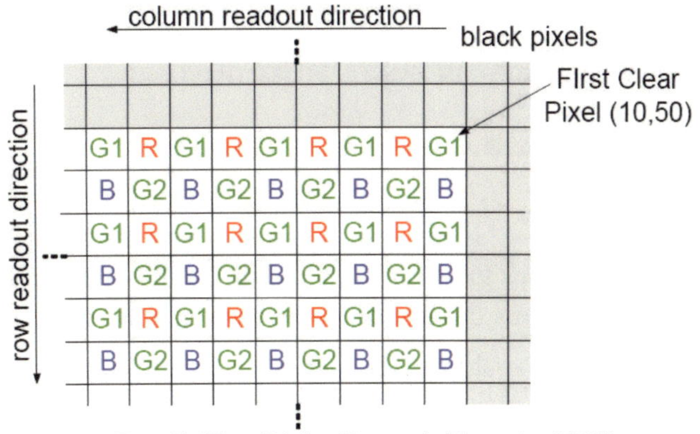

Fig. 11. Pixel Color Pattern Detail (Top Right Corner) (Terasic, 2008)

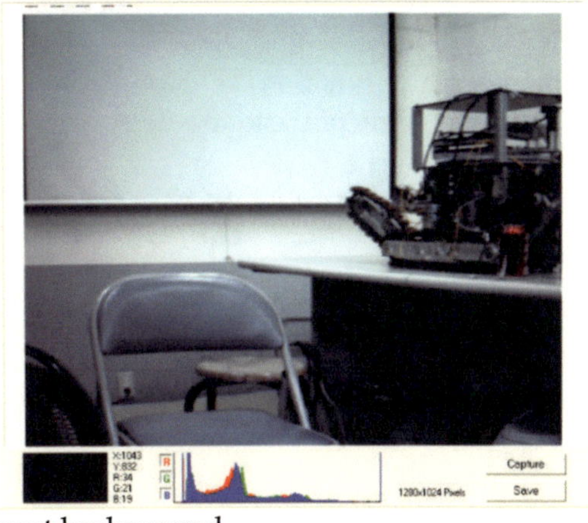

Fig. 12. Image of experiment background

4. Fuzzy Logic Control

A fuzzy logic controller may be viewed as a real-time expert system since it aims to incorporate expert human knowledge in the control algorithm. The fuzzy logic control (FLC) system consists of FI (fuzzification interface), DML (decision making logic), KLB (knowledge base), and DFI (defuzzification interface), shown in Fig. 13. The triangle-shape membership functions of DC bus current I, inclination angle θ_m, and fuzzy output y are shown in Fig. 14, where there are seven linguistic variables, PB (positive big), PM (positive medium), PS (positive small), ZO (zero), NS (negative small), NM (negative medium), and NB (negative big) used in the chapter.

Some of the most successful applications by fuzzy control have been highly related with conventional controllers, such as proportional-integral-derivative (PID) controller. Especially, the PD-like fuzzy control is widely adopted in many applications. In the system, variables of DC bus current and inclination angle are fed back to determine the control action. The inclination angle of the stairs is fixed so that there is little variation on it during motion. In addition, the motion speed of the robot is too slow to need predicting the change of the next states of the sensor signals. As a result, for easily programming, the simplest P control algorithm is employed to achieve the motion control while the robot climbs up and goes down stairs.

The ith fuzzy rule in the fuzzy rule-base system is described as

$$R_i : \text{If } x_1 \text{ is } A_{i1} \text{ and } x_2 \text{ is } A_{i2}, \text{then } y \text{ is } w_i \tag{2}$$

where w_i, x_j, A_{ij}, $j = 1,2$, $i = 1,2,\cdots,n$ are fuzzy output variables, input fuzzy variables and linguistic variables, respectively. Referring to Fig. 15 for ith membership function with isosceles triangle shape, b_{ij} means the length of the base, and a_{ij} stands for the abscissa of the centre of the base. The membership grade of input x_j is calculated by

$$A_{ij}(x_j) = 1 - \frac{2 \cdot |x_j - a_{ij}|}{b_{ij}}, \quad j = 1,2 \tag{3}$$

The bases of triangular membership function keep same for easily programming. By product operation, the membership grade of the antecedent proposition is calculated as

$$\mu_i = A_{i1}(x_1) \cdot A_{i2}(x_2) \tag{4}$$

Then the output will be

$$y = \left(\sum_{i=1}^{n} \mu_i w_i \right) \Big/ \left(\sum_{i=1}^{n} \mu_i \right) \tag{5}$$

Summarily, Table 1 lists the linguistic control rules.

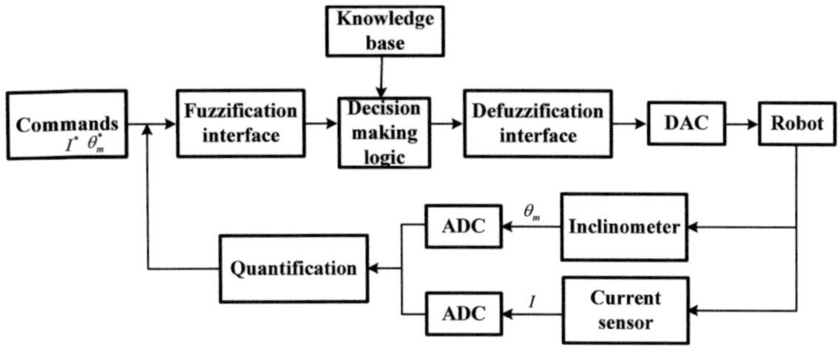

Fig. 13. Block diagram of fuzzy logic control system

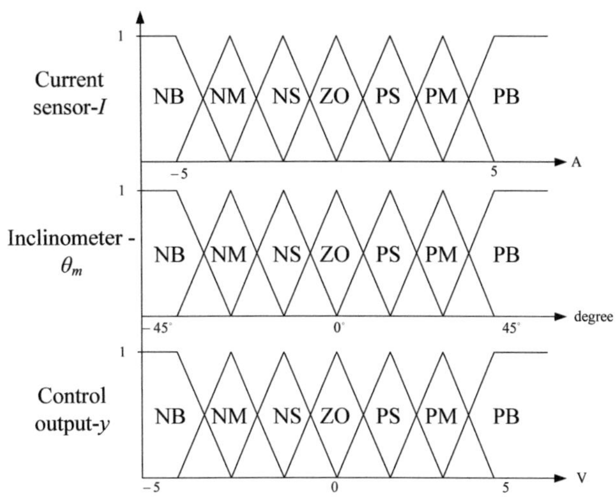

Fig. 14. Membership functions of current sensor, inclinometer, and fuzzy control output

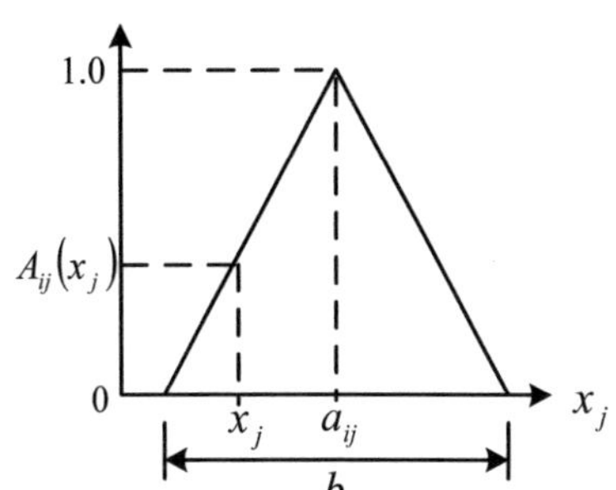

Fig. 15. ith membership function with isosceles triangle shape

5. Experimental Results

The specifications of the stair-climbing robot are as following. The gear ratio and the efficiency of the first (second) gear are $S_1 = 66(S_2 = 20)$ and $\eta_1 = 0.7(\eta_2 = 0.55)$, respectively. The static frictional coefficient of polyurethane rubber blocks is about 0.6. Fig. 16 presents the characteristic curve of an inclinometer in the system. The output voltage depending on the voltage source is almost linear with the inclination angle. The rated specifications of BLDCM are: 200 W, 24 V, 9600 rpm, and $T_e = 0.1336\,Nm$. Since the waveforms of back electromagnetic forces (EMFs) and the armature currents of a BLDCM are trapezoidal alike, not perfectly sinusoidal, the six-step driving algorithm rather than the vector control is adopted on speed control. The popular PI control is adopted for speed regulation.

θ_m \ I	ZO	PS	PM	PB
NB	ZO	PS	PM	PB
NM	ZO	PM	PM	PM
NS	ZO	PB	PS	ZO
ZO	ZO	PB	PS	ZO
PS	ZO	PB	PS	ZO
PM	ZO	PS	PM	PM
PB	ZO	PS	PM	PB

Table 1. Linguistic control rule table

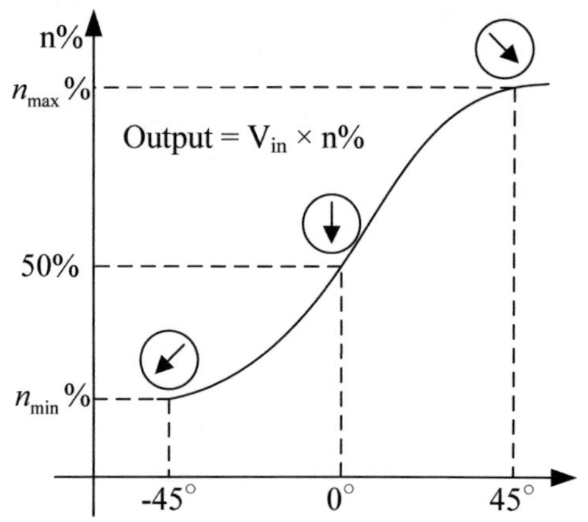

Fig. 16. Characteristic curve of an inclinometer

A preliminary experiment that the unloaded robot climbs up and goes down a gradual stair with the rise of 120 mm and depth of 400 mm ($\theta = 16.7^{\circ}$) by wired control is proceeded. It is firstly easy to check the validness of (1). The results of every motion in Figs. 3 and 5 are shown in Figs. 17 and 18, respectively (Wang & Tu, 2008). It qualifies the designed robot. Then we conduct the second experiment that the robot with loading of one arm moves up and down a steeper stair with the rise of 175 mm and depth of 280 mm ($\theta = 32^{\circ}$) by FLC and (1) still holds. The taped pictures of the experiment and every motion in Figs. 3 and 5 are shown in Figs.19 and 20, respectively. Even lack of any gyroscope, there is little variation in proceeding direction happened during motion due to high friction force between rubber blocks and stairs. The designed robot performs very well during the trip even there are damaged parts on the up-and-down way. The FLC prevents the robot abruptly going down to the ground and damaging itself.

The third experiment contains image processing and arm motion. In order to prevent target damage while clamping, one pressure sensor is installed inside the clamper. The pressure output after calibrating is sent to DSP for reference. Fig. 21 displays the sequentially taped pictures from videos of capturing the cola can and putting it back by the robot arm (Tu, 2009). As the can shifting left or right, the arm can correctly track to the corresponding direction.

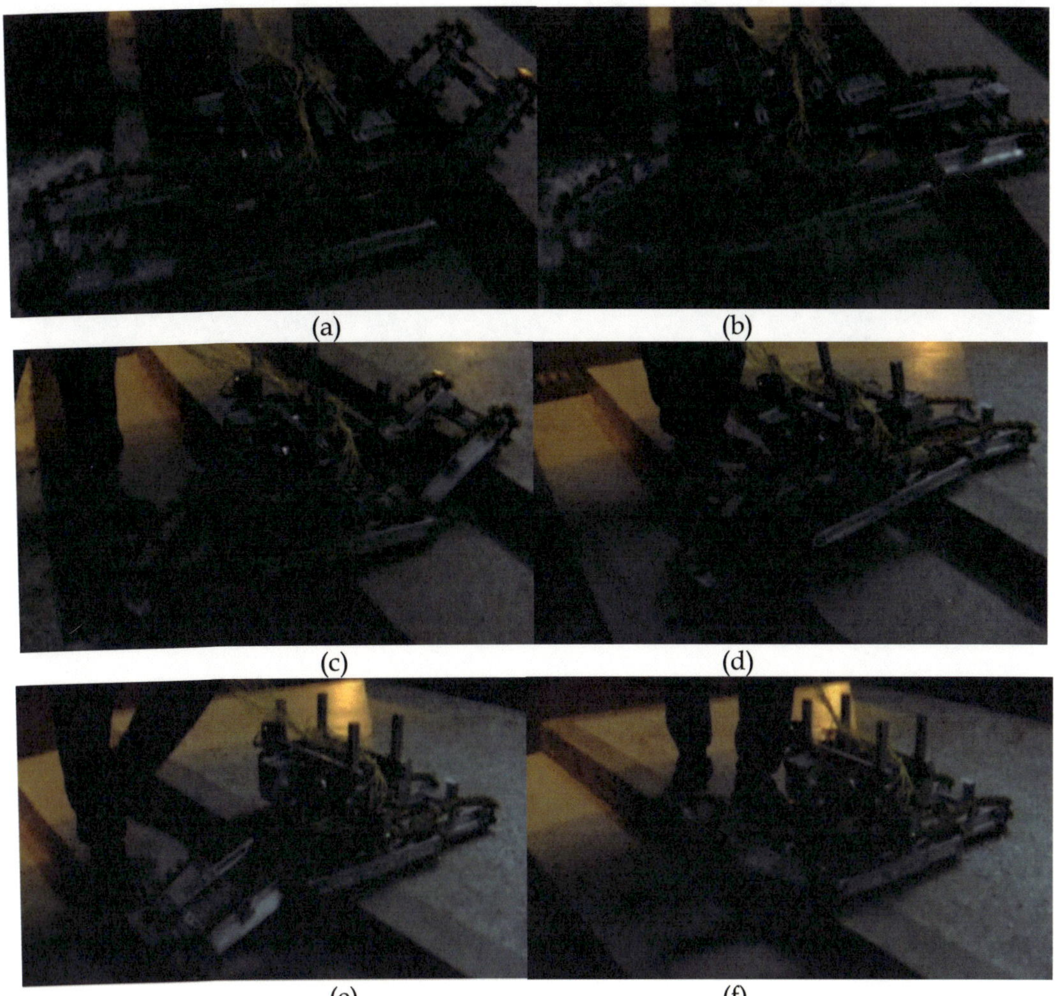

(a) (b) (c) (d) (e) (f)

Fig. 17. Realized motion of climbing up by wired control

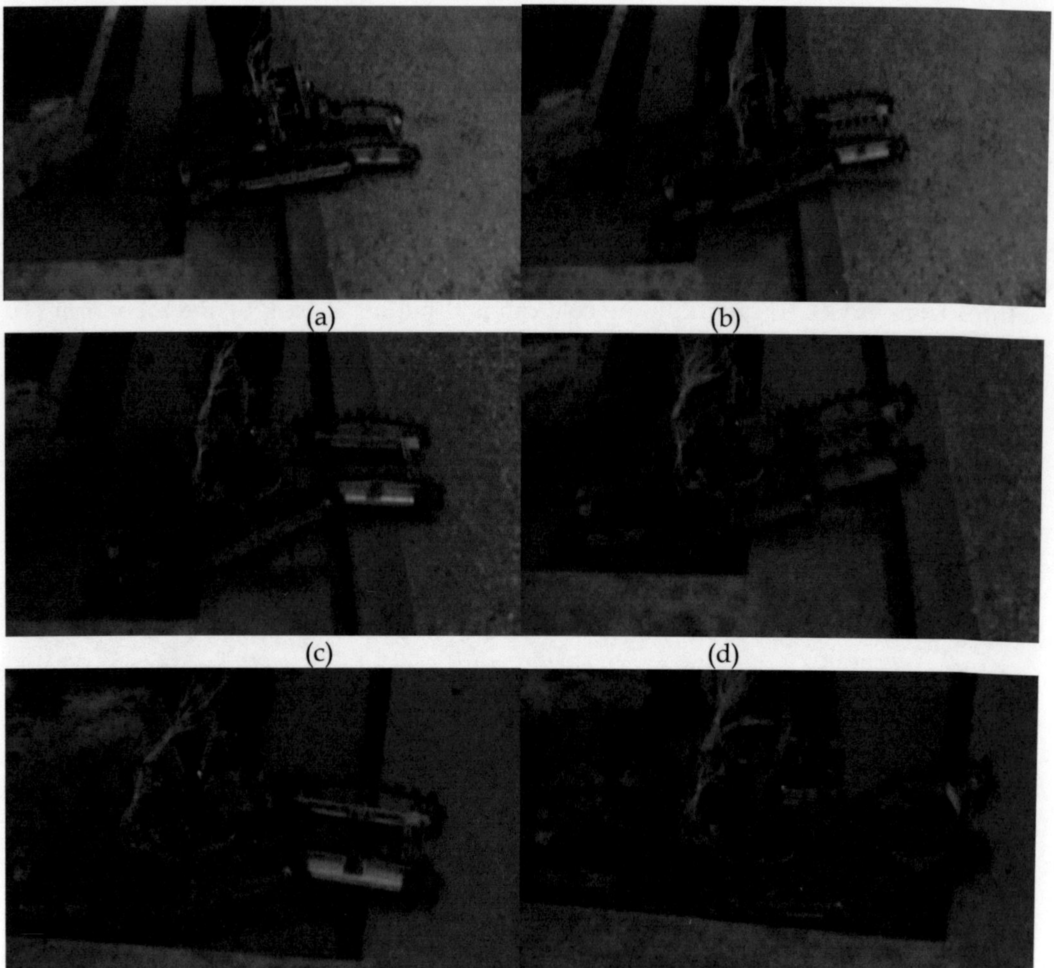

(a) (b) (c) (d) (e) (f)

Fig. 18. Realized motion of going down by wired control

6. Conclusion and Future Work

In the chapter, we have developed a stair-climbing robot to provide service for the elders and completed two walking experiments of moving up and down stairs with the rise/depth of 120/400 mm and 175/280 mm. The third experiment of object tracking, capturing, and loading by the arm have been shown in the taped pictures from videos to verify the proposed design. In fact, we will show the arm may capture the specific object during climbing up and down in the future. In addition, the robot will patrol for security by the CCD camera around the house while more image processing functions are provided.

Fig. 19. Realized motion of climbing up

Fig. 20. Realized motion of going down

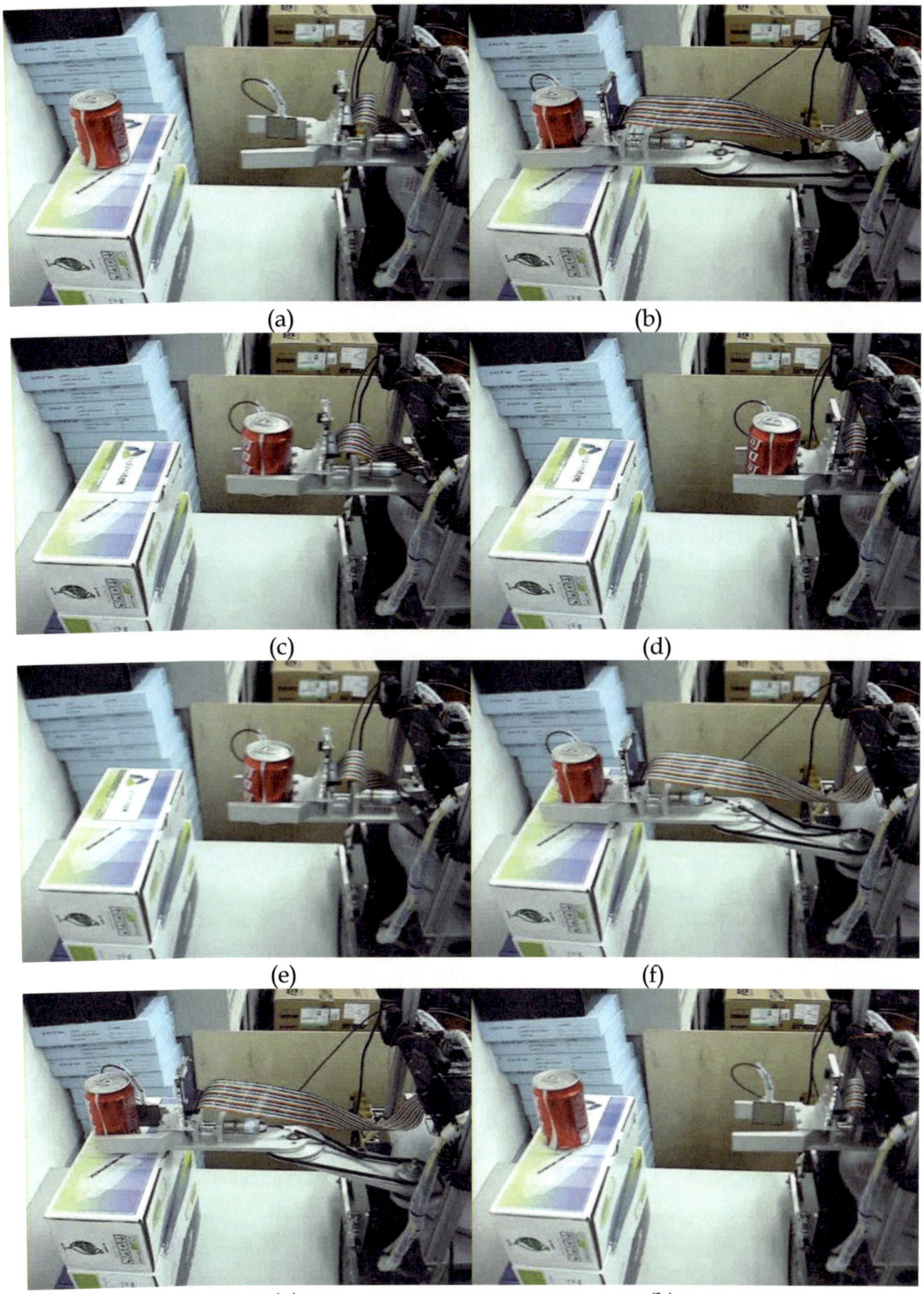

(a) (b) (c) (d) (e) (f) (g) (h)

Fig. 21. The taped pictures of the experimental results of image processing and arm motion

7. References

Benkhalil, A. K.; Sipson, S. S. & Booth, W. (1998). Real-time detection and tracking of a moving object using a complex programmable logic device, *Proc. of IEE Colloquium on Target Tracking and Data Fusion*, pp. 10/1-10/7, Birmingham, UK, 9 Jun.

Chen, P.-B.; Huang, C.-M. & Fu, L.-C. (2004). A robust visual servo system for tracking an arbitrary-shaped object by a new active contour method, *Proceedings of the 2004 American Control Conference*, Vol. 2, pp. 1516-1521, Taipei, Sep. 2-4.

Hamamoto, T.; Nagao, S. & Aizawa, K. (2002). Real-time objects tracking by using-smart image sensor and FPGA, *Proc. of 2002 International Conference on Image Processing*, Rochester, Vol. 3, pp.441-444, New York, USA, 22-25 Sept.

Harada, K.; et al. (2004). Dynamical balance of a humanoid robot grasping an environment, *Proc. of 2004 IEEE/RSJ Int. Conf. on Intelligent Robots and Systems*, Vo. 2, pp. 1167-1173, 28 Sept.-2 Oct.

Independence Technology website, http://www.ibotnow.com/ibot/index.html.

Konuma, Y. & Hirose, S. (2001). Development of the stair-climbing biped robot 'Zero Walker-1', *Proc. of the 19th Annual Conf. of the RSJ*, pp. 851-852

Malima, A.; Ozgur, E. & Cetin, M. (2006). A fast algorithm for vision-based hand gesture recognition for robot control, *Proc. of IEEE 14th Signal Processing and Communications Applications*, pp. 1-4, Antalya, 17-19 April.

Nishiwaki, K.; et al. (2002). Toe joints that enhance bipedal and fullbody motion of humanoid robots, *Proc. of the IEEE ICRA 2002*, Vol. 3, pp. 3105-3110, Washington, DC, USA, 11-15 May.

Park, S. B. (1998). A Motion Detection System Based On A CMOS Photo Sensor Array," *Proc. of IEEE Conference on Image Processing*, Vol. 3, pp. 967 -971, Chicago, USA, 4-7 Oct.

Song, K. T., & Wu, T. Z. (1999). Visual Servo Control of a Mobile Manipulator Using One-dimensional Windows," *Proceedings of IEEE IECON'99*, pp. 547-552.

Sugahara, Y.; Ohta, A.; Hashimoto, K.; Sunazuka, H.; Kawase, M.; Tanaka, C.; Lim, H.-O. & Takanishi, A. (2005). Walking up and down stairs carrying a human by a biped locomotor with parallel mechanism, *Proc. of 2005 IEEE/RSJ International Conf. on Intelligent Robots and Systems*, pp.1489-1494, 2-6 Aug.

Takahashi, Y.; Nagasawa, T.; Nakayama, H.; Hanzawa, T.; Arai, Y.; Nagashima, T.; Hirata, E.; Nakamura, M.; Iizuka, T. & Ninomiya, H. (1998). Robotic assistance for aged people, *Proc. of the 37th SICE Annual Conf.*, pp. 853-858, Japan, 29-31 July

Takita, Y.; Shimoi, N. & Date, H. (2004). Development of a wheeled mobile robot "octal wheel" realized climbing up and down stairs, *Proc. of 2004 IEEE/RSJ International Conf. on Intelligent Robots and Systems*, Vol. 3, pp. 2440-2445, 28 Sept.-2 Oct.

Terasic company, (2008). *THDB-D5M Hardware Specification* and *User Guide.*

Tu, Y.-M. (2009). Design and Implementation of a Stair-Climbing Robot, *Master Thesis*, Department of Electrical Engineering, Southern Taiwan University, Taiwan.

Wang, C.-K.; Cheng, M.-Y.; Liao, C.-H.; Li, C.-C.; Sun, C.-Y. & Tsai, M.-C. (2004). Design and implementation of a multi-purpose real-time pan-tilt visual tracking system, *Proceedings of the 2004 IEEE International Conference on Control Applications*, pp. 1079-1084, Taipei, Sep. 2-4.

Wang M.-S. & Tu, Y.-M. (2008). Design and implementation of a stair-climbing robot," *Proc. of 2008 IEEE International Conference on Advanced Robotics and Its Social Impacts* (ARSO 2008), Taipei, Taiwan.

Evolutionary Multi-Objective Optimization for Biped Walking of Humanoid Robot

Toshihiko Yanase and Hitoshi Iba
The University of Tokyo
Japan

1. Introduction

The recent remarkable progress of robotics research makes advanced skills for robots to solve complex tasks. The divide-and-conquer approach is an intuitive and efficient method when we encounter complex problems. Being a divide-and-conquer approach, the multi-layered system decomposes the problem into a set of levels and each level implements a single task-achieving behaviour. Many researchers employ this approach for robot control system, dividing a complex behaviour into several simple behaviours. For example, Lie et al. developed the Evolutionary Subsumption Architecture, which enables for heterogeneous robots to acquire the cooperative object transferring task (Liu & Iba, 2003).

The autonomous locomotion of humanoid robots consists of following modules: global path planning using given geometrical information, local path planning based on the observation of environment, footstep planning, and whole-body motion generator. Since these modules mainly exchange the information with their neighbours, we can observe that they are hierarchically arranged from the aspect of communication. The parameter settings among these modules are necessary to adapt the system to the targeting environment. The problem involves a number of conflicting objectives such as stability of the robot motion and speed of locomotion.

In this paper, we present a parameter tuning method for multi-layered robot control system by means of Evolutionary Multi-Objective Optimization (EMO). We explore the set of parameters for modules to adapt various kinds of environments. Switching these parameter sets enables us to operate the robots effectively. We developed three modules as the experiment environment: walking pattern generator, footstep planner and path planner. In the experiment, we focused on the footstep planner shown in Fig. 1, which realizes collision-free walking. The parameter setting was manually adjusted in previous researches (Kuffner et al., 2003; Chestnutt et al., 2005). We discuss the conflicting objectives for the optimization footstep planner, and introduce a parameter setting method using EMO. Then we propose a simple rule to use parameter sets obtained by EMO to adapt the footstep planner to both crowded and sparse fields.

The rest of the paper is organized as follows: Section 2 describes our robot control system, Section 3 shows an experiment of the parameter setting of the footstep planner, and Section 4 shows an application using the parameter setting obtained from Section 3 and a comparison with a conventional approach. Section 5 provides the discussion of these experiments, and finally Section 6 concludes with summary and future works.

Fig. 1. Example of Footstep Planning for the humanoid HOAP-1

2. Multi-layered Robot Control System

2.1 System Overview

We implemented a locomotion control system which is able to find an optimal sequence of footsteps from a start location to a goal position without colliding with obstacles. The system consists of three modules, path planner, footstep planner and whole-body motion generator. The path planner generates the geometrical path for the given environmental map. Then footstep planner calculates a sequence of footsteps based on the path information. The whole-body motion generator calculates the trajectories of joint angles which realize the given sequence of footstep without losing the stability of the robot motion. By using these modules cooperatively, the locomotion planning is executed as follows:

1. Plan the entire path from the start position to the goal position by using RRT. (Fig. 2 (a))
2. Place the sub-goals for footstep planner based on the length of path at regular intervals. (Fig. 2 (b))
3. Plan the sequence of footsteps to the sub-goals. (Fig. 2 (c))
4. Repeat Step 3 until the robot reaches the goal position.

In our research, we have used the HOAP-1 (Humanoid for Open Architecture Platform) robot manufactured by Fujitsu Automation. We adjusted the constraints of walking, such as the limitation of landing position and joint angles for HOAP-1. Motions were controlled by specifying the joint angles of the 20 joints of the entire body every 0.002 seconds. The characteristics of the HOAP-1 are noted below: (1) Height: 483 [mm], Weight: 5.9 [kg]. (2) The internal interface between the hardware and software is available for public use. (3) For movable parts, each leg had six degrees of freedom and each arm had four degrees of freedom, for a total of 20 degrees of freedom on the left and right sides.

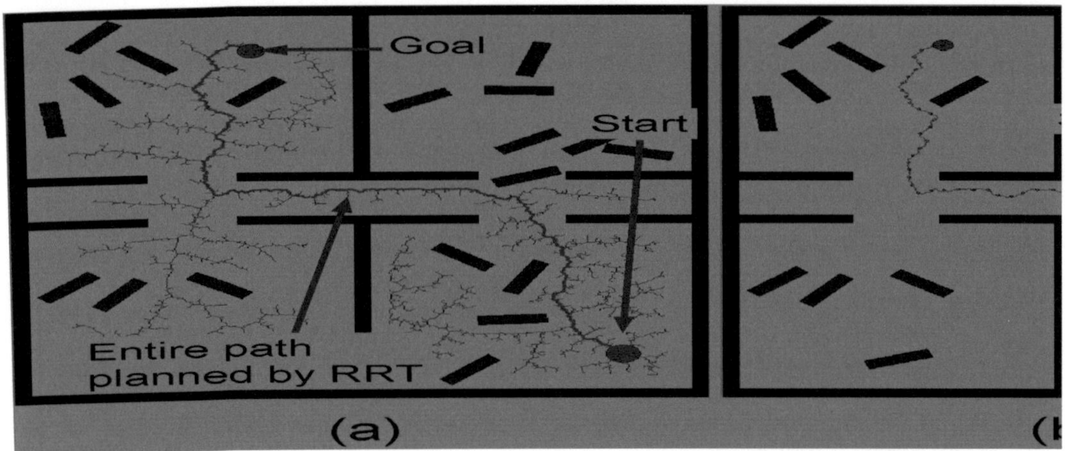

Fig. 2. (a) Path from the initial position to the goal position planned by RRT algorithm. (b) Sub-goals placed on the path. (c) Footstep sequence generated by A* Search algorithm.

2.2 Path Planning

We implemented the path planning module based on the Rapidly-exploring Random Tree (RRT) algorithm (Kuffner et al., 2000). RRT algorithm is a probabilistic motion planner which has found widespread use in the field of robot motion planning (Kamio & Iba, 2006; Zucker et al., 2007). Because RRT offers almost no hypothesis in terms of the problem, we could easily apply it to various problems. In this study, the state of the robot was expressed using 2D information (x coordinate, y coordinate), and we did not deal in another kind of information such as the rotation angle around z axis and the velocity of the robot to simplify the problem. We have to examine the collision of robots and obstacles when we use the path planning. In this experiment, the robot was approximated by the circle whose radius is 20 [cm] in order to detect the conflict between obstacles and the robot.

Before generating a sequence of footsteps which satisfies the planned path, we place the sub-goal for footstep planner. The sub-goal is decided by the given path length because the computational cost for the footstep planning increases with the number of footsteps.

2.3 Footstep Planning

A footstep planner returns an optimal sequence of footstep locations according to the plausible sets of footstep locations defined as shown in Fig. 3. In this research, we employed a footstep planner based on the A* Search Algorithm (Kuffner et al., 2003; Chestnutt et al., 2005). This footstep planner is fast enough to be used for the real environment. This is because the planner descritize the footstep location and explores footstep sequences in order of the estimated cost. Next, we explain the setup of A* Search in this research.

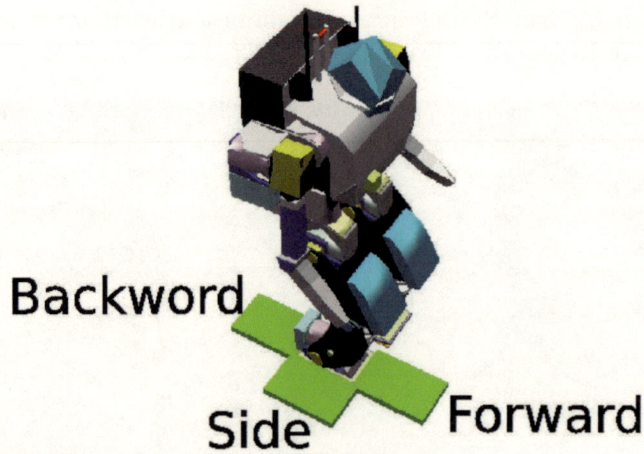

Fig. 3. Example of footsteps of the humanoid HOAP-1

2.3.1 Finite Set of Actions

The A* Search Algorithm uses a finite set of actions for exploration. In case of footstep planning, we define an action as the candidate footstep at the next time step, hence the candidate footstep should be descritized in advance. Fig. 3 shows an example of a set of footsteps which is descritezed into four candidates. The candidate footsteps are placed symmetrically when the robot switch the supporting legs.

2.3.2 Environmental Map

We used two kinds of maps as the experimental environment. At first, we employed a grid map in the experiment of Section 3. Each grid is uniformly-shaped and has information about the existence of obstacles. The other one is the vector map. Obstacles and walls are represented by rectangles in the vector map which is used in Section 4. The collision between footprints and obstacles is detected in both these maps.

2.3.3 Estimation of Cost

The A* Search Algorithm calculates the cost of the entire path by using the cost function and the heuristic estimation. The cost function is the length of footstep sequence from the initial position to the current position. The heuristic estimation calculates the minimum length of the footstep sequence from the current position to the goal position by dividing the distance by the maximum step length.

2.3.4 Termination Criteria

The search is terminated when the footstep sequence reaches the goal position. The search can fail in one of two ways:

- The number of A* search tries exceeded the maximum allowable. In this experiment we limited the maximum number to 3000.
- No more valid successor nodes can be generated. In this case, no collision-free footstep sequence exists using the given discrete set of footstep locations.

2.4 Whole-body Motion Generator

Up to now, many biped robots are successfully developed with the Zero Moment Point (ZMP) concept (Zhu et al., 2004; Nishiwaki et al., 2002; Nishiwaki & Kagami, 2006). In this experiment, one of the ZMP based method, a Fast Generation Method of Motion Pattern that Follows Desired ZMP (Nishiwaki et al., 2002) proposed by Nishiwaki et al. was used as a whole-body motion generator. The features of this method are as follows: (1) realizing the expected feet position accurately and (2) fast enough to used as a real-time module. Since real robots have many performance constraints, the gap between the simulated and real robot becomes a problem. Following performance constraints are taken into consideration in this paper; (1) Existence of the solution for inverse-kinematics of legs, (2) Limitation of joint angle range, (3) Collision between links. We examined the locations of feasible states by using the simulated model of HOAP-1, and stored in the regular grid. The footstep planner uses the regular grid as a reference to detect infeasible footstep locations.

2.5 Problems of Current Biped Walking Control System

The A* Search Algorithm is used for exploring a graph as shown in Fig. 4, which represents the footstep transitions. The node S means the initial state, and each node represents the footstep locations. The red node means the goal state. The gray node means an infeasible footstep which can not be applied at the time step. The number in the footstep location shows the kind of footstep.

Fig. 4 (a) shows a search with the footstep sequence with footstep set {1, 2, 3}, and failed to reach to the goal state. There are two kinds of countermeasures. One is to replace the set of footsteps with {1, 4, 5} as shown in (b), and the other one is to add the footstep {4, 5} as shown in (c). As can be seen from Fig. 4 (b) and (c), both methods successfully reach to the goal state, however, (c) requires more nodes than (b). In the Fig. 4 (a) and (b), three kinds of candidate footsteps are used for the footstep planning, so that the possible search space is a full 3-ary tree. On the other hand, the possible search space of Fig. 4 (c) is full 5-ary tree. The depth of the graph means number of footsteps. The spread of graph changes depending on the number of candidate footsteps. The maximum number of node in the search tree is represented as follow:

$$N_{\max} = \frac{1-k^n}{1-k}, \tag{1}$$

where k is the number of candidate footsteps and n is the length of the footstep sequence. The search space grows significantly as the increase of candidates. For example, in case of n = 10 and k = 3, the maximum number of node is 29524. In case of n = 10 and k = 5, the maximum number of node is 2441406. We have to reduce the number of candidate footstep in order to calculate long footstep sequences with the same amount of search time. In this research, we propose a switching method of footstep sets in order to curb the growth of search space.

We can also observe the side-effect of discretized footstep locations. If we employ only short footsteps, the robot is able to pass through narrow paths. However, such kind of set of footsteps usually requires much computational resources for the footstep planner because short step increases the total steps. On the other hand, the set of landing positions which has only long footsteps reduces the total steps and it increases the feasible path width. In this research, we focused on two requirements, to reduce the computational resources and to minimize the feasible path width.

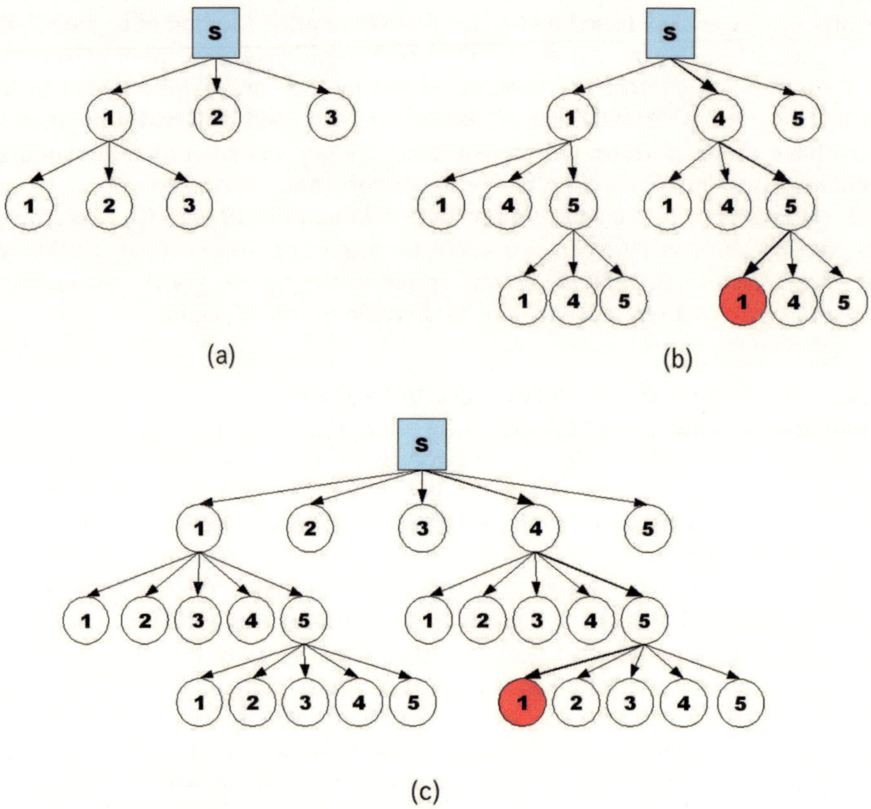

Fig. 4. Search space of A* Search Algorithm. Each node represents a state, which is a footstep location. Gray nodes mean infeasible states. The red nodes mean goal states.

3. Multi-Objective Optimization of Footstep Planner

3.1 Evolutionary Multi-Objective Optimization

The multi-objective optimization problem has multiple criteria to be minimized or maximized. To solve this problem we need to explore a set of solutions called Pareto optima. Evolutionary Multi-Objective Optimization which is one of the methods to solve the multi-objective optimization problem is rapidly developed by many researchers. Representative methods are as follows: Multi-Objective Genetic Algorithm (MOGA) (Fonseca & Fleming, 1993), Non-dominated Sorting Genetic Algorithm II (NSGA-II) (Deb et al., 2002), and Strength Pareto Evolutionary Algorithm 2 (SPEA2) (Zitzler et al., 2001). These methods are widely used in various kinds of research fields (Coello Coello, C. & Lamont, 2004).

3.2 Experimental Setup

We used NSGA-II which is an elitism EMO algorithm for optimization of the discrete set of landing locations. NSGA-II employs the mechanizm to penalize the crowded solutions in order to keep the diversity of solutions. NSGA-II was widely applied for various Real-World Applications (Deb et al., 2002). In this experiment, a chromosome was composed of two real variables. Each variable corresponds to the length of the footstep, forward-backward and sideward.

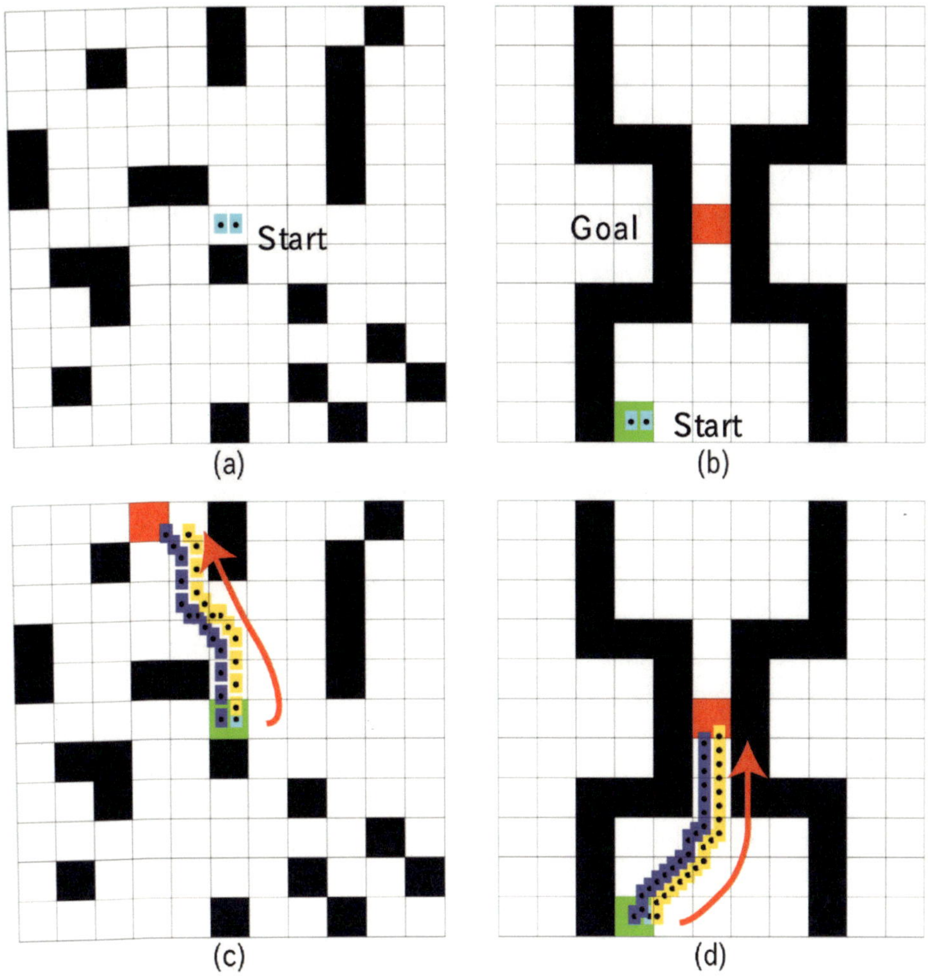

Fig. 5. Environmental maps for optimization. (a) Environment A to evaluate average numbers of A* search tries (b) Environment B to evaluate minimum feasible width of path. (c) and (d) show typical results of footstep planning.

As we mentioned in Section 2, two conflicting objectives exist. The criteria of this optimization were as follows:

- Average numbers of A* search tries
- Minimum feasible width of path

In order to evaluate these two objective functions, we created two environments: Environment A (Fig. 5 left, 220cm ×220cm), Environment B (Fig. 5 right, 220cm ×220cm). With Environment A, we measured average numbers of A* search tries to evaluate the computational cost. We placed the obstacles in Environment A randomly. The start position was settled in the center of the map and the goals were settled in each free space. The footstep planning was carried out from left-top to right-bottom respectively. After evaluating all the free space, we averaged the numbers of A* search tries. With Environment B, we evaluated the second objective. The task is to reach the goal located in the middle of the narrow path. We simply explored the minimum width by using hill-climbing method.

The parameters of NSGA-II are listed in Table 1. The population number was set to 100. The number of generation was set to 70. We employed real-valued mutation based on the uniform distribution as the mutation operator, and the Unimodal Normal Distribution Crossover (UNDX) (Ono et al., 2003) as the crossover operator. The Two parameters of UNDX were determined according to the recommended values in the literature (Ono et al., 2003 ; Ono & Kobayashi, 1997). The crossover rate was 0.9, and mutation rate was 0.01. We assumed that all landing positions in the candidates can transfer to each other because we set the step cycle slow enough to satisfy the constraints.

Parameter Name		Value
Population Size		100
Maximum Generation		70
Crossover	Rate	0.9
	Method	UNDX
	$\delta^2{}_\xi$	0.25
	$\delta^2{}_\eta$	0.06125
Mutation	Rate	0.01
	Method	uniform mutation

Table 1. Parameter setup of NSGA-II

3.3 Results

At 70th generation we observed that most of the individuals became non-dominated solutions. We also applied 10000 times Random Search as a comparison. The nondominated solutions acquired by NSGA-II and Random Search in the typical run were plotted in Fig. 6. The vertical axis means the minimum feasible width of path [mm], and the horizontal axis means the average number of A* search tries. From the Fig. 6, we can observe the trade-off relationship between the two objectives. Both of the method successfully obtained the Pareto solutions, however we can observed the diversity of NSGA-II population is superior to the one of Random Search population.

We used the Ratio of Non-dominated Individuals (RNI) to compare the performance of the two populations. RNI is used for the comparison of two sets of non-dominated individuals S_1 and S_2. At first, the united set of non-dominated individuals S_u is created by combining S_1 and S_2. Then, RNI is calculated by the ratio of S_1 or S_2 among S_u. We say that one set of solutions is closer to the true Pareto-solutions if its RNI is greater than that of the other one. The RNI of NSGA-II was 63% and the RNI of Random Search was 37%. NSGA-II saved the number of evaluations by 30%, and successfully obtained better solutions.

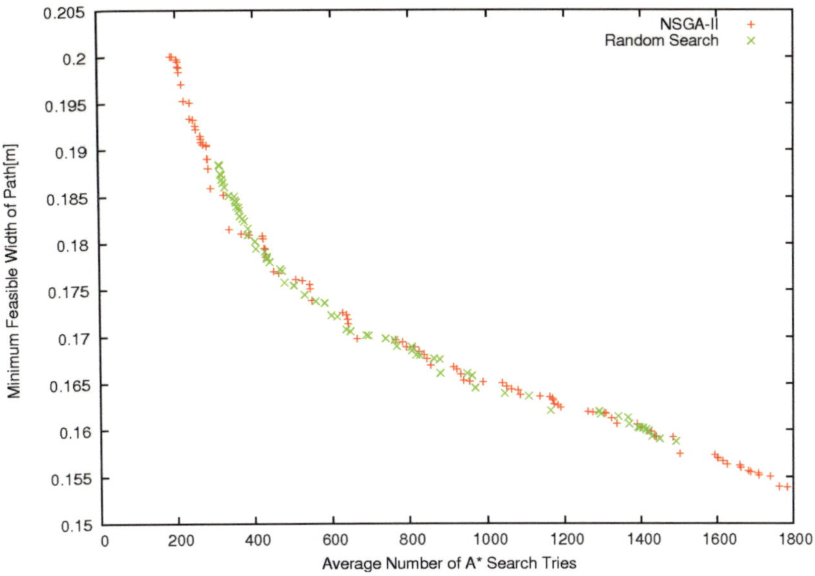

Fig. 6. Pareto-front of NSGA-II and Random Search

4. Switching the Parameter of Footstep Planner

4.1 Experimental Setup

In order to examine the performance of the acquired sets of landing position, we conducted another experiment. The validation environments are shown in Fig. 7. There are four rooms (2[m] ×2[m]) connected by the corridor (4[m]×0.5[m]). The rectangle obstacles (0.15[m] x 0.04[m]) were randomly placed in the environments. We fixed the start position at the right-bottom room and the goal position at the left-top room. The red circles represent the positions of the goals for the successful trials or the sub-goal where the planner failed. After generating the entire path from the initial position to the goal position by using the path planner, we added small obstacles (0.04[m] ×0.04[m]). These small obstacles ignored by the path planner should be avoided by the footstep planner.

4.2 Switching Rule

In order to reduce the number of footsteps, we switched the sets of landing positions. We used two extreme sets of landing positions, which located at both edges of the Pareto front, the set with the smallest average number of A* search tries (Set A) and the set with the smallest minimum feasible width of path (Set B). The Set A was used to reduce the total footsteps, and the Set B redeemed the failures of the plans with Set A. In this experiment, we employed a simple rule to switch the sets of landing positions as follows:

1. Execute the footstep planning by using the Set A.
2. If the planner fails to create the footsteps in step 1, apply the Set B. Otherwise, go to step 1.
3. If the sequence of footsteps is successfully calculated, go to step 1. Otherwise, the plan fails.

4.3 Results

We compared the proposed method to the conventional method which explored the footsteps with a single fixed footstep set, the Set A or the Set B. The performance was evaluated by the total number of footsteps. Fig. 7 shows the experimental environments and a typical result of the path planning and the footstep planning. (a)-(c) in Fig. 7 shows the Environment C. (d)-(e) in Fig. 7 shows the Environment D. The difference between Environment C and D is the placement of the small obstacles. (a)(d) shows the footsteps planned with Set A. (b)(e) shows the footsteps planned with Set B. (c)(f) shows the footsteps planned by the proposed method. The proposed method generated the short sequences of footsteps and the sequence which was able for the robot to pass through the narrow space. On the other hand, the planner with Set A failed at the middle of the map where the small obstacles and the wall formed a narrow path as shown in (a). The planner with Set B successfully passed through the narrow path, however it generated much longer sequences of footstep than the proposed method. The total number of footsteps, averaged over 5 runs are presented in Table 2. The total number of footsteps of the proposed method was about 72.2 % less than that of planner with Set B.

	Planner with Set A	Planner with Set B	Proposed method
Environment C	-	1548	454
Environment D	398	1509	396

Table 2. Total number of footsteps

5. Discussion

5.1 Multi-Objective Optimization of Footstep Planner

As we discussed in section 2, we can observe the trade-off relationship between two performance measures of footstep planner, the computational cost and minimum feasible width of path. In the previous section, we successfully acquired the discrete sets of landing positions for the footstep planner. By using NSGA-II, we saved the number of evaluations by 30%, and successfully obtained better solutions compared to the Random Search.

The trade-off relationship in this experiment is caused by the grain size of descritization of search space, hence this kind of multi-objective problem can be also observed in other problems when we handle the continuous problems by the discrete algorithm.

5.2 Parameter Switching of Footstep Planner

We also introduced a simple rule to switch the sets of landing positions. Using the proposed method, we obtained the 72.2 % shorter sequences of footsteps than that generated by the planner with Set B. We also generated the sequence which was able for the robot to pass through the narrow space where the planner with Set A failed.

Our proposed switching method with two sets of candidates worked efficiently in the environment where only one narrow path exists in the sparse field. In case of the path whose width gradually changes, however, switching more sets of candidates will be more effective than the current method. We can simply change the switching rules and choose appropriate sets of candidate from many of them because we have explored Pareto-optimal solutions in advance.

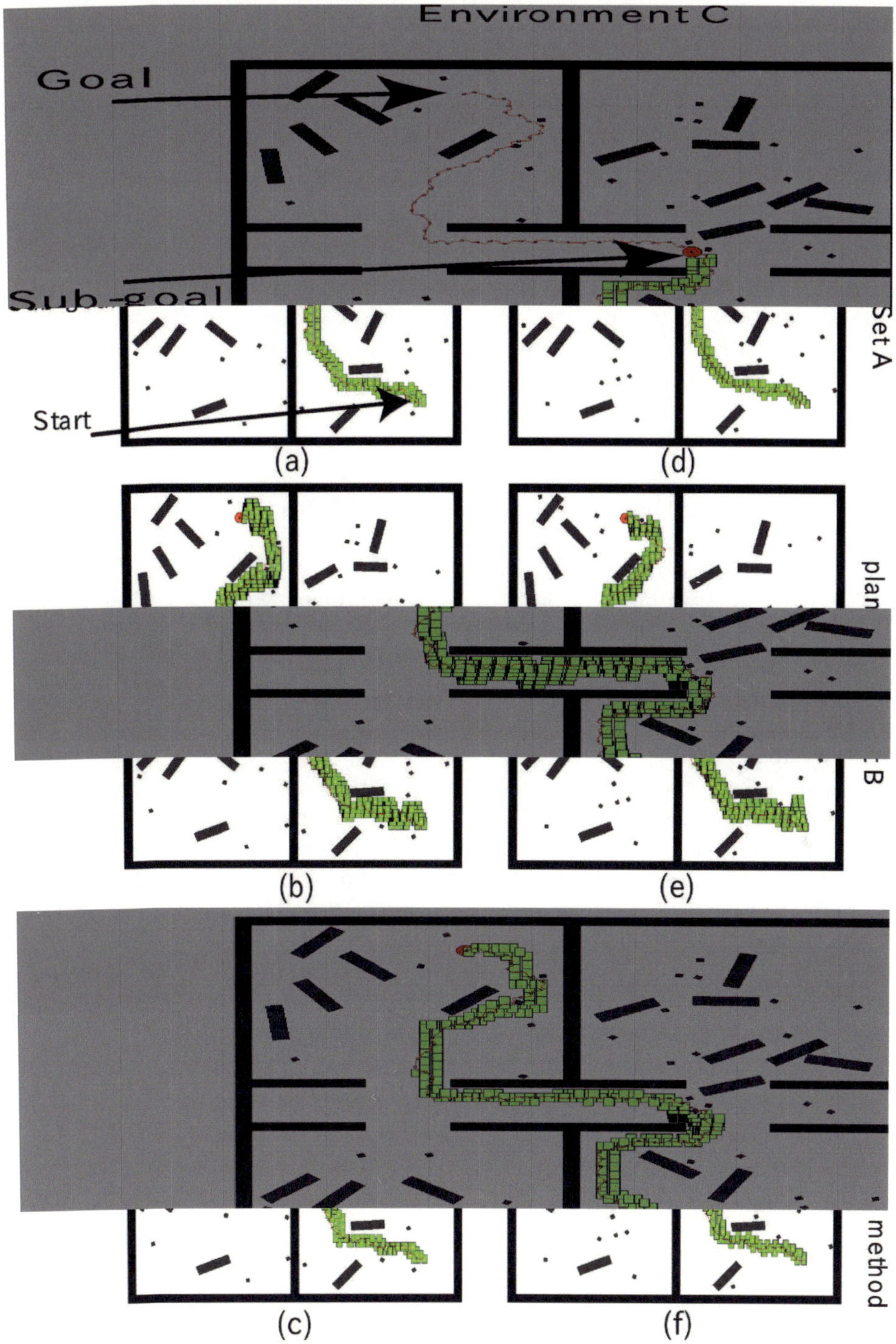

Fig. 7. Environment for validation: (a)(b)(c) show Environment C and typical results. (d)(e)(f) show Environment D and typical results. (a) failed to plan the footstep sequence at the narrow part of the path. (c) changed the set of footsteps at the narrow part of the path.

6. Conclusion and Future Works

We have presented a parameter setting method for a biped walking controller. At first, we introduced the robot control system and discussed the trade-off relationship of parameter settings for the footstep planning module. Next, we applied NSGA-II to acquire a set of parameters for the footstep planner. Then we proposed a simple rule to switch the sets of parameters to obtain the short sequence of footsteps. We also confirmed the proposed method generated the footsteps which was able for the robot to pass through the narrow space.

Our current work focuses on the translation of landing position without taking into consideration the rotation. We plan to optimize both parameters simultaneously to improve the performances of the footstep planner. Also, we intend to conduct experiments applying this system to the humanoid robot in a real environment.

7. References

Chestnutt, J.; Lau, M.; Cheung, G.; Kuffner, J.; Hodgins, J. & Takeo Kanade. (2005). Footstep Planning for the Honda ASIMO Humanoid, *Proceedings of the IEEE International Conference on Robotics and Automation*, pp. 629-634, April, 2005.

Coello Coello, C. & Lamont G. (2004).An Introduction to Multi-Objective Evolutionary Algorithms and Their Applications, In : *Application of Multi-Objective Evolutionary Aogorithms, Advances in Natural Computation*, Vol . 1, pp. 1-28, 2004.

Deb, K.; Pratab, A.; Agrawal, S. & Meyarivan, T. (2002). A fast and elitist multiobjective genetic algorithm: Nsga-II, *IEEE Transaction on Evolutionary Computation*, Vol. 6, No. 2, pp. 182-197, 2002.

Fonseca, C. M. & Fleming, P. J. (1993). Genetic Algorithms for Multiobjective Optimization: Formulation, Discussion and Generalization. *Proceedings of the Fifth International Conference on Genetic Algorithms*, pp 416-423, Morgan Kaufmann, 1993

Kuffner, J. Jr. & LaVall, S. M. (2000). RRT-connect: An efficient approach to single-query path planning, *Proceedings of IEEE International Conference on Robotics and Automation (ICRA '00)*, 2000.

Kamio, S. & Iba, H. (2006). Cooperative object transport with humanoid robots using rrt path planning and re-planning, *Proceedings of the 2006 IEEE/RSJ International Conference on Intelligent Robotics and Systems (IROS2006)*, 2006.

Kuffner, J; Kagami, S.; Nishiwaki, K.; Inaba, M. & Inoue. H. (2003). Online footstep planning for humanoid robots. *Proceedings of IEEE International Conference on Robotics and Automation 2003*, pp. 932-937, 2003.

Liu, H. & Iba, H. (2003). Multi-agent learning of heterogeneous robots by evolutionary subsumption. *Proceedings of the Genetic and Evolutionary Computation Conference (GECCO2003)*, pp. 1715-1728, 2003.

Nishiwaki, K.; Kagami, S.; Kuniyoshi, Y.; Inaba, M. & Inoue, H. (2002). Online generation of humanoid walking motion based on a fast generation method of motion pattern that follows desired zmp. *Proceedings of the 2002 IEEE/RSJ Intl. Conference on Intelligent Robots and Systems*, pp. 2684-2689, 2002.

Nishiwaki, K. & Kagami, S. (2006). High frequency walking pattern generation based on preview control of zmp. *Proceedings of the 2006 IEEE International Conference on Robotics and Automation*, pp. 2667-2672, 2006.

Ono, I. & Kobayashi, S. (1997). A Real-coded Genetic Algorithm for Function Optimization Using Unimodal Normal Distribution Crossover. *Proceedings of 7th Intl. Conference on Genetic Algorithm*, pp. 246-253, 1997.

Ono, I.; Kita, H. & Kobayashi, S. (2003). A Real-coded Genetic Algorithm using the Unimodal Normal Distribution Crossover. In: *Natural Computing Series archive Advances in Evolutionary Computing: theory and applications*, pp. 213-237, 2003.

Zhu, C.; Tomizawa, Y.; Luo, X. & Kawamura, A. (2004). Biped walking with variable zmp, frictional constraint, and inverted pendulum model. *Proceedings of IEEE International Conference on Robotics and Biomimetics 2004. (ROBIO 2004)*, pp. 425-430, 2004.

Zitzler, E.; Laumanns, M. & Thiele, L. (2001). Spea2: Improving the performance of the strengthpareto evolutionary algorithm. *Technical Report 103, Computer Engineering and Commu-nication Networks Lab (TIK)*, Swiss Federal Institute of Technology (ETH) Zurich, 2001.

Zucker, M.; Kuffner, J. & Branicky, M. (2007). Multipartite rrts for rapid replanning in dynamic environments. *Proceedings of IEEE Int. Conference of Robotics and Automation*, 2007.

9

On Adjustable Stiffness Artificial Tendons in Bipedal Walking Energetics

Reza Ghorbani and Qiong Wu
University of Hawaii at Manoa, University of Manitoba
USA, Canada

1. Introduction

Inspired by locomotion in nature, researchers have developed the passive dynamic walking machine principle and applied it to the legged robotics (Coleman & Ruina, 1998; Collins et al., 2001; Garcia, 1999; McGeer, 1990; Wisse, 2004; Wisse & Frankenhuyzen, 2006). The passive dynamic walking machines provide human-like locomotion in legged robots that is more efficient than the precisely joint-angle-controlled robots. On the other hand, tuning the parameters of the passive dynamic walking robots are tricky, time consuming and requires much experimentation. In addition, passive dynamic walking robots still have considerable energy loss through rapid changes in the velocity direction of the center of mass of the robot during collision of the foot with the ground. Precisely joint-angle-controlled bipedal walking robots also undergo significant energy loss caused by rigid collision of the foot with the ground in addition to their common energy dissipation in robot's power systems.

Collision of the foot with the ground during bipedal walking is inevitable which is one of the major sources of the energy loss. Establishing a new technique to reduce this energy loss is a challenging problem which we desire to address in this work by developing the idea of using the adjustable stiffness elastic elements in robot's structure. We believe that the adjustment of the elasticity as a control strategy can significantly improve the energetics of locomotion in bipedal walking robots by reducing the energy loss during the collision phase, which starts with an impact of the heel-strike followed by continuous motion and ends by a second impact at the foot-touch-down. This work , as a first step in this research area, constructively demonstrates the idea through two main efforts. The first effort is to develop the conceptual designs of the adjustable stiffness artificial tendons (ASAT) to show that the idea can be implemented in practice. The second effort is to study the effects of adjustable stiffness elasticity on reducing the energy loss by adding the model of each ASAT into the robot dynamics.

This introductory section reviews the research on legged locomotion which indicates the importance of elasticity in mechanics of locomotion in nature. In human walking, part of the kinetic and potential energy from the body is transiently stored as elastic strain energy during the collision phase and is released later during the rebound phase by elastic recoil (Kuo et al., 2005). This phenomenon greatly reduces the work required from the muscles and lowers the metabolic cost of locomotion (Alexander & Bennett, 1989; Cavagna et al., 1977). The mechanics of elastic recoil were also studied for running and it was found that, the forward kinetic energy of the body's center of mass is in phase with fluctuations in gravitational potential energy (Cavagna et al., 1964). It was also found that, humans and animals most likely store the elastic strain energy in muscles, tendons, ligaments and perhaps even bones, thereby, reducing the fluctuations in total mechanical energy (Cavagna et al., 1964). It has been reported that the leg stiffness influences many kinematic variables such as the stride frequency and the ground contact time (Farley & Gonzalez, 1996; McMahon & Cheng, 1990). Thus, the stiffness of the leg is a key parameter in determining the dynamics of locomotion (Ferris et al., 1998). He and Farley (Farley et al., 1993; He et al., 1991) suggested that the inherent properties of the musculoskeletal system determine an animal's choice of leg stiffness. Their idea was supported by (Roberts et al., 1997) who exposed that the muscles of running turkeys undergo very little change in length during the ground contact. Thus, the tendon may contribute most of the compliance of the muscle-tendon unit and greatly influence the leg stiffness (Alexander, 1988). In addition, adjusting the elasticity of the muscle-tendon unit during human locomotion contributes significantly to its efficiency. Thus, adjusting the stiffness of the robot's structure can be crucial for its energy economy which is studies in this work .

In the context of developing the legged robots, implementation of the adjustable leg stiffness in a running robot has been recommended by researchers to improve the performance on varied terrain (Ferris et al., 1998). Besides allowing the robot to accommodate different surface conditions, the adjustable leg stiffness would permit a robot to quickly adjust its stride length to avoid obstacles on rocky and uneven surfaces. Research is also plentiful in the area of series elasticity. Many of the ideas, problems and solutions of series elasticity related to this work , are initiated and discussed in publications of the MIT leg lab (Howard, 1990; Pratt & Williamson, 1995; Robinson et al., 1999; Williamson, 1995). Beyond the basics, much of the current research in series elasticity addresses topics such as human centered robotics (Zinn et al., 2004) and running robots (Hurst et al., 2004; Hurst & Rizzi, 2004).

Seyfarth developed a simple model of legged locomotion based on compliant limb behavior which is more similar to the human walking behavior (natural walking) than a traditional model of two coupled pendula (Seyfarth, 2000). Geyer also studied the basics of the compliant walking locomotion (Geyer et al., 2002; 2005). Jena walker II was successfully developed at the University of Jena by continuing the research on efficient locomotion using elasticity. However, the stiffness of elastic element in Jena walker II is constant. The electro–mechanical Variable Stiffness Actuation (VSA) motor developed of the University of Pisa is designed for safe and fast physical human/robot interaction in manipulators (Bicchi & Tonietti, 2004). A series elastic actuation system based on the Bowden–cable was developed at the University of Twente, (Veneman et al., 2006) for manipulator robots applications. The idea of controlling the compliance of a pneumatic artificial muscle to reduce the energy consumption of the robot is practically demonstrated, (Vanderborght et al., 2006). Most of the recent research on compliant locomotion is reported by researchers (Geyer, 2005; Ghorbani, 2008; Ghorbani & Wu, 2009a;b).

However, none of the previous research adequately addresses the specific issues of effects of the adjustable stiffness elasticity on reducing the energy loss in bipedal walking robots through a mechanical design approach. This work seeks to fill that gap through the following stages of designing the adjustable stiffness artificial tendons, studying their effects on energetics of bipedal walking robots and investigating the control issues.

The organization of this work is as follows. Section 2 describe three different conceptual designs of ASAT. The OLASAT is selected to continue of studying the energetics. However more information related to the advantages and limitations of each ASAT, the potential applications of ASAT as well as the effects of ASAT on series elastic actuation systems are explained in articles by authors (Ghorbani, 2008; Ghorbani & Wu, 2009a). In order to capture the basic behavior of OLASAT, a simple 2–DOF model of bipedal walking is illustrated in section 3. It also summarized the normalized formulation of the equations of motion of the biped. Section 4 contains the calculation of the energy loss at the foot-touch-down. A controller to automatically adjust the stiffness of OLASAT is proposed in section 5. Then in section 6, computer simulations are carried out to demonstrate the effects of stiffness adjustment of OLASAT on energy efficiency during the single support stance phase. The mathematical model of the bipedal walking is developed in sections 3.1 and 3.2.

2. Conceptual Design and Modeling of ASATs

In this section, three different conceptual designs of ASAT are developed. The conceptual designs have not been fabricated in this project. The first design (section 2.1) is a rotary adjustable stiffness artificial tendon that is a bi-directional tendon able to apply torque in a clockwise as well as a counter clockwise direction. The second design (section 2.2) is a unidirectional linear adjustable stiffness artificial tendon that uses the concept of changing the number of active coils of two series springs. Finally, the third design (section 2.3) is a combination of two offset parallel springs that is an unidirectional tendon. The mathematical model of each tendon is developed. The advantages, limitations of each concept and the potential applications to the development of a compliant actuation system are discussed in (Ghorbani & Wu, 2009a).

2.1 Rotary Adjustable Stiffness Artificial Tendon (RASAT)

The Rotary Adjustable Stiffness Artificial Tendon (RASAT) is specially designed to provide a wide range of the angular stiffness. The schematic of RASAT is illustrated in Fig. 1. In RASAT, a pair of compression springs is intentionally inserted between the two concentric input and output links. Each spring pair consists of a low stiffness spring with a stiffness of K_1 and a high stiffness spring with a stiffness of K_2. The offset between the low and high stiffness springs has a constant value of l. Distance d, of the spring pairs with respect to the center of rotation of the links, is adjustable. In this case, the internal torque T, between the concentric input and output links is calculated from:

$$T = \begin{cases} K_1 d\, x = K_1 d^2 \tan\theta & \frac{l}{d} \geq \tan\theta \\ K_1 d l + d(K_1+K_2)(d\tan\theta - l) & \frac{l}{d} < \tan\theta \end{cases} \tag{1}$$

where θ is the angular displacement between the input and output links, x is the spring deflection. In Equation 1, $\frac{l}{d} > \tan\theta$ represents the situation that only spring 1 is engaged and $\frac{l}{d} < \tan\theta$ is when both springs are engaged. The stiffness of spring 2 is μ times higher than the stiffness of spring 1. Thus:

$$K_2 = \mu K_1 \tag{2}$$

Combining Equations 1 and 2 and converting to the following non-dimensional form:

$$\frac{T}{K_1 d_{max}^2} = \begin{cases} tan\theta(\frac{d}{d_{max}})^2 & \frac{l}{d} > \tan\theta \\ (\mu+1)\tan\theta(\frac{d}{d_{max}})^2 - \mu\frac{l}{d_{max}}(\frac{d}{d_{max}}) & \frac{l}{d} < \tan\theta \end{cases} \tag{3}$$

where d_{max} is the maximum value of distance d. The effects of the distance ratio, $\frac{d}{d_{max}}$, on the output torque index, $\frac{T}{K_1 d_{max}}$, in RASAT are graphically illustrated in Figs. 2 and 3 where $\mu = 5$ and $\frac{l}{d_{max}} = 0.1$. As shown in Fig. 2, by increasing the distance, d, from zero to d_{max}, for a given θ, the torque index, $\frac{T}{K_1 d_{max}}$, increases. This relationship is shown for different θ

while increases from $\theta = 5^o$ to $\theta = 15^o$ with equal steps of 1°. The torque–angular deflection relation in RASAT is graphically illustrated in Fig. 3 for different values of distance indexes $\frac{d}{d_{max}}$. The slope of each curve in Fig. 3 is equivalent to the stiffness of the tendon. As shown in Fig. 3, by decreasing the ratio $\frac{d}{d_{max}}$, from 1 to 0.1 with steps of 0.1, the slopes of curves are reduced significantly. It has been shown in Fig. 3 that the slopes of the curves can be adjusted in a wide range which illustrates the capability of RASAT in adjusting the stiffness in a wide range. Sudden changes in the slopes of the curves in Fig. 3 are caused by engaging the high stiffness spring. Also, the higher the ratio $\frac{d}{d_{max}}$, the sooner the sudden change occurs. The effect of the stiffness ratio of the springs, μ, on the stiffness of RASAT is illustrated in Fig. 4 by assuming $\frac{d}{d_{max}} = 0.8$ and $\frac{l}{d_{max}} = 0.1$. Increasing the μ represents the increasing of the stiffness ratio of spring 2 to spring 1. In Fig. 4, the slope of the curves increases at the turning point that is caused by engaging spring 2 while μ increases from zero to 5 with equal increment of 1.

From the mechanical design point of view, RASAT (Fig. 5a & 5b) is comprised of an input

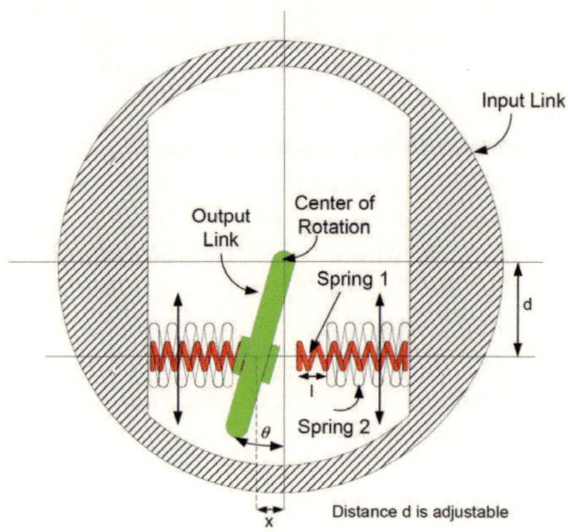

Fig. 1. General schematic of RASAT. A pair of two compression springs (spring 1 and spring 2) with a constant offset, l, are located in each side of the output link. Input and output links are concentric and d, the distance of springs to the center of rotation, is adjustable.

link (Fig. 5d), an output link (Fig. 5c), four springs (not shown in Fig. 5 but is shown in Fig. 1), and the spring positioning mechanism that is installed on the input link as shown in Fig. 5d. Input and output links are concentric and a relative angular displacement between the input and output links, θ, can be measured using a potentiometer installed on the input link (Fig. 5b). Two pairs of parallel helical compression springs configured in an offset are located inside the spring housing. The spring housing is linearly positioned by a non–back drive-able ball screw and a nut, which in turn, is connected to the input link. The ball screw, attached to the input link (Fig. 5d), rotates using a brush-less DC motor. Angular motion of the DC motor is converted to linear motion using a guiding shaft installed at the input link parallel to the ball screw. The distance d (Fig. 1), between the spring housing and the center of rotation can be adjusted using the feedback signal from an encoder installed at the DC motor. A bearing (Fig. 5c) sliding on the output shaft, which is attached to the output link, is pin jointed at the spring

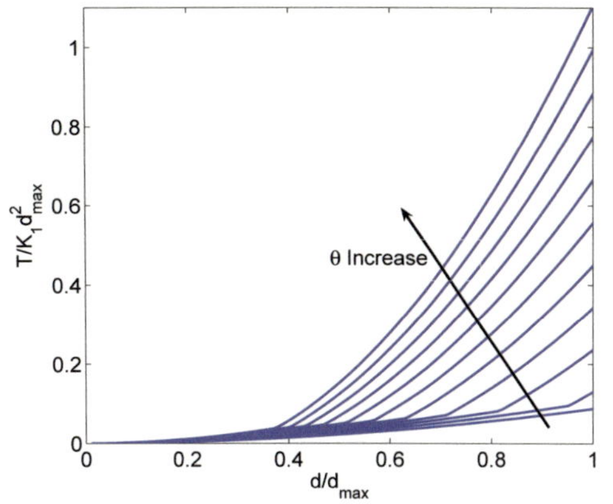

Fig. 2. Effects of increasing $\frac{d}{d_{max}}$ in non–dimensional torque–deformation in RASAT for a constant θ. θ increases in equal steps of 1° from 5° to 15°.

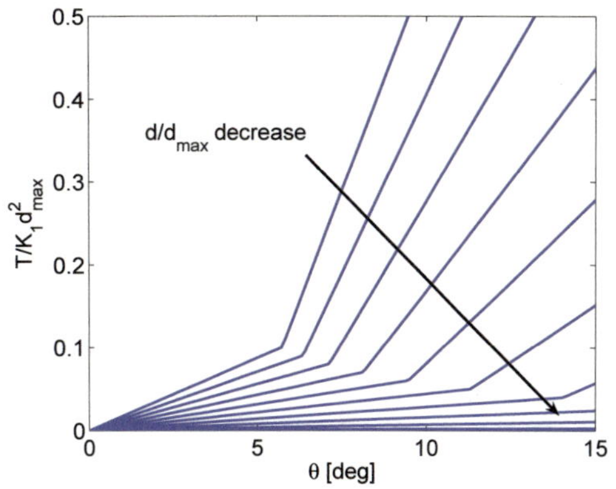

Fig. 3. Each curve shows non–dimensional torque–θ in RASAT for a constant d. $\frac{d}{d_{max}}$ decreases in equal steps of 0.1 from 1 to 0.1.

housing and has sliding motion inside the slot deployed on spring housing. Consequently, with a relative torque between the input and the output links, the bearing slides inside the spring housing and converts the angular motion between the links to the linear motion of the springs. The resultant force caused by the deflection of the springs creates torque through the output shaft via the bearing (Fig. 5c).

2.2 Linear Adjustable Stiffness Artificial Tendon (LASAT)

Linear Adjustable Stiffness Artificial Tendon (LASAT) is an uni–directional compression tendon. LASAT is a series combination of two helical compression springs. A rigid coupler that connects two series springs together is illustrated in Fig. 6. Counterclockwise rotation of the coupler increases the number of active coils in spring 2 with the low stiffness and decreases the number of active coils in spring 1 with the high stiffness; and vice versa for clockwise rotation. Springs 1 and 2 have the stiffnesses of K_{s1} and K_{s2} respectively, which are defined

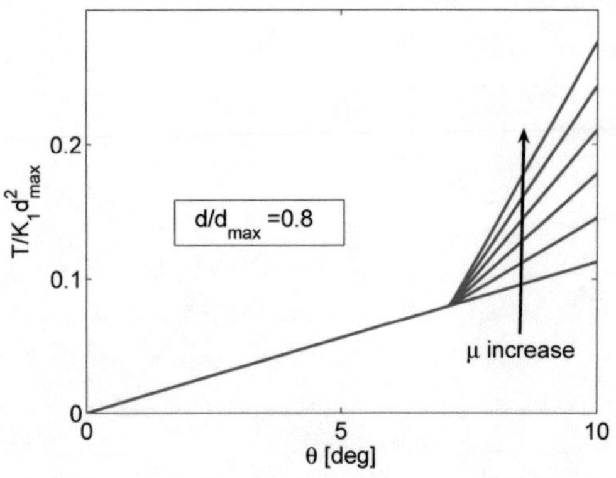

Fig. 4. Effects of increasing μ in non–dimensional torque–θ in RASAT.

by:

$$K_{si} = \frac{P_i}{N_s} \qquad i = 1,2 \tag{4}$$

where parameter 'i' represents the ith spring and number of spring coils, N_s, is assumed to be equal for both springs. P_1 and P_2 are the coil's stiffness of the spring 1 and 2, respectively (Norton, 1999):

$$P_i = \frac{dia_i^4 G_i}{8D_i^3} \qquad i = 1,2 \tag{5}$$

where D_i, dia_i and G_i are the mean coil diameters, wire diameters, the shear modulus of the springs. By changing the position of the coupler, the number of the active coils of spring 1 and spring 2 will be defined by $N_1 = (1-\lambda)N_s$ and $N_2 = \lambda N_s$ respectively, where $0 < \lambda < 1$. The coil's stiffness of spring 1 is assumed ρ times as high as spring 2, thus $P_1 = \rho P_2$. By the above considerations, the effective stiffness of spring 1, K_{a1}, and the effective stiffness of spring 2, K_{a2}, are given by the following Equations:

$$K_{a1} = \frac{K_{s1}}{1-\lambda} \tag{6}$$

$$K_{a2} = \frac{K_{s2}}{\lambda} \tag{7}$$

The resulted stiffness of the series springs, K_{eq}, represents the LASAT stiffness as long as the compression of softer spring is lower than its shut length, Ls, (where the coils are in contact) that is given below:

$$K_{eq} = \frac{K_{a1}K_{a2}}{K_{a1}+K_{a2}} = \frac{\frac{P_1}{(1-\lambda)N_s}\frac{P_2}{\lambda N_s}}{\frac{P_1}{(1-\lambda)N_s}+\frac{P_2}{\lambda N_s}} = \frac{\rho K_{s2}}{1+(\rho-1)\lambda} \tag{8}$$

Thus, the force of the tendon is calculated by the following Equations:

$$F_{LASAT} = \begin{cases} K_{eq} d_{LASAT} & d_{LASAT} \leq Ls\frac{1+(\rho-1)\lambda}{\rho} \\ K_{s2}Ls + K_{a1}(d_{LASAT} - Ls\frac{1+(\rho-1)\lambda}{\rho}) & d_{LASAT} > Ls\frac{1+(\rho-1)\lambda}{\rho} \end{cases} \tag{9}$$

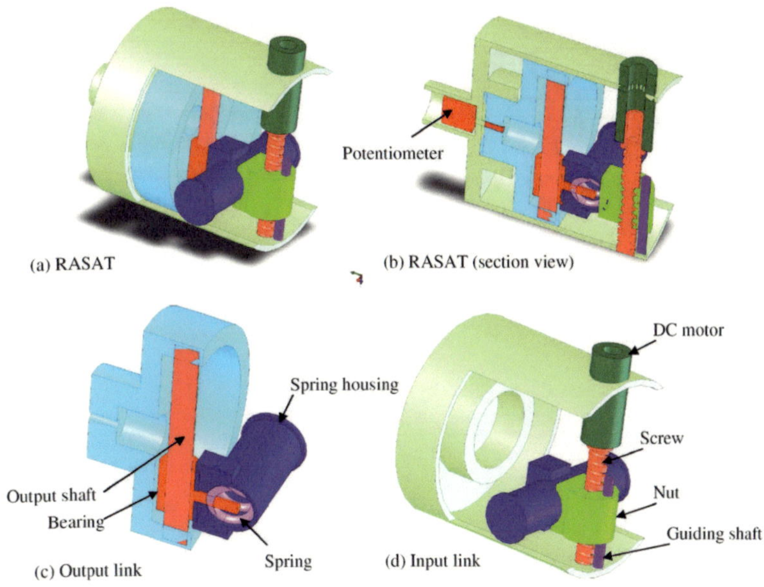

Fig. 5. 3D model of RASAT.

and respectively in its dimension-less form:

$$\frac{F_{LASAT}}{K_{s2}Ls} = \begin{cases} \frac{d_{LASAT}\rho}{Ls(1+(\rho-1)\lambda)} & \frac{d_{LASAT}}{Ls} \leq \frac{1+(\rho-1)\lambda}{\rho} \\ 1 + \frac{\rho}{1-\lambda}\left(\frac{d_{LASAT}}{Ls} - \frac{1+(\rho-1)\lambda}{\rho}\right) & \frac{d_{LASAT}}{Ls} > \frac{1+(\rho-1)\lambda}{\rho} \end{cases} \tag{10}$$

where d_{LASAT} is the deflection of the LASAT and the length $Ls\frac{1+(\rho-1)\lambda}{\rho}$ is the total deflection of the tendon at the instance that spring 2 reaches to the shut length. Fig. 7 illustrates the relationship of the dimension-less resultant stiffness of the LASAT, $\frac{K_{eq}}{K_{s2}}$, to the λ (the ratio of the number of active coils of spring 2 to N_s) for different values of ρ (the ratio of the coil stiffness of the spring 1 to the spring 2). In Fig. 7, each curve corresponds to a constant ρ and the value of ρ increases from 1 to 5 with increments of one. As shown, by increasing λ from zero to one for a constant ρ, the resulted stiffness of LASAT, K_{eq}, decreases.

Fig. 8 shows the relation of dimension-less force index $\frac{F_{LASAT}}{K_{s2}Ls}$, to the dimension-less deflection index $\frac{d_{LASAT}}{Ls}$, when $\rho = 5$ as well as λ varies from 0.1 to 0.9 with equal steps of 0.1. As shown in Fig. 8, there is a discontinuity in the slope of each curve as $\frac{F_{LASAT}}{K_{s2}Ls} = 1$ that is caused by the shut length of spring 2. The slope of the curves before the shut length shown in Equation 8 equals to $\frac{\rho}{1+(\rho-1)\lambda}$. The slope after the shut length equals to $\frac{\rho}{1-\lambda}$. By increasing λ, the slope of each curve before the shut length decreases that is resulted to the softer equivalent spring. On the other hand, the slope of the curve after the shut length increases. In general, helical springs are not acting linearly close to their the shut lengths. Thus, to reduce nonlinear effects on the tendon caused by coil contact and friction at the shut length, the LASAT should be designed in a way to prevent reaching the shut length.

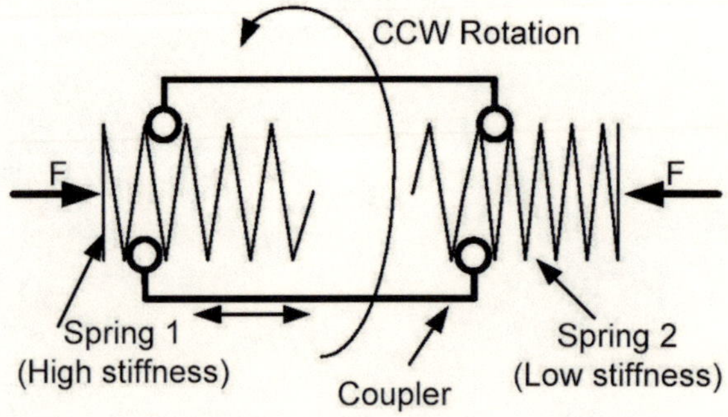

Fig. 6. Schematic of LASAT.

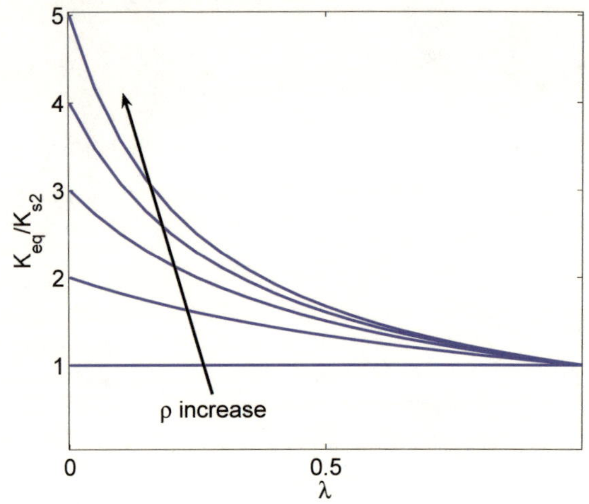

Fig. 7. Non–dimensional relation of stiffness–λ in LASAT before shut length. Each curve corresponds to a constant ρ while ρ increases from 1 to 5 with steps of 1.

From the mechanical design point of view, LASAT is comprised of an input rod, an output rod, two springs and a spring positioning mechanism as shown in Fig. 9. The springs can slide inside the output rod and have the same coil pitch and the mean diameter, but have different wire diameters. The inner diameter of the output rod is assumed to be smaller in the area that contacts with the softer spring than in the area that contacts with the stiffer spring. The output force is directly applied to the low stiffness spring and a notch inside the output rod makes a stopper that prevents the softer spring from reaching to the shut length. The positioning mechanism of the coupler consists of a brush-less DC motor, a spline shaft and a coupling element. The outer surface of the coupler is screw threaded with the lead equal to the spring's coil pitch. The inner surface of the coupler holds a ball spline bush which slides over the spline shaft freely (as shown in Fig. 9). The rotation of the spline shaft by brush-less DC motor transfers to the coupling element by the ball spline. Therefore, the angular motion of the coupling element converts to linear motion and simultaneously changes the number of

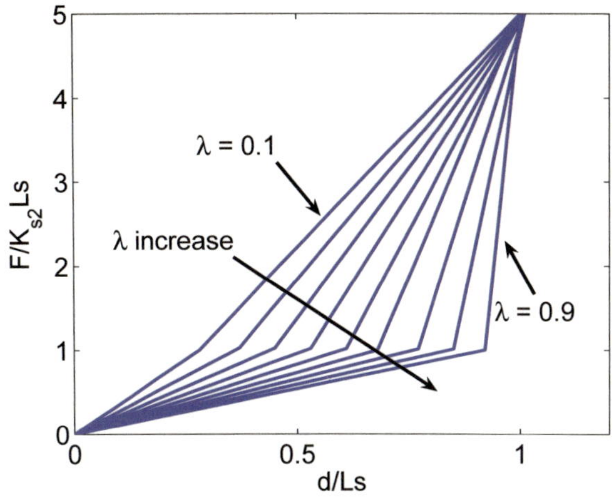

Fig. 8. Non–dimensional graph of force–deformation in LASAT. Sudden changes in slopes of the curves are caused by shut length of spring 2.

the spring coils in each spring. Also, an encoder is installed on the brush-less DC motor to measure the location of the coupling element.

2.3 Offset Location Adjustable Stiffness Artificial Tendon (OLASAT)

The Offset Location Adjustable Stiffness Artificial Tendon (OLASAT) is specially designed to switch the stiffness between two specific values. Here, the artificial tendon is a combination of two parallel springs (spring 1 and spring 2) placed with an offset. As shown in Fig. 10, the offset, a, is the distance between the end points of two springs when the springs are in their neutral lengths. By adjusting the offset using a linear actuator, the deformation requirement which engages spring 2 is changed. The applied force, F_{OLASAT}, of the tendon is a function of the stiffness of spring 1 with a low stiffness (K_{sp1}), spring 2 with a high stiffness (K_{sp2}), the offset (a) and the spring's deflection (d_{OLASAT}) as follows:

$$F_{OLASAT} = \begin{cases} K_{sp1} d_{OLASAT} & d_{OLASAT} < a \\ K_{sp1} a + (K_{sp1} + K_{sp2})(d_{OLASAT} - a) & d_{OLASAT} \geq a \end{cases} \tag{11}$$

The above equation in the dimensionless form appears in Equation(12) where K_{sp2} is replaced by ηK_{sp1}.

$$\frac{F_{OLASAT}}{K_{sp1} a} = \begin{cases} \frac{d_{OLASAT}}{a} & \frac{d_{OLASAT}}{a} < 1 \\ 1 + (1+\eta)(\frac{d_{OLASAT}}{a} - 1) & \frac{d_{OLASAT}}{a} \geq 1 \end{cases} \tag{12}$$

The force-deflection graph of the OLASAT is illustrated in Fig. 11. η is the ratio of the stiffness of spring 2 to that of spring 1 ($\eta = 5$ in Fig. 11). The slopes of the straight lines in Fig. 11 represent the stiffness of OLASAT. The stiffness is suddenly switched from the stiffness of spring 1, K_{sp1}, to the stiffness of two parallel springs, $(\eta + 1)K_{sp1}$, at point $d_{OLASAT} = a$.
From the mechanical design point of view, OLASAT is a uni-directional tendon and consists of an input rod, an output rod, a low stiffness spring and a high stiffness spring, with a positioning mechanism using a ball screw and a nut (as shown in Fig. 12). The low stiffness

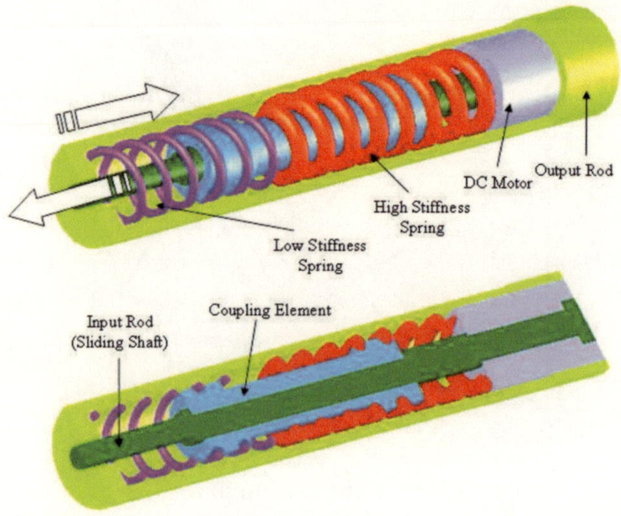

Fig. 9. 3D model of LASAT.

spring is coupled between the input and output rods. The high stiffness spring is connected to the input rod on one side and is free on the other side. A miniature brushless DC motor connected to the ball screw provides the sliding motion of the high stiffness spring over the slot deployed on the input rod, which can adjust the offset between the two springs.

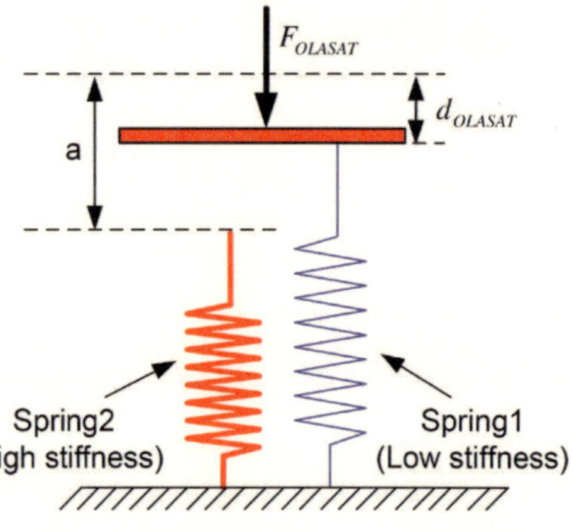

Fig. 10. Schematic of the OLASAT.

3. Bipedal walking gait in the simplified model

A simplified model and the gait cycle of a bipedal walking robot are introduced here. The model offers different advantages. It is simple, and hence decreases the complexity of analysis in energy economy. In addition, it considers the effects of OLASAT and the foot. It also

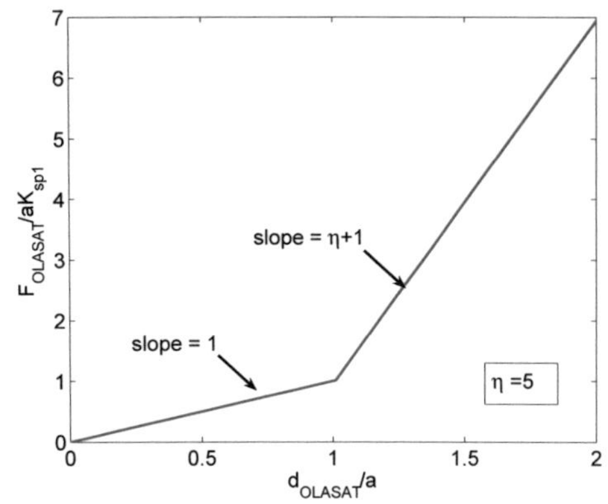

Fig. 11. Non–dimensional force–deformation graph of OLASAT.

includes the double support phase and has the ability to inject energy to the biped. The dynamics of the swing leg is not considered in the model to avoid complexity of the analysis.
In the model, as shown in Fig. 13, a rigid foot with a point mass is pivoted at the ankle joint to a rigid stance leg with a lumped mass at the hip (upper tip of the stance leg). One end of OLASAT is attached to the stance leg and the other end is attached to the foot. A cable and pulley mechanism converts the angular movement of the ankle joint to the linear deformation of the springs in OLASAT. The model also includes a massless linear spring to simulate the force of the trailing leg during the double support stance phase. The linear spring injects the energy to the biped. The input energy through the spring of the trailing leg can be adjusted by either controlling the initial deformation of the spring or adjusting its stiffness. However in this work , only the effects of the stiffness adjustment of OLASAT are studied in the simulation results and the stiffness of the trailing leg spring is taken zero. To simplify the analysis, planar motion and friction-free joints are assumed in the bipedal walking model.
In general, as shown in Fig. 14, the stance phase includes (in both single and double support periods) the collision, the rebound and the preload phases. The collision phase starts with the impact of the heel-strike followed by continuous motion. At the end of the collision, a second impact of the foot-touch-down occurs. Both impacts are assumed to be rigid to rigid, instantaneous and perfectly plastic, which dissipates part of the energy of the biped. In this model, the offset between the two springs of OLASAT, as shown in Fig. 10, can be adjusted to store part of the energy of the biped during the continuous motion of the collision phase and to reduce the impact at the foot-touch-down. The offset is adjusted only once during the swing phase while there is no external load on the foot. Then it remains constant for the following supporting period. The second phase, rebound, is a continuous motion while the foot is assumed stationary on the ground. The stored energy in OLASAT during the collision phase returns to the biped during the rebound phase. The rebound phase ends at midstance (biped upright position). OLASAT is passively loaded during the collision phase and is passively unloaded during the rebound phase. The motion of the biped after midstance is named the preload phase which continues until the heel-strike of the following walking step (Kuo et al., 2005). The kinematics of the heel-strike of the following walking step is specified by step length and the geometry of the robot.
The bipedal walking model in this work consists of a pre-deformed compression linear spring to simulate the force of the trailing leg. The linear spring of the trailing leg is massless with

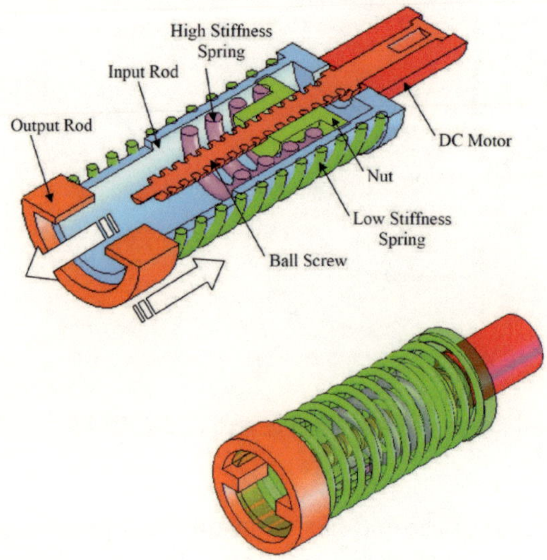

Fig. 12. 3D model of OLASAT.

one end connected to the toe of the foot on the ground and the other to the upper tip of the stance leg as shown in Fig. 13. It is also shown in Fig. 14 by B. The force vector from the compliant trailing leg (F in Fig. 14) is applied on the upper tip of the stance leg until the spring reaches its relaxed length (determining the end of the double support phase). By assuming the trailing leg as an elastic element, the model provides several advantages. The simplicity of dynamic modeling and analysis during impact events and the capability of injecting the external energy are two major advantages.

3.1 Dynamic model of the bipedal walking

The details of the dynamic modeling of the proposed bipedal walking model are given in (Ghorbani, 2008). In this section, the parameters of the simplified model of the bipedal walking on level ground are presented. In Fig. 13, links 1 and 2 are the foot and the stance leg. The values of d_1 and d_2 represent the distance between the center of mass of the foot to the heel and that of the body to the ankle joint respectively. l_1 is the distance between the heel and the ankle joint. l_2 is the distance between the ankle joint and the center of mass of the body which is at the upper end of the stance leg. Thus in the model, $l_2 = d_2$. θ_1 and θ_2 are denoted as the angles of the foot and the stance leg with respect to the horizontal axis as illustrated in Fig. 13. x_h and y_h represent the horizontal and vertical distance between the heel and a reference point O on the ground. In this work, the origin O is defined at the heel of the stance leg. The dimensionless parameters of the model are specified and listed in Table 1. Generalized coordinates of the biped are the horizontal and vertical positions of the heel as

Parameters	β	ψ	ζ	υ	ς	η	ν
Equivalence	$\frac{m_1}{m_2}$	$\frac{l_1}{l_2}$	$\frac{d_1}{l_2}$	$\frac{l_t}{l_2}$	$\frac{K_1 R^2}{m_2 l_2 g}$	$\frac{K_2}{K_1}$	$\frac{K_t l_2}{m_2 g}$

Table 1. Dimensionless Parameters.

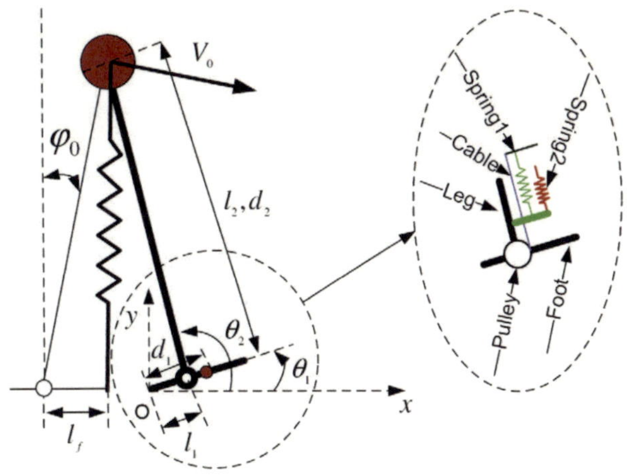

Fig. 13. Bipedal walking model schematic.

well as foot and stance leg angles with respect to the horizontal line which correspond to x_h, y_h, θ_1 and θ_2, respectively. The perpendicular position of the foot to the stance leg is assumed as a neutral position (no force) of OLASAT in this work. The heel is assumed to be pivoted to the ground during the collision phase by assuming enough friction force between the foot and the ground. Dynamic modeling of the bipedal walking, which is detailed in (Ghorbani, 2008), includes the heel-strike, the continuous motion during the collision phase as well as the rebound and the preload phases. The equations of motion in the normalized form with dimensionless parameters can help one to study more efficiently the bipedal walking motion in a generalized form. It also assists in the parametric follow-up study. The section 3.2 presents the normalized form of the equations of motion.

3.2 Equations of motion in normalized form

The dimensionless parameters of the model are specified and listed in Table 1. The equations of motion are normalized by $m_2 l_2^2$, the inertia of the stance leg about the ankle joint. Finally, by replacing the dimensionless parameters into the normalized form of the equations of motion, the normalized form of the equations of motion are arrived at. The normalized form of the dynamics equation at the heel-strike appears below.

$$\begin{bmatrix} \frac{\beta+1}{l_2^2} & 0 & \frac{\beta\zeta+\psi\sin(\theta_1)}{l_2} & \frac{1}{l_2} \\ 0 & \frac{\beta+1}{l_2^2} & \frac{\beta\zeta+\psi\cos(\theta_1)}{l_2} & \frac{1}{l_2} \\ \frac{\beta\zeta+\psi\sin(\theta_1)}{l_2} & \frac{\beta\zeta+\psi\cos(\theta_1)}{l_2} & \beta\zeta^2+\psi^2 & \psi\cos(\Delta\theta) \\ \frac{1}{l_2} & \frac{1}{l_2} & \psi\cos(\Delta\theta) & 1 \end{bmatrix} \begin{bmatrix} -^{HS}\dot{x}^- \\ -^{HS}\dot{y}^- \\ ^{HS}(\dot{\theta}_1^+ - \dot{\theta}_1^-) \\ ^{HS}(\dot{\theta}_2^+ - \dot{\theta}_2^-) \end{bmatrix} = \begin{bmatrix} \frac{^{HS}\hat{\lambda}_1}{m_2 d_2^2} \\ \frac{^{HS}\hat{\lambda}_2}{m_2 d_2^2} \\ 0 \\ 0 \end{bmatrix} \tag{13}$$

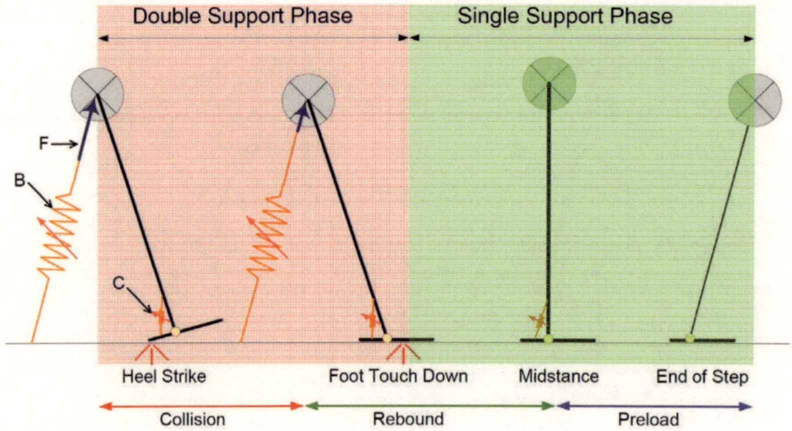

Fig. 14. General Schematic of the bipedal gate.

where $\Delta\theta = \theta_1 - \theta_2$. The normalized form of the equations of motion at the foot-touch-down is

$$\begin{bmatrix} \beta\zeta^2 + \psi^2 & \psi\cos(\Delta\theta) \\ \psi\cos(\Delta\theta) & 1 \end{bmatrix} \begin{bmatrix} -\,{}^{FTD}\dot{\theta}_1^- \\ {}^{FTD}(\dot{\theta}_2^+ - \dot{\theta}_2^-) \end{bmatrix} = \begin{bmatrix} \frac{{}^{FTD}\hat{\lambda}}{m_2 d_2^2} \\ 0 \end{bmatrix} \tag{14}$$

Equation (15) is the normalized counterpart of Equation.

$$\hat{M}(\theta)\ddot{\theta} + \hat{H}(\theta,\dot{\theta})\dot{\theta} + \hat{G}(\theta) + \hat{S}(\theta) = \hat{I}(\theta) \tag{15}$$

where

$$\hat{M} = \begin{bmatrix} \beta\zeta^2 + \psi^2 & \psi\cos(\Delta\theta) \\ \psi\cos(\Delta\theta) & 1 \end{bmatrix} \tag{16}$$

$$\hat{H} = \begin{bmatrix} 0 & \psi\sin(\theta_2 - \theta_1)\dot{\theta_2} \\ \psi\sin(\theta_2 - \theta_1)\dot{\theta_1} & 0 \end{bmatrix} \tag{17}$$

$$\hat{G} = \begin{bmatrix} \frac{\beta\zeta + \psi}{l_2} g\cos(\theta_1) \\ \frac{g}{l_2}\cos(\theta_2) \end{bmatrix} \tag{18}$$

$$\hat{S} = \frac{\varsigma}{l_2} \begin{bmatrix} -g(\Delta\theta - \frac{\pi}{2}) - \gamma\eta g(\Delta\theta - \frac{\pi}{2} - a/R) \\ g(\Delta\theta - \frac{\pi}{2}) + \gamma\eta g(\Delta\theta - \frac{\pi}{2} - a/R) \end{bmatrix} \tag{19}$$

$$\hat{I} = \begin{bmatrix} -\frac{\iota v g(L_0 - L_t)\beta}{l_2^2}\sin(\theta_1)\cos(\delta) + \frac{\iota v g(L_0 - L_t)\beta}{l_2^2}\cos(\theta_1)\sin(\delta) \\ -\frac{\iota v g(L_0 - L_t)\beta}{l_2^2}\sin(\theta_2)\cos(\delta) + \frac{\iota v g(L_0 - L_t)\beta}{l_2^2}\cos(\theta_2)\sin(\delta) \end{bmatrix} \tag{20}$$

The equations of motion during rebound is

$$\ddot{\theta_2} + \frac{g}{2}\cos(\theta_2) + \varsigma\frac{g}{l_2}(\theta_2 - \frac{\pi}{2}) + \gamma\eta\varsigma\frac{g}{l_2}(\theta_2 - \frac{\pi}{2} - \frac{a}{R}) = -\frac{\iota v g(L_0 - L_t)\beta}{l_2^2}\sin(\theta_2)\cos(\delta) + \frac{\iota v g(L_0 - L_t)\beta}{l_2^2}\cos(\theta_2)\sin(\delta) \tag{21}$$

The next section presents the calculations related to energy loss during the foot-touch-down, which is the major source of energy loss in the proposed bipedal walking model.

4. Discussion of the energy loss

The energy loss during the foot-touch-down is one of the major causes of energy reduction in bipedal walking which is reduced by properly adjusting the stiffness of OLASAT. This section studies the key parameters involved in the change in the kinetic energy of the biped, ${}^{FTD}\Delta E = {}^{FTD}E^- - {}^{FTD}E^+$, before and after the foot-touch-down, shedding light on how the stiffness adjustment of OLASAT can reduce the energy loss. ${}^{FTD}\Delta E$ for the model explained in Fig. 13 is given below.

$$\begin{aligned} {}^{FTD}\Delta E = 0.5m_2 l_2^2(({}^{FTD}\dot{\theta}_2^-)^2 - ({}^{FTD}\dot{\theta}_2^+)^2) + 0.5(m_1 d_1^2 + m_2 l_1^2)({}^{FTD}\dot{\theta}_1^-)^2 \\ + m_2 l_1 l_2 \cos({}^{FTD}\theta_2)({}^{FTD}\dot{\theta}_1^-)({}^{FTD}\dot{\theta}_2^-) \end{aligned} \tag{22}$$

On the other hand ${}^{FTD}\dot{\theta}_2^+$ is calculated from the equation of motion detailed in (Ghorbani, 2008)

$${}^{FTD}\dot{\theta}_2^+ = {}^{FTD}\dot{\theta}_2^- + \frac{l_1\cos(\theta_2 - \theta_1)\,{}^{FTD}\dot{\theta}_1^-}{d_2} \tag{23}$$

By substituting of ${}^{FTD}\dot{\theta}_2^+$ from Equation (23) into Equation (22) and after simplification, the following relation is obtained.

$${}^{FTD}\Delta E = 0.5(m_1 d_1^2 + m_2 l_1^2 \sin^2({}^{FTD}\theta_2^-))({}^{FTD}\dot{\theta}_1^-)^2 \tag{24}$$

Masses, lengths of the links and $\sin^2({}^{FTD}\theta_2)$ are all positive. Equation (24) illustrates the direct relation of ${}^{FTD}\Delta E$ with $({}^{FTD}\dot{\theta}_1^-)^2$. It indicates that reducing the magnitude of the angular velocity of the foot immediately before the foot-touch-down can significantly reduce the energy loss of the biped. In addition, Equation (23) illustrates the direct relation of ${}^{FTD}\dot{\theta}_2^+ - {}^{FTD}\dot{\theta}_2^-$ with ${}^{FTD}\dot{\theta}_1^-$ which indicates that by reducing the magnitude of ${}^{FTD}\dot{\theta}_1^-$ the change of angular velocity of the stance leg at the foot-touch-down is reduced.

By taking the time derivative of the position of the center of mass (COM) of the body, ${}^{FTD}Y_{COM}^-$, its velocity is arrived at as given below:

$${}^{FTD}\dot{Y}_{COM}^- = l_1\,{}^{FTD}\dot{\theta}_1^- + l_2\,{}^{FTD}\dot{\theta}_2^- \cos({}^{FTD}\theta_2) \tag{25}$$

${}^{FTD}\dot{\theta}_2^-$ and $\cos({}^{FTD}\theta_2)$ are negative before midstance. The optimum situation ${}^{FTD}\Delta E = 0$ can be achieved by reducing $({}^{FTD}\dot{\theta}_1^-)^2$ to zero. In such a scenario, the velocity of the COM of the body at the foot-touch-down will be upward. This indicates that the direction of the velocity vector of the COM of the body at the heel-strike ${}^{HS}\dot{Y}_{COM}^-$, which is downward, should be reversed to the upward direction at the foot-touch-down during the collision phase to reduce the energy loss. This is made possible by storing part of the kinetic energy of the biped in elastic form during the collision phase. This notion can be reinforced in human walking. Donelan expressed that humans redirect the center of mass velocity during step-to-step transitions not with instantaneous collisions, but with negative work performed by the leading leg over a finite period of time (Donelan & Kuo, 2002; Donelan et al., 2002). These findings serve as the foundation to determine the offset of OLASAT. As a result, the development of an automatic controller to adjust the stiffness of OLASAT is necessary to improve the performance of the bipedal walking which is described in the following section.

5. Design of the stiffness adjustment controller

In general, OLASAT has two major roles during the collision phase. The first role is to compensate the moment about the ankle joint exerted by the gravitational force of the body. The second is to store part of the kinetic energy of the biped. Both of these two roles can reduce $({}^{FTD}\dot{\theta}_1^-)^2$. This section provides a guideline for determining the offset of OLASAT, a, in order to store part of the energy of the biped, thus reducing $({}^{FTD}\dot{\theta}_1^-)^2$, and consequently reducing the energy loss. The development of a controller to satisfy such an optimal condition of $({}^{FTD}\dot{\theta}_1^-)^2 = 0$ can be possible by predicting the dynamics of the bipedal walking in advance. On the other hand, perfectly predicting the dynamics of the biped is not realistic because of the complexity of physical robots. Thus, a controller is developed in this section to estimate the offset of OLASAT without requiring the full knowledge of the system dynamics. To design such a controller, the following assumptions are made in this work.

First, OLASAT is loaded and unloaded passively during the stance phase. Thus for the following walking step, the offset is adjusted during the swing phase of the current walking step while the foot is not in contact with the ground. Second, the feedback signals of the biped are taken to be the angular position, θ_2, and the angular velocity, $\dot{\theta}_2$, of the stance leg. The reason for specifying these two signals as feedback is that the biped is an inverted pendulum during the rebound and the preload phases. Thus, the velocity of the biped at the heel-strike of the following walking step can be determined from the angular velocity of the stance leg at midstance, ${}^{MD}\dot{\theta}_2$. This choice allows enough time to adjust the offset during the swing phase which is important from the practical point of view. Third, the foot is perpendicular to the stance leg immediately before the heel-strike and in such a situation, OLASAT is in the neutral position (with no force). Fourth, the step length is fixed by assuming that the swing leg is perfectly controlled. Fifth, the angular displacement of the stance leg relative to the ground is negligible during the collision phase. This assumption results in the approximation of ${}^{FTD}\theta_2^- = {}^{HS}\theta_2^-$. It ensures that the total deformation of spring 1 in OLASAT is equal to $R\,{}^{HS}\theta_1^-$. ${}^{HS}\theta_1^-$ and ${}^{HS}\theta_2^-$ are the θ_1 and θ_2 immediately before the heel-strike which are known from the walking step length.

The stiffness adjustment controller developed here determines the offset of OLASAT, a, which corresponds to the angular offset of $\frac{a}{R}$ at the pulley of the ankle joint. Here, the maximum angular displacement of the pulley ($\Delta\vartheta$), in which spring 2 is engaged during the collision phase, is determined first to calculate the offset. Before determining the offset, we first discuss the selection of the stiffness of spring 1 of OLASAT. The stiffness of spring 1 must be selected low enough to prevent the leg from bouncing during the collision phase even for minimum bipedal walking speed while spring 2 is not engaged (minimum stiffness of OLASAT). It must also be selected high enough to compensate a portion of the gravitational moment about the ankle joint at the foot-touch-down and also to store part of the kinetic energy of the biped. Next, we explain the procedure of determining $\Delta\vartheta$.

$\Delta\vartheta(n+1)$ is calculated from the feedback loop of the stiffness adjustment controller for the walking step n+1. Walking step n is started from the heel-strike which includes the double support phase and will end immediately before the heel-strike of the step n+1. The inputs of the stiffness adjustment controller are the angular velocity of the stance leg at midstance, ${}^{MD}\dot{\theta}_2(n)$, of the current walking step n, and the stiffness of the trailing leg spring of the following walking step, $K_t(n+1)$. The output of the stiffness adjustment controller is the offset of OLASAT for the walking step n+1. $K_t(n+1)$ can be determined using a speed tracking controller to inject energy to the biped which is not discussed in this work and further information is referred to (Ghorbani, 2008).

Part of the kinetic energy of the biped at the end of the walking step n is stored in the trailing leg spring during the collision phase of the walking step n+1. Based on the results obtained in Section 4, the OLASAT should also store part of the kinetic energy during the collision phase of the walking step n+1. Here, the elastic potential energy, $0.5K_{sp2}R^2(\Delta\vartheta(n+1))^2$, of the walking step n+1 stored in spring 2 is assumed to be proportional to the difference between the kinetic energy of the biped at the end of the walking step n, $0.5m_2(l_2\ {}^{EN}\dot{\theta}_2^-(n)\sin(\varphi_0(n)))^2$, associated with the vertical component of the velocity of the COM of the body, and the elastic potential energy of trailing leg spring, $\frac{1}{2}K_t(\Delta L_{col}^2 - \Delta L_{dss}^2)$, during the collision phase of the walking step n+1 where $\varphi_0(n) = \frac{\pi}{2} - {}^{EN}\theta_2^-(n)$. ${}^{EN}\theta_2^-(n)$ and ${}^{EN}\dot{\theta}_2^-(n)$ are the angle and the angular velocity of the stance leg at the end of the preload phase of the walking step n. ΔL_{dss} and ΔL_{col} are the deformation of the tailing leg spring at the heel-strike and the deformation of the tailing leg spring at the foot-touch-down of the walking step n+1. The following equation describes the above energy relation.

$$\frac{1}{2}K_{sp2}R^2(\Delta\vartheta(n+1))^2 = \frac{1}{2}K_{adjust}(m_2(l_2\ {}^{EN}\dot{\theta}_2^-(n)\sin(\varphi_0(n)))^2 - K_t(n+1)(\Delta L_{col}^2 - \Delta L_{dss}^2)) \tag{26}$$

K_{adjust} is a proportional gain and $\Delta L_{col} = L_0 - {}^{FTD}L_t$, where ${}^{FTD}L_t$ is the length of the trailing leg spring at the foot-touch-down of the walking step n+1 which can be calculated from the kinematics of the biped using the following assumptions. Here, the stiffness of the trailing leg should be limited preventing the right hand side of Equation (26) from having a negative value. The step length and the initial angles of the foot and the stance leg for the walking step n+1 are known values in this work. Preload is also assumed as a free rotating inverted pendulum under gravity. ${}^{EN}\dot{\theta}_2^-(n)$ can be calculated from the angular velocity of the leg at midstance, ${}^{MD}\dot{\theta}_2(n)$, using the following energy relation:

$$0.5m_2l_2^2(\ {}^{EN}\dot{\theta}_2^-(n))^2 = 0.5m_2l_2^2(\ {}^{MD}\dot{\theta}_2^2(n)) + m_2gl_2(1-\cos(\varphi_0(n))) \tag{27}$$

The first term in the right-hand side of the above relation is the kinetic energy of the biped at midstance of the walking step n which is measurable from the feedback signals. It is assumed that the double support phase is ended before the midstance. Thus the injected energy, through the trailing leg spring during the double support phase of the walking step n, is converted to the kinetic energy of the biped which is measured at the midstance. The second term of the right-hand side of Equation (27) is the change in the gravitational potential energy of the biped between the midstance of the walking step n and the heel-strike of the walking step n+1 which can be calculated by assuming a fixed amount for the step length. By calculating ${}^{EN}\dot{\theta}_2^-(n)$ from Equation (27), which results in ${}^{EN}\dot{\theta}_2^-(n) = \sqrt{{}^{MD}\dot{\theta}_2^2(n) + \frac{2g}{l_2}(1-\cos(\varphi_0(n)))}$, and by substituting it into Equation (26), $\Delta\vartheta(n+1)$ is determined to be as follows.

$$\Delta\vartheta(n+1) = \sqrt{\frac{K_{adjust}}{R^2 K_{sp2}}} \sqrt{m_2(l_2 \ {}^{EN}\dot{\theta}_2^-(n)\sin(\varphi_0(n)))^2 - K_t(n+1)(\Delta L_{col}^2 - \Delta L_{dss}^2)} \quad (28)$$

Finally, as mentioned above by assuming that the angular movement of the stance leg relative to the ground is negligible during the collision phase, the offset $\frac{a}{R}(n+1)$ is calculated from the relation below.

$$\frac{a}{R}(n+1) = {}^{HS}\theta_1^-(n+1) - \Delta\vartheta(n+1) \quad (29)$$

The next section presents the simulation results of the bipedal walking motion in different case studies.

6. The study of energetics through simulations

The effects of the adjustable stiffness artificial tendon on the energetics of bipedal walking is studied in this section through computer simulations. In section 6.1, the simulation results of bipedal walking are illustrated during the single support phase for a single walking step. The simulation results show significant improvement in reducing the energy loss by proper adjustment of the stiffness of OLASAT. Then, the simulation results of the robot during single support phase are presented for 5 consecutive walking steps in section 6.2 to compare the results of the two cases of the single spring with best selected stiffness and OLASAT with the well-adjusted stiffness. Finally the controller developed in section 5 is implemented to automatically adjust the offset of OLASAT for reducing energy loss during the collision phase.

6.1 Stance phase of bipedal walking during a single walking step

In this section, a realization of the single support phase of bipedal walking is demonstrated for two different cases of a single spring and OLASAT during a single walking step. The stiffness of the single spring equals the stiffness of the spring 1 in OLASAT. Here, the stiffness of spring 1 in OLASAT is defined through trial and error. The objective was to find a stiffness which prevents the leg bouncing at the FTD for wide range of initial velocities, and which is high enough to store the elastic energy during the collision phase. In order to study the effects of stiffness adjustment on the energy economy of the biped, simulations are performed for the same initial conditions for both cases.

The parameters of the biped in computer simulations are listed in Table 2 in addition to $m_2 = 1kg$ and $l_2 = 1.0m$. The initial conditions of the biped immediately before the heel-strike in

Parameters	β	ψ	ζ	υ	ς	η
Value	0.02	0.05	0.07	0.1	0.5	4

Table 2. Dimensionless Parameters of the simulations.

the simulation are specified as follows: ${}^{HS}\theta_1^- = 15^o$, ${}^{HS}\theta_2^- = 105^o$, $x_h = y_h = 0$, ${}^{HS}\dot{\theta}_1^- = {}^{HS}\dot{\theta}_2^- = 0$ and $\varphi_0 = 11.7^o$. The initial translational velocity V_0 of the biped before the heel-strike has the magnitude of $1.5m/s$ with the horizontal angle of $-\varphi_0$.

Using those initial conditions, the total kinetic energy of the biped before the heel-strike is equal to 1.125 [J]. The stiffness of the trailing leg spring is assumed zero in this section, ($\nu = 0$), to simulate the biped's motion with no external elastic energy input. Note that the dynamics

of the swing leg is not assumed in the model.
Fig. 15 shows the simulation results of the biped in case 1 (single spring with best selected stiffness) and case 2 (OLASAT with well-adjusted stiffness). OLASAT's offset value in case 2 is provided using a trial and error procedure, $\frac{a}{R} = 6.0^o$, which reduces the angular velocity of the foot before the foot-touch-down to decrease the energy loss. Fig. 15a shows the joint angles of the biped θ_1, θ_2 which indicate that the motion of the biped is sustained in both cases. Fig. 15b shows the magnitudes of the velocity vector of the COM of the body. As shown in Fig. 15b, the velocity jump of the COM of the body during the FTD in case 2, which equals 0.01 $\frac{m}{sec}$, is significantly lower than case 1, which equals 0.15 $\frac{m}{sec}$. Consequently, the final velocity of the COM of the body is higher in case 2 compared to case 1. These results indicate that the biped can sustain the motion during the stance phase in both cases. Dimensionless kinetic energy κ, is defined as the ratio of the kinetic energy of the biped to its gravitational potential energy at midstance, $m_2 l_2 g$. Fig. 15c depicts the κ versus time, and the comparison of the results indicates that the kinetic energy loss at the FTD is significantly reduced in case 2, that is by 1%, resulting from the proper adjustment of the OLASAT, compared to the κ in case 1 which is as high as 20%. These results illustrate the effects of proper stiffness adjustment of OLASAT on the energetics of the bipedal walking. The significant reduction in energy loss during the FTD in case 2 results in higher final velocity of the body compared to the case 1. The displacement of the COM of the body during the collision phase together with its velocity vector is shown in Fig. 15d. Fig. 15d shows that the vertical component of the velocity of the body's COM is gradually decreased during the collision phase which can reduce the energy loss at the FTD as discussed in Section 4.
The dimensionless parameter of τ_{ankle} is the ratio of the torque of the ankle joint exerted by OLASAT to $m_2 g l_2$ during the single support stance phase. The graph of τ_{ankle} versus time is given in Fig. 16 for the above two cases. The exerted torque by OLASAT in case 1 is much lower than case 2. As shown in Fig. 16, the graphs include different sudden changes of slopes of the τ_{ankle} in the time period of the stance phase. In case 1, this behavior is caused by the end of the collision phase defined by point B and by the end of the rebound phase defined by F. OLASAT performs differently in case 2 while both of the springs are engaged in part of the time period of the motion. At the beginning of case 2, spring 2 is not engaged. After A in Fig. 16, both springs are engaged and they will remain engaged up to D. The change of slope in case 2 at C occures at the end of the collision phase. Point E is also defined by the end of the rebound phase in case 2. Figure 16 indicates that in OLASAT more energy is stored and released during the stance phase. To continue this study, the next section illustrates the effectiveness of adjusting the stiffness of the OLASAT in consecutive walking steps.

6.2 Bipedal walking simulations in consecutive steps

In the last section, a single step of bipedal walking was realized by implementing the OLASAT at the ankle joint. In this section, the simulation results of the bipedal walking in consecutive steps are presented. The goal is to demonstrate and quantify the effects of adjusting the stiffness of the ankle joint on multiple steps of the bipedal walking. During consecutive walking, a gravitational energy of the mass of the foot is injected to the biped caused by leg's initial angle in each walking step. For the first walking step, the same biped configuration and the same initial conditions as in Section 6.1 are used. The biped is moving on level ground. To demonstrate the kinetic energy loss of the biped in this section, $\Delta\kappa_n$ is defined as the dimensionless ratio of the total kinetic energy loss of the biped after n walking steps to the gravitational potential energy of the biped, $m_2 g l_2$, at midstance.

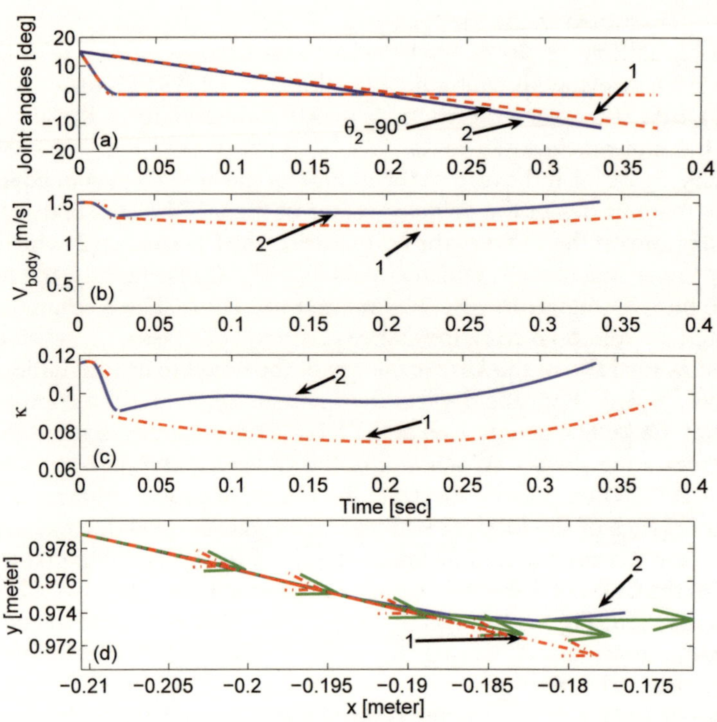

Fig. 15. Simulation results during the single support stance period for both cases of single spring, case1, and OLASAT with well-adjusted stiffness, case 2. (a)- Joint angles vs. time. (b)- Velocity of COM of the body vs time. (c)- Dimensionless kinetic energy of the biped vs time. (d)- Position and velocity vector of COM of the body during the collision phase.

Fig. 17 illustrates the velocity of the COM of the body in 5 consecutive walking steps. It also demonstrates that the motion of the biped is sustained in all steps for all cases. For the case of a single spring, it shows how stepping forward increases the period of each walking step. Reducing the average velocity of the biped is caused by the energy loss of the impact events. On average, for the single spring with the bests stiffness, 8% of the kinetic energy of the biped is dissipated in each step and $\Delta\kappa_5 = 0.0892$. For the case of OLASAT the changes in the step period and velocity of the COM of the body are insignificant as compared with those for the case of the single spring. On average in this case, 0.5% of the kinetic energy of the biped is dissipated in each walking step when $\Delta\kappa_5 = 0.0527$. These results show how the OLASAT, with ability of proper adjustment of the stiffness, significantly improves the energetics of the biped and makes it more efficient. This justifies the developement of an automated stiffness adjustment controller.

Fig. 17 also illustrates the velocity of the COM of the body in 5 consecutive walking steps while the stiffness of OLASAT is adjusted automatically, explained in Section 5. In order to simulate the well-adjusted stiffness, $\frac{a}{R}$ is assumed to be 6.0^o in the first walking step. The proportional gain of the automated stiffness adjustment controller, K_{adjust}, can play an important role in optimally adjusting the offset. Optimal iteration procedure to obtain the best gain remains as future research. Through trial and error simulations, it has been observed

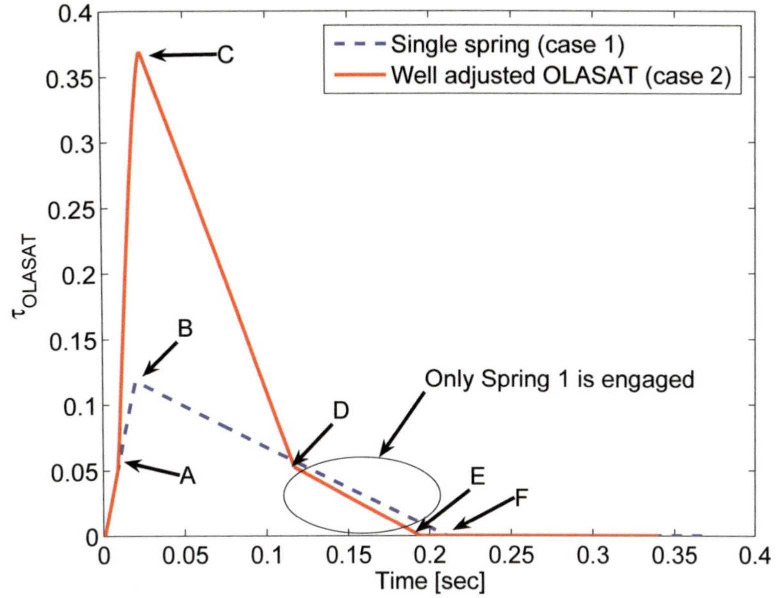

Fig. 16. Dimensionless torque of OLASAT vs. time during single support stance phase for both cases of single spring and OLASAT with well-adjusted stiffness.

that adaptive adjusting of K_{adjust}, as a function of the angular velocity of the stance leg at midstance, provides a better performance of the controller. Thus, in the simulation results of this section, the relation $K_{adjust} = 3l_2^{MD}\dot{\theta}_2$ is assumed. The controller is activated after the first walking step by sensing the angular velocity of the stance leg at midstance. As shown in Fig. 17, the sudden change in the velocity of the COM of the body is significantly reduced with compare to both the single spring and fixed stiffness. The energy loss is also significantly reduced, $\Delta\kappa = 0.052$, compared to case 1 and 2.

7. Conclusions

This work introduced different designs of adjustable stiffness artificial tendons. The modeling and detail conceptual design of each tendon were given. Then the results of the effects of stiffness adjustment of the ankle joint on energetics of the bipedal walking robots were presented. A methodology to reduce the energy loss was presented through three main efforts. In the first effort, one of the adjustable-stiffness artificial tendon, named OLASAT, was selected which is capable of storing and releasing the elastic energy during walking. In the second effort, a simplified model of the bipedal walking robot in the stance phase was developed which consists of a foot, a leg and an OLASAT which is installed parallel to the ankle joint. Such a model was used to compare the effects of a single spring and a well-adjusted stiffness OLASAT on reducing the energy loss during foot-touch-down. As the third effort, a simple controller based on energy feedback was designed to adjust the stiffness of OLASAT. Computer simulations were carried out to compare the energy loss of the biped in the two cases of the single spring and the well-adjusted stiffness OLASAT.

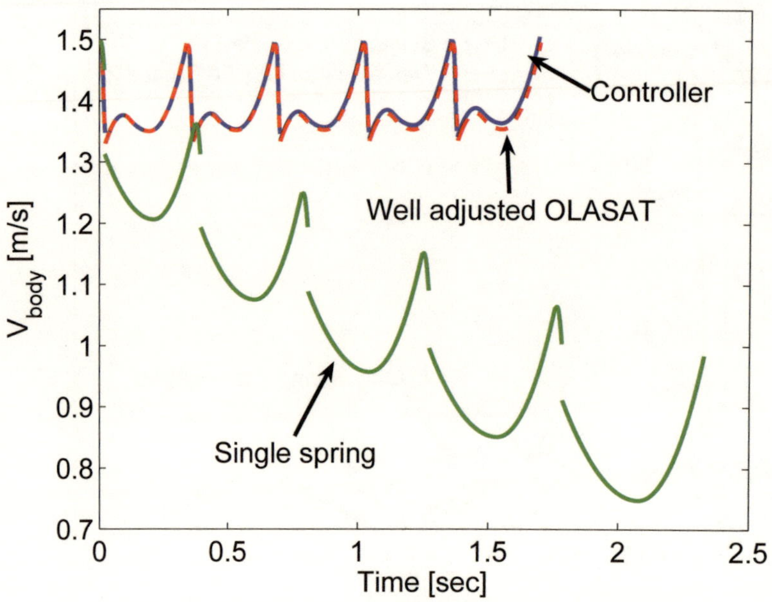

Fig. 17. Velocity of COM of the body vs time during 5 multiple walking steps for three cases of single spring, OLASAT with well-adjusted stiffness and stiffness adjustment controller.

Proper adjustment of the stiffness significantly reduces the kinetic energy loss during the foot-touch-down from 20% (of the single spring) to 1%. Simulation results of the biped during multiple walking steps illustrate that proper stiffness adjustment of OLASAT significantly improves the energetics of the bipedal walking. On average, the kinetic energy loss during multiple walking steps is reduced from 8% to 0.5% in each walking step. Simulation results illustrated that the automated stiffness adjustment controller can successfully reduce the energy loss during the stance phase. In general, the simulation results of this work suggest that various designs of the adjustable stiffness artificial tendons can be included in robot structures to achieve better energetics.

8. References

Alexander, R. M. (1988). *Elastic mechanisms in animal movement*, Cambridge University Press.

Alexander, R. M. & Bennett, M. B. (1989). How elastic is a running shoe?, *New Scientist* **123**: 45–6.

Bicchi, A. & Tonietti, G. (2004). Fast and soft-arm tactics [robot arm design], *IEEE Robotics and Automation Magazine* **11(2)**: 22–33.

Cavagna, G. A., Heglund, N. G. & Taylor, C. R. (1977). Mechanical work in terrestrial locomotion: two basic mechanisms for minimizing energy expenditure, *American Journal of Physiology* **233**: 243–261.

Cavagna, G. A., Saibene, F. P. & Margaria, R. (1964). Mechanical work in running, *Journal of Applied Physiology* **19**: 249–256.

Coleman, M. & Ruina, A. (1998). An uncontrolled toy that can walk but cannot stand still, *Physical Review Letters* **80(16)**: 3658–3661.

Collins, S. H., Wisse, M. & Ruina, A. (2001). A three-dimensional passive-dynamic walking robot with two legs and knees, *International Journal of Robotics Research* **20(7)**: 607–615.

Donelan, J.M. Kram, R. & Kuo, A. (2002). Mechanical work for step-to step transitions is a major determinant of the metabolic cost of human walking, *Journal of Experimental Biology* **205**: 3717–3727.

Donelan, M. J., Kram, R. & Kuo, A. D. (2002). Simultaneous positive and negative external mechanical work in human walking, *Journal of Biomechanical Engineering* **35**: 117–124.

Farley, C. T., Glasheen, J. & McMahon, T. A. (1993). Running springs: speed and animal size, *Journal of Experimental Biology* **185**: 71–86.

Farley, C. T. & Gonzalez, O. (1996). Leg stiffness and stride frequency in human running, *Journal of Biomechanics* **29**: 181–186.

Ferris, D. P., Louie, M. & Farley, C. T. (1998). Running in the real world: adjusting leg stiffness for different surfaces, *The Royal Society B: Biological Sciences* pp. 989–994.

Garcia, M. S. (1999). *Stability, scaling, and chaos in passive dynamic gait models*, PhD thesis, Theoretical and Applied Mechanics, Cornell University.

Geyer, H. (2005). *Simple models of legged locomotion based on compliant limb behavior*, PhD thesis, University of Jena.

Geyer, H., Blickan, R. & Seyfarth, A. (2002). Natural dynamics of spring–like running–emergence of self stability, *5th International Conference on Climbing and Walking Robots* pp. 87–91.

Geyer, H., Seyfarth, A. & Blickhan, R. (2005). Spring–mass running: simple approximate solution and application to gait stability, *Journal of Theoretical Biology* **232(3)**: 315–328.

Ghorbani, R. (2008). *On controllable stiffness bipedal walking*, PhD thesis, Mechanical and Manufacturing Engineering, University of Manitoba.

Ghorbani, R. & Wu, Q. (2009a). Conceptual design of the adjustable stiffness artificial tendons for legged robotics, *Mechanism and Machine Theory* **44(1)**: 140–161.

Ghorbani, R. & Wu, Q. (2009b). On improving bipedal walking energetics through adjusting the stiffness of elastic elements at the ankle joint, *International Journal of Humanoid Robots, In revision* **6(1)**: 23–48.

He, J. P., Kram, R. & McMahon, T. A. (1991). Mechanics of running under simulated low gravity, *Journal of Applied Physiology* **71**: 863–870.

Howard, R. D. (1990). *Joint and actuator design for enhanced stability in robotic force control*, PhD thesis, Massachusetts Institute of Technology.

Hurst, J., Chestnutt, J. & Rizzi, A. (2004). An actuator with physically variable stiffness for highly dynamic legged locomotion, *International Conference on Robotics and Automation* pp. 4662–4667.

Hurst, J. & Rizzi, A. (2004). Physically variable compliance in running, *International Conference on Climbing and Walking Robots* pp. 123–132.

Kuo, A. D., Donelan, J. M. & Ruina, A. (2005). Energetic consequences of walking like an inverted pendulum: step-to-step transitions, *Exercise and Sport Sciences Reviews* **33(2)**: 88–97.

McGeer, T. (1990). Passive dynamic walking, *The International Journal of Robotics Research* **9**: 62–82.

McMahon, T. & Cheng, G. (1990). The mechanics of running: how does stiffness couple with speed, *Journal of Biomechanics* **23 (Suppl. 1)**: 65–78.

Norton, R. L. (1999). *Design of machinery*, Thomas Casson.

Pratt, G. A. & Williamson, M. (1995). Series elastic actuators, *IEEE International Conference on Intelligent Robots and Systems* pp. 399–406.

Roberts, T. J., Marsh, R. L., Weyand, P. G. & Taylor, C. R. (1997). Muscular force in running turkeys: the economy of minimizing work, *Science* **275**: 1113–1115.

Robinson, D. W., Pratt, J. E., Paluska, D. J. & Pratt, G. A. (1999). Series elastic actuator development for a biomimetic walking robot, *Advanced Intelligent Mechatronics* pp. 19–22.

Seyfarth, A. (2000). *Elastically Operating Legs – Strategies and Construction Principles*, PhD thesis, University of Jena.

Vanderborght, B., Verrelst, B., Van Ham, R., Van Damme, M., Lefeber, D., Duran, B. M. Y. & Beyl, P. (2006). Exploiting natural dynamics to reduce energy consumption by controlling the compliance of soft actuators, *The international Journal of Robotics Research* **25(4)**: 343–358.

Veneman, J., Ekkelenkamp, R., Kruidhof, R. & van der Helm, F. C. T. (2006). A series elastic– and bowden–cable–based actuation system for use as torque actuator in exoskeleton-type robots, *The international Journal of Robotics Research* **25(3)**: 261–281.

Williamson, M. M. (1995). *Series elastic actuators*, PhD thesis, Massachusetts Institute of Technology.

Wisse, M. (2004). *Essentials of dynamic walking: analysis and Design of Two-Legged Robots*, PhD thesis, T.U. Delft.

Wisse, M. & Frankenhuyzen, J. V. (2006). Design and construction of mike; a 2d autonomous biped based on passive dynamic walking, *Adaptive Motion of Animals and Machines* pp. 143–154.

Zinn, M., Roth, B., Khatib, O. & Salisbury, J. K. (2004). New actuation approach for human friendly robot design, *The International Journal of Robotics Research* **23(4-5)**: 379–398.

10

Mathematical Modelling and Simulation of Combined Trajectory Paths of a Seven Link Biped Robot

Ahmad Bagheri, Behnam Miripour-Fard and Peiman Naseradin Mousavi
The University of Guilan, Faculty of Engineering
Iran

1. Introduction

The humanoid robot is designed like human for doing various tasks and adapting to human's environment. Next generation humanoids are expected to successfully coexist within human environments. This imposes very difficult challenges to researchers to produce truly dynamic movements to maintain balance under unexpected changes of paths. There are vast majority of studies on single trajectory paths of the biped robot (Bagheri & Mousavi, 2007; Mousavi & Bagheri, 2007) while combined trajectory paths have not received much attention. In this chapter, we have hence focused on generation of combined trajectory paths with the aid of mathematical interpolation. Zarrugh and Radcliffe (1997) have considered a biped robot with respect to a walking pattern by recording human kinematic data while McGeer (1990) have focused on passive walking of a biped robot generated by gravitational and inertial components. Silva and Machado (1999) have focused on actuator power and energy by the adaptation of walking parameters. The stability of the robot is connected to the biped robot's tendency to tip over. Zheng and Shen (1990) have considered a method of gait synthesis with respect to static stability while Chevallereau (1998) have focused on dynamic stability with the aid of a low energy reference trajectory definition. Takanishi et al. (1985), Shih et al. (1990), Hirai et al. (1998), and Dasgupta and Nakamura (1999) have considered the robot dynamic stability with respect to the walking process based on the zero moment point (ZMP) method. With respect to the various conditions of combined trajectory paths, Shih (1997, 1999), Huang et al. (2001) utilized methods to generate the trajectory paths of the robot where they carried out simulation of the combined trajectory paths.

The current chapter focuses on the inverse kinematic and dynamic methods for providing the robot combined trajectory paths in order to obtain a smooth motion of the robot. This procedure avoids the link's velocity discontinuities of the robot in order to mitigate the occurrence of impact effects and also helps to obtain a suitable control process. The main contribution of this work consists of a new method for simulation of a seven link biped robot over combined trajectory paths in order to study the effects of hip height over the torso's modified motion. The process has been performed based on the system given breakpoints and either third-order spline or Vandermonde matrix interpolation method.

The employed methods generate the desired combined trajectory paths to avoid oscillation of the paths because of the high order of the polynomials. The simulations have been carried out for combined trajectory paths. Similar to human gait, the robot's feet make negative, zero and positive angles with the ground.

2. Kinematic modelling

The mathematical interpolation is one of the simplest methods utilized for obtaining the suitable curves with respect to the given break points that the robot must undergo. The process of inverse kinematics with the aid of the defined combined trajectory paths (they have been indicated by the operator) and also solving the nonlinear equations for the robot movement will result in the needed parameters. Parameters such as joint angles will be calculated for use in dynamics and the subsequent controlling equations before the actual calculation of the actuator torques. The actuator torques will be used in the indicated combined trajectory paths. In the current chapter, the simplest mathematical interpolations which are "Vandermonde matrix" and "third-order spline" (Mousavi & Bagheri, 2007; Huang et al., 2001) have been used. The mathematical equations of a biped robot are nonlinear systems of equations which contain complex mathematical relations. In Figs. 1 and 2, the robot sagittal schematic has been presented to indicate the required nodes and the utility of the mentioned curves. With utilization of the specified criteria, the generated paths will be exploited in the designed software to obtain the kinematic parameters.
In general and with respect to Fig. 1, all the needed and important robot parameters are listed as follows:

(a) Hip parameters: The hip parameters include the vertical and horizontal displacements of the joint (z_h, x_h) and the variables which are displayed by Figs. 1 and 2, respectively. The distance between the hip and the fixed coordinate system (which is on the support leg) will be denoted for the instants of the beginning and the end points of the double support phase by x_{ed}, x_{sd}, respectively (Huang et al., 2001).

(b) Foot parameters: During the walking cycle, the horizontal and vertical displacements of ankle joint are displayed by x_a, z_a, respectively. The other parameters are as follows:

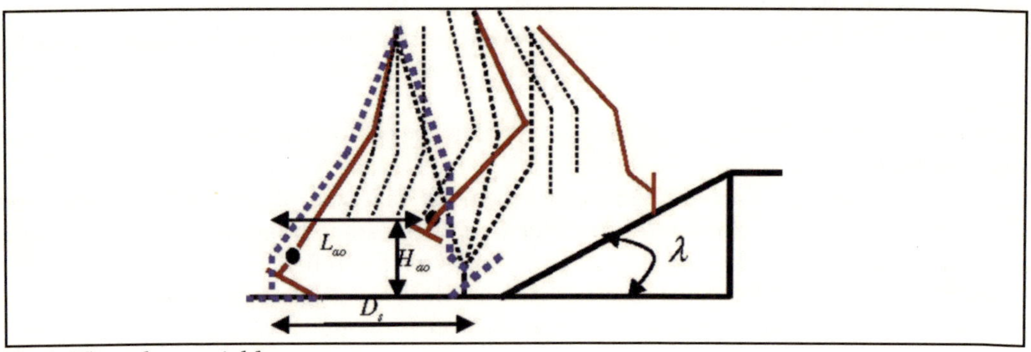

Fig. 1. The robot variables.

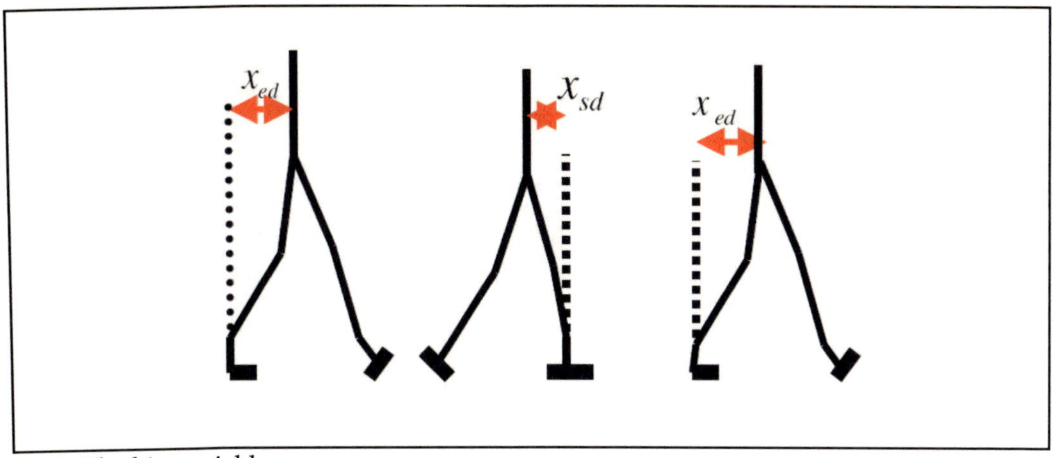

Fig. 2. The hip variables: x_{ed}, x_{sd}.

T_c	total traveling time including single and double support phases
T_d	double support phase time which is regarded as ≈20% of T_c
T_m	the time at which the ankle joint has reached the maximum height during the walking cycle
k	step number
H_{ao}	ankle joint maximum height
L_{ao}	the horizontal distance traveled between the ankle joint and the start point when the ankle joint has reached its maximum height
D_s	step length
q_b, q_f	foot lift angle and contact angle with the level ground
q_{gs}, q_{gf}	the ground initial terrain angles
h_{st}	stair level height
H_{st}	foot maximum height from stair level.

2.1. Foot trajectory interpolation

Five phases are required for calculation of the foot's combined trajectory path as shown in Fig. 3; they are derived as follow:

-First and fifth phases:

$$\theta_a(t) = \begin{cases} -q_{gs} & t = kT_c \\ -q_b & t = kT_c + T_d \\ q_f & t = (k+1)T_c \\ q_{gf} & t = (k+1)T_c + T_d \end{cases} \tag{1}$$

-Second phase:

$$\theta_a(t) = \begin{cases} -q_{gs} & t = kT_c \\ -q_b & t = kT_c + T_d \\ q_f + \lambda & t = (k+1)T_c \\ q_{gf} + \lambda & t = (k+1)T_c + T_d \end{cases} \tag{2}$$

-Third phase:

$$\theta_a(t)=\begin{cases}-q_{gs}+\lambda & t=kT_c\\ -q_b+\lambda & t=kT_c+T_d\\ q_f+\lambda & t=(k+1)T_c\\ q_{gf}+\lambda & t=(k+1)T_c+T_d\end{cases} \tag{3}$$

-Fourth phase:

$$\theta_a(t)=\begin{cases}-q_{gs}+\lambda & t=kT_c\\ -q_b+\lambda & t=kT_c+T_d\\ q_f & t=(k+1)T_c\\ q_{gf} & t=(k+1)T_c+T_d\end{cases} \tag{4}$$

In the current process of the polynomial determination, boundary conditions of the movement play an important role for calculation of the combined trajectory paths. The required boundary conditions are derived with respect to the system requirements. Similar to the human walking process, foot angular velocity ($\dot{\theta}_a$) at the specified moments including the instants of the beginning and the end of foot traveling is equal to zero:

$$\dot{\theta}_a\left[t=kT_c, t=kT_c+T_d\right]=0 \tag{5}$$

With respect to relations (1)-(4), a third-order spline can be utilized to provide the foot trajectory for the combined trajectory paths. Similar processes are exploited to calculate the trajectory paths for the displacements of horizontal and vertical foot traveling as shown below.
The displacement of the horizontal coordinates of the foot for the calculated breakpoints can be obtained as follows:

-First phase (Mousavi, 2006):

$$x_{a\,level\,ground}(t)=\begin{cases}kD_s & t=kT_c\\ kD_s+l_{an}\sin q_b+l_{af}(1-\cos q_b) & t=kT_c+T_d\\ kD_s+L_{ao} & t=kT_c+T_m\\ (k+2)D_s-l_{an}\sin q_f-l_{ab}(1-\cos q_f) & t=(k+1)T_c\\ (k+2)D_s & t=(k+1)T_c+T_d\end{cases} \tag{6}$$

-Second phase:

$$x_{a\,transient11}(t)=\begin{cases}k_{ch}D_s & t=kT_c\\ k_{ch}D_s+l_{an}\sin q_b+l_{af}(1-\cos q_b) & t=kT_c+T_d\\ k_{ch}D_s+L_{ao} & t=kT_c+T_m\\ (k_{ch}+2)D_s+(D_s-l_{ab})\cos\lambda-\cdots & t=(k+1)T_c\\ l_{an}\sin(q_f+\lambda)+l_{ab}\cos(q_f+\lambda) & \\ (k_{ch}+1)D_s+D_s\cos\lambda-l_{an}\sin q_f & t=(k+1)T_c+T_d\end{cases} \tag{7}$$

-Third phase:

$$x_{a\,transient11}(t)=\begin{cases}(k_{ch}+1)D_s & t=kT_c\\ (k_{ch}+1)D_s+l_{an}\sin q_b+l_{af}(1-\cos q_b) & t=kT_c+T_d\\ (k_{ch}+1)D_s+L_{ao}\cos\lambda & t=kT_c+T_m\\ (k_{ch}+1)D_s+(2D_s-l_{ab})\cos\lambda-\cdots & t=(k+1)T_c\\ l_{an}\sin(q_f+\lambda)+l_{ab}\cos(q_f+\lambda) & \\ (k_{ch}+1)D_s+2D_s\cos\lambda-l_{an}\sin q_f & t=(k+1)T_c+T_d\end{cases} \tag{8}$$

-Fourth phase:

$$x_{a\,Dec}(t)=\begin{cases}(k_{ch}+1)D_s+Dec_{st}D_s\cos\lambda-l_{an}\sin\lambda & t=kT_c\\ (k_{ch}+1)D_{ss}+(Dec_{st}D_s+l_{af})\cos\lambda-\cdots & t=kT_c+T_d\\ l_{af}\cos(q_b-\lambda)+l_{an}\cos(q_b-\lambda) & \\ (k_{ch}+1)D_s+(Dec_{ss}+L_{ao})\cos\lambda & t=kT_c+T_m\\ (k_{ch}+1)D_s+((Dec_{st}+2)D_s-l_{ab})\cos\lambda-\cdots & t=(k+1)T_c\\ l_{an}\sin(q_f+\lambda)+l_{ab}\cos(q_f+\lambda) & \\ (k_{ch}+1)D_s+(Dec_{st}+2)D_s\cos\lambda-l_{an}\sin q_f & t=(k+1)T_c+T_d\end{cases} \tag{9}$$

-Fifth phase:

$$x_{a,tran2}(t)=\begin{cases}(k_{ch}+1)D_s+(Dec_{st}+1)D_s\cos\lambda-l_{an}\sin\lambda & t=kT_c\\ (k_{ch}+1)D_{ss}+((Dec_{st}+1)D_s+l_{af})\cos\lambda-\cdots & t=kT_c+T_d\\ l_{af}\cos(q_b-\lambda)+l_{an}\cos(q_b-\lambda) & \\ (k_{ch}+1)D_s+((Dec_{ss}+1)D_s+L_{ao})\cos\lambda & t=kT_c+T_m\\ (k_{ch}+1)D_s+((k_{ch2}-k_{ch1})D_s)\cos\lambda+D_s+\cdots & t=(k+1)T_c\\ l_{ab}\cos(q_f-1)-l_{an}\sin q_f & \\ (k_{ch}+1)D_s+(k_{ch2}-k_{ch1})D_s\cos\lambda+D_s & t=(k+1)T_c+T_d\end{cases} \tag{10}$$

-Six phase:

$$x_{a,tran2}(t)=\begin{cases}(k_{ch}+1)D_s+(Dec_{st}+2)D_s\cos\lambda-l_{an}\sin\lambda & t=kT_c\\ (k_{ch}+1)D_{ss}+((Dec_{st}+2)D_s+l_{af})\cos\lambda-\cdots & t=kT_c+T_d\\ l_{af}\cos(q_b-\lambda)+l_{an}\cos(q_b-\lambda) & \\ (k_{ch}+1)D_s+((k_{ch2}-k_{ch1})D_s)\cos\lambda+L_{ao} & t=kT_c+T_m\\ (k_{ch}+1)D_s+((k_{ch2}-k_{ch1})D_s)\cos\lambda+2D_s+\cdots & t=(k+1)T_c\\ l_{ab}\cos(q_f-1)-l_{an}\sin q_f & \\ (k_{ch}+1)D_s+(k_{ch2}-k_{ch1})D_s\cos\lambda+2D_s & t=(k+1)T_c+T_d\end{cases} \tag{11}$$

-Seven phase:

$$x_{a,\, level\ ground}(t)=\begin{cases}(k_{ch}+1)D_s+((k_{ch2}-k_{ch1})D_s)\cos\lambda+Lev_{st}D_s & t=kT_c\\ (k_{ch}+1)D_{ss}+((k_{ch2}-k_{ch1})D_s)\cos\lambda+Lev_{st}D_s-\cdots & t=kT_c+T_d\\ l_{af}(1-\cos q_b)+l_{an}\cos q_b & \\ (k_{ch}+1)D_s+((k_{ch2}-k_{ch1})D_s)\cos\lambda+Lev_{st}D_s+L_{ao} & t=kT_c+T_m\\ (k_{ch}+1)D_s+((k_{ch2}-k_{ch1})D_s)\cos\lambda+\cdots & t=(k+1)T_c\\ (Lev_{st}+2)D_s+l_{ab}\cos(q_f-1)-l_{an}\sin q_f & \\ (k_{ch}+1)D_s+(k_{ch2}-k_{ch1})D_s\cos\lambda+(Lev_{st}+2)D_s & t=(k+1)T_c+T_d\end{cases} \quad (12)$$

The displacement of the vertical coordinates of the foot for the calculated breakpoints can be obtained as follows:

-First phase (Mousavi, 2006):

$$z_{a,\, level\ ground}(t)=\begin{cases}h_{gs}+l_{an} & t=kT_c\\ h_{gs}+l_{af}\sin q_b+l_{an}\cos q_b & t=kT_c+T_d\\ H_{ao} & t=kT_c+T_m\\ h_{ge}+l_{ab}\sin q_f+l_{an}\cos q_f & t=(k+1)T_c\\ h_{ge}+l_{an} & t=(k+1)T_c+T_d\end{cases} \quad (13)$$

-Second phase:

$$z_{a,\, transient}(t)=\begin{cases}h_{gs}+l_{an} & t=kT_c\\ h_{gs}+l_{af}\sin q_b+l_{an}\cos q_b & t=kT_c+T_d\\ H_{ao} & t=kT_c+T_m\\ (D_s-l_{ab})\sin\lambda+l_{ab}\sin(q_f+\lambda)+l_{an}\cos(q_f+\lambda) & t=(k+1)T_c\\ D_s\sin\lambda+l_{an}\cos\lambda & t=(k+1)T_c+T_d\end{cases} \quad (14)$$

-Third phase:

$$z_{a,\, transient}(t)=\begin{cases}h_{gs}+l_{an} & t=kT_c\\ D_s\sin\lambda+l_{an}\cos\lambda & t=kT_c+T_d\\ L_{ao}\sin\lambda+H_{ao}\cos\lambda & t=kT_c+T_m\\ (2D_s-l_{ab})\sin\lambda+l_{ab}\sin(q_f+\lambda)+l_{an}\cos(q_f+\lambda) & t=(k+1)T_c\\ 2D_s\sin\lambda+l_{an}\cos\lambda & t=(k+1)T_c+T_d\end{cases} \quad (15)$$

-Fourth phase:

$$z_{a,\, dec}(t)=\begin{cases}Dec_{st}D_s\sin\lambda+l_{an}\cos\lambda & t=kT_c\\ (Dec_{st}D_s+l_{af})\sin\lambda+l_{af}\sin(q_f-\lambda)+l_{an}\cos(q_b-\lambda) & t=kT_c+T_d\\ (Dec_{st}D_s+L_{ao})\sin\lambda+H_{ao}\cos\lambda & t=kT_c+T_m\\ ((Decst+2)D_s-l_{ab})\sin\lambda+l_{ab}\sin(q_f+\lambda)+\cdots & t=(k+1)T_c\\ l_{an}\cos(q_f+\lambda) & \\ (Dec_{st}+2)D_s\sin\lambda+l_{an}\cos\lambda & t=(k+1)T_c+T_d\end{cases} \quad (16)$$

-Fifth phase:

$$z_{a,\,tran.2}(t)=\begin{cases}(Dec_{st}+1)D_s\sin\lambda+l_{an}\cos\lambda & t=kT_c\\((Dec_{st}+1)D_s+l_{af})\sin\lambda+l_{af}\sin(q_b-\lambda)+l_{an}\cos(q_b-\lambda) & t=kT_c+T_d\\((k_{ch}-k_{ch1})D_s)\sin\lambda+H_{ao} & t=kT_c+T_m\\((k_{ch}-k_{ch1})D_s)\sin\lambda+l_{ab}\sin(q_f)+l_{an}\cos(q_f) & t=(k+1)T_c\\((k_{ch}-k_{ch1})D_s)\sin\lambda+l_{an} & t=(k+1)T_c+T_d\end{cases}\qquad(17)$$

-Six phase:

$$z_{a,\,tran.2}(t)=\begin{cases}(Dec_{st}+2)D_s\sin\lambda+l_{an}\cos\lambda & t=kT_c\\((Dec_{st}+2)D_s+l_{af})\sin\lambda+l_{af}\sin(q_b-\lambda)+l_{an}\cos(q_b-\lambda) & t=kT_c+T_d\\((k_{ch2}-k_{ch1})D_s)\sin\lambda+H_{ao} & t=kT_c+T_m\\((k_{ch}-k_{ch1})D_s)\sin\lambda+l_{ab}\sin(q_f)+l_{an}\cos(q_f) & t=(k+1)T_c\\((k_{ch2}-k_{ch1})D_s)\sin\lambda+l_{an} & t=(k+1)T_c+T_d\end{cases}\qquad(18)$$

-Seven phase:

$$z_{a,\,level\,2ground}(t)=\begin{cases}((k_{ch2}-k_{ch1})D_s)\sin\lambda+l_{an} & t=kT_c\\((k_{ch2}-k_{ch1})D_s)\sin\lambda+l_{af}\sin(q_b)+l_{an}\cos(q_b) & t=kT_c+T_d\\((k_{ch2}-k_{ch1})D_s)\sin\lambda+H_{ao} & t=kT_c+T_m\\((k_{ch}-k_{ch1})D_s)\sin\lambda+l_{ab}\sin(q_f)+l_{an}\cos(q_f) & t=(k+1)T_c\\((k_{ch2}-k_{ch1})D_s)\sin\lambda+l_{an} & t=(k+1)T_c+T_d\end{cases}\qquad(19)$$

-Eighth phase:

$$z_{a,\,stair,descending}(t)=\begin{cases}((k_{ch2}-k_{ch1})D_s)\sin\lambda+l_{an} & t=kT_c\\((k_{ch2}-k_{ch1})D_s)\sin\lambda+l_{af}\sin(q_b)+l_{an}\cos(q_b) & t=kT_c+T_d\\((k_{ch2}-k_{ch1})D_s)\sin\lambda+H_{ao} & t=kT_c+T_m\\((k_{ch}-k_{ch1})D_s)\sin\lambda+l_{ab}\sin(q_f)+l_{an}\cos(q_f)+h_s & t=(k+1)T_c\\((k_{ch2}-k_{ch1})D_s)\sin\lambda+l_{an}+h_s & t=(k+1)T_c+T_d\end{cases}\qquad(20)$$

-Nine phase:

$$z_{a,\,stair,descending}(t)=\begin{cases}((k_{ch2}-k_{ch1})D_s)\sin\lambda+l_{an} & t=kT_c\\((k_{ch2}-k_{ch1})D_s)\sin\lambda+l_{af}\sin(q_b)+l_{an}\cos(q_b) & t=kT_c+T_d\\((k_{ch2}-k_{ch1})D_s)\sin\lambda+H_{ao} & t=kT_c+T_m\\((k_{ch}-k_{ch1})D_s)\sin\lambda+l_{ab}\sin(q_f)+\cdots & t=(k+1)T_c\\l_{an}\cos(q_f)+2h_s & \\((k_{ch2}-k_{ch1})D_s)\sin\lambda+l_{an}+2h_s & t=(k+1)T_c+T_d\end{cases}\qquad(21)$$

-Tenth phase:

$$z_{a,stair,descending}(t)=\begin{cases}((k_{ch2}-k_{ch1})D_s)\sin\lambda+l_{an}-St_{con}h_s & t=kT_c\\ ((k_{ch2}-k_{ch1})D_s)\sin\lambda+l_{af}\sin(q_b)+\cdots & t=kT_c+T_d\\ l_{an}\cos(q_b)+St_{con}h_s & \\ ((k_{ch2}-k_{ch1})D_s)\sin\lambda-(St_{con}+1)h_s+H_s & t=kT_c+T_m\\ ((k_{ch2}-k_{ch1})D_s)\sin\lambda+l_{ab}\sin(q_f)+\cdots & t=(k+1)T_c\\ l_{an}\cos(q_f)+(St_{con}+2)h_s & \\ ((k_{ch2}-k_{ch1})D_s)\sin\lambda+l_{an}-(St_{con}+2)h_s & t=(k+1)T_c+T_d\end{cases} \quad (22)$$

In all the obtained relations, l_{ab}, l_{af} and l_{an} indicate the foot configuration as displayed in Fig. 4. H_s, h_s and St_{con} indicate the stair height, foot's maximum height measured from the stair level and the step number of the robot over the stair. The trajectory path of the hip follows the above utilized procedure with respect to walking of the robot phases (Mousavi & Bagheri, 2007). The applicable constraints of the ankle and hip joints have been discussed in (Mousavi & Bagheri, 2007).

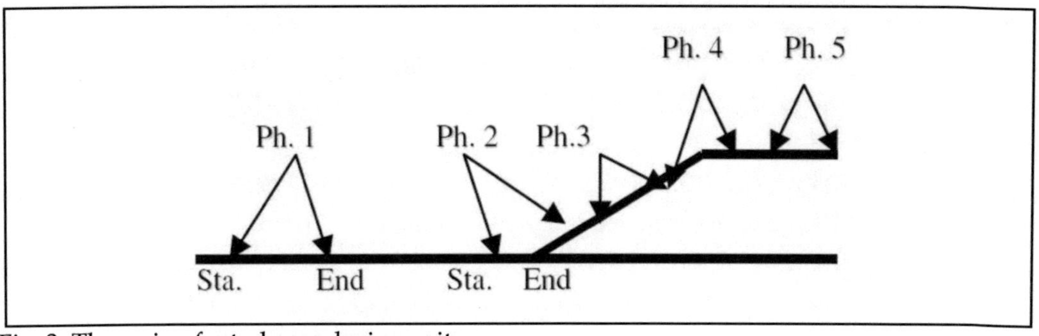

Fig. 3. The swing foot phases during gait.

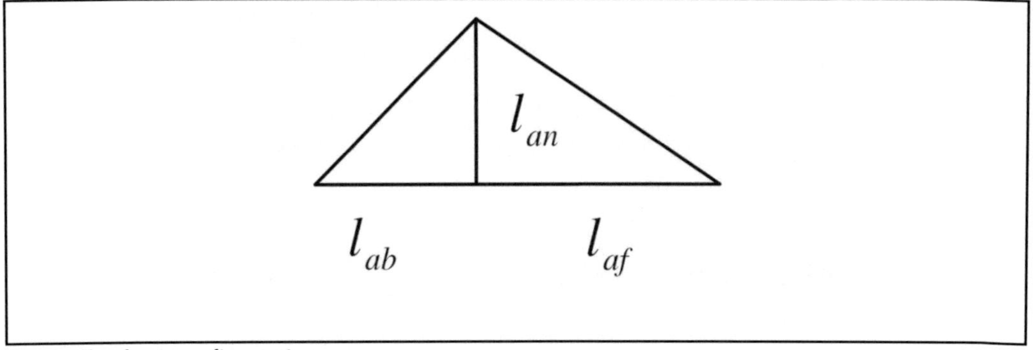

Fig. 4. The foot configuration.

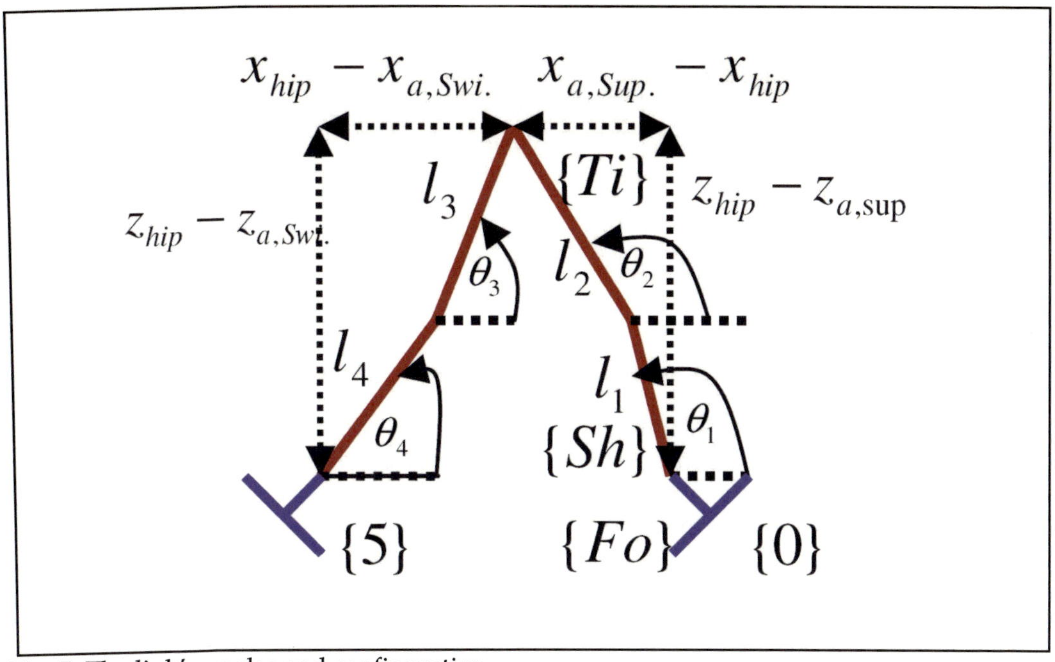

Fig. 5. The link's angles and configuration.

Now, the kinematic parameters will be obtained with respect to the above mentioned combined trajectory paths combined with the domain of the nonlinear equations (see Fig. 5). The nonlinear equations can be obtained as follows:

For support legs:

$$\begin{aligned} l_1 \cos(\pi - \theta_1) + l_2 \cos(\pi - \theta_2) = a \\ l_1 \sin(\pi - \theta_1) + l_2 \sin(\pi - \theta_2) = b \end{aligned} \tag{23}$$

For swing legs:

$$\begin{aligned} l_3 \cos(\theta_3) + l_4 \cos(\theta_4) = c \\ l_3 \sin(\theta_3) + l_4 \sin(\theta_4) = d \end{aligned} \tag{24}$$

where

$a = x_{a,sup} - x_{hip}$

$b = z_{hip} - z_{a,sup}$

$c = x_{hip} - x_{a,swing}$

$d = z_{hip}, z_{a,swing}$

With the aid of the written programs and designed software, the above nonlinear equations based on the gait parameters are solved and also the link's angles are obtained. The kinematic parameters of the robot for single phase of the walking can be found in (Mousavi & Bagheri, 2007; Mousavi, 2006).

3. Dynamic investigations

With the biped's motion an important stability criteria (in similarities to the human gait) is defined using the zero moment point (ZMP). The ZMP is a point on the ground about which the sum of all the moments around is equal to zero. The ZMP formula is written as follows (Huang et al., 2001):

$$x_{zmp} = \frac{\sum_1^n m_i (g\cos\lambda + \ddot{z}_i)x_i - \sum_1^n m_i (g\sin\lambda + \ddot{x}_i)z_i - \sum_1^{n\ i} I_i \ddot{\theta}_i}{\sum_1^n m_i (g\cos\lambda + \ddot{z}_i)} \tag{25}$$

where $\ddot{x}_i, \ddot{z}_i$ are the vertical and horizontal acceleration of the mass center of link (i) with respect to the fixed coordinate system (which is on the support foot). $\ddot{\theta}_i$ is the angular acceleration of link (i) obtained from the interpolation process and k denotes the slope of the surface. Principally, two types of ZMP are defined: (a) moving ZMP and (b) fixed ZMP.
The moving ZMP of the robot is similar to that for the human gait (Mousavi & Bagheri, 2007). In the fixed type, the ZMP position is restricted through the support feet or the user's selected areas. Consequently, the significant torso's modified motion is required for stable walking of the robot. For the process here, the software has been designed to find the target angle of the torso for providing the fixed ZMP position automatically. In the designed software, q_{torso} shows the deflection angle of the torso determined by the user or calculated by the auto detector module of the software. Note that in the auto detector, the torso's motion needed for obtaining the mentioned fixed ZMP will be extracted with respect to the desired ranges. The desired ranges include the defined support feet area by the users or is determined automatically by the designed software. Note that the most affecting parameters for obtaining the robot's stable walking are the hip's height and position. By varying the parameters with an iterative method for x_{ed}, x_{sd} (Huang et al., 2001) and choosing the optimum hip height, the robot control process with respect to the torso's modified angles and the mentioned parameters can be performed. To obtain the joint's actuator torques, Lagrange equations (John, 1989) have been used at the single support phase as follows:

$$\tau_i = H(q)\ddot{q} + C(q,\dot{q})\dot{q} + G(q_i),$$

where i = 0, 2, . . ., 6 and H, C, G are mass inertia, coriolis and gravitational matrices of the system which can be written as follows:

$$H(q) = \begin{bmatrix} h_{11} & h_{12} & h_{13} & h_{14} & h_{15} & h_{16} & h_{17} \\ h_{21} & h_{22} & h_{23} & h_{24} & h_{25} & h_{26} & h_{27} \\ h_{31} & h_{32} & h_{33} & h_{34} & h_{35} & h_{36} & h_{37} \\ h_{41} & h_{42} & h_{43} & h_{44} & h_{45} & h_{46} & h_{47} \\ h_{51} & h_{52} & h_{53} & h_{54} & h_{55} & h_{56} & h_{57} \\ h_{61} & h_{62} & h_{63} & h_{64} & h_{65} & h_{66} & h_{67} \end{bmatrix} \qquad C(q,\dot{q}) = \begin{bmatrix} c_{11} & c_{12} & c_{13} & c_{14} & c_{15} & c_{16} & c_{17} \\ c_{21} & c_{22} & c_{23} & c_{24} & c_{25} & c_{26} & c_{27} \\ c_{31} & c_{32} & c_{33} & c_{34} & c_{35} & c_{36} & c_{37} \\ c_{41} & c_{42} & c_{43} & c_{44} & c_{45} & c_{46} & c_{47} \\ c_{51} & c_{52} & c_{53} & c_{54} & c_{55} & c_{56} & c_{57} \\ c_{61} & c_{62} & c_{63} & c_{64} & c_{65} & c_{66} & c_{67} \end{bmatrix}$$

$$G(q) = \begin{bmatrix} G_1 \\ G_2 \\ G_3 \\ G_4 \\ G_5 \\ G_{TOR} \end{bmatrix}$$

The most important point of the double support phase signifies the occurrence of the impact between the swing leg and the ground. Due to presence of the reaction force of the ground, Newton's equations must be employed for determination of the reaction force applied through the double support phase ((Huang et al., 2001; Lum et al., 1999; Eric, 2003). The method of (Huang et al., 2001) for simulation of the ground reaction force has been used for the inverse dynamics. Now, we have chosen an impeccable method involved slight deviations for dynamical analysis of the robot included the Lagrangian and Newtonian relations. The components of the matrices are complex and the detailed mathematical relations can be found in (Mousavi, 2006).

l_{Sh}	l_{Ti}	l_{To}	l_{an}	l_{ab}	l_{af}
0.3m	0.3m	0.3m	0.1m	0.1m	0.13m
m_{Sh}	m_{Th}	m_{To}	m_{Fo}	D_s	T_c
5.7kg	10kg	43kg	3.3kg	0.5m	0.9s
T^d	T_m	H_{ao}	L_{ao}	x_{ed}	x_{sd}
0.18s	0.4s	0.16m	0.4m	0.23m	0.23m
g_{gs}	g_{gf}	H_{min}	H_{max}	h_s	H_s
0	0	0.60m	0.62m	0.1m	0.15m
I_{shank}		I_{tight}		I_{torso}	I_{foot}
0.02kgm^2		0.08kgm^2		1.4kgm^2	0.01kgm^2
k_{Ch}		k_{Ch1}		k_{Ch2}	
2		5		7	

Table 1. The simulated robot specifications

The following conditions must be applied during the simulation process:

$if\ k \le k_{ch1}\ and\ k = k_{ch2} + 1 \qquad \lambda = 0$

$if\ k > k_{ch1}\ and\ k < k_{ch2} + 1 \qquad \lambda = The\ user\ defined\ value$

Dec_{st} the number of robot's steps over the slope

k_{Ch} the number of steps that the robot changes during the walking process from the ground to slope

k_{Ch1} the number of steps that the robot changes during the walking process from slope to the ground

k_{Ch2} the number of steps that the robot changes during the walking process from the ground to stair

The ranges of the fixed ZMP are selected with respect to the descending and ascending surfaces respectively as follows:

$-0.05m \le X_{zmp} \le 0.26$

$-0.1m \le X_{zmp} \le 0.26$

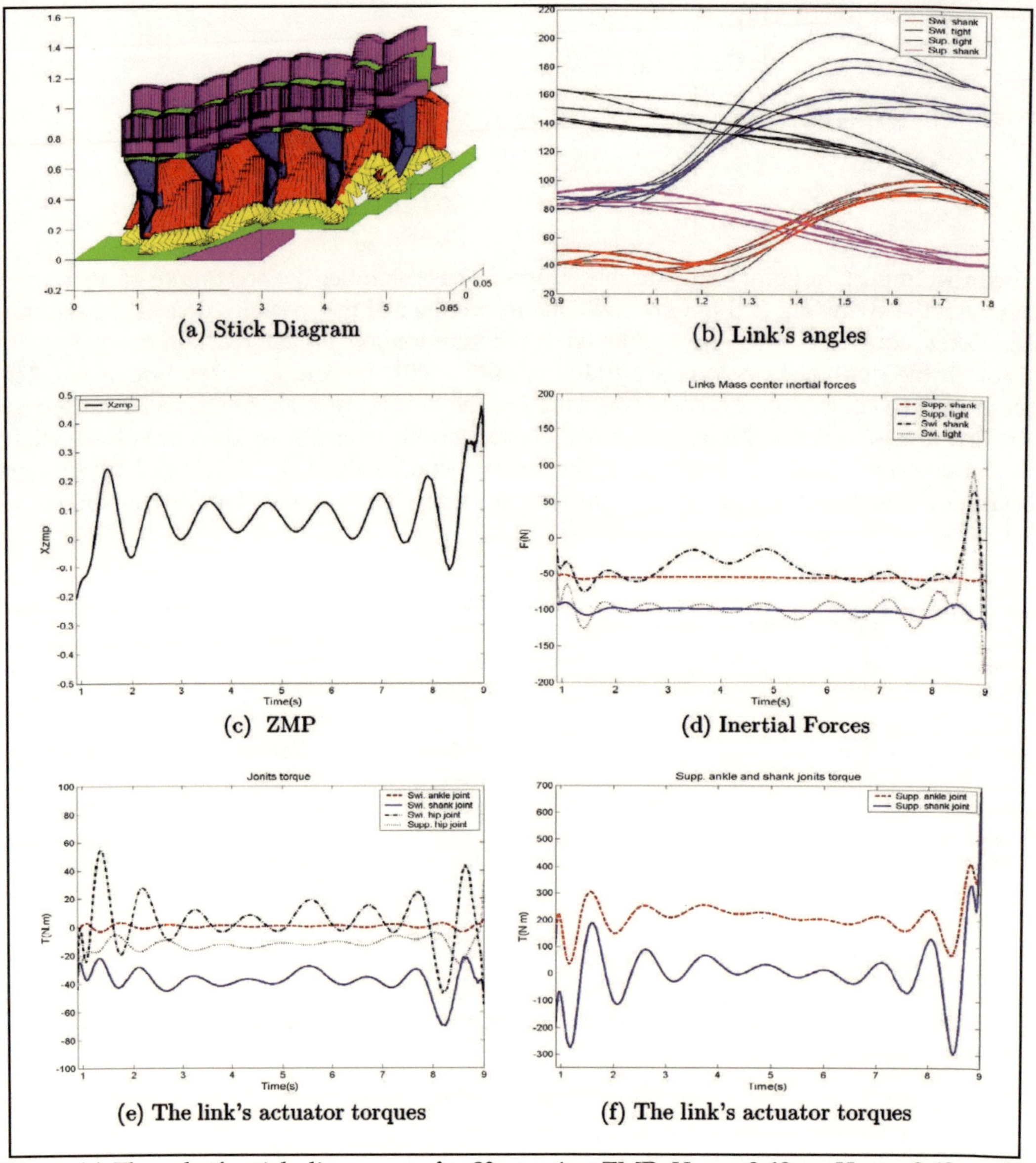

Fig. 6. (a) The robot's stick diagram on λ= 8°, moving ZMP, H_{min} = 0.60 m, H_{max} = 0.62 m; (b) the Link's angles during combined trajectory paths; (c) the moving ZMP diagram in nominal gait which satisfies stability criteria; (d) Inertial forces: (—) supp. thigh, (- - -) supp. shank, (…) swing thigh, (_._.) swing shank; (e) joint's torques: (—) swing shank joint, (- - -) swing ankle joint, (…) supp. hip joint, (_._.) swing hip joint; (f) joint's torques: (-) supp. Ankle joint, (- - -) supp. shank joint.

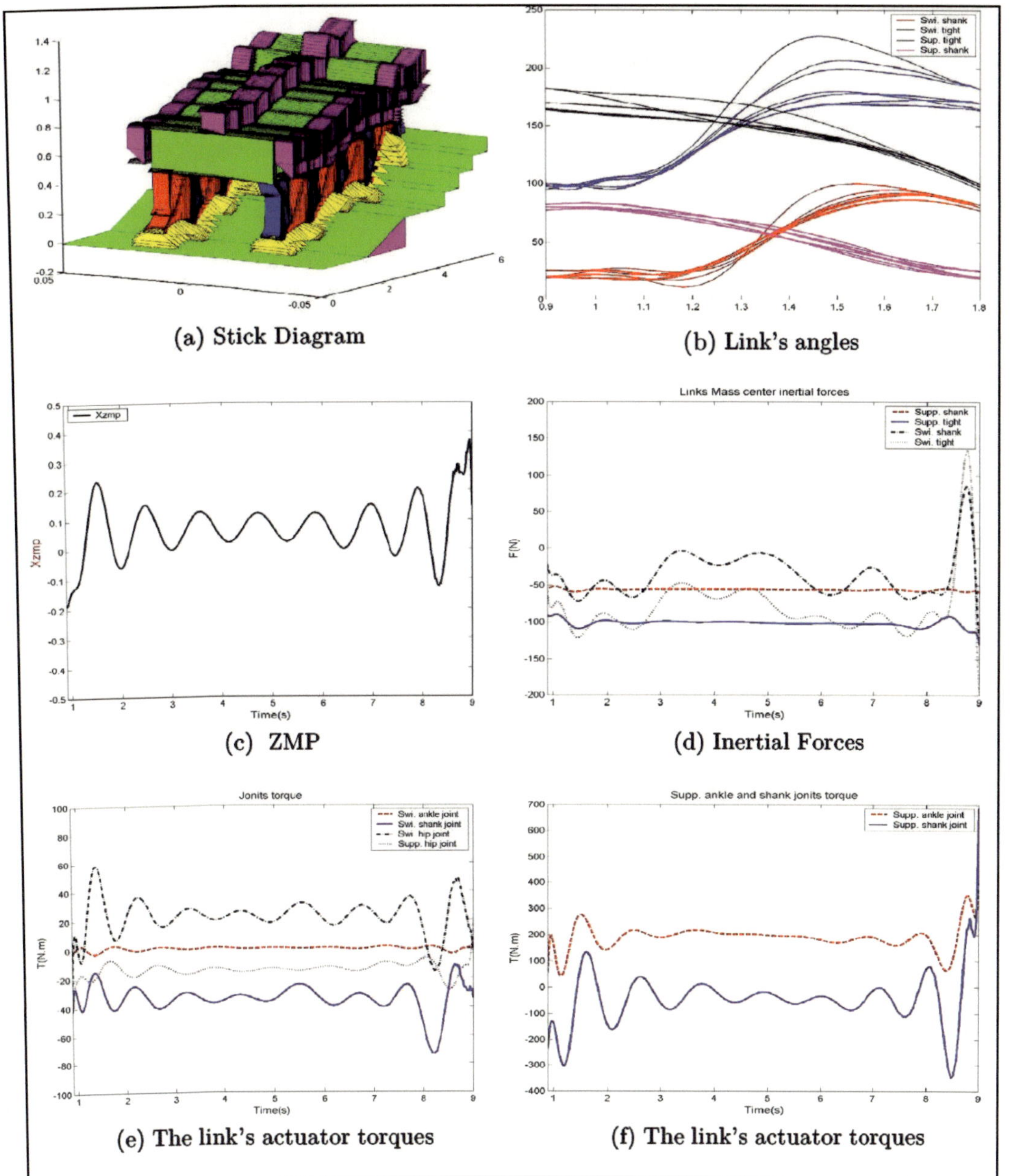

Fig. 7. (a) The robot's stick diagram on λ= 8°, moving ZMP, H_{min} = 0.5 m, H_{max} = 0.52 m. (b) The Link's angles during combined trajectory paths. (c) The moving ZMP diagram in nominal gait which satisfies stability criteria. (d) Inertial forces: (—) supp. thigh, (- - -) supp. shank, (…) swing thigh, (_._.) swing shank. (e) Joint's torques (—) swing shank joint, (- - -) swing ankle joint, (…) supp. hip joint, (_._.) swing hip joint. (f) Joint's torques: (—) supp. ankle joint, (- - -) supp. shank joint.

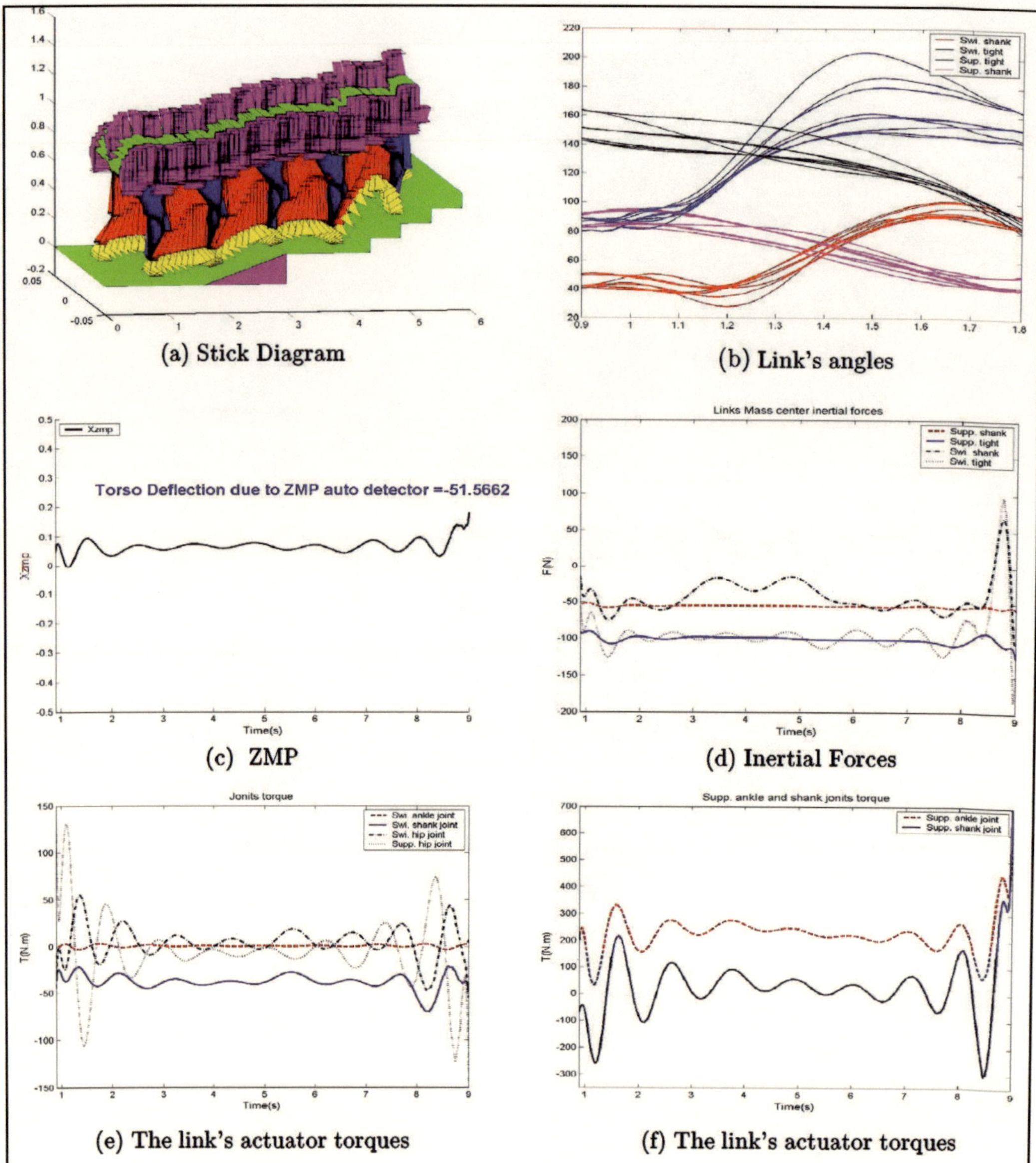

Fig. 8. (a) The robot's stick diagram on λ= 8°, fixed ZMP, H_{min} = 0.6 m, H_{max} = 0.62 m. (b) The Link's angles during combined trajectory paths. (c) The fixed ZMP diagram in nominal gait which satisfies stability criteria. (d) Inertial forces: (—) supp. thigh, (- - -) supp. shank, (...) swing thigh, (_._.) swing shank. (e) Joint's torques (—) swing shank joint, (- - -) swing ankle joint, (...) supp. hip joint, (_._.) swing hip joint. (f) Joint's torques: (—) supp. ankle joint, (- - -) supp. shank joint.

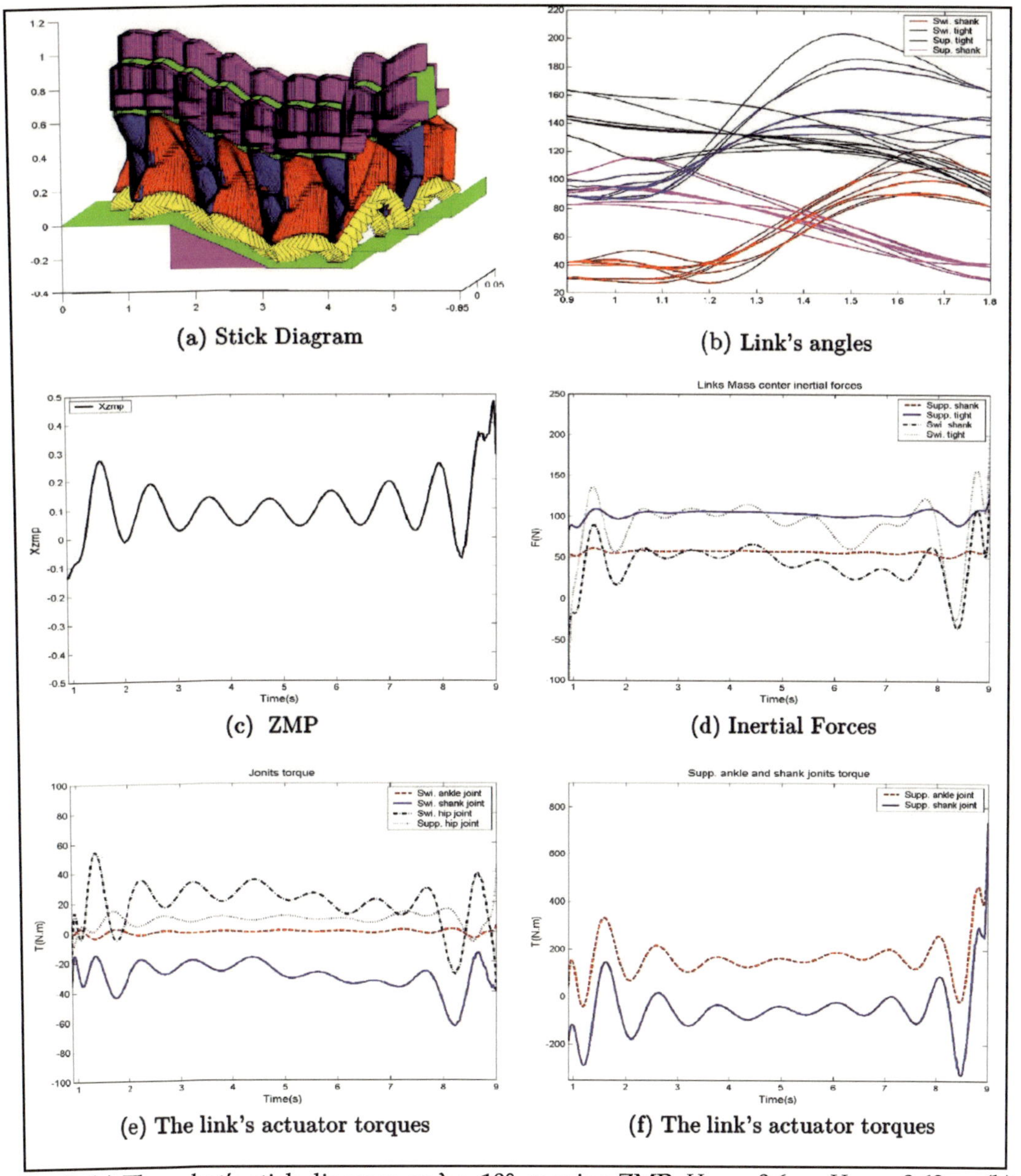

Fig. 9. (a) The robot's stick diagram on λ= -10°, moving ZMP, H_{min} = 0.6 m, H_{max} = 0.62 m. (b) The Link's angles during combined trajectory paths. (c) The moving ZMP diagram in nominal gait which satisfies stability criteria. (d) Inertial forces: (—) supp. tight, (- - -) supp. shank, (...) swing thigh, (_._._) swing shank. (e) Joint's torques: (—) swing shank joint, (- - -) swing ankle joint, (...) supp. hip joint, (_._._) swing hip joint. (f) Joint's torques: (—) supp. ankle joint, (- - -) supp. shank joint.

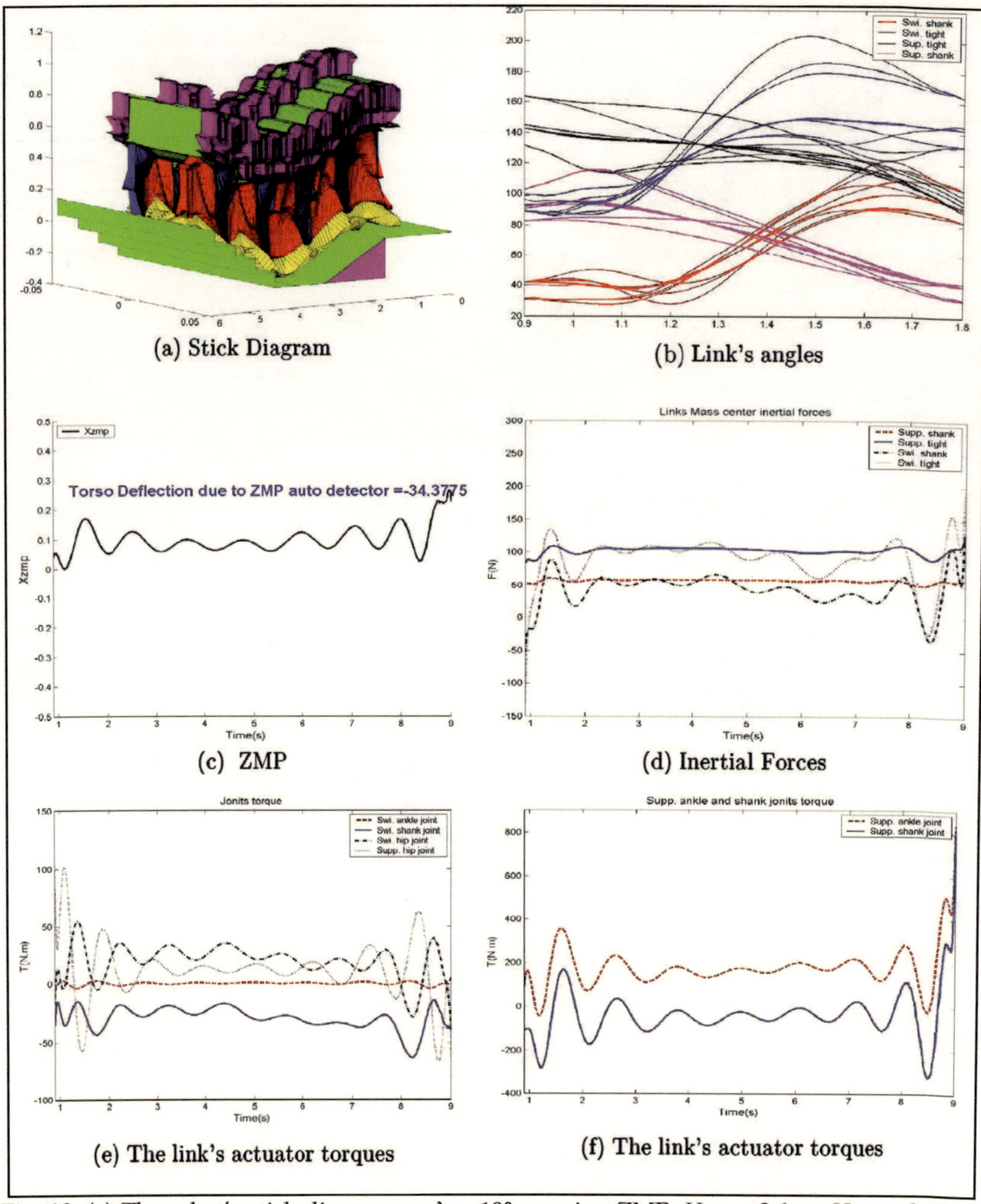

Fig. 10. (a) The robot's stick diagram on λ= -10°, moving ZMP, H_{min} = 0.6 m, H_{max} = 0.62 m. (b) The Link's angles during combined trajectory paths. (c) The fixed ZMP diagram in nominal gait which satisfies stability criteria. (d) Inertial forces: (—) supp. thigh, (- - -) supp. shank, (…) swing thigh, (_._._.) swing shank. (e) Joint's torques: (—) swing shank joint, (- - -) swing ankle joint, (…) supp. hip joint, (_._._.) swing hip joint. (f) Joint's torques: (—) supp. ankle joint, (- - -) supp. shank joint.

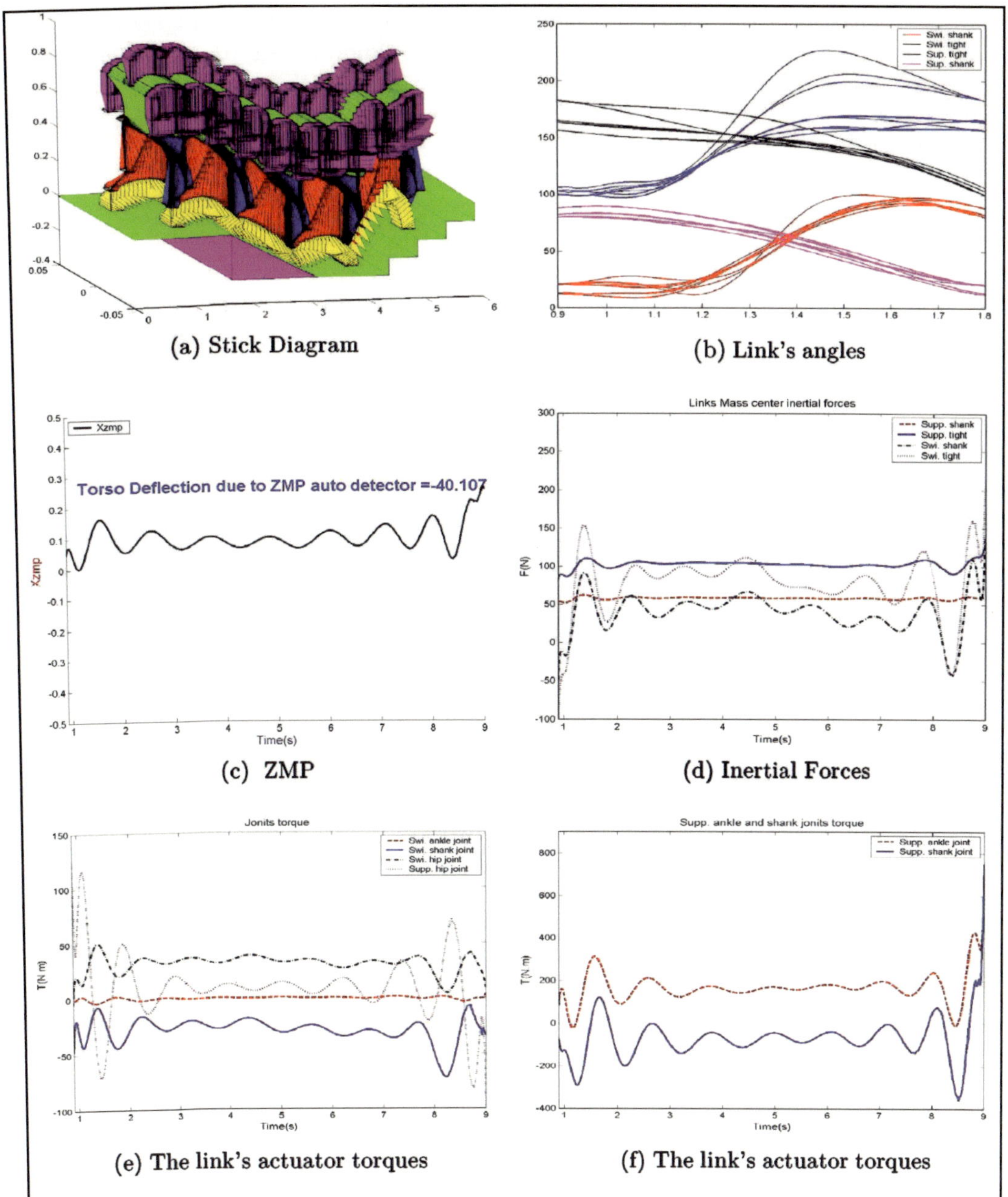

Fig. 11. (a) The robot's stick diagram on λ= -10°, fixed ZMP, H_{min} = 0.5 m, H_{max} = 0.52 m. (b) The Link's angles during combined trajectory paths. (c) The fixed ZMP diagram in nominal gait which satisfies stability criteria. (d) Inertial forces: (—) supp. thigh, (- - -) supp. shank, (...) swing thigh, (-.-.-) swing shank. (e) Joint's torques: (—) swing shank joint, (- - -) swing ankle joint, (...) supp. hip joint, (-.-.-) swing hip joint. (f) Joint's torques: (—) supp. ankle joint, (- - -) supp. shank joint.

In the designed software, these methods are used to simulate the robot including AVI (audio and video interface) files for each identified condition by the users. Differentiating and also using the mathematical methods in the program, the angular velocities and accelerations of the robot's links are calculated to use in the ZMP, Lagrangian and Newtonian equations Table 1.

4. Simulation results

For the described process, the software has been designed based on the cited mathematical methods for simulation of a seven link biped robot. Because of the very high precision of third-order spline method, this method has been applied to calculate the trajectory paths of the robot. The result is 14,000 lines of program in the MATLAB/SIMULINK environment for simulation and stability analysis of the biped robot. By choosing the type of the ZMP in the Fixed and Moving modes, stability analysis of the robot can be judged easily. For the fixed type of ZMP, the torso's modified motion has been regarded to be identical with respect to various phases of the robot's motion. The results have been displayed in Figs. 6–11. Figs. 6–8 present the combined trajectory paths for nominal and non-nominal (with changed hip heights from nominal values) walking of the robot over ascending surfaces. Figs. 9–11 present the same types of walking process over descending surfaces. Both ZMPs have been displayed and their effects on the joint's actuator torques are presented. The impact of swing leg and the ground has been included in the designed software (Huang et al., 2001; Lum et al., 1999; Hon et al., 1978).

5. Conclusion

In this chapter, simulation of combined trajectory paths of a seven link biped robot over various surfaces has been presented. We have focused on generation of combined trajectory paths with the aid of mathematical interpolation. The inverse kinematic and dynamic methods have implemented for providing the robot combined trajectory paths in order to obtain a smooth motion of the robot. This procedure avoids the link's velocity discontinuities of the robot in order to mitigate the occurrence of impact effects and also helps to obtain a suitable control process. The sagittal movement of the robot has been investigated while 3D simulations of the robot are presented. From the presented simulations, one can observe important parameters of the robot with respect to stability treatment and optimum driver torques. The most important factor is the hip height measured from the fixed coordinate system. As can be seen from Fig. 7f, the support knee needs more actuator torque than the value of the non-nominal gait (with lower hip height measured from the fixed coordinate system). This point can be seen in Figs. 8f and 10f. This is due to the robot's need to bend its knee joint more at a lower hip position. The role of the hip height is considerable over the torso's modified motion for obtaining the desired fixed ZMP position. With respect to Figs. 10c and 11c, the robot with the lower hip height needs more modified motion of its torso to satisfy the defined ranges of ZMP by the users. The magnitude of the torso's modified motion has drastic effects upon the control process of the robot. Assuming control process of an inverse pendulum included a stagnant origin will present relatively sophisticated control process for substantial deflection angle of pendulum. Note that the torso motion in a biped (as an inverted pendulum) includes both the rotational and translational movements which will complicate the process of control. Consequently, the lower torso's modified motion is desirable which can be derived from the higher hip height. Therefore, the hip height plays an important role in both the stability and optimum actuator torques of the joints. Meanwhile, the higher hip height will avoid the link's velocity discontinuities.

6. References

Bagheri, A. & Mousavi, P. N. (2007). Dynamic Simulation of Single and Combined Trajectory Path Generation and Control of A Seven Link Biped Robot, In: *Humanoid Robots New Developments,* Armando Carlos de Pina Filho, (Ed.), 89-120, Advanced Robotics Systems International and I-Tech, ISBN 978-3-902613-00-4, Vienna Austria.

Chevallereau, C.; Formal'sky, A. & Perrin, B. (1998). Low Energy Cost Reference Trajectories for a Biped Robot, in: Proc. IEEE Int. Conf. Robotics and Automation, 1998, pp. 1398–1404.

Dasgupta, A. & Nakamura, Y. (1999). Making Feasible Walking Motion of Humanoid Robots from Human Motion Capture Data, in: Proc. IEEE Int. Conf. Robotics and Automation, pp. 1044–1049.

Hirai, K; Hirose, M.; Haikawa, Y. & Takenaka, T. (1998). The Development of Honda Humanoid Robot, in: Proc. IEEE Int. Conf. Robotics and Automation, pp. 1321–1326.

Huang, Q.; Yokoi, K.; Kajita, S.; Kaneko, K.; Arai, H.; Koyachi, N. & Tanie, K. (2001). Planning Walking Patterns For A Biped Robot, IEEE Trans. Robot. Automat. 17 (3).

Hon, H.; Kim, T. & Park, T. (1978). Tolerance Analysis of a Spur Gear Train, in: Proc. Third DADS Korean User's Conf, pp. 61–81.

John, J. G. (1989) *Introduction to Robotics: Mechanics and Control*, Addison-Wesley.

Lum, H. K.; Zribi, M. & Soh, Y. C. (1999). Planning and Contact of A Biped Robot, *Int. J. Eng. Sci*. 37 -1319–1349.

McGeer, T. (1990). Passive walking with knees, in: Proc. IEEE Int. Conf. Robotics and Automation, pp. 1640–1645.

Mousavi, P. N. (2006). Adaptive Control of 5 DOF Biped Robot Moving on a Declined Surface, M.S. Thesis, Guilan University.

Mousavi, P. N. & Bagheri, A. (2007). Mathematical Simulation of A Seven Link Biped Robot on Various Surfaces and ZMP Considerations, *Applied Mathematical Modelling*, vol. 31/1, Elsevier, pp. 18–37.

Shih, C. L; Li, Y. Z.; Churng, S.; Lee, T. T. & Cruver, W. A. (1990). Trajectory Synthesis And Physical Admissibility For A Biped Robot During The Single Support Phase, in: Proc. IEEE Int. Conf. Robotics and Automation, pp. 1646–1652.

Shih, C. (1997). Gait Synthesis For A Biped Robot, *Robotica,* 15, 599–607.

Shih, C. L. (1999). Ascending And Descending Stairs For A Biped Robot, IEEE Trans. Syst. Man. Cybern. A 29 (3) 255–268.

Silva, F.M. & Machado, J. A. T. (1999). Energy Analysis during Biped Walking, in: Proc. IEEE Int. Conf. Robotics and Automation, pp. 59–64.

Takanishi, A.; Ishida, M.; Yamazaki, Y. & Kato, I. (1985). The Realization of Dynamic Walking Robot WL-10RD, in: Proc. Int. Conf. Advanced Robotics, pp. 459–466.

Westervelt, E. R. (2003). Toward A Coherent Framework for the Control of Plannar Biped Locomotion, A Dissertation Submitted in Partial Fulfilment of the Requirements for the Degree of Doctor of Philosophy, (Electrical Engineering Systems), the University of Michigan.

Zarrugh, M. Y. & Radcliffe, C.W. (1979). Computer Generation of Human Gait Kinematics, *J. Biomech.* 12, 99–111.

Zheng, Y. F. & Shen, J. (1990). Gait Synthesis for the SD-2 Biped Robot to Climb Sloping Surface, IEEE Trans. Robot. Automat. 6, 86–96.

11

Bipedal Walking Control based on the Assumption of the Point-contact: Sagittal Motion Control and Stabilization

Tadayoshi Aoyama[1], Kosuke Sekiyama[1],
Yasuhisa Hasegawa[2] and Toshio Fukuda[1]
[1]Nagoya University, [2]University of Tsukuba
Japan

1. Introduction

In the age of an aging society, the prospective role of robots is turning gradually from just working machines to do monotonous work in a factories to partners who support human life. In recent years, a lot of autonomous humanoid robots have been actually realized (Hirai et al. (1998); Kaneko et al. (2008)). These robots can walk on two legs stably by means of the control based on ZMP (Zero Moment Point). ZMP (Vukobratovic & Borovac (2004)), the indicator of the stability of biped walking, is a point on the floor where the torque generated by both inertial and gravitational forces becomes zero. That is, using ZMP-based control to realize stable walking makes sense, thus a number of researches of ZMP-based control have been presented (Nishiwaki et al. (2002); Takanishi et al. (1985)). However, in terms of the practical use of humanoid robots, these controllers based on ZMP have a problem in terms of the run-time of the battery since ZMP-based method does not take advantage of the robot inherent dynamics.

In order to achieve natural and energy efficient biped walking, many control methods based on robot dynamics had been proposed up to this day. As one of such methods, some researchers presented the control methods to take advantage of robot dynamics directly by use of point-contact state between a robot and the ground (Furusho & Sano (1990); Goswami et al. (1997); Grishin et al. (1994); Kuo (1999); Nakanishi et al. (2004); Ono et al. (2004)). Miura et al. produced the point-contact biped robot like stilt and realize dynamic walking by means of stabilizing control to change the configuration at foot-contact (Miura & Shimoyama (1984)). Kajita et al. proposed the control and stabilizing method based on the conserved quantity derived by designing the COG trajectory parallel to the ground (Kajita et al. (1992)). Chevallereau presented the control to converge robot dynamics on optical trajectory by introducing the virtual time (Chevallereau (2003)). Grizzle and Westervelt et al. built the controller by use of the virtual holonomi constraint of joints named *virtual constraint* realize stable dynamic walking by means of the biped robot with a torso (Grizzle et al. (2001); Westervelt et al. (2004)).

As one of point-contact methods, Doi et al. proposed Passive Dynamic Autonomous Control (PDAC) previously (Doi et al. (2004b)). PDAC expresses the robot dynamics as an one-dimensional autonomous system based on the two concepts: 1) point-contact 2) *virtual con-*

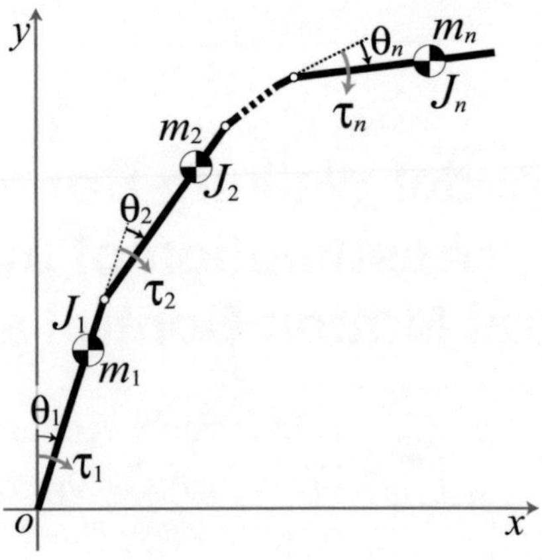

Fig. 1. Mechanical model of the serial n-link rigid robot. θ_i and τ_i are the angle and the torque of ith joint respectively. m_i and J_i are the mass and the moment of inertia of ith link respectively.

straint (proposed by Grizzle and Westervelt et al. (Grizzle et al. (2001); Westervelt et al. (2004))). In this chapter, we design the sagittal motion controller by applying PDAC to sagittal motion. In addition, we find the convergence domain of the proposed controller and prove the stability by the Liapunov Theory. Finally, 3-D dynamic walking based on the robot inherent dynamics is realized by coupling the sagittal motion proposed in this chapter and lateral motion proposed previously (Doi et al. (2004a)).

2. Passive Dynamic Autonomous Control

2.1 Converged dynamics

As mentioned previously, PDAC is base on the two concepts, i.e. point-contact and *virtual constraint*. Point-contact denotes that a robot contacts the ground at a point, that is, the first joint is passive. *virtual constraint* was defined by Grizzle and Westervelt et al. (Grizzle et al. (2001); Westervelt et al. (2004)) as a set of holonomic constraints on the robot's actuated DoF parameterized by the robot's unactuated DoF. Assuming that PDAC is applied to the serial n-link rigid robot shown in Fig. 1, these two premises are expressed as follows:

$$\tau_1 = 0, \tag{1}$$

$$\begin{aligned}\boldsymbol{\Theta} &= [\theta_1, \theta_2, \cdots, \theta_n]^T = [f_1(\theta), f_2(\theta), \cdots, f_n(\theta)]^T \\ &:= \boldsymbol{f}(\theta),\end{aligned} \tag{2}$$

where θ is the angle around the contact point in the absolute coordinate system, that is, $\theta_1 = f_1(\theta) = \theta$.

The dynamic equations of this model are given by

$$\frac{d}{dt}\left(\boldsymbol{M}(\boldsymbol{\Theta})\dot{\boldsymbol{\Theta}}\right) - \frac{1}{2}\frac{\partial}{\partial \boldsymbol{\Theta}}\left(\dot{\boldsymbol{\Theta}}^T \boldsymbol{M}(\boldsymbol{\Theta})\dot{\boldsymbol{\Theta}}\right) - G(\boldsymbol{\Theta}) = \boldsymbol{\tau}, \tag{3}$$

where $\boldsymbol{M}(\boldsymbol{\Theta}) := [\boldsymbol{m}_1(\boldsymbol{\Theta})^T, \boldsymbol{m}_2(\boldsymbol{\Theta})^T, \cdots, \boldsymbol{m}_n(\boldsymbol{\Theta})^T]^T$, $\boldsymbol{\Theta} := [\theta_1, \theta_2, \cdots, \theta_n]^T$, $\boldsymbol{G}(\boldsymbol{\Theta}) := [G_1(\boldsymbol{\Theta}), G_2(\boldsymbol{\Theta}), \cdots, G_n(\boldsymbol{\Theta})]^T$, $\boldsymbol{\tau} := [\tau_1, \tau_2, \cdots, \tau_n]^T$, $\frac{\partial}{\partial \boldsymbol{\Theta}} = [\frac{\partial}{\partial \theta_1}, \frac{\partial}{\partial \theta_2}, \cdots, \frac{\partial}{\partial \theta_n}]^T$.

Since in this model the dynamic equation around the contact point has no term of the Coriolis force, it is given as

$$\frac{d}{dt}\left(\boldsymbol{m}_1(\boldsymbol{\Theta})^T\dot{\boldsymbol{\Theta}}\right) - G_1(\boldsymbol{\Theta}) = \tau_1. \tag{4}$$

By differentiating Eq. (2) with respect to time, the following equation is acquired,

$$\dot{\boldsymbol{\Theta}} = \frac{\partial \boldsymbol{f}(\theta)}{\partial \theta}\dot{\theta} = \left[\frac{\partial f_1(\theta)}{\partial \theta}, \frac{\partial f_2(\theta)}{\partial \theta}, \cdots, \frac{\partial f_n(\theta)}{\partial \theta}\right]^T \dot{\theta}. \tag{5}$$

Substituting Eqs. (1), (2), and (5) into Eq. (3) yields the following dynamic equation,

$$\frac{d}{dt}\left(M(\theta)\dot{\theta}\right) = G(\theta), \tag{6}$$

where

$$M(\theta) := \boldsymbol{m}_1\left(\boldsymbol{f}(\theta)\right)^T \frac{d\boldsymbol{f}(\theta)}{d\theta} \tag{7}$$

$$G(\theta) := G_1\left(\boldsymbol{f}(\theta)\right). \tag{8}$$

By multiplying both sides of Eq. (6) by $M(\theta)\dot{\theta}$ and integrating with respect to time, the dynamics around the contact point is obtained as follows:

$$\int \left(M(\theta)\dot{\theta}\right)\frac{d}{dt}\left(M(\theta)\dot{\theta}\right) dt = \int M(\theta)G(\theta)\dot{\theta}\, dt \tag{9}$$

$$\Longleftrightarrow \frac{1}{2}\left(M(\theta)\dot{\theta}\right)^2 = \int M(\theta)G(\theta)\, d\theta. \tag{10}$$

Therefore, the whole robot dynamics is expressed as the following one-dimensional autonomous system (that is, the phase around contact point),

$$\dot{\theta} = \frac{1}{M(\theta)}\sqrt{2\int M(\theta)G(\theta)\, d\theta} \tag{11}$$

$$:= \frac{1}{M(\theta)}\sqrt{2(D(\theta)+C)} \tag{12}$$

$$:= F(\theta). \tag{13}$$

In this chapter, we term Eqs. (12) and (13) Converged dynamics.

2.2 PDAC Constant

Since Converged dynamics is autonomous, in addition, independent of time, it is considered as a conservative system. The integral constant in right side of Eq. (10), C, is a conserved quantity, which is termed PDAC Constant. Its value is decided according to the initial condition (as for biped walking, the state just after foot-contact), and kept constant during a cycle of motion. Thus, it is possible to stabilize the motion by keeping PDAC Constant at certain value.

The dimension of PDAC Constant is equal to the square of angular momentum and has relevance to it. As is well know, assuming that the robot shown in Fig. 1 is placed on its side, the angular momentum around contact point is conserved since there is no effect of gravitational force on the robot dynamics. In this condition, it is clear that $M(\theta)\dot{\theta}$(angular momentum)$= \sqrt{2C}$ from Eq. (12), since $G(\boldsymbol{\Theta}) = 0$ in Eq. (3) hence $D(\theta) = 0$. Note that, although angular momentum is not conserved in the condition which robot dynamics is affected by the gravitational force, PDAC Constant is conserved since it includes the effects of the gravitation. This chapter demonstrates the convergence of PDAC Constant by the Lyapunov theory and proves the stability of walking.

3. 3D biped walking

In this section, control architecture of 3-D biped walking is summarize.

3.1 Sagittal motion

3.1.1 3-link model

For the sake of simplicity, in this chapter upper body of a robot is not moved, hence the 3-link model as shown in Fig. 2 is employed. The dynamic equation of this model is described as Eq. (3) and that of the ankle joint of the stance leg is Eq. (4) where $n = 3$. The left-hand side of Eq. (4) is described as follows:

$$\begin{aligned}
M_{11}(\boldsymbol{\Theta}) &= J_1 + J_2 + J_3 + m_1 a_1^2 + m_2 l_1^2 + m_2 a_2^2 - 2m_2 a_2 l_1 \cos\theta_2 \\
&\quad + m_3 l_1^2 + m_3 a_3^2 + 2m_3 a_3 l_1 \cos(\gamma - \theta_3), \\
M_{12}(\boldsymbol{\Theta}) &= -J_2 - m_2 a_2^2 + m_2 a_2 l_1 \cos\theta_2, \\
M_{13}(\boldsymbol{\Theta}) &= -J_3 - m_3 a_3^2 - m_3 a_3 l_1 \cos(\gamma - \theta_3), \\
G_1(\boldsymbol{\Theta}) &= (m_1 a_1 + m_2 l_1 + m_3 l_1) g \sin\theta_1 + m_2 g a_2 \sin(\theta_2 - \theta_1) \\
&\quad + m_3 g a_3 \sin(\theta_1 + \gamma - \theta_3),
\end{aligned}$$

where $\mathbf{m}_1(\boldsymbol{\Theta}) = [M_{11}(\boldsymbol{\Theta}), M_{12}(\boldsymbol{\Theta}), M_{13}(\boldsymbol{\Theta})]$.

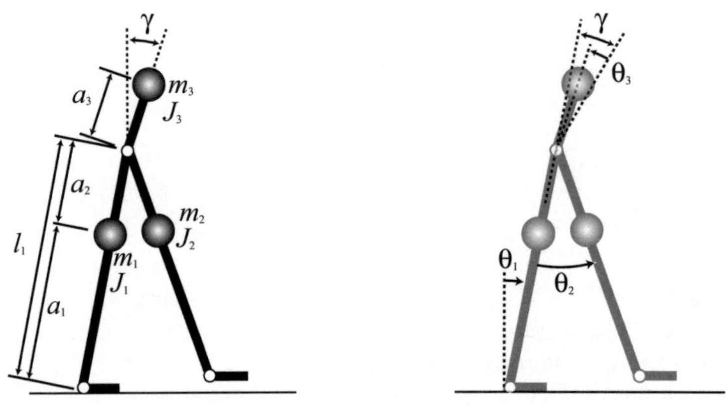

Fig. 2. 3-link model in the sagittal plane. m_i, J_i, l_i and a_i are the mass, the moment of inertia, the length of link and the distance from the joint to the link COG of link i respectively. γ is the angle of the forward tilting. In the right figure, θ_1, θ_2 and θ_3 are the ankle angle of the stance leg, the angle from the stance leg to the the swing leg, the angle to swing the trunk up respectively.

3.1.2 Constraints of sagittal joints

Constraints are simply designed as follows:

- The angle of the torso is constant.
- The swing leg is symmetrical to the stance leg.

That is,

$$\theta_1 = f_1(\theta) = \theta, \tag{14}$$
$$\theta_2 = f_2(\theta) = 2\theta, \quad \text{and} \tag{15}$$
$$\theta_3 = f_3(\theta) = \theta. \tag{16}$$

From Eqs. (14)-(16) and (1), Eq. (6) is

$$\begin{aligned} M_s(\theta) &= J_1 - J_2 + m_1 a_1^2 + m_2 l_1^2 - m_2 a_2^2 + m_3 l_1^2 + m_3 a_3 l_1 \cos(\gamma - \theta) \\ &:= E_1 + E_2 \cos(\gamma - \theta), \qquad (17) \\ G_s(\theta) &= (m_1 a_1 + m_2 l_1 + m_2 a_2 + m_3 l_1) g \sin\theta + m_3 g a_3 \sin\gamma \\ &:= E_3 + E_4 \sin\theta. \qquad (18) \end{aligned}$$

Thus,

$$\begin{aligned} &\int M_s(\theta) G_s(\theta) d\theta \\ &= \int \left(E_1 + E_2 \cos(\gamma - \theta)\right)\left(E_3 + E_4 \sin\theta\right) d\theta \\ &= E_2 E_4 \left(\frac{\sin\gamma}{2}\theta - \frac{\cos(2\theta - \gamma)}{4} \right) + E_1 E_3 \theta \\ &\quad + E_2 E_3 \sin(\theta - \gamma) - E_1 E_4 \cos\theta + C_s \\ &:= D_s(\theta) + C_s, \qquad (19) \end{aligned}$$

where C_s is the integral constant, which is PDAC Constant of the sagittal motion. From Eq. (12), Converged dynamics in the sagittal plane is

$$\begin{aligned} \dot{\theta} &= \frac{1}{M_s(\theta)} \sqrt{2(D_s(\theta) + C_s)} \qquad (20) \\ &:= F_s(\theta). \qquad (21) \end{aligned}$$

Note that it is necessary that γ is decided so that $M_s(\theta) > 0$ in order to avoid singular point. Generally speaking, as for humanoid robots and biped robots, $E_1 > E_2$ since $l_1 > a_3$. Thus, we assume $M_s(\theta) > 0$ below.

3.1.3 Foot-contact model

Regarding foot-contact, it is assumed that the ground is perfectly inelastic collision and occurred for a moment similarly to previous works (Goswami et al. (1997); Grizzle et al. (2001); Kuo (1999); Westervelt et al. (2004)). That is, the angular momentum around the contact point is conserved before and after foot-contact.
Fig. 3 shows the angle and length of the inverted pendulum at foot-contact. Here, consider the foot-contact at the end of kth step, i.e. at the beginning of $k+1$th step. Denoting the angular

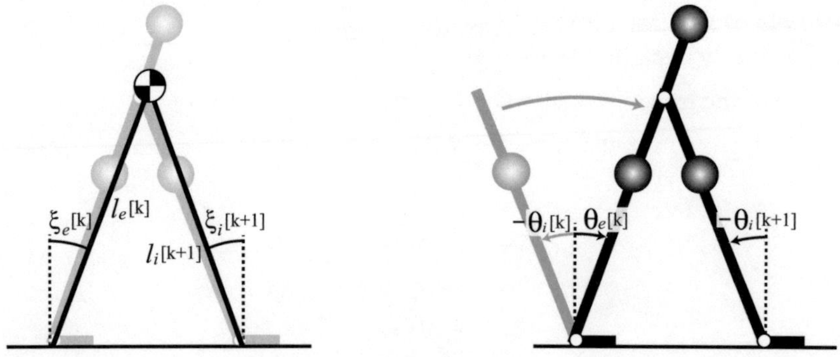

Fig. 3. Parameters at foot-contact. $l_e[k]$ and $\xi_e[k]$ are the length and inclination of the inverted pendulum which connects the ankle of support leg and robot COG before impact at the end of kth step. $l_i[k+1]$ and $\xi_i[k+1]$ are those after impact. $\theta_e[k]$ and $\theta_i[k+1]$ are the angles around the contact point before and after impact.

velocity of ankle joint of the rear leg at foot-contact as $\dot{\theta}_e[k]$, the following equation is derived from Eq. (12):

$$\dot{\theta}_e[k] = \frac{1}{M_s(\theta_e[k])}\sqrt{2(D_s(\theta_e[k]) + C_s[k])}, \tag{22}$$

where $C_s[k]$ denotes PDAC Constant of kth step.
Since the torso angle is constant and COG is not rotated, the angular velocity of ankle joint of the fore leg at foot-contact, $P_i[k+1]$, is described as follows:

$$\begin{aligned} P_i[k+1] &= m_t l_e[k] l_i[k+1] \dot{\theta}_e[k] \cos(\xi_e[k] + \xi_i[k+1]) \\ &= \frac{m_t l_e[k] l_i[k+1] \cos(\xi_e[k] + \xi_i[k+1])}{M_s(\theta_e[k])}\sqrt{2(D_s(\theta_e[k]) + C_s[k])} \\ &:= h[k]\sqrt{2(D_s(\theta_e[k]) + C_s[k])}, \end{aligned} \tag{23}$$

where $m_t = m_1 + m_2 + m_3$.
Since the angular velocity around the passive joint is

$$P = M_s(\theta)\dot{\theta},$$

PDAC Constant after foot-contact, $C_s[k+1]$, is represented as

$$\begin{aligned} C_s[k+1] &= \frac{1}{2}P_i[k+1]^2 - D_s(\theta_i[k]) \\ &= h[k]^2 C_s[k] + h[k]^2 D_s(\theta_e[k]) - D_s(\theta_i[k]) \\ &:= s_1[k] C_s[k] + s_2[k]. \end{aligned} \tag{24}$$

3.1.4 Sagittal motion period

In order to satisfy the condition of constant step-length, it is necessary to control the lateral motion so that lateral foot-contact period matches sagittal one. Since sagittal dynamics is expressed as an one-dimensional autonomous dynamics, it is possible to calculate the sagittal foot-contact period by integrating sagittal Converged dynamics with time as follows:

$$T_s = \int_{{}_i\theta^+}^{{}_{i+1}\theta^-} \frac{1}{F_s(\theta)}\, d\theta. \tag{25}$$

In next section, we design the lateral motion and build the controller satisfying the synchronization between lateral and sagittal motion.

3.2 Lateral motion control

3.2.1 Lateral motion

In this section, the lateral motion is composed by means of PDAC. In phase(A), a robot starts to turn over toward its swing-leg-side and is accelerated by gravitation from the tilting position at a standstill on the stance-leg-side to foot-contact. In phase(B), after foot-contact, a robot is got up toward the tilting position at a standstill by the energy obtained in phase(A).

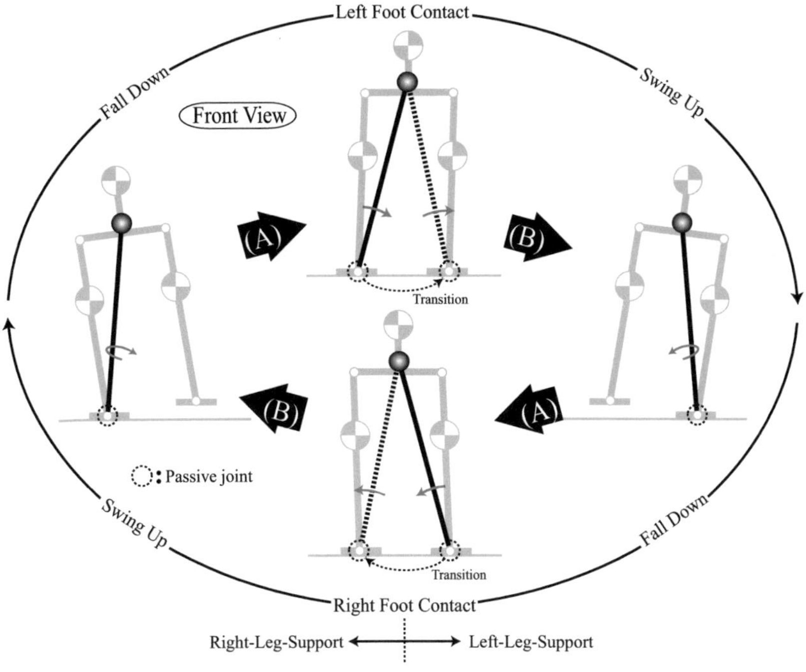

Fig. 4. The lateral motion of lateral-based walk (front view). The inverted pendulum falls off in phase(A) and swing up in phase(B)

3.2.2 Constraint of lateral joints

The dynamic equation of an inverted pendulum is described as follows:

$$\frac{d}{dt}\left((ml^2 + J)\dot{\phi}\right) = mgl\sin\phi. \tag{26}$$

Multiplying both side of this equation by $(ml^2 + J)\dot{\phi}$ and integrating with respect to time yields the following equations,

$$\begin{aligned}
&((ml^2 + J)\dot{\phi})\frac{d}{dt}((ml^2 + J)\dot{\phi}) = mgl(ml^2 + J)\dot{\phi}\sin\phi \\
\Longleftrightarrow\ &\frac{1}{2}((ml^2 + J)\dot{\phi})^2 = \int mgl(ml^2 + J)\dot{\phi}\sin\phi\, dt \\
\Longleftrightarrow\ &\dot{\phi} = \frac{1}{ml^2 + J}\sqrt{2\int mgl(ml^2 + J)\dot{\phi}\sin\phi\, dt}.
\end{aligned} \tag{27}$$

Here, we decide the *virtual constraint*, that is, pendulum length l is described as the function of ϕ. It is clear that the right side of Eq. (27) can be integrated if $f(\phi)$ is a polynomial equation. Thus in this chapter, $f(\phi)$ is decided as follows:

$$l = f(\phi) \tag{28}$$

$$= a\phi^2 + b\phi + c, \tag{29}$$

where a, b, and c are determined so as to satisfy the conditions described below. At first, the conditions of pendulum length at the beginning and ending of phase(A) and phase(B) introduce the following four equations:

$$f^A(\phi_0) = l_0, \tag{30}$$

$$f^A(\phi_1) = l_0 + \Delta l, \tag{31}$$

$$f^B(-\phi_2) = l_0 - \Delta l, \tag{32}$$

$$f^B(-\phi_3) = l_0, \tag{33}$$

where upper-suffixes denote the differentiation of phases.
In addition, the pendulum motion is designed so that the angular velocity of robot joints is not discontinuous, that is, the velocity along pendulum is zero,

$$\frac{\partial f^A}{\partial \phi}(\phi_1) = 0, \quad \text{and} \tag{34}$$

$$\frac{\partial f^B}{\partial \phi}(-\phi_2) = 0. \tag{35}$$

From Eq. (30)-(35), a and b, c in each phase are decided.
Finally, the phase around contact point (phase of passive joint) is obtained from Eqs. (27) and (28) as follows:

$$\begin{aligned}
\dot{\phi} &= \frac{1}{mf(\phi)^2 + J}\sqrt{2\int mgf(\phi)(mf(\phi)^2 + J)\sin\, d\phi} \\
&:= \frac{1}{M_l(\phi)}\sqrt{2(D_l(\phi) + C_l)} \\
&:= F(\phi).
\end{aligned} \tag{36}$$

Lastly, the value to lift up pelvis, Δl, is determined. Δl is necessary to be decided so that Eq. (36) satisfies the initial condition of phase(A) and the end condition of phase(B), that is,

$$\begin{aligned} & F_l^A(\phi_0) = F_l^B(-\phi_3) = 0 \\ \Longleftrightarrow \quad & \frac{\sqrt{2(D_l^A(\phi_1) - D_l^A(\phi_0))}}{M_l^A(\phi_1)}\cos(\phi_1+\phi_2) = \frac{\sqrt{2(D_l^B(-\phi_2) - D_l^B(-\phi_3))}}{M_l^B(-\phi_2)}, \end{aligned} \tag{37}$$

where upper suffixes denote the differentiation of phases. Δl is so small that it is possible to find the appropriate value satisfying Eq. (37) by use of the quadratic approximation.

3.2.3 Control of lateral period

Next we design the period controller of the lateral motion described in the previous subsection. The period of lateral motion is decided by the amplitude of pendulum motion, that is, the period is long if the amplitude is large and it is short if the amplitude is small. In this chapter, the desired period is realized by controlling the lateral amplitude.

Assuming that the pendulum angle at the transition from phase(B) to phase(A) is ϕ_3, the motion period T can be found properly by the following calculation

$$\int_{-\phi_3}^{-\phi_2} \frac{1}{F_B(\phi)} d\phi + \int_{\phi_3}^{\phi_1} \frac{1}{F_A(\phi)} d\phi = T. \tag{38}$$

However, it is not easy to solve this equation for ϕ_3. The pendulum extension is so small that the desired amplitude is decided approximately by use of the model of inverted pendulum, length of which is not variable, as follows:

$$\phi_3 = \frac{\phi_c}{\cosh\left(\sqrt{\frac{g}{l_0}}\frac{T}{2}\right)}, \tag{39}$$

where, ϕ_c is the pendulum angle in the standing posture, i.e. the pendulum angle at the foot-contact under the condition of $\Delta l = 0$.

4. Stability proof of the sagittal motion

4.1 Constraint of constant step-length

In this work, the step-length is fixed at constant value in order to stabilize walking. Under such condition, it is clear that the following is held:

$$\theta_e[j] = -\theta_i[j] = \arcsin\left(\frac{\lambda_d}{2l_1}\right) := \theta_c = \text{const.}, \tag{40}$$

where $j \in \mathbf{N}$ and $0 \le \theta_c < \frac{\pi}{2}$. Since the torso angle, γ, is kept constant, $\xi_e[j]$ and $\xi_i[j]$, $l_e[j]$, $l_i[j]$ are also all constant similarly. Hence, in Eq. (23),

$$h[j] := H = \text{const.}$$

is held. Besides, in Eq. (24),

$$\begin{aligned} s_1[j] &= h[j]^2 = H^2 := S_1 = \text{const.} \\ s_2[j] &= h[j]^2 D_s(\theta_e[j]) - D_s(\theta_i[j]) \\ &= H^2 D_s(\theta_c) - D_s(-\theta_c) := S_2 = \text{const.} \end{aligned}$$

are also held.

4.2 Requisite to perform walking continuously

We consider the requisite to generate walking. In terms of practicality, let γ be $0 < \gamma < \frac{\pi}{2}$. In order to perform walking continuously, $\dot{\theta} > 0$ is required at all times. Considering $M_s(\theta) > 0$, this condition is equivalent to $P = M_s(\theta)\dot{\theta} > 0$. Since

$$\begin{aligned}\frac{dD_s(\theta)}{d\theta} &= M_s(\theta)G_s(\theta)\\ &= (E_1 + E_2\cos(\gamma - \theta))(E_3 + E_4\sin\theta),\end{aligned}$$

and also since $E_3 > 0$ and $E_4 > 0$, in can be seen that $D_s(\theta)$ i.e. the angular velocity of passive joint, P, is minimum when

$$\theta = \arcsin\left(-\frac{E_3}{E_4}\right) := \hat{\theta},$$

and that it decreases monotonically on $-\frac{\pi}{2} < \theta < \hat{\theta}$ and increases monotonically on $\hat{\theta} < \theta < \frac{\pi}{2}$. Since $\frac{1}{2}P^2 = D_s(\theta) + C_s$, the condition discussed above, $P > 0$, is described as below,

$$P > 0 \Longleftrightarrow C_s > -D_s(\hat{\theta}) := \hat{C}_s. \tag{41}$$

Therefore, from Eq. (24), the requisite to perform walking continuously is found as follows:

$$S_1\hat{C}_s + S_2 > \hat{C}_s. \tag{42}$$

Next, we argue the state that dynamics of walking is converged on a unique trajectory, i.e. the equilibrium state. On such condition, PDAC Constant of every step is converged on constant value. That is,

$$C_s[k] = C_s[k+1] := C_s^*.$$

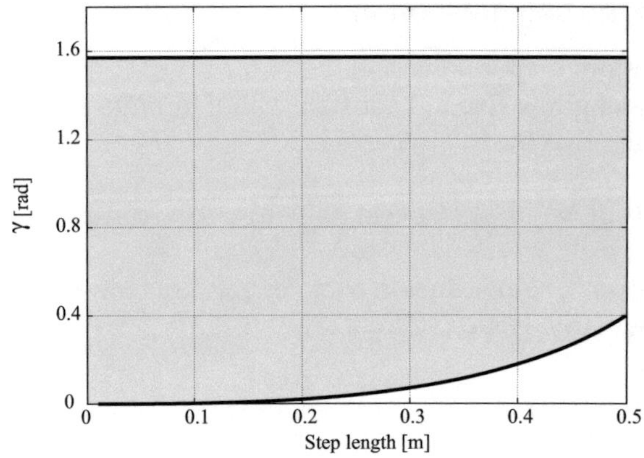

Fig. 5. Condition of θ and γ in order to generate continual walking

Hence, from Eq. (24),

$$\begin{aligned} & C_s^* = S_1 C_s^* + S_2 \\ \Longleftrightarrow \quad & S_2 = C_s^*(1 - S_1) \end{aligned} \tag{43}$$

is held. Substituting Eq. (43) into Eq. (42),

$$(1 - S_1)(C_s^* - \hat{C}_s) > 0 \tag{44}$$

is obtained. From Eq. (41), it is clear that

$$C_s^* > \hat{C}_s. \tag{45}$$

From Eqs. (44) and (45), the requisite to perform walking continuously, Eq. (42), is described as follows:

$$1 - S_1 > 0. \tag{46}$$

As for our robot (Gorilla Robot Ⅲ (Kajima et al. (2004)): Fig. 9, Table 1), Eq. (46) is the range shown in Fig. 5. Although we assume $\gamma < \frac{\pi}{2}$ in Fig. 5, actual upper bound is decided according to the limitation of robot's specification such as the torque to swing a leg forward or to keep a torso angle at constant value.

4.3 Proof of stability

Lastly, we prove the stability by Liapunov Theory using the conditions found above. The error between actual C_s and convergent value is defined as

$$\delta C_s := C_s^* - C_s. \tag{47}$$

The following positive definite function, V, is defined,

$$V = (\delta C_s)^2.$$

Since $V(0) = 0$ and $V > 0$ ($\delta C_s \neq 0$), it is apparent that V is positive definite. From Eqs. (24), (43), and (47), finite difference of V is

$$\begin{aligned} \Delta V &= V[k+1] - V[k] \\ &= \delta C_s[k+1]^2 - \delta C_s[k]^2 \\ &= (\delta C_s[k+1] + \delta C_s[k])(\delta C_s[k+1] - \delta C_s[k]) \\ &= -(2C_s^* - C_s[k+1] - C_s[k])(C_s[k+1] - C_s[k]) \\ &= -(2C_s^* - S_1 C_s[k] - S_2 - C_s[k])(S_1 C_s[k] + S_2 - C_s[k]) \\ &= -\left(2C_s^* - S_1 C_s[k] - (1 - S_1)C_s^* - C_s[k]\right)\left(S_1 C_s[k] + (1 - S_1)C_s^* - C_s[k]\right) \\ &= -(1 + S_1)(1 - S_1)(C_s^* - C_s[k])^2 \\ &= -(1 + S_1)(1 - S_1)\Delta C_s[k]^2. \end{aligned} \tag{48}$$

Since it is clear that $1 + S_1 = 1 + H^2 > 0$ from Eq. (41) and that $1 - S_1 > 0$ from Eqs. (46) and (48) is

$$\Delta V = 0 \quad (\delta C_s = 0). \tag{49}$$

In addition,

$$\Delta V < 0 \quad (\delta C_s \neq 0) \tag{50}$$

is held. From Eq. (49) and (50), ΔV is negative definite. Therefore, the equilibrium point, C_s^*, is asymptotically stable in the range shown in Fig. 5.

4.4 Simulation

Next, stability proof described in the previous subsection is confirmed by the numerical simulation. Fig. 6 is the phase portrait of θ and alteration in terms of time. From these figures, the convergence of the sagittal motion can be ascertained. On this simulation, step-length is 0.18[m], the torso angle is $\gamma = 0.035$[rad].

Fig. 7 depicts the return maps of PDAC Constant and the angular velocity of passive joint at foot-contact. These figures show that the sagittal dynamics has a sole stable fixed point.

Finally, in order to confirm that the stability of the sagittal motion is independent of step-length, we perform the simulation of the various step-length. Fig. 8 is the graph of the return map of the angular volocity of passive joint at foot-contact v.s. step-length. From this figure, it can be confirmed that the sagittal dynamics is stable regardless of step-length.

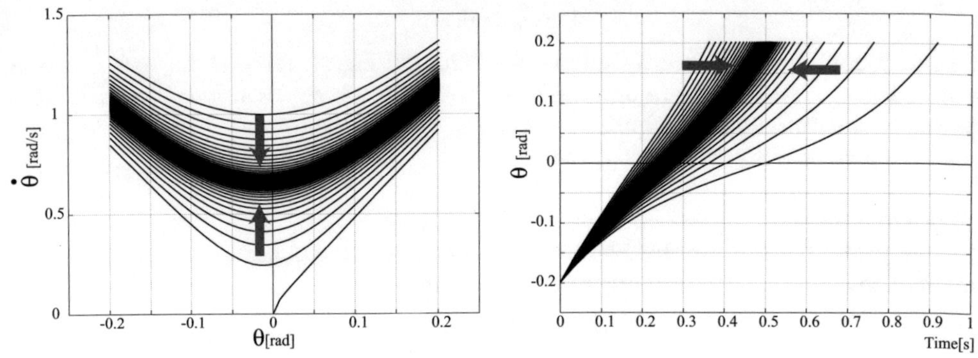

Fig. 6. Simulation results of sagittal stabilization. (Left) Phase portrait of the sagittal dynamics. (Right) θ trajectory vs. time.

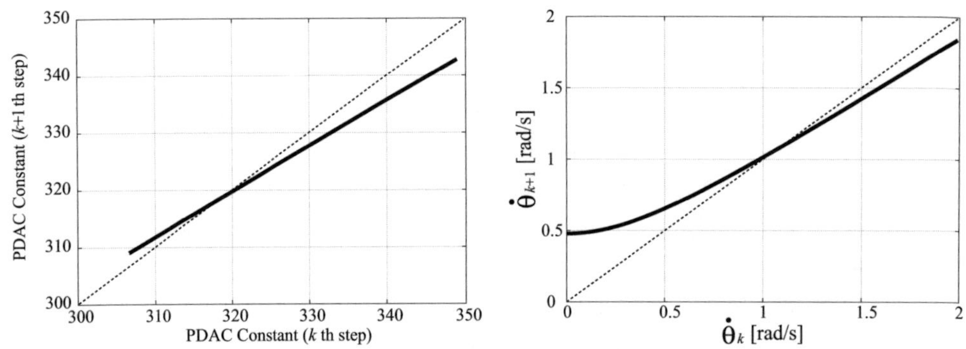

Fig. 7. Simulation results of sagittal stabilization. (Left) Return map of PDAC Constant. (Right) Return map of $\dot{\theta}$ at foot-contact.

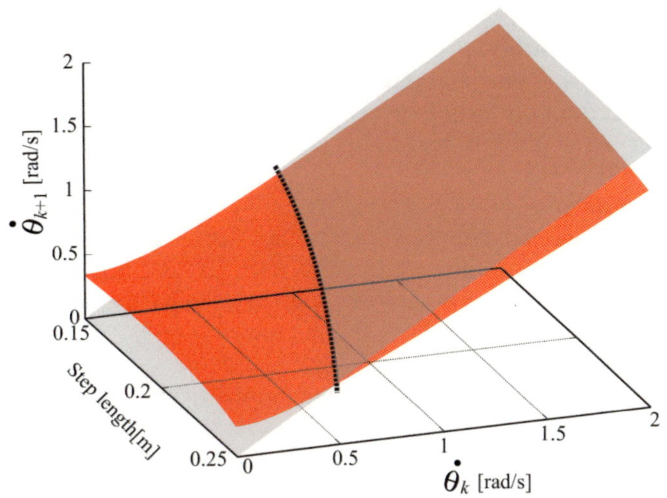

Fig. 8. Return map of $\dot{\theta}$ at foot-contact vs. step-length

Table 1. Link parameters of Gorilla Robot Ⅲ

Mass[kg]	link1	m_1	2.618
	link2	m_2	3.451
	link3	m_3	15.143
COG position	link1	a_1	0.23
	link2	a_2	0.28
	link3	a_3	0.22
Moment of inertia	link1	J_1	0.042
	link2	J_2	0.070
	link3	J_3	0.047

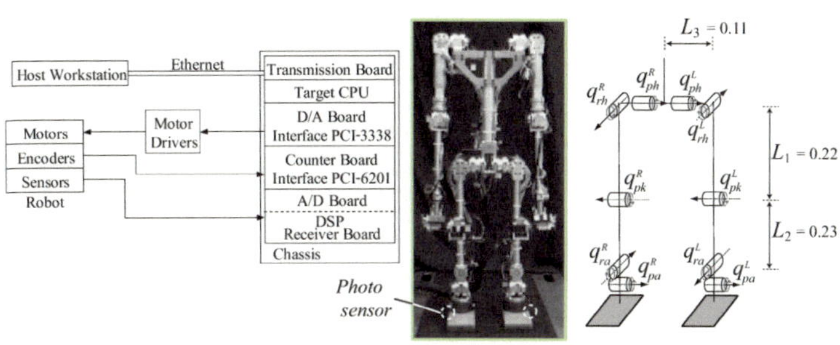

Fig. 9. Gorilla Robot Ⅲ (about 1.0[m] height, 22.0[kg] weight, 24 DOF)

5. Experiment

In order to ascertain the validity of proposed method, we conducted the experiment. Both sagittal controller and lateral one are employed simultaneously and 3-D dynamic walking is realized. Note that, in the following experiment, estimated viscous torque is applied to the ankle joints of stance-leg in order to realize the passive joint virtually.

5.1 Experimental Setup

Fig. 9 depicts our experimental setup and our robot "Gorilla Robot Ⅲ (Kajima et al. (2004))". This robot is driven by 24 AC motors of 20-30W with 100-200 times of speed reduction by harmonic gears. It has the photo sensors on its soles to perceive foot-contact.

5.2 Experimental results

In this experiment, the robot bends its knee joint of the swing leg so as to prevent the foot being in friction with the ground immediately after foot-contact on the assumption that the effect of knee bending on the robot dynamics can be neglected. The foot of the swing leg is actuated so as to be kept parallel to the ground. The experiment was performed on the basically level and flat ground which has maximum 1.0[cm] irregularity without discontinuities.

In this chapter, the desired step-length is given to be gradually increased within initial 5 steps up to 0.15[m] and after 5 steps it is fixed at 0.15[m]. As a result of experiment, the dynamic and natural walking is realized over 25 steps. The step-length is about 0.15[m] and the walking velocity is about 0.23[m/s]. Fig. 10 shows the snapshots of the PDAC walking at 1st, 5th, 9th, 12th, 14th, 16th step respectively. The angle and angular velocity of the lower body joints are depicted in Fig. 11 and Fig. 12. As shown in these figures, the smooth dynamics motion is realized periodically.

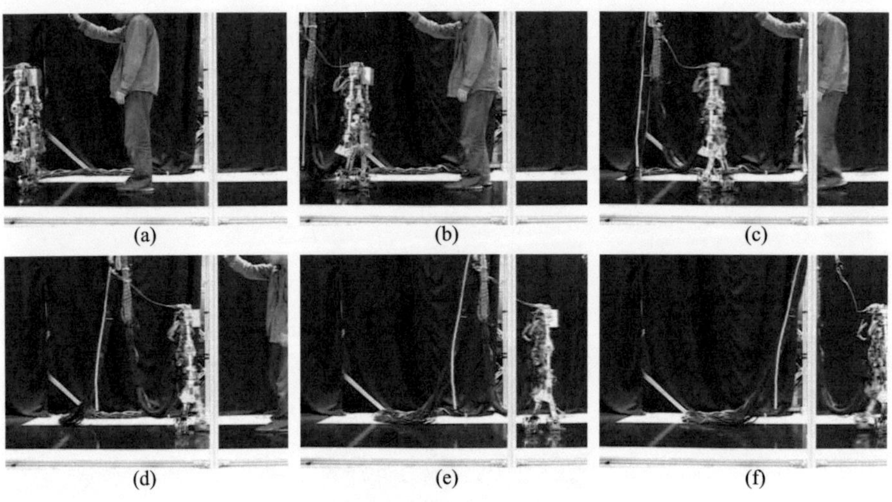

Fig. 10. Snapshots of the walking experiment. Each figure shows the snapshots at (a)1st (b)5th (c)9th (d)12th (e)14th (f)16th step.

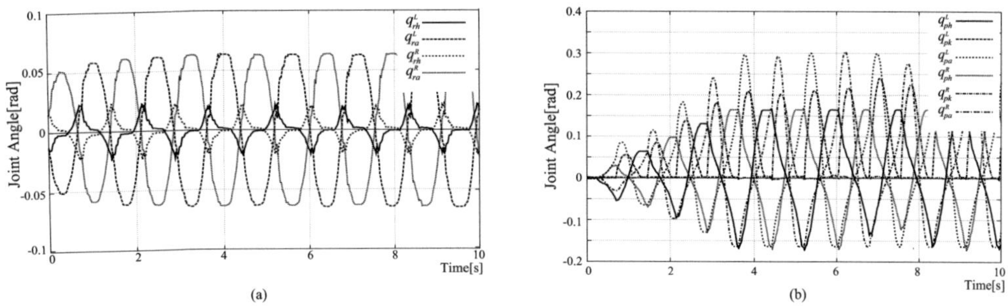

Fig. 11. Angle of the joints (a) in the lateral plane (b) in the sagittal plnae

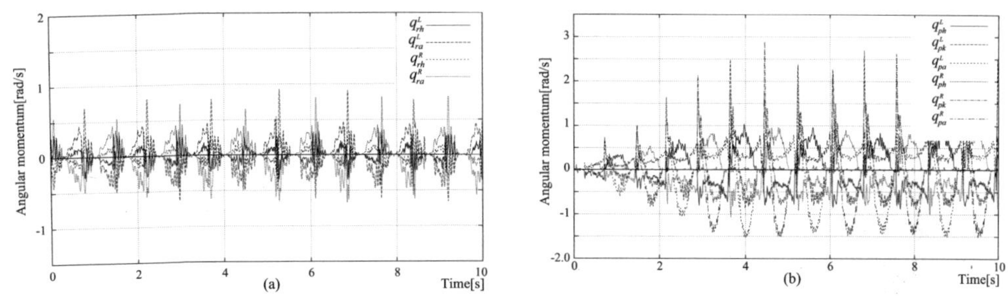

Fig. 12. Angular Velocity of the joints (a) in the lateral plane (b) in the sagittal plnae

6. Conclusion

This chapter designed the sagittal motion controller of biped walking and proved the stability of sagittal motion designed by means of PDAC. The proof was conducted by the Liapunov Theory, and the convergence domain was also investigated. We confirmed the correctness of proof by numerical simulation. Finally, a three-dimensional dynamic walking whose step-length is about 0.15[m] and velocity is about 0.23[m/s] was realized by coupling the sagittal motion and lateral motion. The proof in this chapter handled the stability of 2-D dynamics, thus the future work is to propose the walking controller achieving the stability of 3-D dynamics, and prove the 3-D stability of dynamic walking.

7. References

Chevallereau, C. (2003). Time-Scaling Control for an Underactuated Biped Robot. *IEEE Transactions on Robotics Automation*, Vol.19, No.2, pp.362-368, 2003.

Doi, M.; Hasegawa, Y. & Fukuda, T. (2004a). Passive trajectory control of the lateral motion in bipedal walking. *Proceedings of IEEE International Conference on Robotics and Automation*, pp.3049-3054, New Orleans, May 2004.

Doi, M.; Hasegawa, Y. & Fukuda, T. (2004b). Passive Dynamic Autonomous Control of Bipedal Walking. *Proceedings of IEEE-RAS/RSJ International Conference on Humanoid Robots*, Paper no.72, Los Angeles, November 2004.

Furusho, J. & Sano, A. (1990). Sensor-based Control of a Nine-Link Biped. *The International Journal of Robotics Research*, Vol.9, No.2, pp.83-98, 1990.

Goswami, A.; Espiau, B. & Keramane, A. (1997). Limit cycles in a passive compass gait biped and passivity-mimicking control laws. *Autonomous Robots*, Vol.4, No.3, pp.273-286, 1997.

Grishin, A. A.; Formal'sky, A. M.; Lensky, A. V. & Zhitomirsky, S. V. (1994). Dynamic Walking of a Vehicle With Two Telescopic Legs Controlled by Two Drives. *The International Journal of Robotics Research*, Vol.13, No.2, pp.137-147, 1994.

Grizzle, J. W.; Abba, G. & Plestan, F. (2001). Asymptotically Stable Walking for Biped Robots: Analysis via Systems with Impulse Effects. *IEEE Transactions on Automatic Control*, Vol.46, No.1, pp.56-64, 2001.

Hirai, K.; Hirose, M.; Haikawa, Y. & Takenaka, T. (1998).The Development of Honda Humanoid Robot. *Proceedings of the IEEE International Conference on Robotics and Automation*, pp.1321-1326, Leuven, May 1998.

Kajima, H.; Doi, M.; Hasegawa, Y. & Fukuda, T. (2004). A study on brachiation controller for a multi-locomotion robot -realization of smooth, continuous brachiation-. *Advanced Robotics*, vol.18, no.10, pp.1025-1038, 2004.

Kajita, S.; Yamaura, T. & Kobayashi, A. (1992). Dynamic walking control of a biped robot along a potential conserving orbit. *IEEE Transactions on Robotics Automation*, Vol.8, No.4, pp.431-438, 1992.

Kaneko, K.; Harada, K.; Kanehiro, F.; Miyamori, G. & Akachi, K. (2008). Humanoid Robot HRP-3. *Proceedings of the IEEE/RSJ International Conference on Intelligent Robots and Systems*, pp.2471-2478, Nice, September 2008.

Kuo, A. D. (1999). Stabilization of Lateral Motion in Passive Dynamic Walking. *The International Journal of Robotics Research*, Vol.18, No.9, pp.917-930, 1999.

Miura, H. & Shimoyama, I., (1984). Dynamic Walking of a biped. *The International Journal of Robotics Research*, Vol.3, No.2, pp.60-74, 1984.

Nakanishi, J.; Morimoto, J.; Endo, G.; Cheng, G.; Schaal, S. & Kawato, M. (2004). A Framework for Learning Biped Locomotion with Dynamical Movement Primitives, *Proceedings of IEEE-RAS/RSJ International Conference on Humanoid Robots*, Paper no.81, Los Angeles, November 2004.

Nishiwaki, K.; Kagami, S.; Kuniyoshi, Y.; Inaba, M. & Inoue, H. (2002). Online Generation of Humanoid Walking Motion based on a Fast Generation Method of Motion Pattern that Follows Desired ZMP. *Proceedings of the International Conference on Intelligent Robots and Systems*, pp.2684-2689, Lausanne, September 2002.

Ono, K.; Furuichi, T. & Takahashi, R. (2004). Self-Excited Walking of a Biped Mechanism With Feet. *The International Journal of Robotics Research*, Vol.23, No.1, pp.55-68, 2004.

Takanishi, A.; Ishida, M.; Yamazaki, Y. & Kato, I. (1985). The Realization of Dynamic Walking Robot WL-10RD. *Proceedings of International Conference on Advanced Robotics*, pp.459-466, Tokyo, September 1985.

Vukobratovic, M. & Borovac, B. (2004). ZERO-MOMENT POINT-THIRTY FIVE YEARS OF ITS LIFE. *International Journal of Humanoid Robotics*, vol.1, no.1, pp.157-173, 2004.

Westervelt, E. R.; Buche, G. & Grizzle, J. W. (2004). Experimental Validation of a Framework for the Design of Controllers that Induce Stable Walking in Planar Bipeds. *The International Journal of Robotics Research*, Vol.24, No.6, pp.559-582, 2004.

12

Simulated Regulator to Synthesize ZMP Manipulation and Foot Location for Autonomous Control of Biped Robots

Tomomichi Sugihara
Kyushu University
Japan

1. Introduction

Biped robots have potentially high mobilities. Since they have similar morphologies with the humans' lower limbs, they are expected to grow up into the robots which can travel wherever humans can. To make biped robots catch up with such an expectation still needs to resolve many challenges.

The two bases of pedipulation, namely, the legged motion control are the indirect reaction force manipulation (Fujimoto et al., 1998; Furusho & Masubuchi, 1986; Mitobe et al., 1995; Vukobratović et al., 1970) and the discontinuous grounding foot location (Kajita & Tani, 1995; Miura & Shimoyama, 1984; Raibert et al., 1984; Witt, 1970). The former is necessary to transport the center of mass (COM) of the floating multibody system, which is not mechanically connected to the inertial frame. The latter is required to reform the supporting region of the system, which determines the limitation of the physically available reaction forces. Each has a strongly nonlinear property, so that the synthesis of them is still an open problem.

A major solution against it is to refer the motion trajectory which is defined as a function of time. It can coordinate the fullbody motion involving foot location and COM transportation rather easily with physical feasibility represented by ZMP (Vukobratović & Stepanenko, 1972) and geometric constraints such as collision avoidance taken into account. Many successful biped walkers based on this approach have been appeared (Hirai et al., 1998; Huang et al., 2001; Kajita et al., 2003; Löffler et al., 2003; Nagasaka et al., 1999; 2004; Sugihara & Nakamura, 2005; Takanishi et al., 1988). Such time-slaved controls, however, are not robust against various extrinsic events. While they work in situations where sufficient knowledge about the environment and the task is given in advance, they are less promising in fields of action with many uncertainties. It is desired that the control system is designed as an autonomous system, namely, a system which does not explicitly depend on time.

Some COM control methods by manipulating the reaction force or ZMP in realtime were proposed (Fujimoto et al., 1998; Mitobe et al., 1995; Sugihara et al., 2002). They focused on the COM control under a given supporting condition, and the foot location strategies were out of the scope. Some stepwise-stable autonomous biped controllers in accordance with the point-foot contact were also proposed (Gubina et al., 1974; Kajita et al., 1992; Westervelt et al., 2004). The application to realistic robots which support themselves on their soles has been a future work. Passive dynamic walking (Collins et al., 2001; McGeer, 1990; Yamakita et al., 2000) is

another approach to design an autonomous biped controller by utilizing an inherent stability of discretized biped dynamics. It stands on the ideally perfect plastic collision between the robot and the ground, and thus, has a low stabilizing ability.
This paper proposes a control to synthesize the above ZMP manipulation control and the foot location in a consistent manner. We design a regulator based on the approximate dynamical model of a biped robot, focusing on a simple relationship between COM and ZMP. In this stage, the feasible area where ZMP can exist is unbounded against the physical constraint. In this sense, we call it the simulated regulator. When the desired ZMP is located out of the supporting region, it is modified to be within the actual region. The robot is controlled in such a way that the real ZMP tracks the desired ZMP. Simultaneously, the supporting region is deformed by a foot replacements to include the original desired ZMP in the future. The regulator gains are decided by the pole assignment method in order to give COM a slow mode and ZMP a fast mode explicitly, which matches the role of each foot. Since both the ZMP manipulation and the foot location originate from the identical simulated regulator, a totally consistent control system is made up. In addition, it is shown that a cyclic walk is automatically generated without giving a walking period explicitly by coupling the support-state transition and the goal-state transition. It does not assume a periodicity of the motion trajectory, and hence, seamless starting and stopping can be achieved.

2. Simulated COM–ZMP regulator

2.1 Linearized biped system and simulated regulator

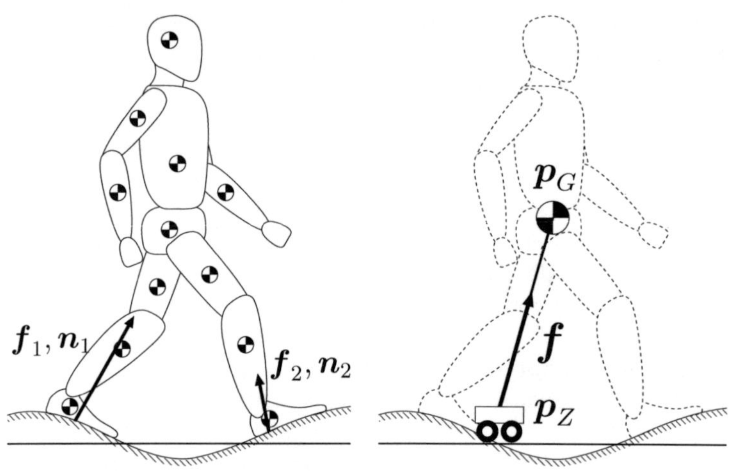

(a)Precise anthropomorphic model (b)Inverted pendulum metaphorized model

Fig. 1. Approximately mass-concentrated biped model.

The strict equation of motion of a biped robot takes a complicated form with tens of degrees-of-freedom. Here, we assume that an effect of the moment about COM is smaller enough to be neglected than that about ZMP due to the movement of COM. Then, the macroscopic behavior of the legged system is represented by the motion of COM. The equation of motion in horizontal direction of a biped model with such a mass-concentrated approximation as

Fig. 1(B) is expressed as follows:

$$\ddot{x} = \omega^2 (x - x_Z) \tag{1}$$

$$\ddot{y} = \omega^2 (y - y_Z), \tag{2}$$

where $\boldsymbol{p}_G = [\, x \;\; y \;\; z \,]^{\mathrm{T}}$ is the position of COM, and $\boldsymbol{p}_Z = [\, x_Z \;\; y_Z \;\; z_Z \,]^{\mathrm{T}}$ is ZMP. ω is defined as:

$$\omega^2 \equiv \frac{\ddot{z} + g}{z - z_Z} \left(\geq 0 \right), \tag{3}$$

where g is the acceleration of gravity, and z_Z is the ground level, which is known. z, x and y axes are aligned along the gravity, the forward and the leftward directions, respectively. Eq.(1) and (2) imply that COM can be controlled via manipulation of ZMP.

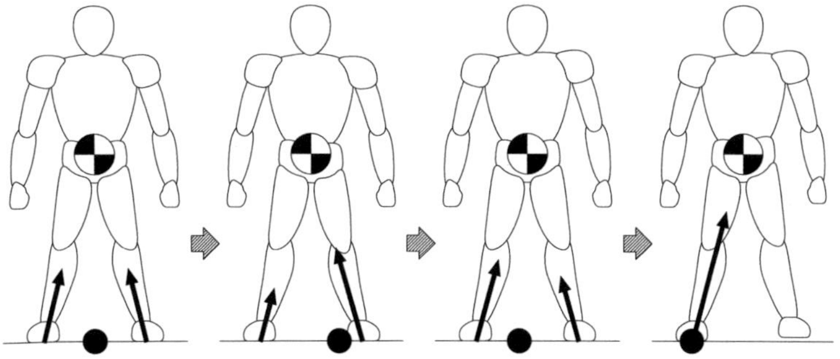

Fig. 2. Coupled movement of ZMP and COM in the ground-kick in the double support phase. ZMP travels fast between the feet to overtake COM.

The coupled movement of ZMP and COM is not simple. Let us consider a case where the robot lifts up one foot from the both-standing state, for example. Note that, in such situations, a conventional distinction between swing foot and stance foot does not make sense any longer, since neither feet are swinging. However, they are obviously different from each other in terms of function. In this paper, the foot to be the swing foot is called *kicking foot*, and that to be the stance foot is called *pivoting foot*, instead.
The sequence is illustrated by Fig. 2. ZMP is required to be within the pivoting sole at the end of the phase in order to detach the kicking foot off the ground, while it moves into the sole of kicking foot in the initial phase in order to accelerate COM towards the pivoting foot. Namely, ZMP initially moves oppositely against the direction of the desired COM movement, and overtakes COM during the motion. The fact that the biped robot is a non-minimum-phase-transition system as well as the inverted pendulum underlies the requirement of such a complex manipulation of ZMP. In addition, ZMP travels faster than COM between the feet in the double support phase, as ZMP depends on the acceleration of the robot. Both modes of COM and ZMP movement are desired to be explicitly designed in accordance with the locations of feet. Then, we include ZMP in the state variable and regard the ZMP rate as the input. The linearized state equation is represented as follows:

$$\dot{\boldsymbol{x}} = \boldsymbol{A}\boldsymbol{x} + \boldsymbol{b}u, \tag{4}$$

where the motion along x-axis is only considered from the isomorphism of Eq.(1) and (2), and:

$$\boldsymbol{x} \equiv \begin{bmatrix} x \\ \dot{x} \\ x_Z \end{bmatrix}, \boldsymbol{A} \equiv \begin{bmatrix} 0 & 1 & 0 \\ \omega^2 & 0 & -\omega^2 \\ 0 & 0 & 0 \end{bmatrix}, \boldsymbol{b} \equiv \begin{bmatrix} 0 \\ 0 \\ 1 \end{bmatrix}, u \equiv \dot{x}_Z,$$

respectively. In the above equation assumed that the vertical movement of COM is slower enough to regard as $\omega \simeq$ const. than the horizontal movement. The ZMP rate is decided based on the state feedback around the referential state ${}^{ref}\boldsymbol{x}$.

$$u = \boldsymbol{k}^{\mathrm{T}}({}^{ref}\boldsymbol{x} - \boldsymbol{x}). \tag{5}$$

The gain $\boldsymbol{k}$ is designed by the pole assignment method so as to embed a faster mode explicitly into ZMP movement than the mode of COM. The motion along y-axis is dealt with as well. In this stage, we don't constrain ZMP in the supporting region, so that the system is not necessarily physically consistent. In this sense, let us call it *the simulated ZMP* and represent it by $\tilde{\boldsymbol{p}}_Z = [\, \tilde{x}_Z \;\tilde{y}_Z \; z_Z \,]^{\mathrm{T}}$. As long as $\tilde{\boldsymbol{p}}_Z$ is within the supporting region, the actual desired ZMP ${}^d\boldsymbol{p}_Z$ is set for the same position with $\tilde{\boldsymbol{p}}_Z$.

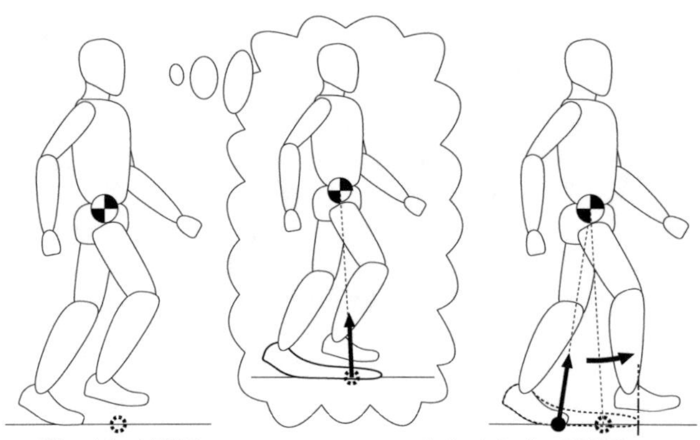

Fig. 3. The concept of the simulated regulator. When the simulated ZMP $\tilde{\boldsymbol{p}}_Z$ lies out of the supporting region, the desired ZMP ${}^d\boldsymbol{p}_Z$ is set for the proximity to the supporting region. At the same time, the swing foot is relocated to deform the supporting region so as to include $\tilde{\boldsymbol{p}}_Z$ in the future.

Fig. 3 illustrates the idea of the proposed control. The situation where $\tilde{\boldsymbol{p}}_Z$ lies out of the supporting region means that COM cannot be provided with the desired acceleration under the current supporting condition. In order to compromise this inconsistency between the desired control and the acceptable control, the following two maneuvers are required. One is to take a physically-feasible acceleration which is the nearest to the desired value by setting the desired ZMP ${}^d\boldsymbol{p}_Z$ for the proximity of $\tilde{\boldsymbol{p}}_Z$ to the supporting region as Fig. 4 depicts. The motion continuity at the moment of landing is held by resetting the simulated ZMP $\tilde{\boldsymbol{p}}_Z$ for the originally desired ZMP ${}^d\boldsymbol{p}_Z$. This idea has been already proposed by the authors (Sugihara et al., 2002). The other is to deform and expand the supporting region so as to include $\tilde{\boldsymbol{p}}_Z$ in the future, which is described in the following section.

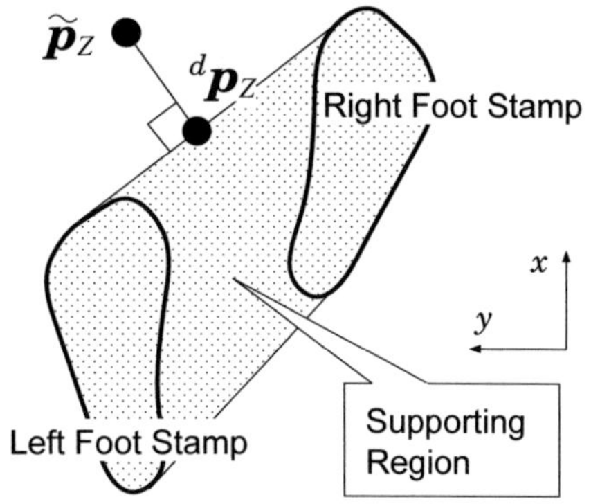

Fig. 4. Substitution of $\tilde{p}_Z$ for ${}^d p_Z$ to match the actual supporting region.

2.2 Foot location control based on simulated ZMP

The deformation of the supporting region is achieved via the relocation of stance feet. Suppose ZMP is within the pivoting sole. Let us define that $\boldsymbol{p}_S = [\, x_S \;\; y_S \;\; z_S \,]^{\mathrm{T}}$ and $\boldsymbol{p}_K = [\, x_K \;\; y_K \;\; z_K \,]^{\mathrm{T}}$ are the tip positions of the pivoting foot and the kicking foot, respectively. They correspond to the positions of the stance foot and the swing foot during the single support phase, respectively. We decide the desired position of the foot ${}^d\boldsymbol{p}_K = \left[{}^d x_K \;\; {}^d y_K \;\; {}^d z_K \right]^{\mathrm{T}}$ by the following procedure.

The COM acceleration which the simulated regulator requires (called *the simulated COM acceleration*, hereafter), and the desired COM acceleration which conforms to the actual supporting condition (called *the desired COM acceleration* in short, hereafter) are defined by the relative COM locations with respect to the simulated ZMP $\tilde{\boldsymbol{p}}_Z$ and the originally desired ZMP ${}^d\boldsymbol{p}_Z$, respectively. The necessity of a relocation of grounding feet arises in case where the desired COM acceleration is inconsistent with the simulated COM acceleration. It is judged with respect to x- and y-axes independently. ${}^d x_K$ is defined as follows:

$$
{}^d x_K = \begin{cases} \lambda_x \tilde{x}_Z + (1 - \lambda_x) x_S & (\text{for } \iota_x < 0) \\ x_K & (\text{for } \iota_x \geq 0) \end{cases} \tag{6}
$$

$$
\iota_x \equiv (x - \tilde{x}_Z)(x - {}^d x_Z), \tag{7}
$$

where λ_x is a constant to define the step magnitude ($\lambda_x > 1$). The above rule means that the robot puts its swing foot on the place where the desired COM acceleration orients to the same direction with the simulated COM acceleration, if they direct counterwards to each other.

For the motion in y-axis, ${}^d y'_K$ is firstly computed from the designed $\lambda_y (> 1)$ as well. Then, it is converted to ${}^d y_K$ by the following rule in order to avoid the self-collision between both feet:

$$
{}^d y_K = \bar{y} + \frac{1}{2} \left\{ {}^d y'_K - \bar{y} \pm \sqrt{({}^d y'_K - \bar{y})^2 + a} \right\}, \tag{8}
$$

where $+$ is chosen for the left leg for the double sign, while $-$ for the right leg, and $\bar{y}$ is the inner boundary of the swing foot. The above function has a profile as shown in Fig. 6. A

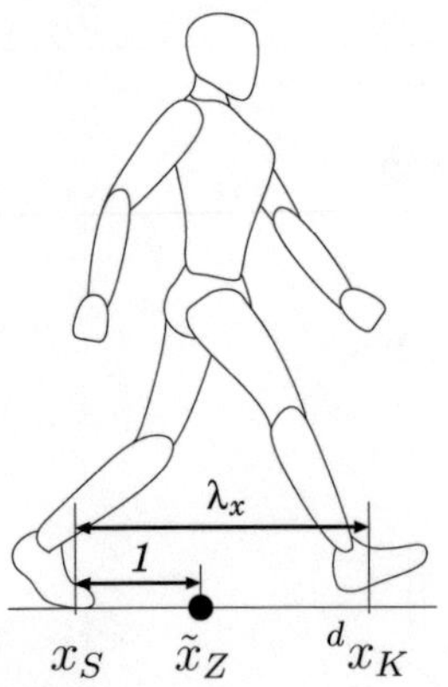

Fig. 5. Step ratio λ_x to cover simulated ZMP in the future.

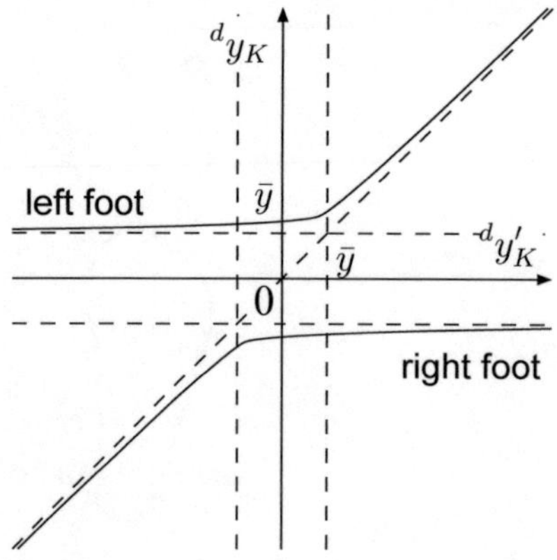

Fig. 6. Foot location transformation in y-axis for self-collision avoidance.

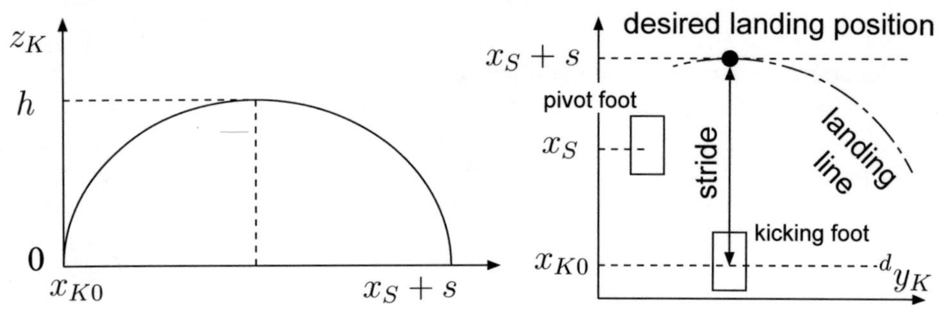

Fig. 7. Spatial foot trajectory (left) in xz-plane (right) in xy-plane.

smaller constant a makes the curve approach to the asymptotic lines with the break point $(^d y'_K, {}^d y_K) = (\bar{y}, \bar{y})$.

Suppose the initial position of the swing foot is $\boldsymbol{p}_{K0} = [\ x_{K0}\ \ y_{K0}\ \ z_{K0}\ ]^{\mathrm{T}}$, and the lift height of the swing foot $^d z_K$ is defined as:

$$^d z_K = 2h\sqrt{\theta(1-\theta)} \tag{9}$$

$$\theta \equiv \min\left\{ \frac{(^d x_K - x_{K0})^2 + (^d y_K - y_{K0})^2}{|x_S - x_{K0} + s|}, 1 \right\}. \tag{10}$$

It generates a spatial trajectory which carries the swing foot along a half ellipsoid with a height h as the leftside of Fig. 7, and makes it land on a circle with the center (x_{K0}, y_{K0}) and the radius $x_S - x_{K0+s}$, the bird's-eye view of which is depicted in the right side of Fig. 7; it lands to the point with a stride $x_S - x_{K0} + s$ from the initial position as long as $^d y_K = y_{K0}$ is ensured.

The above procedure does not guarantee the time continuity of $^d \boldsymbol{p}_K$, so that it might jump largely at the moment when ZMP travels to the pivoting sole, or when the relative COM location with respect to the simulated ZMP comes in the opposite side of that with respect to

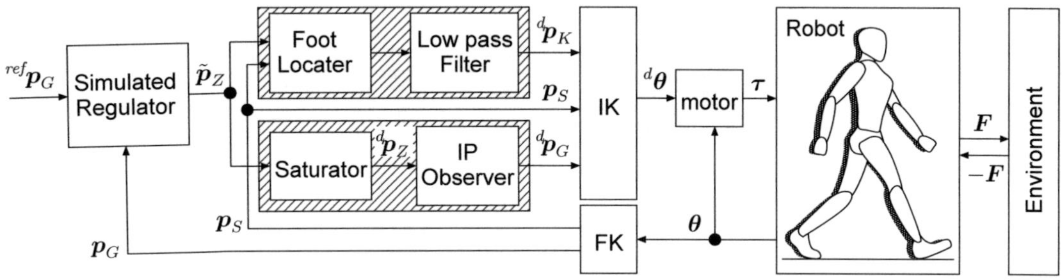

Fig. 8. Block diagram of the proposed biped control system with the simulated regulator.

the desired ZMP, for instance. Then, the time sequence of ${}^{d}\boldsymbol{p}_K$ is smoothened by second-order low-pass filters, for example.

Fig. 8 is a block diagram of the proposed control system described above. 'IP Observer' in the figure shows a subsystem which outputs the desired COM position ${}^{d}\boldsymbol{p}_G$ equivalent to the desired ZMP ${}^{d}\boldsymbol{p}_Z$ (Sugihara et al., 2002). One can note that both the COM controller with ZMP manipulation and the foot relocation controller branch from the identical simulated regulator and join in the inverse kinematics solver (the motion rate resolver).

3. Autonomous walk by coupled goal-state/support-state transition

Suppose the referential COM position is ${}^{ref}\boldsymbol{p}_G = \begin{bmatrix} {}^{ref}x & {}^{ref}y & {}^{ref}z \end{bmatrix}^{\mathrm{T}}$, the referential state of the simulated regulator in x-axis is ${}^{ref}\boldsymbol{x} = \begin{bmatrix} {}^{ref}x & 0 & {}^{ref}x \end{bmatrix}^{\mathrm{T}}$. The control in the previous section yields a step motion automatically by locating ${}^{ref}\boldsymbol{p}_G$ out of the supporting region on purpose. This property is utilized to achieve an autonomous continual walk by coupling the referential goal state transition and the supporting state transition, namely, by repeating to set ${}^{ref}\boldsymbol{p}$ out of the supporting region after the supporting region is deformed so as to include ${}^{ref}\boldsymbol{p}_G$ by the stepping. More concretely, ${}^{ref}x$ is defined by the following equation for a given s and the position of pivoting foot x_S in x-axis:

$$ {}^{ref}x = x_S + rs, \tag{11} $$

where r is a positive coefficient ($0 < r < 1$). In cases where the robot changes the orientation, x- and y-axes are again realigned with respect to the moving direction, and the desired COM position is computed with the above Eq.(11).

4. Simulation

We verified the proposed control via a simulation with an inverted pendulum model whose mass was concentrated at the tip. The length of the pendulum was 0.27[m], which fits to the robot "mighty" (Sugihara et al., 2007) shown in Fig. 9. Note that the robot mass does not affect the behavior of the inverted pendulum. The both sole were modelled as rectangles with the length 0.055[m] to the toe edge, 0.04[m] to the heel edge, and 0.035[m] to each side. The state feedback gains were designed by the pole assignment method. The poles were -3, -6 and -10 with respect to x-axis, and -2.5, -25 and -30 with respect to y-axis. The other control

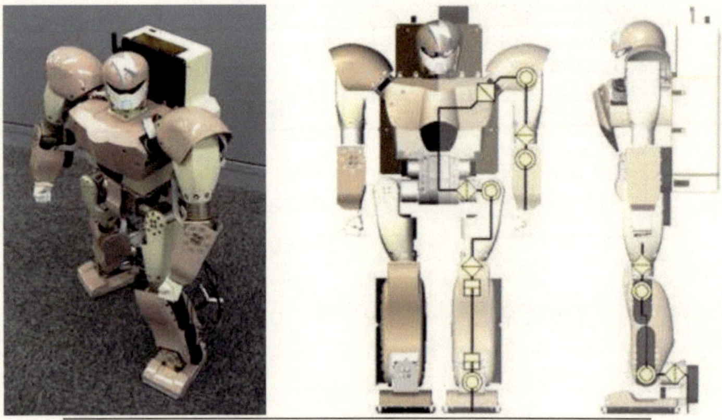

Height:	580 [mm]
Weight:	6.5 [kg]
Number of joints:	20 (8 for arms,12 for legs)

Fig. 9. External view and specifications of the robot "mighty".

parameters were set for $\lambda_x = 2$, $\lambda_y = 3$, $a = 0.001$, $r = 0.9$ and $h = 0.01$[m], respectively. The desired swing foot position was smoothened by a second-order low-pass filter $\frac{1}{(0.02s+1)^2}$. The initial state was set for $(x, y) = (0, 0)$ and $(\dot{x}, \dot{y}) = (0, 0)$. The initial stance position of the left and the right feet were $(0, 0.045)$ and $(0, -0.045)$, respectively. From the first to the sixth step, the stride s was set for 0.3[m], and the referential COM position was automatically updated by the method described in section 3. Immediately after landing the sixth step, the referential COM position was settled at the midpoint of both feet.

The loci of the referential COM position $({}^{ref}x, {}^{ref}y)$, the actual COM (the tip point of the inverted pendulum)(x, y), the simulated ZMP position $(\tilde{x}_Z, \tilde{y}_Z)$, the actually desired ZMP position $({}^{d}x_Z, {}^{d}y_Z)$, the referential feet positions $({}^{d}x_L, {}^{d}y_L)$, $({}^{d}x_R, {}^{d}y_R)$ and the filtered positions of them (x_L, y_L), (x_R, y_R) are plotted in Fig. 10. It is seen that an almost cyclic continual walk was achieved without giving a walk period explicitly by an alternation of the supporting-region deformation via the pedipulation and the goal-state transition. In this example motion, the simulated ZMP and the actually desired ZMP in y-axis always coincided with each other, so that a sideward stepping was not resulted. The difference of COM and ZMP modes particularly appear in the movement along y-axis. The given pole to design feedback gains set the time-constant of the sideward kicking for about 0.1[s], which contributed to ensure about 60% of duty ratio of the swinging phase. Fig. 11 zooms a part of Fig. 10 from $t = 0 \sim 1.5$. ${}^{d}x_Z$ differs from $\tilde{x}_Z$ in $t \simeq 0.4 \sim 0.5$, $t \simeq 0.9 \sim 1.0$ and $t \simeq 1.4 \sim 1.5$. ${}^{d}x_Z$ in those terms are thought to be saturated at the toe edge of the supporting sole. $\tilde{x}_Z$ is synchronized at $t \simeq 0.5, 1.0$ when the swing foot lands on the ground, and the continuity of ZMP is held. ${}^{d}x_L$ and ${}^{d}x_R$ discontinuously jump at $t \simeq 0.15, 0.75, 1.25$ which are thought to be times when the ZMP reaches the pivoting sole. In spite of that, x_L and x_R keep continuous, thanks to the low-pass filters. The robot responded to the sudden stop of the reference at $t \simeq 3.0$ without bankruptcy. Fig. 12 shows some sequential snapshots of a motion of the inverted pendulum. The red ball and the green ball in the figure indicate the referential COM position and the simulated ZMP position, respectively. The magenta area is the supporting region composed from the grounding sole.

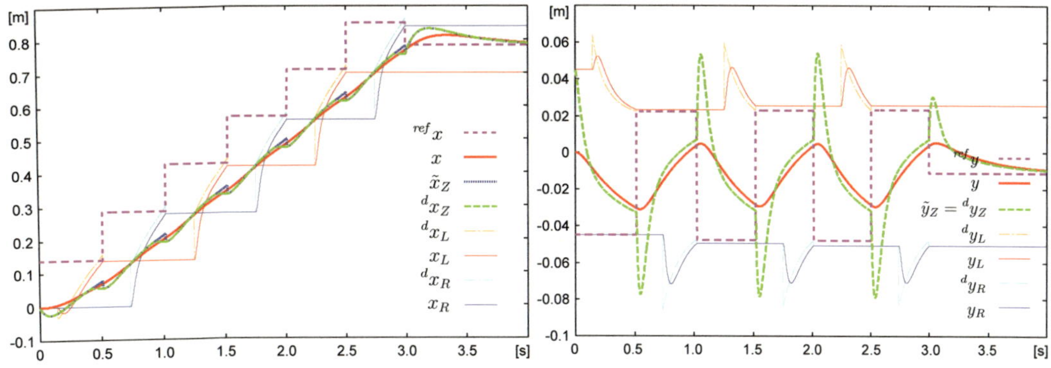

(a) Loci of COM, ZMP and feet in x-axis (b) Loci of COM, ZMP and feet in y-axis

Fig. 10. Resulted loci of COM, ZMP and feet.

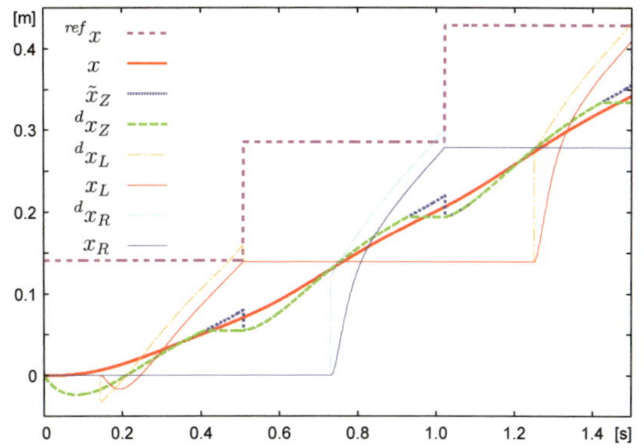

Fig. 11. Zoomed loci of COM, ZMP and feet.

Fig. 13 shows snapshots of the synthesized robot motion computed by the above result and the inverse kinematics. Note that the fullbody dynamics is not considered.

5. Conclusion

We developed an autonomous biped controller, in which the ZMP manipulation under the current support condition and the pedipulation to deform the future support region were synthesized. Both are based on an identical simulated regulator, so that they are integrated into the total control system without any conflicts. Since the simulated regulator involves ZMP in the state variable, it is possible to give a slow mode to COM and a fast mode to ZMP, which is accommodated to the current choice of stance and kicking feet, explicitly by the pole assignment method.

The autonomous controller is promising to improve the system robustness against extrinsic events and uncertainties in the environment. The next short-term issues are to verify the absorption performance of perturbations and to examine the adaptability against rough terrains.

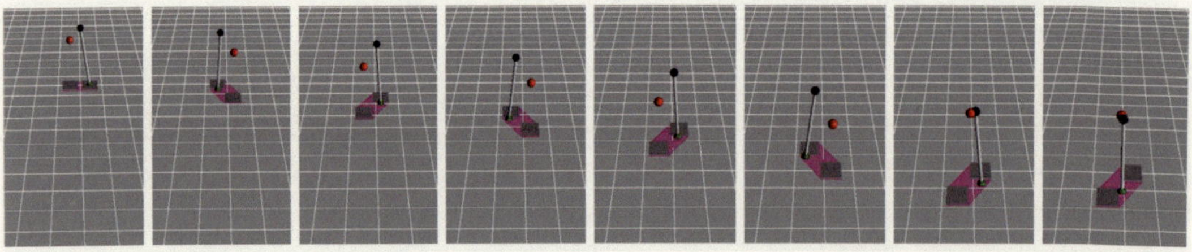

Fig. 12. Snapshots of an inverted pendulum motion controlled by the proposed method.

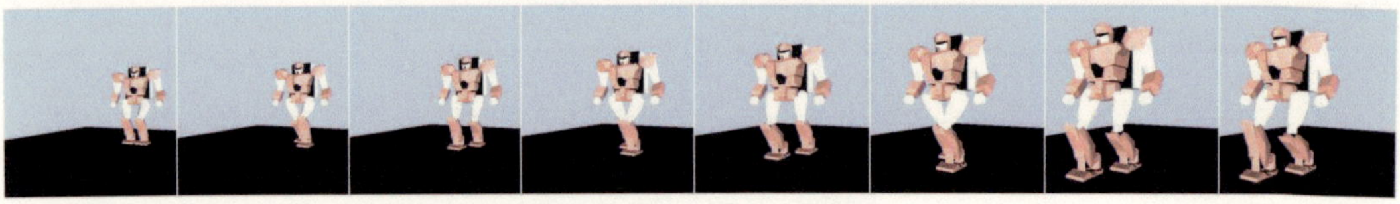

Fig. 13. Snapshots of a walking motion replayed by mighty.

This work was supported in part by Grant-in-Aid for Young Scientists (B) #20760170, Japan Society for the Promotion of Science and by "The Kyushu University Research Superstar Program (SSP)", based on the budget of Kyushu University allocated under President's initiative.

6. References

Collins, S. H., Wisse, M. & Ruina, A. (2001). A Three-Dimensional Passive-Dynamic Walking Robot with Two Legs and Knees, *The International Journal of Robotics Research* **20**(7): 607–615.

Fujimoto, Y., Obata, S. & Kawamura, A. (1998). Robust Biped Walking with Active Interaction Control between Foot and Ground, *Proceedings of the 1998 IEEE International Conference on Robotics & Automation*, pp. 2030–2035.

Furusho, J. & Masubuchi, M. (1986). Control of a Dynamical Biped Locomotion System for Steady Walking, *Transactions of the ASME, Journal of Dynamic Systems, Measurement, and Control* **108**: 111–118.

Gubina, F., Hemami, H. & McGhee, R. B. (1974). On the dynamic stability of biped locomotion, *IEEE Transactions on Bio-Medical Engineering* **BME-21**(2): 102–108.

Hirai, K., Hirose, M., Haikawa, Y. & Takenaka, T. (1998). The Development of Honda Humanoid Robot, *Proceeding of the 1998 IEEE International Conference on Robotics & Automation*, pp. 1321–1326.

Huang, Q., Yokoi, K., Kajita, S., Kaneko, K., Arai, H., Koyachi, N. & Tanie, K. (2001). Planning Walking Patterns for a Biped Robot, *IEEE Transactions on Robotics and Automation* **17**(3): 280–289.

Kajita, S., Kanehiro, F., Kaneko, K., Fujiwara, K., Harada, K., Yokoi, K. & Hirukawa, H. (2003). Biped Walking Pattern Generation by using Preview Control of Zero-Moment Point, *Proceedings of the 2003 IEEE International Conference on Robotics & Automation*, pp. 1620–1626.

Kajita, S. & Tani, K. (1995). Experimental Study of Biped Dynamic Walking in the Linear Inverted Pendulum Mode, *Proceedings of the 1995 IEEE International Conference on Robotics & Automation*, pp. 2885–2819.

Kajita, S., Yamaura, T. & Kobayashi, A. (1992). Dynamic Walking Control of a Biped Robot Along a Potential Energy Conserving Orbit, *IEEE Transactions on Robotics and Automation* **8**(4): 431–438.

Löffler, K., Gienger, M. & Pfeiffer, F. (2003). Sensor and Control Design of a Dynamically Stable Biped Robot, *Proceedings of the 2003 IEEE International Conference on Robotics & Automation*, pp. 484–490.

McGeer, T. (1990). Passive Dynamic Walking, *The International Journal of Robotics Research* **9**(2): 62–82.

Mitobe, K., Mori, N., Aida, K. & Nasu, Y. (1995). Nonlinear feedback control of a biped walking robot, *Proseedings of the 1995 IEEE International Conference on Robotics & Automation*, pp. 2865–2870.

Miura, H. & Shimoyama, I. (1984). Dynamic Walk of a Biped, *The International Journal of Robotics Research* **3**(2): 60–74.

Nagasaka, K., Inaba, M. & Inoue, H. (1999). Walking Pattern Generation for a Humanoid Robot Based on Optimal Gradient Method, *Proceedings of 1999 IEEE International Conference on Systems, Man, and Cybernetics*, pp. VI–908–913.

Nagasaka, K., Kuroki, Y., Suzuki, S., Itoh, Y. & Yamaguchi, J. (2004). Integrated Motion Control for Walking, Jumping and Running on a Small Bipedal Entertainment Robot, *Proceedings of the 2004 IEEE International Conference on Robotics and Automation*, pp. 3189–3914.

Raibert, M. H., Jr., H. B. B. & Chepponis, M. (1984). Experiments in Balance with a 3D One-Legged Hopping Machine, *The International Journal of Robotics Research* **3**(2): 75–92.

Sugihara, T. & Nakamura, Y. (2005). A Fast Online Gait Planning with Boundary Condition Relaxation for Humanoid Robots, *Proceedings of the 2005 IEEE International Conference on Robotics & Automation*, pp. 306–311.

Sugihara, T., Nakamura, Y. & Inoue, H. (2002). Realtime Humanoid Motion Generation through ZMP Manipulation based on Inverted Pendulum Control, *Proceedings of the 2002 IEEE International Conference on Robotics & Automation*, pp. 1404–1409.

Sugihara, T., Yamamoto, K. & Nakamura, Y. (2007). Hardware design of high performance miniature anthropomorphic robots, *Robotics and Autonomous System* **56**(1): 82–94.

Takanishi, A., Egusa, Y., Tochizawa, M., Takeya, T. & Kato, I. (1988). Realization of Dynamic Walking Stabilized with Trunk Motion, *ROMANSY 7*, pp. 68–79.

Vukobratović, M., Frank, A. A. & Juričić, D. (1970). On the Stability of Biped Locomotion, *IEEE Transactions on Bio-Medical Engineering* **BME-17**(1): 25–36.

Vukobratović, M. & Stepanenko, J. (1972). On the Stability of Anthropomorphic Systems, *Mathematical Biosciences* **15**(1): 1–37.

Westervelt, E. R., Buche, G. & Grizzle, J. W. (2004). Experimental Validation of a Framework for the Design of Controllers that Induce Stable Walking in Planar Bipeds, *The International Journal of Robotics Research* **24**(6): 559–582.

Witt, D. C. (1970). A Feasibility Study on Automatically-Controlled Powered Lower-Limb Prostheses, *Report*, University of Oxford.

Yamakita, M., Asano, F. & Furuta, K. (2000). Passive Velocity Field Control of Biped Walking Robot, *Proceedings of the 2000 IEEE International Conference on Robotics & Automation*, pp. 3057–3062.

13

Nonlinear $\mathcal{H}_\infty$ Control Applied to Biped Robots

Adriano A. G. Siqueira, Marco H. Terra and Leonardo Tubota
University of São Paulo, São Carlos, São Paulo
Brazil

1. Introduction

Researches on biped robots have been the focus of many universities and industries around the world Hirukawa et al. (2004). The challenges in the study of this kind of system, beyond, of course, the stability of the walking, are strongly coupled with nonlinearities and with the occurrence of discontinuities in the joint variables when the swing foot touches the ground.
Control strategies for biped robots can be classified in two categories according to the degree of actuation of the joints: control strategies based on the Zero Moment Point (ZMP), which is considered a fully actuated system, and control strategies based on the dynamic passive walking, which the few (or even none) degrees of actuation are used.
The ZMP criterion to evaluate biped robot stability was initially proposed by Vukobratovic & Juricic (1969). Shortly, the ZMP is the point in the ground where the resultant of reaction moments are null. If the ZMP is inside the support polygon, defined by the contact points of the robot with the ground, the walking is stable. In Huang et al. (2001), it is presented one of many trajectory generators for biped robots taking into account the ZMP criterion. Specific points of the ankle and hip trajectories are defined according to the desired step length and duration. The minimization of a functional related to the ZMP gives the desired trajectories, which, tracked trough the actuation of all joints, generate a stable movement of the biped robot.
On the other hand, the movement of passive walking robots is created by the action of the gravity in an inclined surface, without any joint actuation. In this case, when the foot of the swing leg touches the ground there is energy spending which is compensated by the ground slope. A passive walking occurs when the joint positions and velocities at the final of the step are identical to the initial conditions. The movement generated can be characterized as periodic movement or, in terms of nonlinear systems analysis, a limit cycle.
The problems related to passive dynamic walking are more complex than to the static case, however, the dynamic walking gives higher velocities, bigger energy efficiency, and a smooth and anthropomorphic movement. A scientific research on these mechanisms starts with McGeer (1990). McGeer shows that a fully no actuated biped robot, and hence no controllable, can present a stable walking on an inclined surface for some dynamic parameter selections. In Asano et al. (2005) the passive walk is used in order to define a control law that uses the energy trajectory as trajectory reference, which is time independent (standard trajectories which depend on state variables are time dependent) and it assures the robustness of the system. This is the motivation to consider the passive walk as the best model in terms of energy consumption.

However, the basin of attraction of the limit cycle for passive walking is generally small and sensitive to disturbances and ground slope variations. For example, in Goswani et al. (1998), bifurcations with period duplication occur when the ground slope varies from 3° to 5°. In Spong & Bhatia (2003), it is shown that the walking can be made slope invariant for a bidimensional biped robot without knees and torso, and with actuators on the hip and ankles, considering a controller based on gravity compensation. It is also shown that the basin of attraction and the robustness can be increased using an energy-based controller. In Bhatia & Spong (2004), this procedure is extended to a bidimensional biped robot with knees and torso. Gravity compensation and energy-based controllers are also presented in Asano & Yamakita (2001) for passive walking robots.
In this chapter, it is used a control strategy for passive walking robots to increase the robustness against external disturbances based on nonlinear $\mathcal{H}_\infty$ controllers. They are based on linear parameter varying (LPV) representation of the biped robot. This kind of approach has been applied to fully actuated and underactuated robot manipulators Siqueira & Terra (2002; 2004). The objective, as in Bhatia & Spong (2004), is to apply the robust controller to leads back the robot to the basin of attraction of the limit cycle. We consider that the robot leaves it due to the disturbances. In this way, the period of application of the controller is minimal and the main characteristic of passive walking (energy save) is preserved.
The biped robot considered here, in addition to knees and torso of the robot presented in Bhatia & Spong (2004), has feet. For biped robots with feet, the time instant when the foot rotates around the toes can be determined by the Foot Rotation Indicator (FRI), described in Goswani (1999). The FRI is a point in the ground surface where the resultant ground reaction force would have to act to keep the foot stationary. In Choi & Grizzle (2005), the authors consider a biped robot with feet and use the FRI as control variable in the procedure to find the zero dynamics of the system proposed in Westervelt et al. (2003). Here, the FRI is also used as control variable to provide to the biped robot an anthropomorphic walking.
This chapter is organized as follows: in Section 2, the dynamic model of the biped robot, considering the foot rotation, the knee strike, and the ground collision, is presented; in Section 3.2, the robust dynamic walking control strategy using the nonlinear $\mathcal{H}_\infty$ control is proposed; and in Section 4, the results obtained from simulation are shown.

2. Dynamic Model of 2D Biped Robots

In this section we present the dynamic model of a planar walking robot, including the swing and ground collision phases. These phases are characterized, respectively, by having the robot only one foot or both feet in contact with the ground. Consider a biped robot with a torso, knees and feet, Figure 1. The overall system has nine degrees of freedom (DOF), corresponding to two DOF of the Y-Z plane, three DOF of each leg, and one DOF of the torso. The biped robot moves over a planar surface with inclination ϕ with relation to the inertial coordinate system.
The robot configuration is described by the angles: q_1 (foot of the stance leg), q_2 (stance leg), q_3 (torso), q_4 (thigh of the swing leg), q_5 (shank of the swing leg), and q_6 (foot of the swing leg). The generalized coordinate q_1 is defined as absolute value with relation to the horizontal axis Y and the remaining angles are defined as relative value to the previous link. The parameters mass (m), length (l), center of mass (c), and inertia momentum with relation to the center of mass (I) of each link are shown in Table 1, where the indexes $i = (1, \cdots, 6)$ are related to the foot of the stance leg, stance leg, torso, thigh of the swing leg, shank of the swing leg, and foot of the swing leg, respectively. The parameters in Table 1 correspond to the values of the

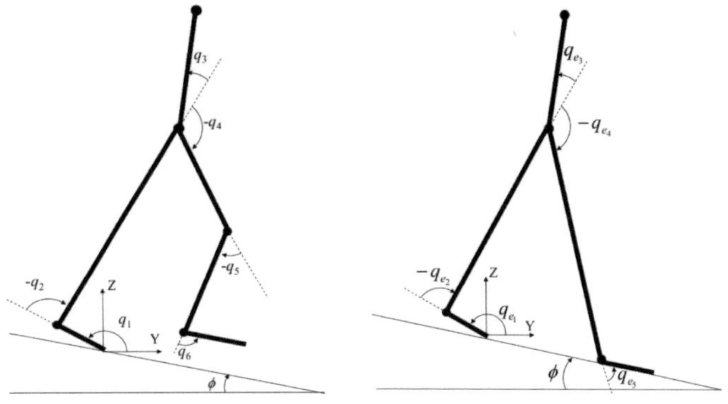

(a) (b)

Fig. 1. a) Bidimensional biped robot with feet. b) Ground collision model.

bidimensional experimental robot RABBIT described in Chevallereau et al. (2003), except the values related to the feet, since RABBIT has no feet. The robot motion in the sagittal plane is divided in two phases: the swing phase and the ground collision phase.

2.1 Swing phase

The swing phase is considered in this chapter as consisting of three sub-phases: before the foot rotation, after the foot rotation and before the knee strike, and after the knee strike. During the first sub-phase, before the foot rotation, the foot link is parallel to the ground and its angular velocity is zero. The change from this phase to the phase of foot rotation around the toe occurs when the Foot Rotation Indicator (FRI) point is outside the foot support area.
The FRI definition was first described in Goswani (1999), as the point in the ground surface where the resultant ground reaction force would have to act to keep the foot stationary. The FRI has some important properties for the control of a biped robot:

- The FRI indicates the occurrence of foot rotation and its direction;
- The location of the FRI indicates the amount of unbalanced moment on the foot;
- The minimum distance between the support polygon and the FRI is a measurement of the stability margin, since that the FRI is inside of the support polygon.

In other words to keep the foot stationary, the FRI must remain inside the foot support area. The FRI incorporates the biped robot dynamics, differently from the ground projection of the center of mass, that represents a static characteristic. The FRI is also different from the center of pressure, known as the zero moment point (ZMP). The ZMP is a point in the ground surface where the resultant ground reaction force is actually applied. Hence, the ZMP never leaves the foot support area, whereas the FRI does so. When the point of application of the resultant ground reaction force is inside the foot support area, the FRI and ZMP are coincident. In a

Parameter	Value	Unit	Parameter	Value	Unit
m_1	$0,5$	kg	c_1	$0,05$	m
m_2	10	kg	c_2	$0,28$	m
m_3	20	kg	c_3	$0,2$	m
m_4	$6,8$	kg	c_4	$0,163$	m
m_5	$3,2$	kg	c_5	$0,128$	m
m_6	$0,5$	kg	c_6	$0,05$	m
l_1	$0,1$	m	I_1	$0,329$	kg.m^2
l_2	$0,8$	m	I_2	$2,06$	kg.m^2
l_3	$0,625$	m	I_3	$1,42$	kg.m^2
l_4	$0,4$	m	I_4	$0,899$	kg.m^2
l_5	$0,4$	m	I_5	$0,878$	kg.m^2
l_6	$0,1$	m	I_6	$0,329$	kg.m^2

Table 1. Biped Robot Parameters.

three-dimensional biped robot, the coordinates of the FRI in the X-Y plane can be obtained as:

$$x_{FRI} = \frac{m_1 x_{c_1} g + \sum_{i=2}^{n} m_i x_{c_i}(\ddot{z}_{c_i} + g) - \sum_{i=2}^{n} m_i z_{c_i}\ddot{x}_{c_i} - \sum_{i=2}^{n} (\dot{H}_{c_y})_i}{m_1 g + \sum_{i=2}^{n} m_i(\ddot{z}_{c_i} + g)},$$

$$y_{FRI} = \frac{m_1 y_{c_1} g + \sum_{i=2}^{n} m_i y_{c_i}(\ddot{z}_{c_i} + g) - \sum_{i=2}^{n} m_i z_{c_i}\ddot{y}_{c_i} - \sum_{i=2}^{n} (\dot{H}_{c_x})_i}{m_1 g + \sum_{i=2}^{n} m_i(\ddot{z}_{c_i} + g)}, \tag{1}$$

where g is the gravity acceleration, x_{c_i} and y_{c_i} are the position coordinates of the center of mass of the link i, $\ddot{x}_{c_i}$, $\ddot{y}_{c_i}$ and $\ddot{z}_{c_i}$ are the acceleration coordinates of the center of mass of the link i and $(\dot{H}_{c_x})_i$ and $(\dot{H}_{c_y})_i$ are the derivative of the angular momentum around the center of mass of the link i. As we are working with a planar biped robot in this chapter, if y_{IRP} in Eq. 1 is above 0, the foot rotation occurs.

The foot rotation occurs in the human being due to the application of a torque around the ankle joint at the end of the foot stance phase. Trying to reproduce this natural behavior on the biped robot considered in this chapter, a torsional spring is introduced on the ankle joint of the stance leg (joint 2), as shown in the Figure 2.

The function describing the torque applied by the spring, τ_M, with relation to the ankle joint displacement, q_2, is given by

$$\tau_M = K_0 - K_M q_2, \tag{2}$$

where $K_0 > K_M > 0$ are control design parameters. They are adjusted to set the foot rotation instant during the step, Palmer (2002). Before the foot rotation, the spring is compressed and

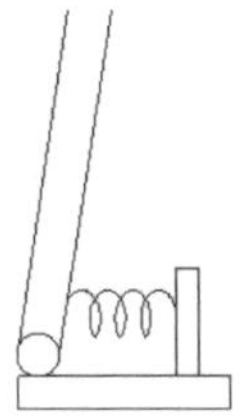

Fig. 2. Representation of a torsional spring in the ankle joint.

stores some part of the dynamic walking energy. During the foot rotation, the stored potential energy is transformed into kinematic one, keeping the dynamic walking.
During the two sub-phases before the knee strike, the dynamic equations of the system are given by:

$$M(q)\ddot{q} + C(q,\dot{q})\dot{q} + g(q) = \tau, \tag{3}$$

where $q = [q_1 \ q_2 \ q_3 \ q_4 \ q_5 \ q_6]^T$, $M(q) \in \Re^{6\times 6}$ is the inertia matrix, $C(q,\dot{q}) \in \Re^{6\times 6}$ is the Coriolis/centripetal matrix, $g(q) \in \Re^6$ is the gravitational vector, and $\tau = [\tau_1 \ \tau_2 \ \tau_3 \ \tau_4 \ \tau_5 \ \tau_6]^T$ is the applied torque vector, τ_i $(i = 1,\cdots,6)$ are the torques in the toe of the stance leg, in the ankle of the stance leg, between the stance leg and the torso, between the stance leg and the thigh of the swing leg, in the knee of the swing leg, and in the ankle of the swing leg, respectively. The shank of the swing leg can move around the knee only to the posterior part of the leg, that is, $q_5 < 0$.
At the instant of the impact, the knee displacement vanishes, $q_5 = 0$. The joint positions do not change during the impact, but the velocities suffer an abrupt change. This event can be modeled by conservation of the angular momentum with relation to the stance leg:

$$M(q^+)\dot{q}^+ = M(q^-)\dot{q}^- - J_i^T\lambda_i, \tag{4}$$

where λ_i are the impulsive forces generated by the impact, q^+ and q^- are the values of q just before and after the knee strike, and $J_i = [0 \ 0 \ 0 \ 0 \ 1 \ 0]$ is the Jacobian matrix generated by the constraint $J_i\dot{q}^+ = 0$. Since $M(q^+) = M(q^-)$, because $q^+ = q^-$.
From (4), the joint velocities after the impact are computed as function of the velocities just before the impact and of the constraint forces as $\dot{q}^+ = \dot{q}^- - M(q)^{-1}J_i^T\lambda_i$. From the above considerations, the following relation is obtained:

$$J_i\dot{q}^+ = 0 = J_i\dot{q}^- - J_iM(q)^{-1}J_i^T\lambda_i,$$

which results in $\lambda_i = X_i^{-1}J_i\dot{q}^-$ where $X_i = J_iM(q)^{-1}J_i^T$.
After the knee strike, the swing leg remains straight due the constraint forces on the knee generated by a mechanism that locks the joint ($q_5 = 0$). The dynamic equations of the system, considering the constraint forces, are given by:

$$M(q)\ddot{q} + C(q,\dot{q})\dot{q} + g(q) = \tau - J_r^T\lambda_r, \tag{5}$$

where $J_r = [0 \ 0 \ 0 \ 0 \ 1 \ 0]$. The constraint forces are given by $\lambda_r = -X_r^{-1}J_rM(q)^{-1}(C(q,\dot{q})\dot{q} + g(q))$ where $X_r = J_rM(q)^{-1}J_r^T$.

2.2 Ground Collision Phase

The collision of the foot of the swing leg with the ground can be modeled as a contact between two rigid bodies. It is considered that the heel and the toe of the foot impact the ground at the same instant and the foot of the stance leg is rotated. Also, the system is modeled as a biped robot without knees, Figure 1. In this case, there are seven DOF (two DOF on the Y-Z plane, two DOF for each leg and the DOF of the torso). The dynamic equation is given by:

$$M_e(q_e)\ddot{q}_e + C_e(q_e,\dot{q}_e)\dot{q}_e + g_e(q_e) = \tau_e + \delta F_{ext}, \tag{6}$$

where $q_e = [q_{e_1}\ q_{e_2}\ q_{e_3}\ q_{e_4}\ q_{e_5}\ y\ z]^T$, y and z are the position coordinates of the toe of the stance leg, and δF_{ext} are the external forces acting on the robot in the contact points.
It is assumed that the collision occurs during an infinitesimal period of time. The external forces in the impact moment can be represented by impulses. These impulsive forces generate an instantaneous change of joint velocities, but the positions remain the same. In this case the torques given by the actuators are not impulsive and can be neglected. Under these conditions, integrating (6) during the impact duration we obtain:

$$M_e(q_e)(\dot{q}_e^+ - \dot{q}_e^-) = F_{ext}, \tag{7}$$

where F_{ext} are the impulsive forces, and $\dot{q}_e^+$ and $\dot{q}_e^-$ are the joint velocities just after and before the impact, respectively. The joint positions do not change during the impact, then $q_e^+ = q_e^-$.
It is also assumed that the stance leg lifts from the ground without interaction, no external forces are applied at its contact point. In this case, the external forces, F_{ext}, contain only the forces acting on the heel and toe of the foot of the swing leg. Let Ψ_c and Ψ_d be the vectors of position coordinates of the heel and the toe, respectively. Hence:

$$F_{ext} = [E_c^T(q_e)\ E_d^T(q_e)] \begin{bmatrix} F_{c_T} \\ F_{c_N} \\ F_{d_T} \\ F_{d_N} \end{bmatrix}, \tag{8}$$

where

$$E_c(q_e) = \frac{\partial \Psi_c}{\partial q_e}, \quad E_d(q_e) = \frac{\partial \Psi_d}{\partial q_e}, \tag{9}$$

and F_{c_T}, F_{c_N}, F_{d_T}, and F_{d_N} are the tangential and normal forces applied on the heel and on the toe. An additional set of four equations is obtained from the condition that the heel and the toe of the swing leg can not move after the impact, $d/dt\Psi_c(q_e) = (\partial\Psi_c/\partial q_e)\dot{q}_e^+ = 0$ and $d/dt\Psi_d(q_e) = (\partial\Psi_d/\partial q_e)\dot{q}_e^+ = 0$, that is:

$$E_c(q_e)\dot{q}_e^+ = 0, \quad E_d(q_e)\dot{q}_e^+ = 0. \tag{10}$$

From (7) and (10):

$$\begin{bmatrix} M_e(q_e) & -E_c(q_e)^T & -E_d(q_e)^T \\ E_c(q_e) & 0 & 0 \\ E_d(q_e) & 0 & 0 \end{bmatrix} \begin{bmatrix} \dot{q}_e^+ \\ F \end{bmatrix} = \begin{bmatrix} \dot{q}_e^- \\ 0 \end{bmatrix}, \tag{11}$$

with $F = [F_{c_T}\ F_{c_N}\ F_{d_T}\ F_{d_N}]^T$. The value of $\dot{q}^+$ is expressed in terms of $\dot{q}_e^+$ by:

$$\dot{q}^+ = [\dot{q}_{e_1}^+\ \dot{q}_{e_2}^+\ \dot{q}_{e_3}^+\ \dot{q}_{e_4}^+\ 0\ \dot{q}_{e_5}^+]^T. \tag{12}$$

After the impact, the swing leg becomes the stance leg and vice-versa. Hence, a coordinate transformation is needed to reinitialize the System (3). Considering $x = [q^T \ \dot{q}^T]^T$, the transformation can be given by

$$x = \Delta(x), \tag{13}$$

where

$$\Delta(x) = \begin{bmatrix} \pi \\ -q_6 \\ (q_1 + q_2 + q_3 - \pi + q_6) \\ (-2\pi - q_4) \\ 0 \\ -q_2 \\ 0 \\ -\dot{q}_6 \\ (\dot{q}_1 + \dot{q}_2 + \dot{q}_3 + \dot{q}_6) \\ -\dot{q}_4 \\ -\dot{q}_2 \end{bmatrix}. \tag{14}$$

3. Robust Dynamic Walking Control

The main problem on passive walking consists on finding a stable limit cycle without actuation of the joints for a given ground slope. However, stable limit cycles cannot be found for the biped robot considered in this paper without a loop of control be applied to the torso. In this way, PD controllers are used to keep the torso in a fixed position with relation to the horizontal line, that is, the absolute angle, $\theta_4 = q_1 + q_2 + q_3$, is controlled. The foot of the swing leg must also be controlled to reach the ground surface parallel to it at the impact instant to guarantee that the toe and the ankle touch the ground in the same time. In this case, the absolute angle $\theta_6 = q_1 + q_2 + q_4 + q_5 + q_6$, must be zero, $\theta_6^d = 0$. The PD controllers are given by:

$$\begin{aligned} \tau_4 &= K_{P_4}(\theta_4^d - \theta_4) - K_{D_4}\dot{\theta}_4, \\ \tau_6 &= K_{P_6}(\theta_6^d - \theta_6) - K_{D_6}\dot{\theta}_6, \end{aligned} \tag{15}$$

where K_{P_4}, K_{D_4}, K_{P_6}, and K_{D_6} are proportional and derivative gains, and q_4^d and q_6^d are the desired joint position for the torso and the foot of the swing leg, respectively.

In order to guarantee the passive walking at any ground slope, including zero inclination, a gravity compensation control developed by Spong & Bhatia (2003) is used. The key idea is to consider the existence of an initial condition $x(\phi_0) = [q_1(0) \ q_2(0) \ \cdots \ q_6(0) \ \dot{q}_1(0) \ \cdots \ \dot{q}_6(0)]^T$ such that, with only the PD controllers described above, a limit cycle is found for ground slope ϕ_0. Now suppose that the desired slope is ϕ. Let $\beta = \phi_0 - \phi$, R_β be the rotation defined by angle β and $P(\theta)$ the potential energy, where θ is the absolute joint position vector. In the robot configuration space the representation of angle β is given by

$$R_\beta(\theta) = \begin{bmatrix} \theta_1 - \beta \\ \vdots \\ \theta_6 - \beta \end{bmatrix}. \tag{16}$$

Now, defining $P_\beta(\theta) = P(R_\beta(\theta))$ and

$$g_\beta = \frac{\partial P_\beta(\theta)^T}{\partial \theta}, \tag{17}$$

the feedback control law is given by

$$\tau = B^{-1}(g(\theta) - g_\beta(\theta)). \tag{18}$$

This control law allows the biped robot to perform a limit cycle for ground slope ϕ, considering as initial condition $x(\phi) = [q_1(0) + \beta \; q_2(0) \; \cdots \; q_6(0) \; \dot{q}_1(0) \; \cdots \; \dot{q}_6(0)]^T$.

3.1 Nonlinear $\mathcal{H}_\infty$ control

In this work, nonlinear $\mathcal{H}_\infty$ control is applied to biped robots to increase the basin of attraction of the limit cycle and the robustness against disturbances. The proposed controller is based on design procedures developed for LPV systems. This approach provides a systematic way, using solutions based on Linear Matrix Inequalities (LMI), to design controllers that satisfy the $\mathcal{L}_2$ gain condition Wu et al. (1996). The nonlinear dynamics of the biped robot is represented as a LPV system whose parameters are in function of the state ($\rho(x)$), namely, quasi-LPV representation.

Consider the system:

$$\begin{bmatrix} \dot{x} \\ z_1 \\ z_2 \end{bmatrix} = \begin{bmatrix} A(\rho) & B_1(\rho) & B_2(\rho) \\ C_1(\rho) & 0 & 0 \\ C_2(\rho) & 0 & I \end{bmatrix} \begin{bmatrix} x \\ w \\ u \end{bmatrix}, \tag{19}$$

where x is the state vector, u is the control input vector, w is the external input vector, z_1 and z_2 are the output variables, and ρ is the parameter varying vector.

To present $\mathcal{H}_\infty$ performance the closed loop system need to satisfy the $\mathcal{L}_2$ gain condition given by:

$$\int_0^T \|z\|^2 \, dt \leq \gamma^2 \int_0^T \|w\|^2 \, dt. \tag{20}$$

If it is possible to find a solution $X(\rho) > 0$ to the following set of LMIs:

$$\begin{bmatrix} E(\rho) & X(\rho)C_1^T(\rho) & B_1(\rho) \\ C_1(\rho)X(\rho) & -I & 0 \\ B_1^T(\rho) & 0 & -\gamma^2 I \end{bmatrix} < 0 \tag{21}$$

where

$$E(\rho) = -\sum_{i=1}^{m} \pm \left(\nu_i \frac{\partial X}{\partial \rho_i} \right) - B_2(\rho)B_2^T(\rho) + \widehat{A}(\rho)X(\rho) + X(\rho)\widehat{A}(\rho)^T$$

and $\widehat{A}(\rho) = A(\rho) - B_2(\rho)C_2(\rho)$, then, the closed loop system has $\mathcal{L}_2$ gain $\leq \gamma$ under the state feedback law

$$u = -(B_2(\rho)X^{-1}(\rho) + C_2(\rho))x.$$

Hence, it is need to solve a set of parametric LMIs, (21), that is an infinitesimal convex optimization problem. Fortunately, a practical scheme using basis functions for $X(\rho)$ and griding the parameter set was developed in Wu et al. (1996) to solve this problem.

The quasi-LPV representation we are considering in this chapter for biped robots is shown in the following. The hybrid control strategy proposed in the next section considers that the nonlinear $\mathcal{H}_\infty$ control is applied in the two sub-phases of the movement before the knee strike. For the sub-phase before the foot rotation, the system is considered fully-actuated since the first joint value (q_1) is fixed. It is considered the following dynamic equation with the inclusion of a disturbance vector d, corresponding to external disturbance and parametric uncertainties:

$$M_{aa}(q_a)\ddot{q}_a + C_{aa}(q_a,\dot{q}_a)\dot{q}_a + g_a(q_a) = \tau_a + d, \tag{22}$$

where $q_a = [q_2 \;\; q_3 \;\; q_4 \;\; q_5 \;\; q_6]^T$ are the active joints, and $M_{aa}(q_a)$, $C_{aa}(q_a,\dot{q}_a)$, $g_a(q_a)$, and τ_a are the corresponding partitions of the dynamic matrices.
Define the state as the composition of the trajectory tracking error and its derivative as:

$$\tilde{x} = \begin{bmatrix} \dot{q}_a - \dot{q}_a^d \\ q_a - q_a^d \end{bmatrix} = \begin{bmatrix} \dot{\tilde{q}}_a \\ \tilde{q}_a \end{bmatrix}. \tag{23}$$

From (22) and (23), the biped robot state space equation is given by

$$\dot{\tilde{x}} = A(q_a,\dot{q}_a)\tilde{x} + B_1 w + B_2 u, \tag{24}$$

with

$$A(q_a,\dot{q}_a) = \begin{bmatrix} -M_{aa}^{-1}(q_a)C_{aa}(q_a,\dot{q}_a) & 0 \\ I & 0 \end{bmatrix}, \quad B_1 = \begin{bmatrix} I \\ 0 \end{bmatrix},$$
$$B_2 = \begin{bmatrix} I \\ 0 \end{bmatrix}, \quad w = M_{aa}^{-1}(q_a)d,$$
$$u = M_{aa}^{-1}(q_a)(\tau_a - M_{aa}(q_a)\ddot{q}_a^d - C_{aa}(q_a,\dot{q}_a)\dot{q}_a^d - g_a(q)).$$

The applied torque is computed as

$$\tau_a = M_{aa}(q_a)(\ddot{q}_a^d + u) + C_{aa}(q_a,\dot{q}_a)\dot{q}_a^d + g_a(q_a). \tag{25}$$

For the sub-phase after the foot rotation and before the knee strike, the system is underactuated since no torque is applied to the first joint and it is free. In this work, it is proposed to control only the active joints in this phase, considering the dynamic coupling with first joint and its dynamics. The following dynamic equation is used to design the quasi-LPV controller:

$$\bar{M}(q)\ddot{q}_a + \bar{C}(q,\dot{q})\dot{q}_a + \bar{D}(q,\dot{q})\dot{q}_1 + \bar{g}(q_a) = \tau_a + \bar{d}. \tag{26}$$

where $\bar{M} = M_{aa} - M_{a1}{M_{11}}^{-1}M_{1a}$, $\bar{C} = C_{aa} - M_{a1}{M_{11}}^{-1}C_{1a}$, $\bar{D} = C_{a1} - M_{a1}{M_{11}}^{-1}C_{11}$, $\bar{g} = g_a - M_{a1}{M_{11}}^{-1}g_1$, and $\bar{d}$ are the external disturbances and parametric uncertainties. Following the steps described above, a quasi-LPV representation of the biped robot, similar to (24), can be found for this sub-phase of the movement.

3.2 Hybrid Control

The PD controllers and the gravity compensation controller provide a stable walking for different ground slopes for initial positions close to the limit cycle. However, the basin of attraction of a limit cycle must be large enough to provide the system robustness against disturbances, parametric uncertainties and terrain variations. The controllers presented above are combined to produce a robust dynamic walking Bhatia & Spong (2004).

The minimum distance between the current position to the limit cycle is computed and compared to an adjustable value C, corresponding to the limit cycle basin of attraction. The minimum distance is given as

$$\bar{d} = min_i||q - q_r(i)||, \tag{27}$$

where $q_r(i)$ are discrete points on the limit cycle.
If the distance of the initial position limit cycle is lower than C, the system converges to the limit cycle. If the trajectory leaves the basin of attraction, that is, the value of $\bar{d}$ is greater than C, a control strategy based on the nonlinear $\mathcal{H}_\infty$ controller is used to come back the system to the basin of attraction as shown in Figure 3.

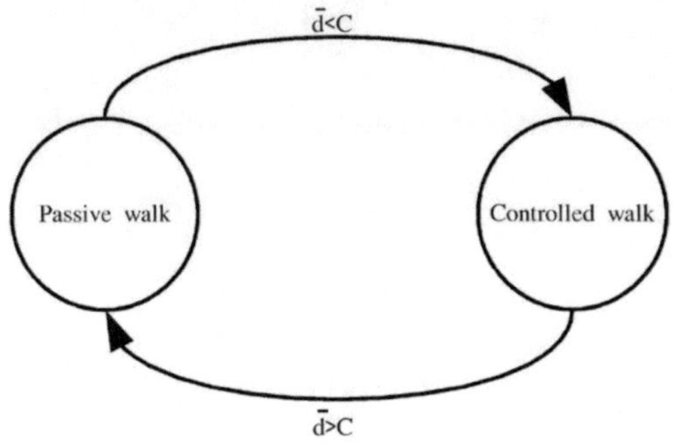

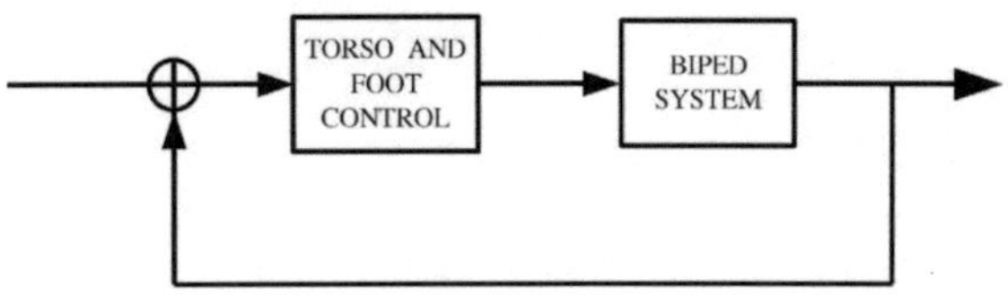

(a) Passive Diagram

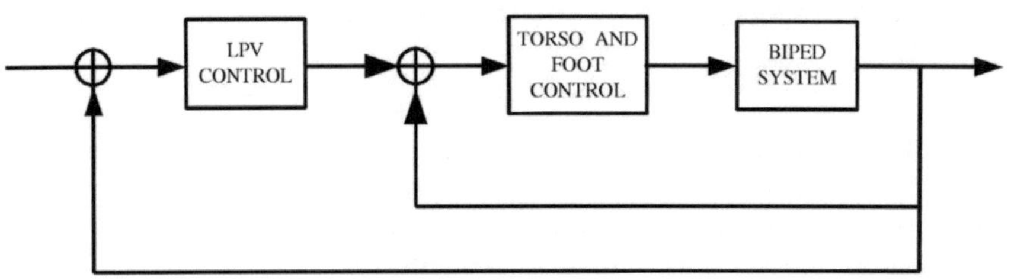

(b) Controlled Diagram

Fig. 3. Hybrid Control Strategy.

The changing point between the nonlinear $\mathcal{H}_\infty$ controller and when there is not trajectory tracking controllers (gravity compensation and PD ones) must be chosen to guarantee a smooth and natural walking. Considering the walking cycle, it is chosen the instant when

the knee strike as been this point. At this moment the swing leg is ahead the stance leg. The desired trajectory, q^d, between the position where the biped leaves the basin of attraction and the position of the knee strike is defined in this work as a 5-th polynomial, with duration corresponding to the remaining time to the impact.
Since the third sub-phase (after the knee strike and before the ground collision) is very short in the limit cycle, the nonlinear $\mathcal{H}_\infty$ controller is applied only before the knee strike. If the biped robot leaves the basin of attraction after the knee strike, only the PD controllers are used until the ground collision, and then, the nonlinear $\mathcal{H}_\infty$ controller is applied from this point until the next knee strike.

4. Results

In this section, it is presented the limit cycle for the bidimensional biped robot with feet described in Section 2 and the simulated results obtained from the implementation of the proposed hybrid control strategy. Using the Newton-Raphson method and considering a ground slope $\phi = 3°$, the initial conditions to the limit cycle is given by

$$x(\phi) = \begin{bmatrix} \pi & -1.30 & -0.55 & -3.89 & 0 & 1.52 & 0 & -1.22 & 2.09 & 0.34 & 0.34 & -1.33 \end{bmatrix}^T.$$

The values used to find the solution, and the values used thorough this section, are

$$K_{P_4} = 150, \quad K_{D_4} = 100, \quad K_{P_6} = 80 \;\text{ e }\; K_{D_6} = 10,$$

for the gains of the PD controllers, and $\theta_4^d = 80°$ and $\theta_6^d = 0°$, for the desired joint positions. The reference position to the trunk, θ_4^d, was selected closer to the human trunk position for a walking ahead movement. The foot rotation instant is properly adjusted through K_0 and K_M of the torsional spring on the ankle (2). For this case they were chosen as

$$K_0 = -10 \quad \text{and} \quad K_M = 30.$$

With the above values, it is found a walking cycle close to the human being with the foot rotation occurring after the swing leg to pass the normal to the ground surface and before the knee strike, which occurs just before the ground collision of the swing leg. The joint trajectories as function of the time are shown in Figure 4 and the limit cycles for the joints are shown in Figure 5, where each number in the limit cycles represents a pose of the robot during the step. These points are described below and shown in Figure 6, where the green leg is the support one and the red leg is the stance one:

1. Beginning of the limit cycle: support leg in front of the swing leg;
2. Intermediary point of the swing phase, before the foot rotation: support foot parallel to the ground;
3. Beginning of the foot rotation around the toe;
4. Intermediary point of the swing phase, before the knee strike: angle of the shank is negative ($q_5 < 0$);
5. Knee strike: angle of the shank is zero and the swing leg is straight;
6. Collision of the swing foot with the ground: support leg in the back of the swing leg.

The limit cycle found is stable. This characteristic can be verified computing the eigenvalues of the Jacobian matrix, J, obtained at the last Newton-Raphson method iteration. It is observed that the eigenvalues of J are all within the unitary circle of the complex plane, showing the stability. However, the maximum absolute value of the eigenvalues of J is given by: $max|\lambda(J)| = 0,99411$, that is, the limit cycle has a low convergence rate.

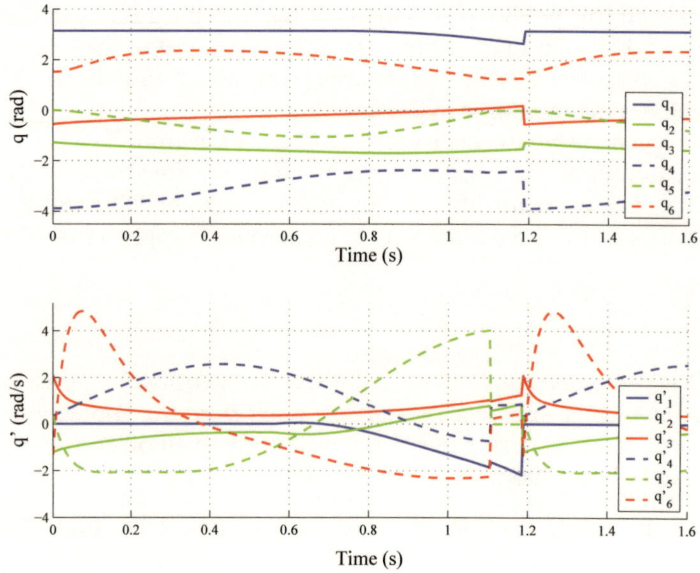

Fig. 4. Joint position and velocity time variations.

4.1 Nonlinear $\mathcal{H}_\infty$ Control

To apply the algorithm described in Section 3.1, the parameters $\rho(\tilde{x})$ are chosen as been the states representing only the position errors of the joints 2, 3, and 4 ($\widetilde{q}_2, \widetilde{q}_3$, and $\widetilde{q}_4$, respectively). This simplification is considered to alleviate the computational effort to find the solution for the set of LMIs (21). This is possible since the influence of the other states is small on the computation of the dynamic matrix $A(q_a, \dot{q}_a)$. Figure 7 shows the value of $||A(q_a, \dot{q}_a)||$ for a variation of each joint within its operation range. Hence, $m = 3$ and:

$$\rho(\widetilde{x}) = [\widetilde{q}_2 \;\; \widetilde{q}_3 \;\; \widetilde{q}_4]^T.$$

In order to find the solution for the LMIs, the compact set where the parameters vary is defined as $\rho \in [-0.3, 0.3]rad \times [-0.4, 0.4]rad \times [-0.8, 0.8]rad$ and divided in ($L = 3$). Also the parameter variation rate is bounded by $|\dot{\rho}| \leq [1{,}5 \;\; 2{,}0 \;\; 3]rad/s$. These values are obtained from the limit cycle. The basis functions for $X(\rho)$ are selected as $f_1(\rho(\widetilde{x})) = 1$ $f_2(\rho(\widetilde{x})) = cos(\widetilde{q}_2)$, $f_3(\rho(\widetilde{x})) = cos(\widetilde{q}_3)$ e $f_3(\rho(\widetilde{x})) = cos(\widetilde{q}_4)$. The best attenuation level found was $\gamma = 1.5$.

4.2 Initial Conditions with Zero Velocities

Since the limit cycle has small basin of attraction, the biped robot cannot reach the limit cycle on its own from initial conditions with zero velocities. To do this, it is used the control strategy proposed in Section 3.2. The nonlinear $\mathcal{H}_\infty$ control is applied from the initial conditions to knee strike, considered as desired final position. After the knee strike, the PD and gravity compensation controllers are used. If at the beginning of the next step the initial conditions are again outside the basin of attraction, the procedure is repeated. To illustrate the effectiveness of the control strategy, consider the following initial conditions

$$x_0 = [\pi \;\; 1.3 \;\; -0.6 \;\; -3.68 \;\; 0 \;\; 1.84 \;\; 0 \;\; 0 \;\; 0 \;\; 0]^T.$$

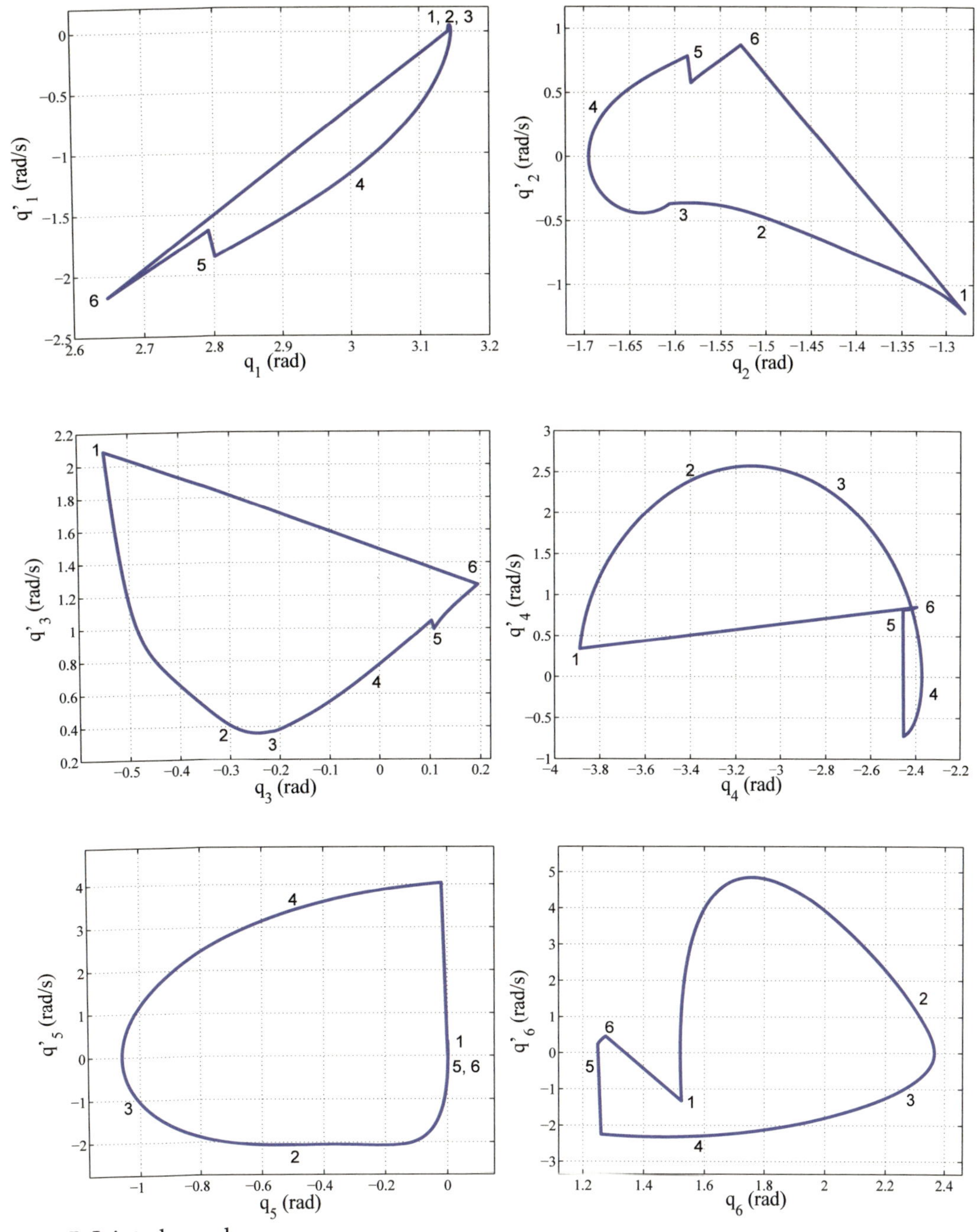

Fig. 5. Joint phase planes.

These initial conditions correspond to the legs be close to the normal to the ground surface and the feet be parallel to it. Figures 8 and 9 show the limit cycle trajectory for all joints and the joint position and velocity time variations. With the hybrid control strategy, the biped robot

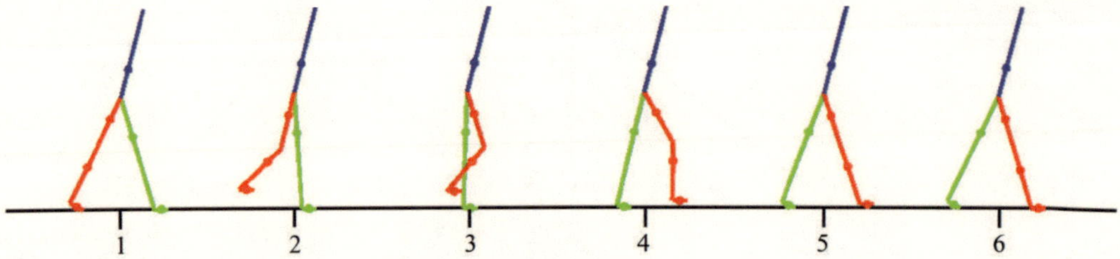

Fig. 6. Ilustration of the step in different moments.

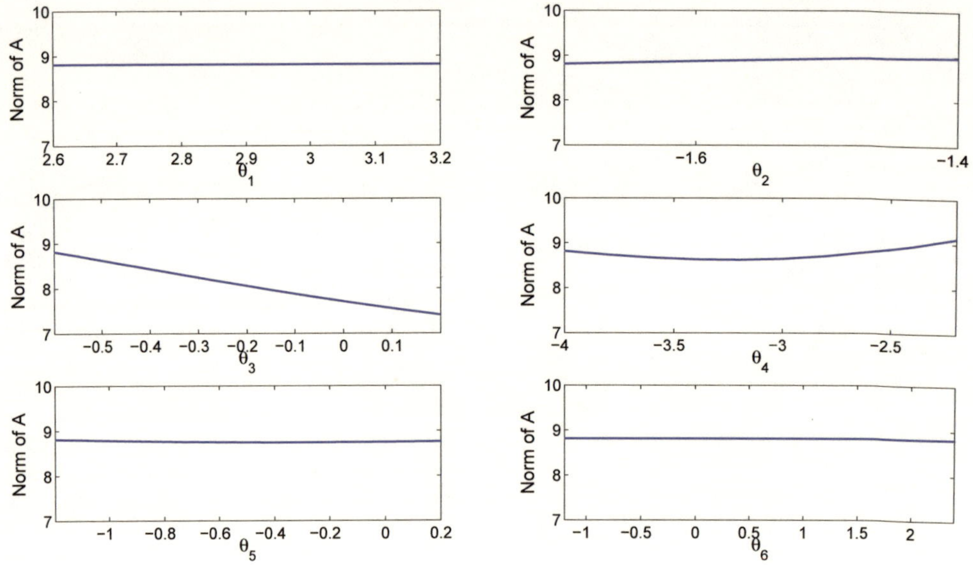

Fig. 7. Variation of $||A||$ with respect to the states.

reached the limit cycle after three steps, with a smooth trajectory from the initial conditions to the first knee strike. This example shows the basin of attraction of the limit cycle was increased for beyond the initial conditions with zero velocities with the application of the nonlinear $\mathcal{H}_\infty$ controller.

4.3 Disturbance Rejection

As seen before, the basin of attraction is small when only the PD and gravity compensation controllers are acting. Hence, even little disturbances can generate gait instability. In this section, the hybrid control strategy is used to give robustness against disturbances. In the example shown, the biped robot starts from the limit cycle initial condition, $x(\phi)$, with only the PD and gravity compensation controllers. External disturbances composed of normal and sine functions, as described in (28), are applied to the variable robot joints. At $t_f = 0{,}1$ s, occurs the maximum peak of the disturbance.

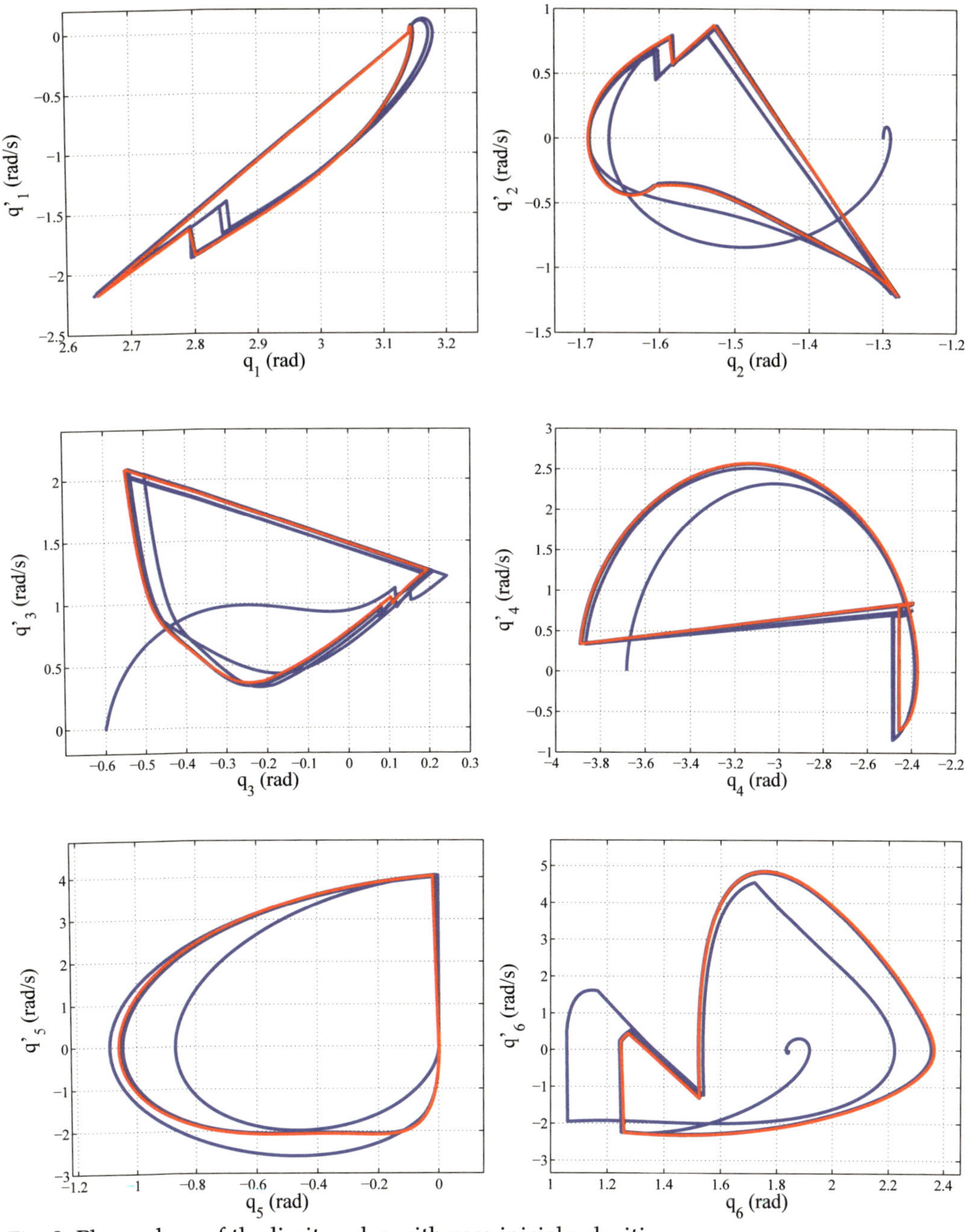

Fig. 8. Phase plane of the limit cycles with zero inicial velocities.

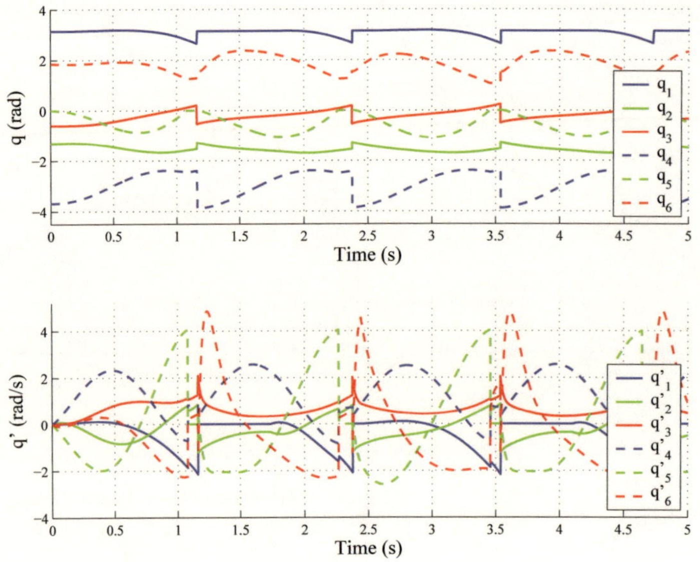

Fig. 9. Joint position and velocities with zero inicial velocities.

$$d = \begin{bmatrix} 0 \\ 0,24e^{\frac{-(t-t_f)^2}{2\sigma^2}} sin(7,2\pi t) \\ 0,18e^{\frac{-(t-t_f)^2}{2\sigma^2}} sin(5,4\pi t) \\ 0,12e^{\frac{-(t-t_f)^2}{2\sigma^2}} sin(3,6\pi t) \\ 0,012e^{\frac{-(t-t_f)^2}{2\sigma^2}} sin(3,6\pi t) \\ 0,012e^{\frac{-(t-t_f)^2}{2\sigma^2}} sin(3,6\pi t) \end{bmatrix} \quad (28)$$

The value of C, the constant that indicates the basin of attraction size, is empirically obtained from simulations with different initial conditions. In this example, $C = 0,15$. Figures 10 and 11 show the results obtained. With the disturbance application, the robot leaves the basin of attraction at the beginning of the second step at $t = 1,25$ s, instant where the control changes to the nonlinear $\mathcal{H}_\infty$ controller. It is then computed the joint desired trajectory considering as desired final position the knee strike position, with duration time corresponding to the remaining time to the impact. Note that after the knee strike at the second step, the robot does not leave any more the basin of attraction and converges to the limit cycle. Hence, the nonlinear $\mathcal{H}_\infty$ controller leads back the robot trajectory to the basin of attraction in only one step, showing the robustness of the adopted control strategy. Note also from Figure 11 that the trajectory performed by the robot during the nonlinear $\mathcal{H}_\infty$ control application is close to the limit cycle trajectory.

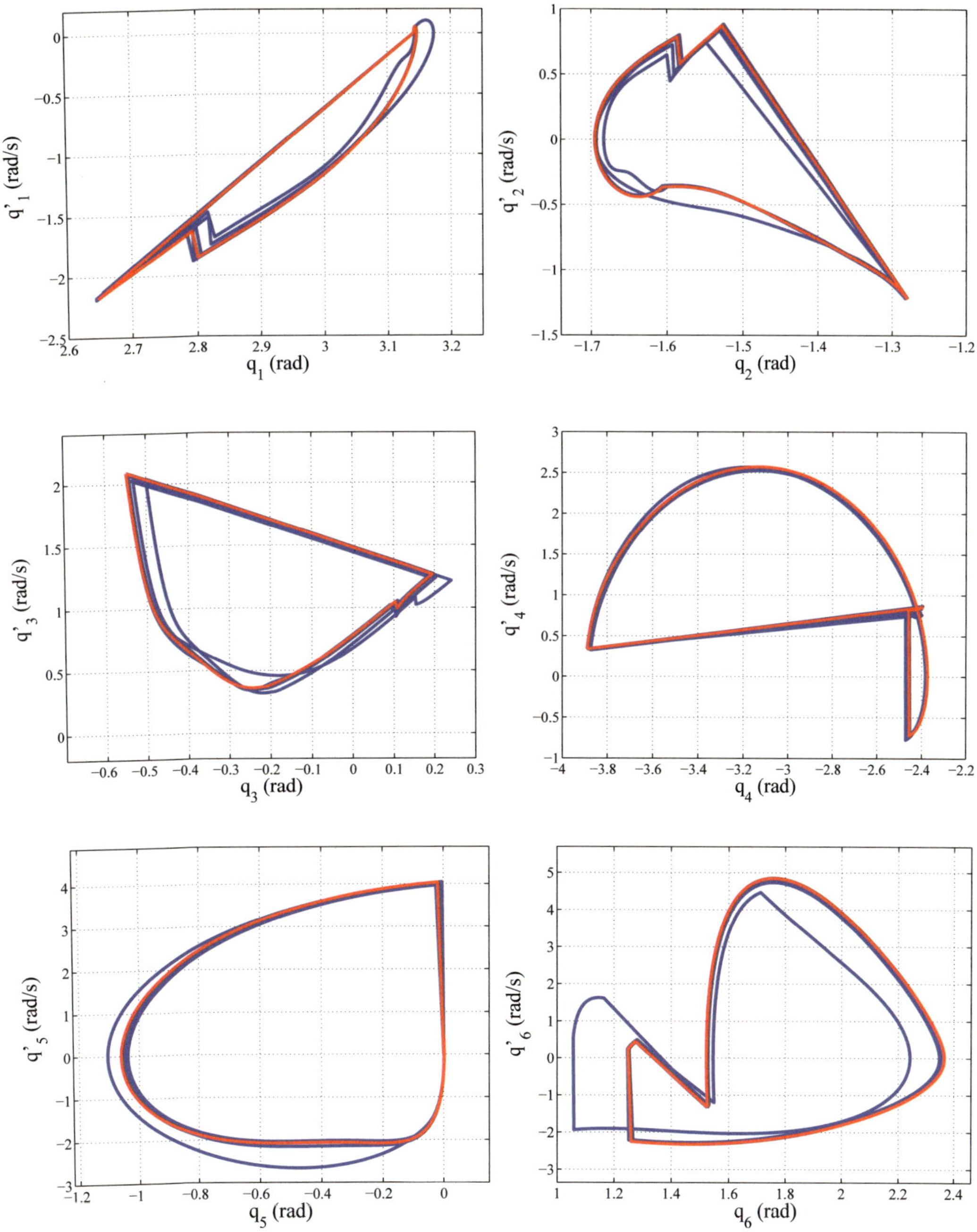

Fig. 10. Phase plane with disturbance rejection.

5. Conclusions

In this chapter, a stable limit cycle is found for a biped robot with torso, knees, and feet. However, the basin attraction is extremely small. To increase it, a control strategy based on

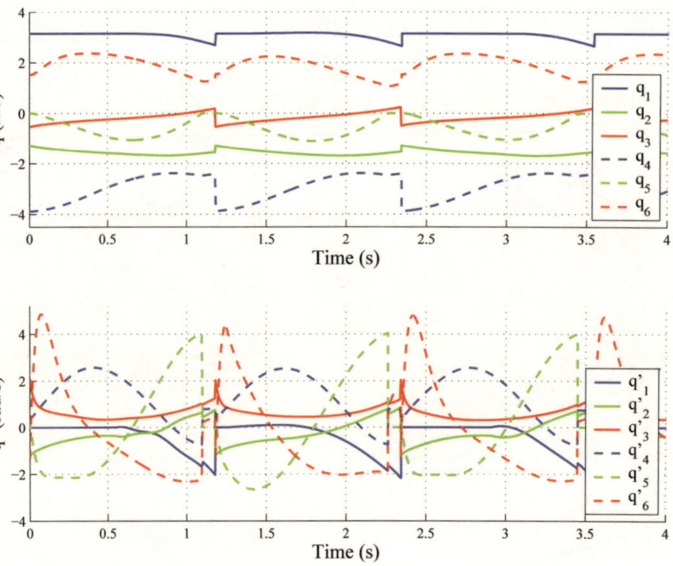

Fig. 11. Joint position and velocity time variations with disturbance rejection.

nonlinear $\mathcal{H}_\infty$ control and on a suitable choice of the switch point between the passive walking and the controlled trajectory is proposed. The results show that the basin of attraction is increased with the adopted strategy, for example, to beyond the initial conditions with zero velocities. In addition, system robustness is verified with the fast convergence to the stable gait, only one step, after the application of external torque disturbances.

6. References

Asano, F., Luo, Z.-W. & Yamakita, M. (2005). Biped gait generation and control based on a unified property of passive dynamic walking, *IEEE Transactions on Robotics* **21**(4): 754–762.

Asano, F. & Yamakita, M. (2001). Virtual gravity and coupling control for robot gait synthesis, *Transactions on Systems, Man and Cybernetics* **31**(6): 737–745.

Bhatia, G. & Spong, M. W. (2004). Hybrid control for smooth walking of a biped with knees and torso, *Institute of Electrical and Eletronics Engineers Conference on Control Applications*, Taipei, Taiwan.

Chevallereau, C., Abba, G., Aoustin, Y., Plestan, F., Westervelt, E. R., Canudas-de Wit, C. & Grizzle, J. W. (2003). Rabbit: A testbed for advanced control theory, *IEEE Control Systems Magazine* **23**(5): 57–79.

Choi, J. H. & Grizzle, J. W. (2005). Planar bipedal walking with foot rotation, *Proceedings of the 2005 American Conrtol Conference (ACC)*, Portland, Oregon, USA.

Goswani, A. (1999). Foot rotation indicator (fri) point: A new gait planning tool to evaluate postural stability of biped robots, *Proceedings of the 1999 IEEE International Conference on Robotics and Automation (ICRA)*, Detroit, Michigan, USA.

Goswani, A., Thuilot, B. & Espiau, B. (1998). A study of the passive gait of a compass-like biped robot, *International Journal of Robotics Research* **17**(12): 1282–1301.

Hirukawa, H., Kanehiro, F., Kaneko, K., Kajita, S., Fujiwara, K., Kawai, Y., Tomita, F., Hirai, S., Tanie, K., Isozumi, T., Akachi, K., Kawasaki, T., Ota, S., Yokoyama, K., Handa, H., Fukase, Y., Maeda, J., Nakamura, Y., Tachi, S. & Inoue, H. (2004). Humanoid robotics platforms developed in hrp, *Robotics and Autonomous Systems* **48**(4): 165 – 175. Humanoids 2003.

Huang, Q., Yokoi, K., Kajita, S., Kaneko, K., Arai, H., Koyachi, N. & Tanie, K. (2001). Planning walking patterns for a biped robot, *IEEE Transactions on Robotics and Automation* **17**(3): 280–289.

McGeer, T. (1990). Passive dynamic walking, *International Journal of Robotics Research* **9**(2): 62–82.

Palmer, M. L. (2002). *Sagittal Plane Characterization of Normal Human Ankle Function Across A Range Of Walking Gait Speeds, M.S. Thesis*, Massachusetts Institute of Technology, Department of Mechanical Engineering.

Siqueira, A. A. G. & Terra, M. H. (2002). Control of underactuated manipulators using nonlinear $\mathcal{H}_\infty$ techniques, *Proceedings of the 2002 IEEE Conference on Decision and Control (CDC)*, Las Vegas, Nevada, USA.

Siqueira, A. A. G. & Terra, M. H. (2004). Nonlinear and markovian $\mathcal{H}_\infty$ controls of underactuated manipulators, *IEEE Transactions on Control Systems Technology* **12**(6): 811–826.

Spong, M. W. & Bhatia, G. (2003). Further results on passivity based control of bipedal locomotion, *International Conference on Intelligent Robots and Systems*, Las Vegas, USA.

Vukobratovic, M. & Juricic, D. (1969). Contribution to the synthesis of biped gait, *IEEE Transactions on Bio-Medical Engineering* **BM-16**(1): 1–6.

Westervelt, E. R., Grizzle, J. W. & Koditschek, D. E. (2003). Hybrid zero dynamics of planar biped walkers, *IEEE Transactions on Automatic Control* **48**(1): 42–56.

Wu, F., Yang, X. H., Packard, A. & Becker, G. (1996). Induced $\mathcal{L}_2$-norm control for LPV systems with bounded parameter variation rates, *International Journal of Robust and Nonlinear Control* **6**(9-10): 983–998.

Method to Estimate the Basin of Attraction and Speed Switch Control for the Underactuated Biped Robot

Yantao Tian[1,3], Limei Liu[1,2], Xiaoliang Huang[1], Jianfei Li[1] and Zhen Sui[1,3]
1.School of Communication Engineering, Jilin University, Changchun,130025, China
2.Department of Mathematics,Changchun Taxation College, Changchun, 130117, China
3.Key Laboratory of Bionics Engineering, Ministry of Education, Jilin University, Changchun, China

1. Introduction

The biped robots have higher mobility than conventional wheeled robots, especially when moving on rough terrains, up and down slopes and in environments with obstacles. The geometry of the biped robot is similar to the human beings, so it is easy to adapt to the human life environment and can help the human beings to finish the complex work. With the development of the society, the needs for robots to assist human beings with activities in daily environments are growing rapidly. Therefore, a large number of researches have been done on the bipedal walking.

The dynamic system of the biped robot is a nonlinear hybrid dynamic system, which consists of continuous differential equations and discrete events dynamic maps. Therefore, this system is a complex nonlinear system. The most effective way of analyzing the global properties of the nonlinear system is probably the straightforward numerical evaluation to compute the motions and then to infer some global properties from the numerical results. It has been reported that the passive biped robot has weak tolerance for large disturbances. The basin of attraction is widely used as a measure for the disturbance rejection for the biped robots, and it is a total set of state variables from which the walker can walk successfully (Ning, L. et. al., 2007). The larger the size of the basin of attraction is, the stronger the stability is. Therefore, more and more researchers have studied the methods to compute the basin of attraction for the biped robot. The cell mapping method was proposed to compute the basin of attraction for the simplest walking model with point feet and the planar model with round feet (Schwab, A.L. & Wisse, M., 2001); (Ning, L. et. al., 2007). The results of experiments show that this method is effective; however, it is time-consuming for multidimensional state space (Zhang, P., et. al., 2008). Based on the bionics study, most humanoid robot control methods are in terms of the basic principles and characteristics of hominine gait. A robotic simulacrum potentially can be very useful. The passive biped robot can walk down along the slope only by inertial and gravitational force. But this passive walking has weak robust and stability. The basin of attraction of the simplest walker can only tolerate a deviation of 2% from the fixed point (Schwab, A.L. & Wisse, M., 2001). In order to improve the stability of the biped robot and expand the size of the basin of attraction, most researchers have designed the controller on the biped robot. Powered robots based on the concept of the passive biped robot can walk on a level floor by ankle push-off (Tedrake, R., et. al., 2004) or hip actuation(Wisse, M., 2004). S.H. Collins exploited the robot with only ankles actuators (Collins, S.H & Ruina A., 2005). M. Wisse exploited the robot with pneumatic actuators (Wisse, M. & Frankenhuyzen, J.van, 2003). Ono proposed the self-excited control with hip joint (Ono, K. et. al., 2004). Since the number of the input torques of these robots is less than the freedom degree, they are called the underactuated biped robot. Compared with traditional biped robots such as Asimo, the underactuated biped robot has higher energy efficiency (Garcia, M., et. al., 1998). Goswami have carried out the extensive simulation analyses of the stability of the underactuated biped walker. However, the biped robots are expected not only to walk steadily, but also to walk fast. How to accelerate the biped walking has attracted a number of researchers during the last years. Energy shaping control law was proposed by Mark Spong for the non-linear hybrid system. J.K. Holm and others applied the law to two passive-dynamic bipeds: the compass-like biped and a simple biped with torso (Jonathan, K. Holm, et. al., 2007). As the compensation for the self-gravity effects, the robot can get different speeds and different stable limit cycles. The angular velocity is changed with changing the gravity compensation coefficient; but step length can not be changed. Based on this study, we eventually develop a method to accelerate the speed of the kneed biped robot and analyze the changes of potential energy and kinetic energy during this process.

The chapter is organized as follows. In section 2, poincaré-like-alter-cell-to-cell mapping method is presented for estimating the basin of attraction of the biped robot. This method is based on the theories of the cell mapping and the point-to-point mapping. Based on the theory of the cell mapping, a method to find the fixed point of Poincaré map for the biped robot is proposed. The basin of attraction for the biped robot with knees is estimated with this method. The effects of parameters variation on the basin of attraction are discussed. Simulations and experiments will be introduced. In section 3, the speed switch control is introduced for the biped robot with knees, and the transformation of potential energy and kinetic energy is analyzed in the control process. The relationship between the control parameters and the forward speed is obtained by simplifying and analyzing the model of the kneed passive walker. In section 4, the conclusion will be presented.

2. Method to estimate the basin of attraction for the biped robot

In this section, we introduce a new method to estimate the basin of attraction of the biped robot. This method is called Poincaré-like-alter-cell-to-cell mapping method, which is guided by the method proposed by (Liu, L. et. al., 2008). Poincaré-like-alter-cell-to-cell mapping method can not only be used to estimate the basin of attraction of the biped robot, but also can be used to estimate the fixed point of the Poincaré map. And then, the effects of the configurable parameters on the basin of attraction are discussed. In experiments, a kneed biped robot with point feet is used; and the effect on the basin of attraction is obtained with the variation of the mass ratio between the thigh and the shank. Results show that the size of the basin of attraction is enlarged with increasing the ratio.

2.1 Cell mapping

We will introduce some concepts and terminology of the cell mapping. Firstly, a domain of interest $\Omega \subset R^N$ in the state space is chosen. Let the coordinate axis of a state variable x_i $(i=1,2,\cdots,N)$ of this domain be divided into a large number of intervals with an interval size h_i. The total number of the intervals on the x_i-axis of the domain of interest is denoted by n_i. The interval Z_i of the x_i-axis is defined to be one which contains all x_i satisfying $a_i + (Z_i - 1)\cdot h_i \le x_i \le a_i + Z_i \cdot h_i$, where a_i is the smallest value of the x_i-axis of this domain. Z_i is a positive integer. An N-tuple Z_i $(i=1,2,\cdots,N)$ is called a cell vector of the state space and is denoted by z. All the cell vectors constitute a cell space, and the total number of cells is $\prod_{i=1}^{N} n_i$. A point $x = (x_1, x_2, \cdots, x_N)$ belongs to a cell $z = (Z_1, Z_2, \cdots, Z_N)$, if x_i belongs to a cell Z_i for all i. Each cell is now considered as an entity and the state space is regarded as a collection of cells. So with this procedure, the continuous state space is replaced by a discrete cell space. The evolution of the cell $z(n)$ can be described by $z(n+1) = C(z(n))$, that is $Z_i(n+1) = C(Z_i(n))$ $(i=1,2,\cdots,N)$, where C map a set of positive integers to a set of positive integers. Obviously, this mapping is called a cell-to-cell mapping, or a cell mapping.

Let C^m denote the cell mapping C applied m times with C^0 understood to be the identity mapping. A sequence of K distinct cells $z^*(j)(j=1,2,\cdots,K)$ that satisfy

$$\begin{cases} z^*(m+1) = C^m(z^*(1)),\ m = 1,2,\cdots,K-1 \\ z^*(1) = C^K(z^*(1)) \end{cases} \tag{1}$$

are said to form a periodic motion of period K. They are called a P-K motion. And each of its elements $z^*(j)$ is called a periodic cell of period K or simple a P-K cell.

A cell z^* is called equilibrium cell, if it satisfies $z^* = C(z^*)$.

A cell z is said to be transient cell which is "r-step removed from a P-K motion", if r is the minimum positive integer such that $C^r(z) = z^*(j)$, where $z^*(j)$ is one of the P-K cells of that P-K motion. In other words, z is mapped after r-steps into one of the P-K cells of the P-K motion and any further mapping will lock the evolution of the system in the P-K motion. The set of all cells which are r-steps removed from a P-K motion is called the "r-step domain of attraction" for that P-K motion.

For most physical problems once the state variable exceeds a certain scope of the domain of interest, none is interested in the further evolution of the variable. If the range of the state variable exceeds the ones that are interested, then the cell lying in it is called sink cell. Once the cell is sink cell, none is interested in its further evolution. That is to say that the region outside the domain of interest constitutes a collection of the sink cells.

It is obviously that the total number of cells is always finite, although the total number could be usually huge. The evolution of the system starting with any regular cell z can lead to only three possible outcomes: periodic cell, sink cell and transient cell (Hsu, C.S., 1980).

2.2 Method to find the fixed point of Poincaré map for the biped robot

On the basis of the theory of the fixed point of the Poincaré map, we have analyzed the stability of the biped robot. Starting from the given point, if the gaits converge to the limit cycle that starts from the fixed point, we can say that the given point is a stable initial state and the robot can walk stably. But it is difficult to find the fixed point of the Poincaré map. The common method is Newton-Raphson iteration method. However, the initial values of the iteration have to be guessed in experience. If the initial values are not close enough to the fixed point, the iteration can not converge. Coleman and Garcia have made failure in finding the stable fixed point for 3D model with the Newton-Raphson iteration method (Garcia, M., 1999; Coleman, M.J., 1998).

Based on the theory of cell mapping, we propose a new method to find the fixed point of the Poincaré map. Firstly, we choose an initial condition state space at random. And the initial condition state space will be subdivided into cell states with feasible interval sizes. Then we obtain periodic cells under the cell mapping. All center points $x^{(d)}$ of the periodic cells are selected together. Each center point $x^{(d)}$ of the periodic cell is looked as the inital point of iteration. The iteration evolution of $x^{(d)}$ is as follows: let $x' = x^{(d)}$, $x'' = P(x')$, where P is a Poincaré map. Let $x' = x''$, and calculate the above Poincaré map repeatedly. $x'' = P(x')$ is a fixed point of the Poincaré map of the biped robot, if $|P(x') - x'| \le \varepsilon$. That is to say $x'' = P(x')$. The effective initial value of iteration can be obtained easily with this method. This method is still effective in the multidimensional state space.

2.3 Poincaré-like-alter-cell-to-cell mapping method

The system that is considered in this section is

$$\begin{cases} \dot{x} = f(x) & x \notin S \\ x^{+} = H(x) & x \in S \\ S = \{x | r(x) = 0\} \end{cases}$$

The steps of estimating the basin of attraction for the biped robot are listed as follows:

Step1. The state space is divided into a discrete cell space.

A domain of interest $\Omega \subset R^N$ in the state space is chosen. The coordinate axis of the state variable x_i $(i=1,2,\cdots,N)$ of this domain is divided into a large number of intervals with an interval size $h1_i$. The total number of the intervals is denoted by n_i. The interval Z_i of the x_i-axis is defined to be one which contains all x_i satisfying

$$a_i + (z_i - 1) \cdot h1_i \le x_i < a_i + Z_i \cdot h1_i \tag{2}$$

Where a_i is the smallest value of the x_i-axis of this domain. So the cell vector z of this domain is denoted by an N-tuple Z_i $(i=1,2,\cdots,N)$. All the cell vectors constitute a cell space, and the total number of cells is $\prod_{i=1}^{N} n_i$.

Step2. The cell space is classified by the evolution of the cell.

(1)Let $n=1$, choosing a cell $z(n)$ ($Z_i(n)$ $(i=1,2,\cdots,N)$), $x^{(d)}(n)$ is its center point, where $x_i^{(d)}(n)=a_i+(z_i(n)-1/2)\cdot h1_i$. $z(n)$ is called an original cell.

(2) Let $x_0=x^{(d)}(n)$ is the initial point of $\dot{x}=f(x)$ at $t=0$. Then the trajectory of this equation is calculated with initial point x_0. The original cell is a sink cell, if $x(t_e)$ satisfies $r(x(t_e))=\varnothing$, and the evolution of cell is stopped. If $r(x(t_e))=0$, $x^+=H(x(t_e))$. Then $z(n+1)=C(z(n))$, that is

$$Z_i(n+1)=C_i(z_i(n))=Int\left\{\frac{\left(x_i^+-x_i\right)}{h1_i}+\frac{1}{2}\right\} \tag{3}$$

where $Int(y)$ denotes the largest integer which is not bigger than y. One can obtain the new cell $z(n+1)$.

(3) If the state value in which $z(n+1)$ lies exceeds the domain of interest of the state space, the original cell is a sink cell and the evolution of cell is stopped. Otherwise $z(n)$ and $z(n+1)$ are compared. The evolution of the cell is ended, if $z(n)=z(n+1)$, and the original cell is an equilibrium cell. When $z(n)\neq z(n+1)$, let $x_0=x^+$ and one does repetitive operation of (2) and (3), until the new cell is equal to one of the cells that get from the evolution of $z(n)$ previously. In the end, the original cell is a periodic cell or a transient cell that is r-step removed from a P-K motion.

Every cell of the cell space must carry out the procedure of the evolution.

Step3. Almost basin of attraction is obtained

Every sink cell is divided into many intervals with an interval size $h2_i$ $(i=1,2,\cdots,N)$. Every equilibrium cell and periodic cell and transient cell are divided into a lot of intervals with an interval size $h3_i$ $(i=1,2,\cdots,N)$, where $h3_i$ is much smaller than $h2_i$. Then the total number of cells in the cell space is much larger. All center points of the cells are picked out. They constitute an almost basin of attraction of the biped robot.

Step4. Filter step-obtaining the basin of attraction

Since the division of the cell space affects the accuracy of the results, we set this step. Let the fixed point be a reference point. Every point $x^{(d)}$ of the almost basin of attraction is imposed $x''=P(x')$ repeatedly. The point belongs to the basin of attraction, if it is close to the fixed point under the calculations.

2.4 Basin of attraction for the biped robot with knees

2.4.1 Model of the biped robot with knees

In this section, the goal is to estimate the basin of attraction for the biped robot with knees with the Poincaré-like-alter-cell-to-cell mapping method. Here we focus on the biped robot which could go down incline by using potential energy. This robot does not have a torso and consists of two point feet and two legs that are connected at the hip joint. Each leg has a thigh and a shank connected at a passive knee joint that has a knee stopper. By the knee stopper, an angle of the knee rotation is restricted like the human knee. The thigh and the shank of the swing leg are assumed to be kept straight by the knee stopper during a period from the knee collision to the end of the heelstrike. Fig. 1 shows the diagram of the model of the biped robot with knees (Vanessa, F.& Hsu Chen, 2007). Table 1 lists the physical parameters and the values in simulation.

Symbols	Parameter	Value in Simulation
mH	The hip mass [kg]	0.5
m1	The shank mass [kg]	0.05
m2	The thigh mass [kg]	0.5
a1	Length between the heel and the shank COG of the swing leg [m]	0.375
b1	Length between the knee and the shank COG of the swing leg [m]	0.125
a2	Length between the knee and the thigh COG of the stance leg [m]	0.175
b2	Length between the hip and the thigh COG of the stance leg [m]	0.325
L	Leg length [m]	0.5
ϕ	The slope of ground [radian]	0.0504
α	Interleg angle [radian]	0.4772

Table 1. The physical parameters and the values in simulation

The entire step cycle is divided into four processes:

(1) The stance leg straightens out and the knee is locked, just like a single link. While the swing leg with unlock knee comes forward, just like two links connected by a frictionless joint. This stage is called unlocked swing stage.

(2) When the swing leg straightens out, the knee of swing leg is locked. The kneestrike occures. The impact takes place instantaneously.

(3) After the kneestrike, the knee of swing leg remains locked and the system switchs to the double-link pendulum dynamics. Therefore, this stage is just like the swing stage of the compass gait model. This stage is called locked swing stage.

(4) The locked-knee swing leg hits the ground. The premises underlying this stage are that: the impact takes place instantaneously; the impact of the swing leg with the ground is assumed to be inelastic and without sliding; the tip of the support leg is assumed not to be slip, and the robot behaves as a ballistic double-pendulum.

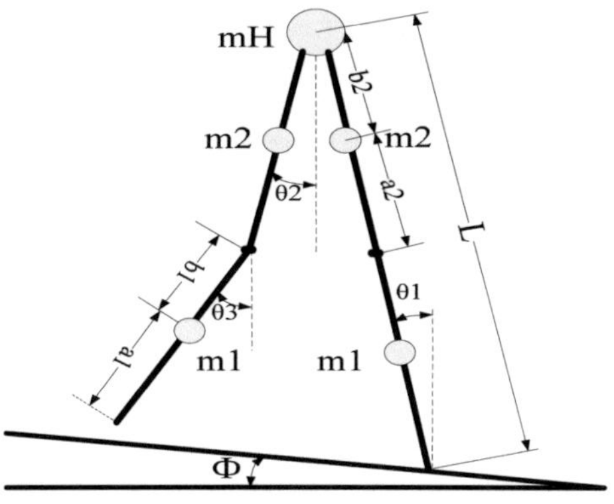

Fig. 1. Model of the biped robot with knees (Vanessa, F. & Hsu Chen, 2007)

Figure 2 shows the diagram of the four stages of the entire step cycle. Equations of the entire step cycle are shown in (Zhang, P. et. al., 2009).

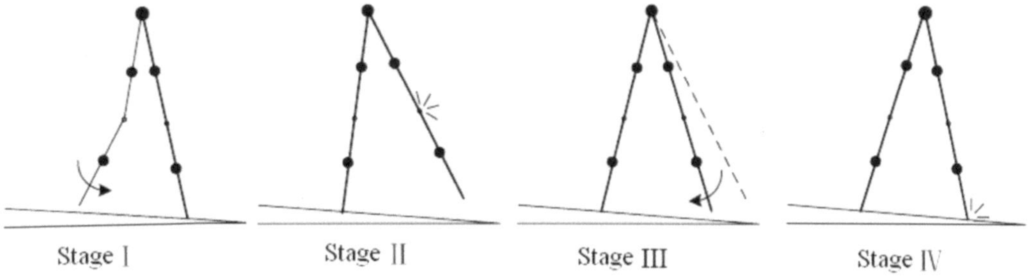

Fig. 2. Diagram of the four stages of the entire gait cycle

2.4.2 Finding the fixed point of the Poincaré map for the biped robot with knees

A passive biped robot with knees is chosen to do experiment. The values of parameters are listed in Table 1 and all of the input torques are zero. The instant just after heelstrike is defined as the Poincar é section. Let the initial condition state spaces be respectively $[0,0.8]\times[-2,0]\times[-2,0]\times[-5,10]\times[-3,0]\times[-3,0]$ and $[-0.8,0]\times[-1,2]\times[-1,2]\times[-5,10]\times[-3,0]\times[-3,0]$. Each state space is subdivided into five equal division. The fixed point is found to be [0.1882 -0.2890 -0.2890 -1.1090 -0.0571 -0.0571] by using the method proposed in this chapter, though the initial condition state spaces are different. Figure 3 presents a limit cycle for the thigh of the swing leg starting from this fixed point. In figure 3, the instantaneous angle velocity changes from the kneestrike and heelstrike are expressed as the straight lines in the limit cycle, while the angles remain the same. Figure 4 shows that the gaits of the biped robot will converge to this limit cycle within a few steps, if the initial state starts slightly away from this fixed point. [0.1982 -0.2890 -0.2890 -0.0590 -0.0571 -0.0571] is marked as a blue star, and the fixed point is marked as a red star. Therefore, the biped robot with knees can walk stably. This experiment shows that the method to find the fixed point of Poincaré map for the biped robot is effective and the result of the method does not rely on the initial condition state space. The initial state of the iterations is not to be guessed.

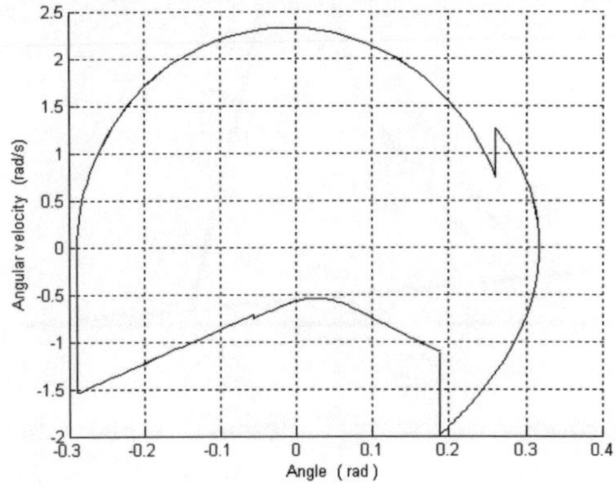

Fig. 3. Limit cycle of the thigh of the swing leg starting from [0.1882 -0.2890 -0.2890 -1.1090 -0.0571 -0.0571]

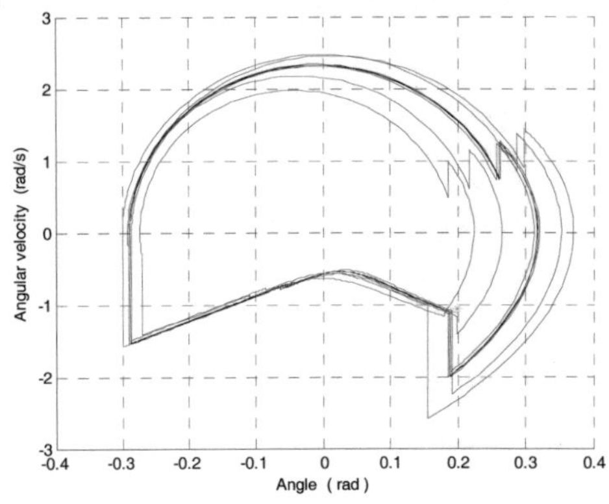

Fig. 4. Limit cycle of the thigh of the swing leg starting from [0.1982 -0.2890 -0.2890 -0.0590 -0.0571 -0.0571]

2.4.3 Estimate the basin of attraction for the biped robot with knees

In simulations, the basin of attraction of the biped robot with knees is estimated with the Poincar é -like-alter-cell-to-cell mapping method. The instance just after heelstrike is set to be the Poincar é section, so the state space satisfies $\theta_2 = \theta_3; \dot{\theta}_2 = \dot{\theta}_3$. In order to reduct dimensions, the interleg angle is fixed to be the fixed point's interleg angle. That is to say $\theta_1 - \theta_2$ be equal to the fixed point's interleg angle. The values of parameters in simulation are listed in table1. The feasible state space is set as $\theta_1 \in [0, 0.4772], \theta_2 \in [-0.4772, 0], \dot{\theta}_1 \in [-4, 1], \dot{\theta}_2 \in [-5, 11]$. Each state space is subdivided into 80000 cells. Every periodic cell and transient cell are divided into 20 cells, and every sink cell is divided into 4 cells. Figure 5 shows the sections of this basin of attraction. In order to ensure accuracy, time-consuming is inevitable for the cell mapping method. Therefore, compared with the cell mapping method, the advantages of Poincaré-like-alter-cell-to-cell mapping method are that this method is more accuracy and saves time.

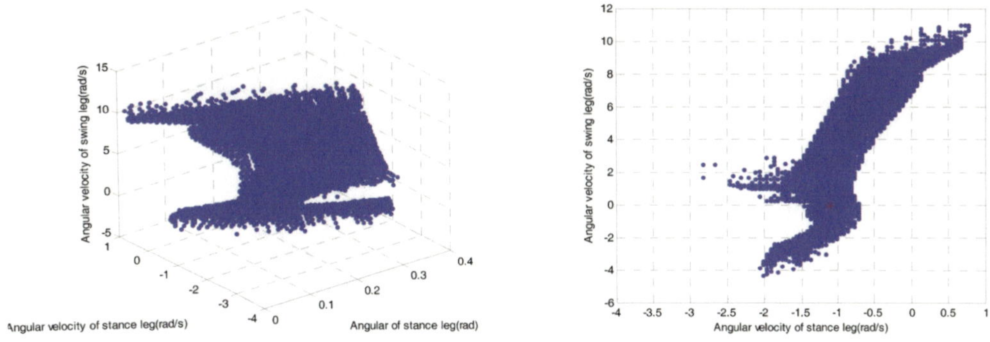

Fig. 5. The sections of the basin of attraction of the biped robot with knees

2.4.4 Effect on the basin of attraction with parameters variation

In this section, we will do research in the effect on the basin of attraction of the biped robot with knees, when the mass ratio in each leg is varied. Let total mass of each leg be 0.55 [kg], and μ denotes the ratio between the mass of the thigh and the mass of the shank. Figure 6 shows the variations of the basin of attraction of the biped robot with knees, when μ is increased. It presents that the size of the basin of attraction becomes larger with increasing μ . Most of the points lying in the basin of attraction assemble in the neighborhood of the fixed points. Since the basin of attraction of the biped robot with knees is a collection of the initial state points that lead to the perpetual walking, the size of the basin of attraction determines the disturbance rejection of the stable gaits. From the results of this simulation, it is proved that the greater μ is, the stronger roboustness is. And it further proved that the greater the ratio between the mass of the thigh and the mass of the shank was, the more stable the walker became ((Vanessa, F.& Hsu Chen, 2007).

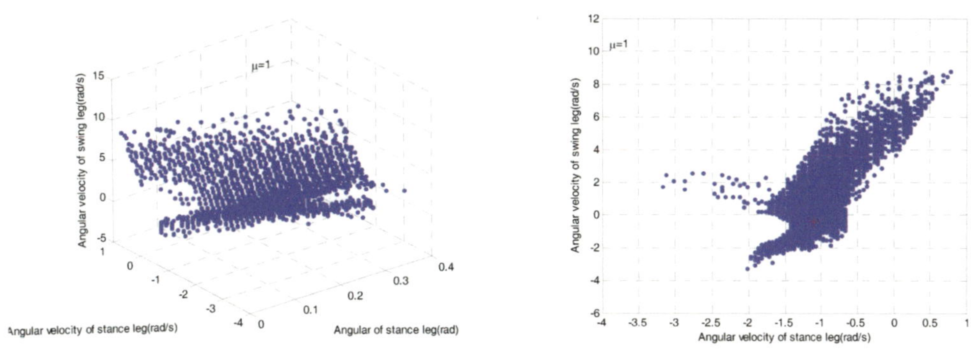

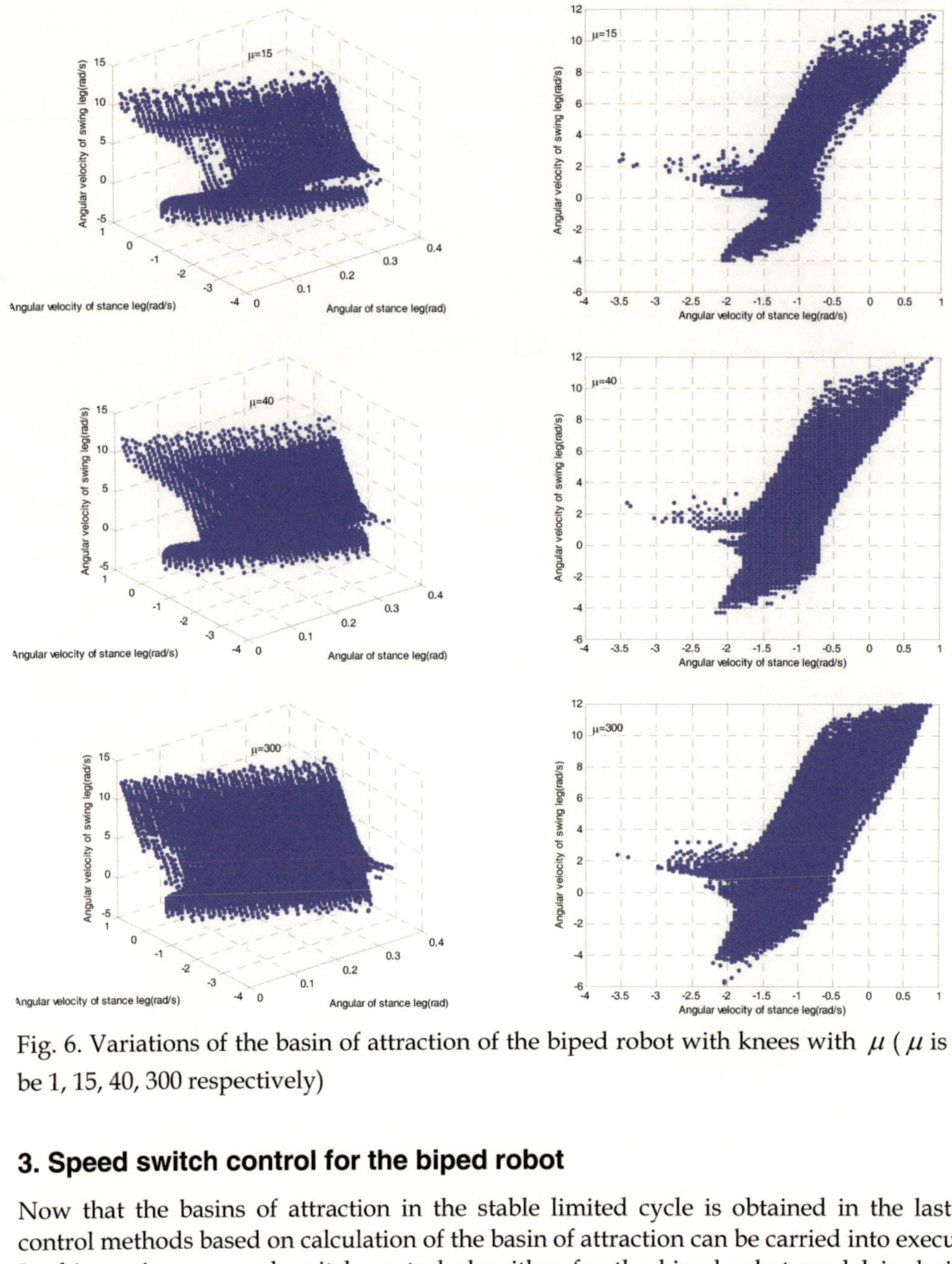

Fig. 6. Variations of the basin of attraction of the biped robot with knees with μ (μ is set to be 1, 15, 40, 300 respectively)

3. Speed switch control for the biped robot

Now that the basins of attraction in the stable limited cycle is obtained in the last part, control methods based on calculation of the basin of attraction can be carried into execution. In this section, a speed switch control algorithm for the biped robot model is designed, based on the energy shaping theory and the estimate of the basin of attraction, to accelerate the dynamic walking and regulating the speed of walking when the parameters are varied. In order to keeping the gaits stable in accelerating process, we design a switch rule based on distinguishing the position of the switch point in the phase space. If the switch point lies in the common pats of the basins of attraction which is belong to the stable limited cycle in different speeds, the walking speed can be adjusted only by changing the grave parameter to objective value directly. Otherwise, a transition function based on the equational constraint condition is necessary to be constructed.

First of all, the functional relationship between the control parameters and the forward speed is analyzed, which is used to construct the speed switch control algorithm. Then the speed swith control algorithm is introduced in detail and the effects on the forward speed and the walking gaits under control are studied. In the end, the trends of the kinetic energy and the potential energy are analyzed with control parameters variations during the controlled walking.

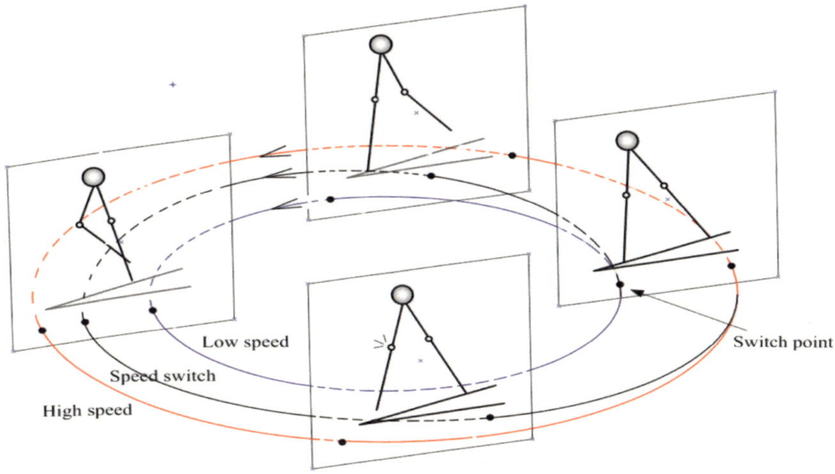

Fig. 7. Respective postures of the biped robot with knees during controlled walking stage

3.1 Stable walking and gravity parameter

3.1.1 Relationship between forward speed and gravity parameter

Energy shaping control law was proposed by Mark Spong for the non-linear hybrid system. Holm and others applied the law to the simple biped with torso (Jonathan K. Holm. et. al., 2007).

Based on this method, we calculate the relationship between the average speed $\overline{v}$ and gravity parameter f ,and showed in the eqution (4)

$$\overline{v} \approx \sqrt{\frac{f\, g_0}{L}} \frac{L(\sin(\theta_1(0)) + \sin(\theta_2(0)))}{\pi} \tag{4}$$

where f is defined as the gravity parameter, which is the proportional coefficient of the gravity: $f = g / g_0$, $g_0 = 9.8 m/s^2$, $\overline{v}$ is the average speed of each gait cycle, $\theta_1(0)$ is the initial angle values of the stance leg and $\theta_2(0)$ is the initial angle values of the swing leg. Also we can get the different stable limit cycles that means different walking speeds due to the different initial values with the compensation for the self-gravity effects as below.

Fig. 9 shows that the limit cycle stretches in the vertical direction with increasing the value of f. This experiment indicates that the angular velocity is changed, but the step length holds in line when changing f. Fig.10 shows the curve of $(\theta_1, \theta_2, \theta_3)$ and $(\dot{\theta}_1, \dot{\theta}_2, \dot{\theta}_3)$ for the different value of f.

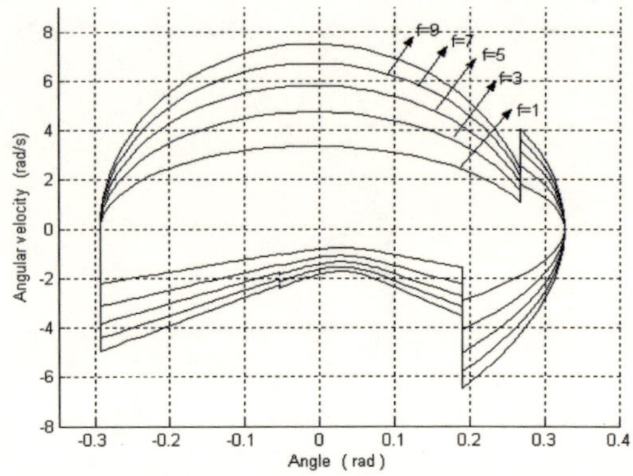

Fig. 9. The limit cycle corresponding to different value of f (f is 1, 3, 5, 7, 9 respectively)

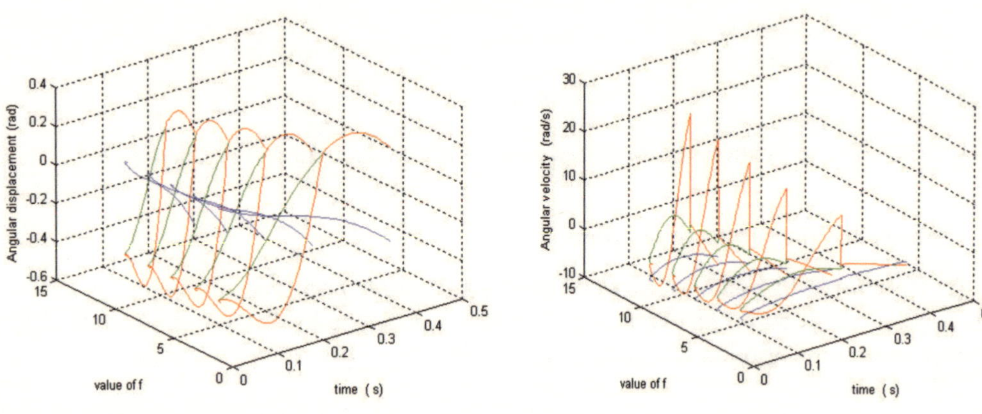

Fig. 10. Angular displacement and angular velocity corresponding to different value of f (f is 1, 3, 5, 7 respectively)

3.1.2 Energy analysis for different gravity parameter

Expressions of the kinetic energy and the potential energy for the biped robot with knees are showed as the equation (5) and (6).

$$\begin{aligned} P(\theta) = [&(m_H + m_1 + m_2) g L \cos\theta_1 + m_1 g a_1 \cos\theta_1 \\ &+ m_2 g (l_1 + a_2) \cos\theta_1 - (m_2 b_2 + m_1 l_2) g \cos\theta_2 \\ &- m_1 g b_1 \cos\theta_3] f(t) \end{aligned} \tag{5}$$

$$\begin{aligned} K(\theta,\dot{\theta}) = & \frac{1}{2} m_H L^2 \dot{\theta}_1^2 + \frac{1}{2}(m_1 a_1^2 + m_2(l_1 + a_2)^2)\dot{\theta}_1^2 \\ & + \frac{1}{2} m_1 (b_1^2 \dot{\theta}_3^2 + l_2^2 \dot{\theta}_2^2 + L^2 \dot{\theta}_1^2) + m_1 b_1 l_2 \dot{\theta}_3 \dot{\theta}_2 \cos(\theta_3 - \theta_2) \\ & - m_1 b_1 L \dot{\theta}_3 \dot{\theta}_1 \cos(\theta_1 - \theta_3) - m_1 l_2 L \dot{\theta}_1 \dot{\theta}_2 \cos(\theta_1 - \theta_2) \\ & + \frac{1}{2} m_2 (b_2^2 \dot{\theta}_2^2 + L^2 \dot{\theta}_1^2 - 2 b_2 L \dot{\theta}_1 \dot{\theta}_2 \cos(\theta_1 - \theta_2)) \end{aligned} \quad (6)$$

With the analysis of energy evaluation, it can be seen that the mechanical energy as well as the kinetic and potential energy changes in a similar manner. From Fig. 11, at the swing stage, mechanical energy is conserved, showd nearly as level lines in the fingure. The curves of kinetic energy with different f have similar variation, which simply decrease to the minimum value and then rise up, compensating the energy loss at kneestrike. The same trend can also be seen in the figure of potential energy.

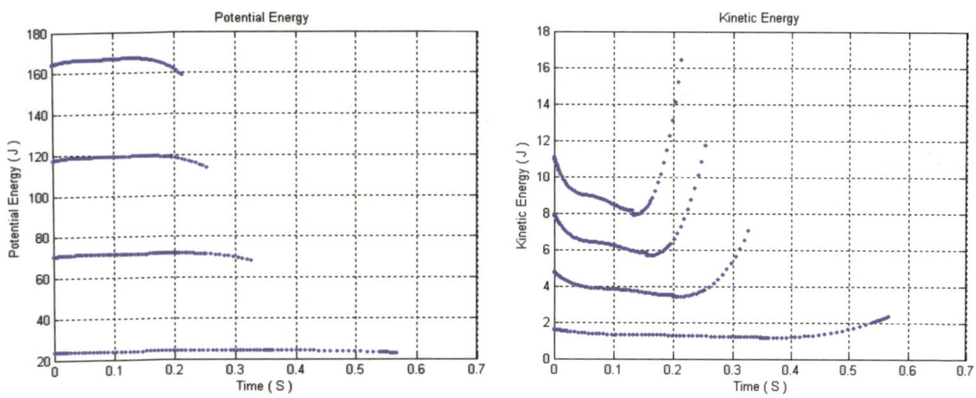

Fig. 11. The variation of potential energy and the variation of kinetic energy corresponding to different value of f (f is 1, 3, 5, 7 respectively)

3.2 Construction of the control law

From the last part, we can get the basin of the attraction for different stable limit cycles, which represents different walking speeds. If we want to switch the gait from one stable limit cycle into another, we should determine whether the switch point (the initial position under control) lies in the common part of the two basins of attraction or not. In general, the point of heelstrike is chosen as the switch point and the changes of touch sensors are examined as the controlling signals. Also this piont is the fixed point in the stable limited cycle as we know from section 2.

Since different speeds corresponds to the different gravity parameters f , the speed switch control method is mainly based on the gravity parameter. The speed switch controller is designed as

$$\begin{aligned} J_r^T l_r + B(q)u &= (1 - f(t))g(q) \\ &= (1 - f(t))\,(g_1, g_2, g_3)^T \end{aligned} \quad (7)$$

When the controller is carried out, we only need to adjust the value of f .

Let $x_0 = (q_0, \dot{q}_0)^T$, the basin of attraction related to f_i is denoted as follows (Liu, L.., et. al., 2008)

$$U_i = \left\{ x_0 \middle| \begin{array}{l} \exists\ T_0 > 0, \\ \text{s.t}\ ,\ P(T_0, 0, x_0, f_i) = x^* \end{array} \right\} \text{(i=1,2)} \quad (8)$$

Where f_i is the gravity parameter, and x^* is the fixed point of the Poincaré return map for the biped robot with knees, that is $P(x^*) = x^*$.

(1) If the switch point x_0 lies in the common part of U_1 and U_2 , f can be turn to f_2 from f_1 directly. Since the switch point in the basin of attraction U_2 , the stable gait in the high speed related to f_2 will be obtained again after a few steps.

(2) If the switch point x_0 does not lie in the common part of U_1 and U_2 , we construct a transition function $f(t)$ to finish the switch. Considering $f(t) = a_0 + a_1 t + a_2 t^2$, a_0 a_1 and a_2 are the coefficients parameters. Table 2 gives some necessary signals to estimate the coefficient parameters.

f_i	The initial value of f
f_e	The final value of f
t_f	The transition time
T	The stable gaits period

Table 2. The signals used to deduce $f(t)$

The transition function $f(t)$ should satisfy the conditions as follows

$$\begin{cases} f_t(0) = f_i \\ f_t(t_{-f}) = f_e \\ \int f_t(0)dt = 0 \\ \int f_t(t_{-f})dt = T \end{cases} \quad (9)$$

In our simulation experiment, the coefficient of $f(t)$ are listed as follows($f_i = f_1, f_e = f_2$)

$$a_0 = f_1$$

$$a_1 = \frac{1}{t_{-f}}\left(f_2 - a_0 - a_2 t_{-f}{}^2\right)$$

$$a_2 = \frac{1}{2 \times t_{-f}{}^3}\left(6T - 6a_0 t_{-f} - 3a_1 t_{-f}{}^2\right)$$

The transition function of $f(t)$ is designed as follows:

$$f(t)=\left\{9\left(\frac{f_1+f_2}{t_{-f}}-\frac{2T}{t_{-f}^{\ 3}}\right)^2 t^4+12\left(\frac{3T}{t_{-f}^{\ 2}}-\frac{2f_1+f_2}{t_{-f}}\right)\left(\frac{f_1+f_2}{t_{-f}}-\frac{2T}{t_{-f}^{\ 3}}\right)t^3\right.$$
$$+\left[4\left(\frac{3T}{t_{-f}^{\ 2}}-\frac{2f_1+f_2}{t_{-f}}\right)^2+6\left(\frac{f_1+f_2}{t_{-f}}-\frac{2T}{t_{-f}^{\ 3}}\right)\tau_1\right]t^2 \tag{10}$$
$$\left.+4\left(\frac{3T}{t_{-f}^{\ 2}}-\frac{2f_1+f_2}{t_{-f}}\right)f_1t^2+f_1^{\ 2}\right\}^{1/2}$$

3.3 Simulation experiments

In order to verify the validity of the speed switch control for the biped robot with knees, a model has been described by visual software. Fig. 12 shows the simulation model. The values of the structure parameters have been given in Table 1.

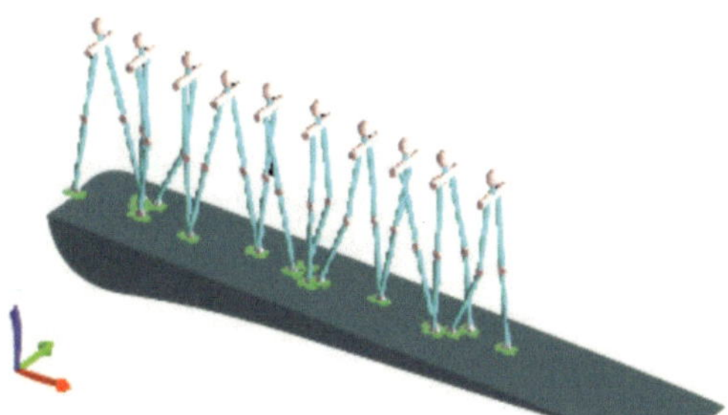

Fig. 12. Visual simulation model of dynamic biped robot with knees

We choose the fixed ponit of the gait cycle as the switch initial position of our algorithm. Fig.13 shows the basins of the attraction in the fixed points of the limit cycles, which are obtained by the method when the gravity parameter is equal to 2 and 3 respectively. We named the basin (red part) U_2 when f =2, and the basin (blue part) U_3 when f =3. We can see the common parts of the two basins clearly. And the green points is the fixed points, which are used for the beginning points of the switch process.

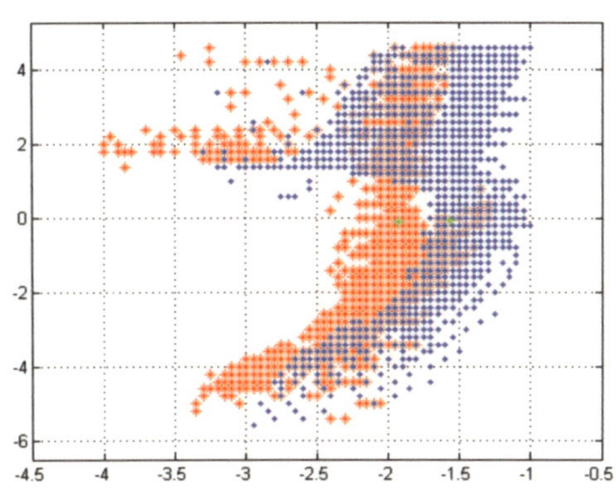

Fig. 13. The basins of attraction for the biped robots with different gravity parameters

The parameters of transition function are set as $f_1=2$, $f_2=3$, $f_{-f}=0.5$ and T=1.1, which means the speed can change from 1.3466 times to 2.1130 times of the original speed within 0.5s, and the variable of transition function begins with f_1 =2 and ends with f_2 =3.

Fig. 14 lists the variation of the hip angular velocity in one step. Fig. 15 shows the change process of the hip angular velocity with changing $f(t)$, which indicates the angular velocity began to accelerate in the second step, and then the biped robot walked fast and stably after the fifth step.

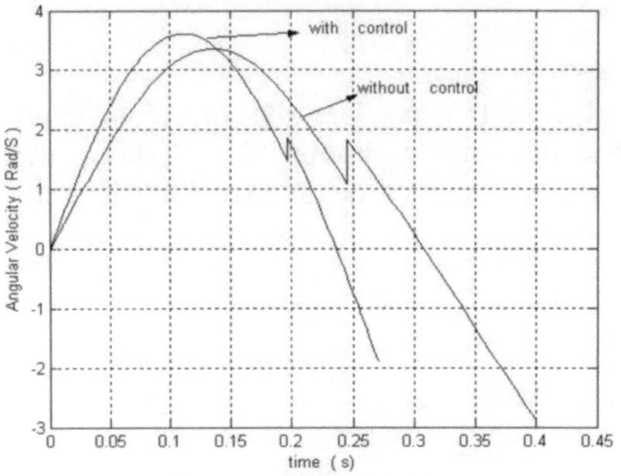

Fig. 14. The hip angular velocity with $f(t)$ compare to the hip angular velocity without $f(t)$ in one step

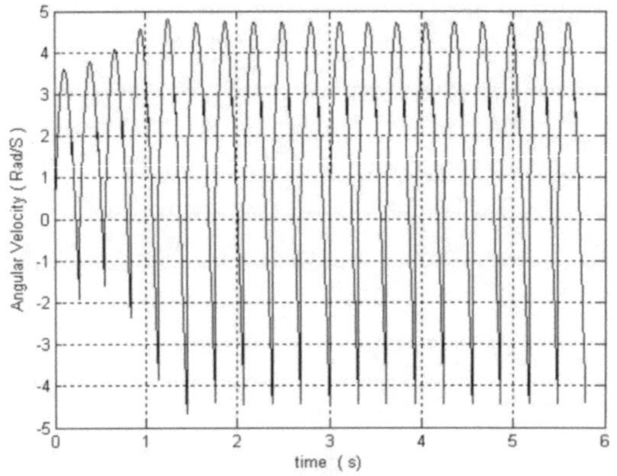

Fig. 15. The hip angular velocity corresponding to $f(t)$ in several steps (begin with f_1 =2 and end with f_2 =3)

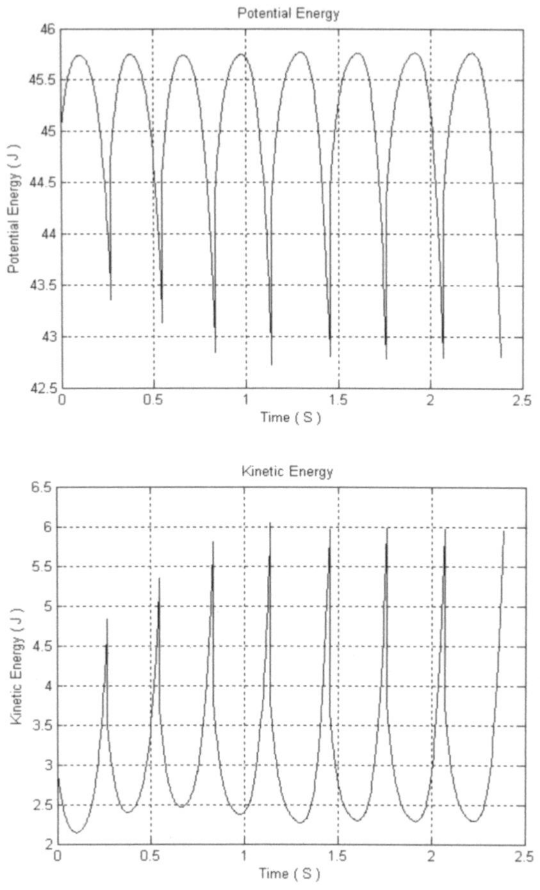

Fig. 16. The variation of potential energy and kinetic energy using the transition function $f(t)$ ($f(t)$ transits from 2 to 3)

The energy transformation is different from original when the forward speed is controlled. Fig.16 show the potential energy and kinetic energy with different value of f . Obviously, with increasing of the values of control parameter, the speed of the kinetic energy and the potential energy transformation raises proportionally. However, the variable range of the kinetic energy is larger than potential energy, because that the energy supplied by the control torques compensates the kinetic energy and accelerates the dynamic walking. With the gravity potential energy shaping, the kneed biped trajectory converges to a high energy limit cycle in one to three steps depending on the transition function $f(t)$. Fig.16 present the variation process with the energy explanation variation graph, which is indicated that the trajectory after each step moves closer to the limit cycle on which the energy is high and stable.

4. Conclusion

In this chapter, we introduce a new method to estimate the basin of attraction for the biped robot and a speed switch control algorithm to change its walking speed. The method which is called Poincaré-like-alter-cell-to-cell mapping proposed in this chapter can also be used as a tool to estimate the basin of attraction of the hybrid system. Then a research about the law of controlling the walking speed based on energy shaping and the estimate of the basin of attraction for the underactuated kneed dynamic walker is carried out, which can change the walking speed smoothly. Simulation tests are done to verify the validity of these methods.

We make a brief conclusion about the main works as follows
1) The robustness of the biped robot can be justified by the size of the basin of attraction that can be obtained by Poincaré-like-alter-cell-to-cell mapping method.
2) Poincaré-like-alter-cell-to-cell mapping method has more accuracy, and it is not time-consuming.
3) Fixed point of the Poincaré map for the biped robot can be easily found with our method which is also used as an initial parameter in the proposed control algorithm.
4) Based on this speed switch control, the speed of the dynamic walking robot can be adjusted easily.
5) The description of the kinetic energy and potential energy and the visual simulation model can be used as effective tools for the analysis and control of our model.

The work in our near future study as below
1) Using this method to estimate the stability of more complex dynamic models, especially the dynamic walking robot whose locomotion is in the three dimensional space. And obtaining its stable initial state is a tough work.
2) Optimizing the speed switch algorithm and the switch function to obtain the time-optimized and energy-optimized trajectory in the controlled gait.
3) Researching for new mathematical tools to give the basin of the attraction more elaborated description. Meanwhile reducing the computational work in solving process is also a pressing problem.

5. Acknowledgment

This work is supported by National High Technology Research and Development Program of China (863 Program), grant No.2006AA04Z251; National Natural Science Foundation of China, grant No.60974067; the Science and Technology Development Program of Jilin Province, grant No: 20070524.

6. References

Ning, L.; Junfeng, L. & Tianshu, W.(2007). The effects of parameter variation on the basins of attraction of passive walking models. Institute of Electrical and Electronics Engineers Computer Society, Piscataway, NJ 08855-1331, United States, Harbin, China, pp. 1908-1913

Schwab, A.L. & Wisse, M. (2001). Basin of attraction of the simplest walking model. In: 18th Biennial Conference on Mechanical Vibration and Noise. American Society of Mechanical Engineers, Pittsburgh, PA, United States, pp. 531-539

Zhang, P.; Tian, Y., Liu, Z. & Yang, S. (2008). Further Research and Comparison of Gaits for Compass-like Biped and Kneed Passive Dynamic Walker. 2008ICIRA-International Conference on intelligent Robotics and Application.

Tedrake, R.; Zhang, T.W. & Fong, M.F., Seung, H.S. (2004). Actuating a simple 3D passive dynamic walker. In: 2004 IEEE International Conference on Robotics and Automation. IEEE Press, Piscataway, New Orleans, LA, United States, pp.4656-4661

Wisse, M. (2004). Three additions to passive dynamic walking; actuation, an upper body, and 3D stability. In: 2004 4th IEEE-RAS International Conference on Humanoid Robots. IEEE Press, New York, Santa Monica, CA, United States, pp. 113-132

Collins, S.H & Ruina A. (2005). A bipedal walking robot with efficient and human-like gait, Proceeding of IEEE International Conference of Robotics and Automation, Barcelona, Spain, Bipedal Locomotion I, CD-ROM No. 1935

Wisse, M. & Frankenhuyzen, J.van (2003). Design and construction of MIKE; a 2d autonomous biped based on passive dynamic walking, Proceeding of Conference of Adaptive Motion of Animals and Machines, Kyoto, Japan, Analysis & Control of Bipedal Locomotion, CD-ROM

Ono, K.; Furuichi, T. & Takahashi, R. (2004). Self-Excited Walking with Feet, International Journal of Robotics Research, Vol. 23, No. 1, pp. 55-68

Garcia, M.; Chatterjee, A. & Ruina, A. (1998). Speed, efficiency, and stability of small-slope 2-D passive dynamic bipedal walking. In: Proceedings of the 1998 IEEE International Conference on Robotics and Automation. Part 3 (of 4). IEEE Press, Piscataway, Leuven, Belgium, pp. 2351-2356

Jonathan K. Holm; Dongjun Lee & Mark W. Spong (2007). Time-Scaling Trajectories of Passive-Dynamic Bipedal Robots, Pro IEEE Int. Conf. Robotics and Automation, Rome, Italy, pp. 3603-3608.

Liu, L.; Tian, Y. & Huang, X. (2008). A Method to Estimate the Basin of Attraction of System with Impulse Effects: Application to the Biped Robot, International Conference on Intelligent Robotics and Applications, Wuhan, China.

Hsu, C.S.(1980) A Theory of Cell-to-Cell Mapping Dynamical Systems. Journal of Applied Mechanics, vol. 47, pp. 931-939

Garcia, M.(1999). Stability, scaling, and chaos in passive-dynamic gait models. Ithaca, New York: Cornell University

Coleman, M.J. (1998). A stability study of a three-dimensional passive-dynamic model of human gait. Ithaca, New York: Cornell University

Vanessa, F. & Hsu Chen (2007). Passive Dynamic Walking with Knees: A Point Foot Model. Submitted to the Department of Electrical Engineering and Computer Science on February 2, pp. 21-28

Zhang, P.; Zhang, D., Tian, Y. & Liu, Z. (2009). Dynamic Modeling and Stability Analysis of Passive Biped Robot. Journal of Beijing University of Technology, vol. 35, No. 2, pp. 258-263

15

Zappa, a Compliant Quasi-Passive Biped Robot with a Tail and Elastic Knees

Félix Monasterio-Huelin and Álvaro Gutiérrez
Universidad Politécnica de Madrid
Spain

Fernando J. Berenguer
Fundación Prodintec
Spain

1. Introduction

Bipeds are complex hybrid dynamical systems mixing both continuous and discrete-event phenomena (Hurmuzlu et al., 2004). The main characteristic of biped walkers is the abrupt kinematic change between the swing phase and the stance phase together with dynamical impacts. The main problem is how to achieve a rhythmical or periodical walk. Another difficulty of these robots is their high power requirement and, consequently, high energy consumption, which limits their autonomy. It can be attributed to the high number of actuated joints, and because energetic studies are not typically considered during the movement's planning.
Different studies focusing on the construction of locomotion controllers for completely actuated legged robots are found in the literature. These studies are mainly oriented to solve trajectory generation problems for the active control centered approach. In (Boeing & Bräunl, 2004) the authors show there is an example of this methodology showing that an impedance control is better than the computed torque method. However, the conventional approach has been questioned by researchers inspired by biomechanical models (Kuo, 2007; McMahon, 1984). The discovery of McGeer about passive dynamic walking by building a biped without any motors or sensors (McGeer, 1990), opens the doors to a new design concept based on morphological considerations. Moreover, the interaction between morphology and control is a topic of actual researches and debates in robotics (Pfeifer & Bongard, 2007). Therefore, exploiting the intrinsic passive dynamics has many advantages compared to the classical two-part methodology (trajectory generation and active controller). Two of them are the energy consumption reduction (the energetic cost is produced in step-to-step transitions) and the simplicity of control (low computational cost). Nevertheless, some theoretical and practical studies (Collins & Ruina, 2005) show that it is difficult to achieve the complex dynamics exhibited by humans and animals taking into account only the properties of simple passive-dynamic walking, but might help in the design of walking robots. In (Fumiya & Pfeifer, 2006) the authors show how to exploit the above mentioned passive properties of biped robots with the incorporation of sensors.

Typically, bipedal walking models assume rigid body structures. On the other hand, elastic materials seem to play an essential role in nature (Alexander, 2005). Therefore, spring-damper elements have been proposed as an initial solution for the compliant leg concept (Fumiya et al., 2009; Geyer & Seyfarth, 2006). These solutions can be extended with adaptable compliance (inverse of stiffness) mechanisms (Ham et al., 2007).
A mechanism for a real biped robot with a tail (i.e. *Zappa*) was proposed in (Berenguer & Monasterio-Huelin, 2006) and improved in (Berenguer & Monasterio-Huelin, 2008). *Zappa* is a robot able to walk in flat floors of different positive and negative slopes (Gutiérrez et al., 2008). The tail is the only actuated limb that imposes an oscillatory movement of variable frequency. By using this simple idea, it is possible to achieve an inverted pendulum trajectory for the legs. In this work, we propose a structural modification to *Zappa*, by adding elastic knees to the robot. This solution offers two main advantages: *(i)* a greater maneuverability and *(ii)* more compliant gaits.
The paper is organized as follows. In Section 2 the robot kinematics are fully described. In section 3 we propose a new controller applied to the robot's tail. Moreover, we study the robot energy consumption, and focus on the robot initialization, as a restriction to the walking performance. Section 4 describes the robot parameters along with performance graphs. Finally, Section 5 summarizes the most relevant conclusions and suggests future research directions.

2. The mechanism of the biped robot and the gait pattern

2.1 Kinematic Description

The previous implementation of *Zappa* (see Figure 1) had two legs, each composed of a four bars mechanism, four hip bars and a tail. The four bar mechanism includes a two bars femur, a one-bar foot and one hip's bar. The hip is composed of four bars, two of them joining the two bars femur of each leg, and the others two joining the legs. The tail is attached to one bar of the hip. The frontal plane is the ZY plane, the sagittal one is the XY plane and the transverse one is the ZX. To talk about the mechanism, we describe the following nomenclature as an example: A represents the line of the ankles, and AL and AR the left and right A lines respectively. When needed we will write AFL and ABL for the front and back of AL bars respectively. Figures 2a, 2b and 3a show the disposition of all bars and rotational joints. The foot has four possible contact points (C) with the floor (or ground, G) as depicted in Figure 3b. To start walking, it is enough to move the tail around the Y axis and to fix a spring to the ABL and ABR ankles. In Zappa this movement is managed by a DC servomotor with a position sensor controlling the tail and an extensional spring holding the back femur and the foot. Nevertheless, in simulations we use a torsional spring for the ankles and a controller to generate torque for the tail joint, as explained in Section 2.2.
The superior bar in the hip is added to avoid independent rotations of the legs around the X axis (see Figure 4). This bar is redundant in the stance phase. In general, this bar fixes a kinematic constraint to the hip angle:

$$q_{1R} - q_{1L} = \pi \tag{1}$$

This constraint could be relaxed adding a extensional spring ($q_5 = x_5$). The advantage of relaxing this constraint is to gain maneuverability to facilitate the rotation of the biped around the Y axis, and to avoid the sliding of the feet in the stance phase. In this work, we do not study the different configurations. On the other hand, the spring has been adjusted to generate enough force to avoid this balancing.

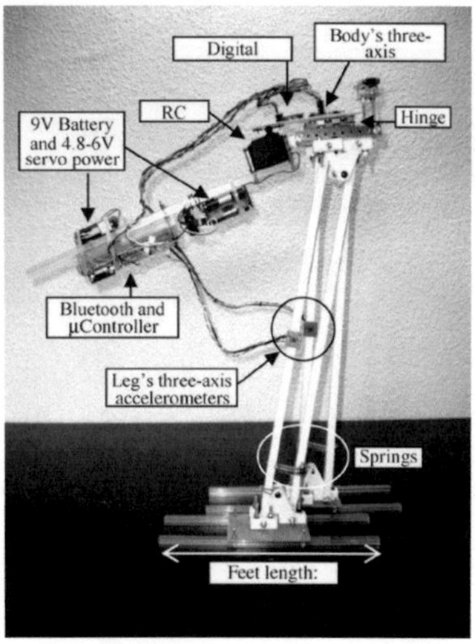

Fig. 1. Zappa biped robot. Prototype without knees.

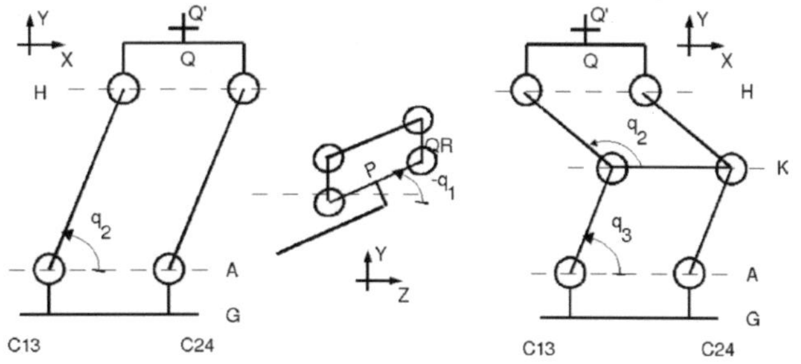

Fig. 2. Sagittal plane of a leg (a) without and (c) with knee. (b) Frontal plane of the hip bars and the tail.

For the stance phase there is a strong kinematic constraint that is avoided by producing a slight sliding of the feet in the Z direction, as depicted in Figure 5,

$$l_2 sin(q_{2R}) + l_1 sin(q_1) = l_2 sin(q_{2L}) \tag{2}$$

In the particular situation of $q_1 = 0$, the sole solution is $q_{2R} + q_{2L} = \pi$ or $q_{2R} = q_{2L}$. Therefore, the center of the hip bars will always lie in the midpoint between the feet. Furthermore, the distance between the feet depends on one ankle angle: $L = 2l cos(q_2)$.

To avoid the sliding, we propose to relax the latter kinematic constraint adding a knee. To do this, we have divided the femur in two parts, from now on femur and tibia, as depicted in Figure 2c (q_2). If the knee bar is rigid, each leg femur bars are parallel, and so are the tibia bars. The second kinematic constraint has been broken, but the pentagonal kinematic closed chain imposes a new constraint:

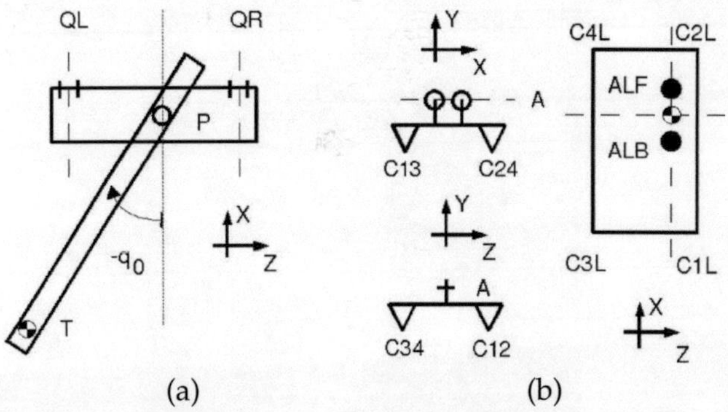

Fig. 3. (a) Transverse plane of the Central Hip and the Tail. (b) Three planes of a foot: joints with the leg and contact points with the floor.

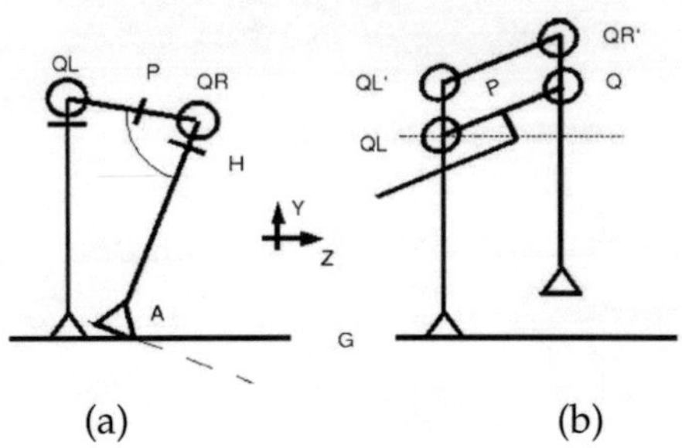

Fig. 4. Hip superior bar added to avoid oscillations around the X axis.

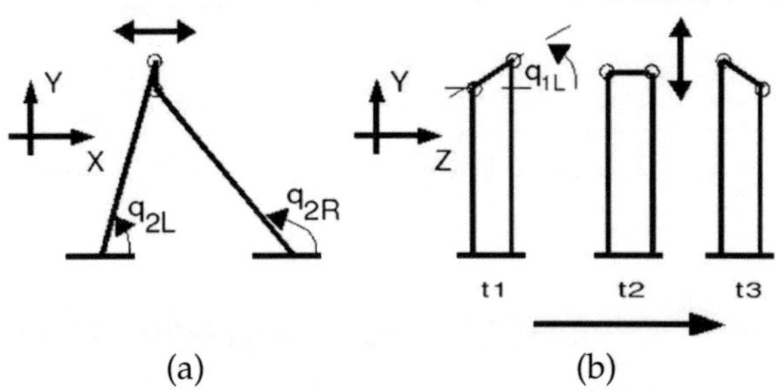

Fig. 5. (a) A possible configuration during the stance phase. (b) A possible time sequence t_1-t_3 to show the needed feet sliding in Z direction for the robot without knees.

$$l_3 sin(q_{3R}) + l_2 sin(q_{2R}) = l_3 sin(q_{3L}) + l_2 sin(q_{2L}) \tag{3}$$

As shown in Figure 6, a robot with knees is able to move from posture I to posture II without modifying the distance L and without sliding. Moreover, we observe point O changes its relative position and the height of the hip to the ground also varies. Therefore, this way of moving is more compliant.

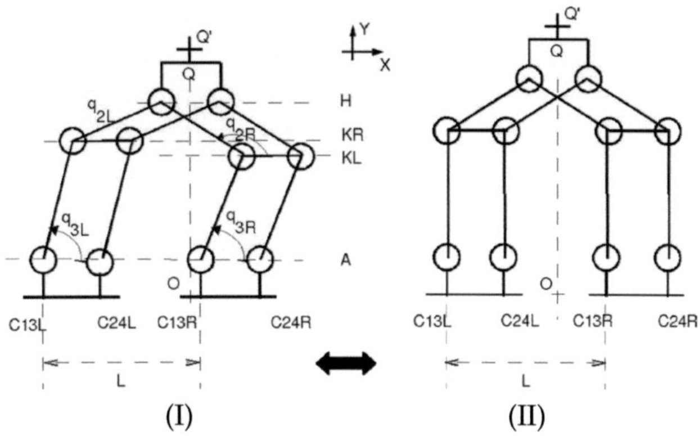

Fig. 6. A possible movement of the robot with knees which avoids the sliding in the Z axis.

Both in the swing and the stance phase the feet are parallel to each other, both in the biped with and without knees. However, if we add elastic knees to the robot, we are able to avoid this third kinematic constraint, obtaining interesting properties. For example, with the feet completely aligned the sole solution is the one shown in Figure 7a. With only one foot on the floor it is possible that the other leg is in the posture of Figure 7b. This mechanism adds two new degrees of freedom (dof) to the robot (q_4) (one translational dof for each leg).

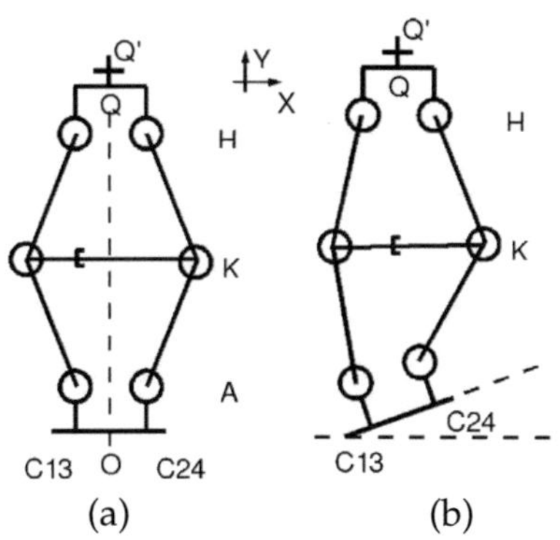

Fig. 7. Two possible postures for the robot with elastic knees.

2.2 Gait Description

As shown in Figure 2c, the springs are rotational in the ankles (ABL, ABR, q_3) and in the knee/femur (joining the femur and the knee bar; KBL, KBR, q_2). In the knee bar (line K, $q_4 = x_4$) and in the superior hip bar (line Q', $q_5 = x_5$) the springs are extensional. They are modelled as linear torque or force generators according to the equations:

$$\tau_i = -k_i(q_i - q_i^0) - b_i\dot{q}_i, \quad i = 2,3 \tag{4}$$

$$F_i = -k_i(x_i - x_i^0) - b_i\dot{x}_i, \quad i = 4,5 \tag{5}$$

where q_i^0 and x_i^0 are the i-spring equilibrium position, k_i the i-spring constant and b_i the i-viscous damping coefficient.

Selecting the springs is far from a trivial task as shown in (Gutiérrez et al., 2008). In that work, the authors studied a variable tuning for the ankle spring in the biped without knees using evolutionary optimization techniques. The fitness function was the travelled distance in a specified time and at different slopes of the floor. A mechanism that is able to change the spring parameters in real time was proposed in (Ham et al., 2007) using two servomotors (in general it is possible to generate the spring torque or force with only one DC motor and an electrical current controller). In the present work we have empirically obtained constant parameters by means of computer simulations.

One important question, not mentioned before, is related to the reaction forces between the feet and the ground. This question is extremely important in walking robots because of the strokes that occur when changing from the swing phase to the stance phase, and because they are variables that show the biped stability.

The total reaction force ($\mathbf{F}_R$) of the robot satisfies the following vectorial equation,

$$\mathbf{F}_R = M(\ddot{\mathbf{cm}} - \mathbf{g}) \tag{6}$$

where $\ddot{\mathbf{cm}}$ is the acceleration of its center of mass referred to some inertial coordinates system, $\mathbf{g}$ the gravitational acceleration and M the mass of the robot. The Newton equation of a foot is

$$\mathbf{f}_R + \mathbf{f}_{AB} + \mathbf{f}_{AF} + m_F\mathbf{g} = m_F\mathbf{a}_F \tag{7}$$

where $\mathbf{f}_R$ is the reaction force of one leg, m_F the mass of the foot, $\mathbf{f}_{AB}$ and $\mathbf{f}_{AF}$ the forces exerted by the back femur (or tibias, for the robot with knees) and frontal femur respectively on the foot (ankles AB and AF).

Hence, $\mathbf{F}_R = \mathbf{f}_{RL} + \mathbf{f}_{RR}$. When the foot is on the floor $\mathbf{a}_F = \mathbf{0}$. In the other case $\mathbf{f}_R = \mathbf{0}$. Therefore, measuring $\mathbf{a}_F$ is enough to know if one leg is on the floor except in the case of elastic knees, because the foot may oscillate around the Z axis. Simulations made in this work consider that the foot is in the air if all the reaction forces at each contact point in the three directions of space (x,y,z) are zero. Figure 8 shows the timing of the moments of contact of feet with the ground (represented as zero) overlapped with the reaction forces of each foot.

For the computer simulations, we have modelled the contact point between the feet and the floor as attractive forces that depend on the velocity of each foot contact point for the X and Z direction. For the Y direction, a stiffness factor has been added. This factor depends on the position only when position and velocity are negative (Berenguer & Monasterio-Huelin, 2007a).

Most figures that follow have been overlapped with this timing to properly interpret them.

The functioning of the biped robot is as follows: when $q_0 = 0$ the robot remains standing. For $\mu(t)$ around $\pi/2$ the robot rises a leg while the other remains in the floor. In other words, the tail works as a dynamic counterweight varying q_1. Because of the torque applied to the tail joint, the moment produced in the legs creates a pendular oscillation in both legs. This oscillation is produced around the ankle (A) for the foot on the floor and around the hip (H) for the foot in the air. This oscillation depends on all the springs of the robot. The potential energy stored in the springs acts as an energy generator allowing a compliant walking.

To be more concise, we can distinguish a clear sequence of movements of point P (tail joint) during the turn of the tail. The swing phase begin lifting one leg when $q_0 = \pi/2 - \mu$ until $\pi/2$, and provokes that $\dot{P}_x > 0$. From $\pi/2$ until $q_0 = \pi/2 - \mu'$ the swing phase continues until the foot makes contact with the ground; hence $\dot{P}_x > 0$. In the stance phase $\dot{P}_x$ changes

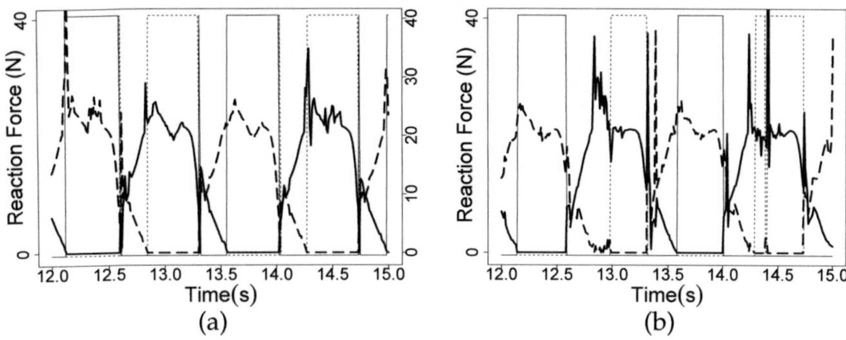

Fig. 8. Left (solid) and Right (dashed) feet reaction forces for a robot (a) without knees and (b) with elastic knees.

its sign until $q_0 = -\pi/2 + \mu$, repeating the cycle. Figure 9 shows this sequence. In Figure 10 the limit cycle ($\dot{q}_2$ versus q_2 for the robot without knees, $\dot{q}_3$ versus q_3 and $\dot{q}_2$ versus q_2 for the robot with elastic knees) of the biped gait pattern is shown. We define the limit cycle as a sequence of steps used as an index of whole stability (Hobbelen & Wisse, 2007) in contrast to the ZMP (Vukobratovic et al., 1990) that imposes local stability at every instant in time.
In Figure 9 we show the P_y kinematic equation when all the four contact points of the foot are in the floor. However, this is not the general case when the robot has elastic knees because P_y depends on the angle between the foot and the floor in the stance phase. Therefore, the biped robot with elastic knees walks raising the heel (line C13 in Figure 3b) of the swinging leg and falls on it at the end of a step (in a similar way to human walking).

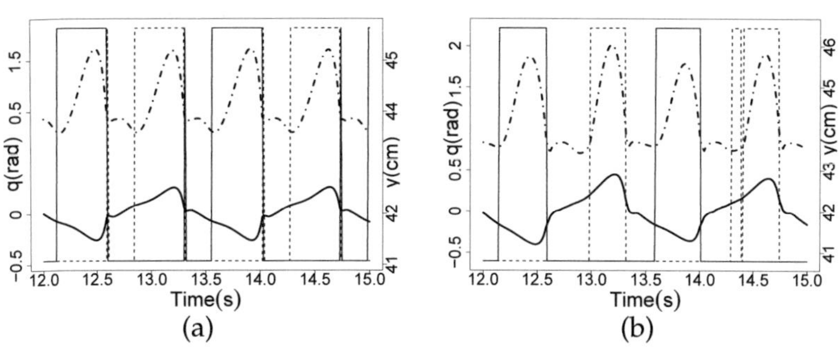

Fig. 9. q_{1L} (solid) and P_y (dotdashed) of the biped (a) without knees (in the stance phase $P_{yL} = l_1 \, |sin(q_{1L})| + l_2 sin(q_{2L})$) and (b) with elastic knees (in the stance phase $P_{yL} = l_1 \, |sin(q_{1L})| + l_2 sin(q_{2L}) + l_3 sin(q_{3L})$ with no elastic knees, as explained in the text).

The angular position and torque of the Knee/Femur spring and the linear position and force of the Knee spring are shown in Figure 11.

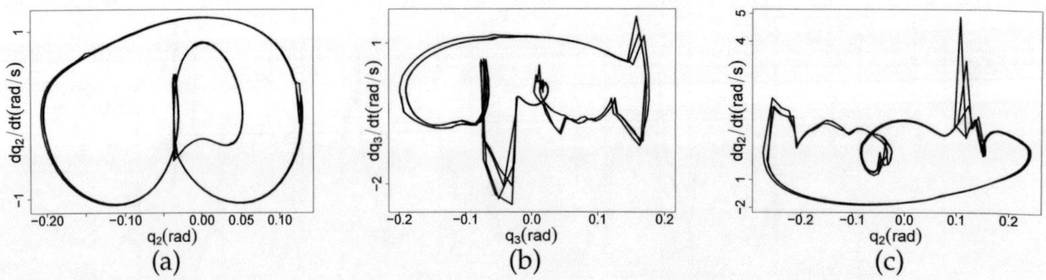

Fig. 10. Limit cycle of the biped (a) without knees. (b) $\dot{q}_3$ versus q_3 and (c) $\dot{q}_2$ versus q_2 with elastic knees.

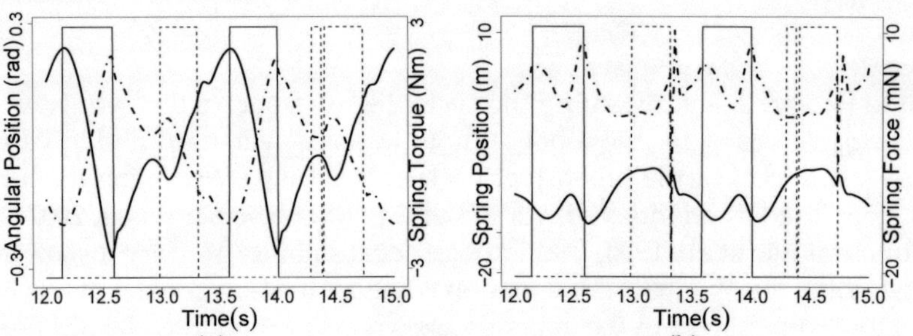

Fig. 11. (a) Left (dotdashed) and Right (solid) femurs torques of a robot with elastic knees. (b) Left (dotdashed) and Right (solid) knees forces of a robot with elastic knees.

3. Tail control, power consumption and stability of the biped

To achieve a regular walking gait the tail has to be controlled adequately. This can be done imposing a sinusoidal movement to the tail (q_{0r}). However, because all initial conditions are zero (the robot is with both feet together) and because an abrupt change could make the biped to fall, a special starting procedure is needed. The idea is to begin the walk with short strides before imposing a sinusoidal oscillation of constant frequency f. In (Berenguer & Monasterio-Huelin, 2007b) chirp functions were used to study the dependency of this starting procedure as a function of frequency variation. In this work, we use a chirp function to switch to the sinusoidal by imposing continuity conditions in position, velocity and acceleration. For these reasons, there is only one solution depending on two parameters: (i) the time at which the switch occurs (t_f) and (ii) the number of periods the chirp must perform (N).

$$q_{0r} = \begin{cases} \pi/2sin(at^2), & t < t_f \\ \pi/2sin(2\pi f(t-t_f) + (2N-1)\pi/2), & t \geq t_f \end{cases} \quad (8)$$

where $a = (2N-1)\pi/(2t_f^2)$. Figure 12 shows the reference signal for the tail.
To control the tail, we set up this chirp function as the reference signal. The tail satisfies the following two vectorial Newton-Euler equations:

$$\mathbf{f}_T + m_T\mathbf{g} = m_T\mathbf{a}_T \quad (9)$$

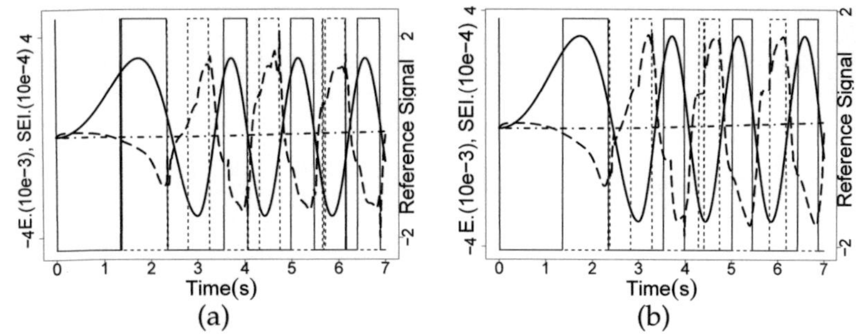

Fig. 12. Reference signal (solid) (q_{0r}) with $t_f = 3s$, $N = 2$, error signal (dashed) and squared error integral (dotdashed) for a PD with $k_p = 300$ and $k_D = 20$. (a) Without knees and (b) with elastic knees.

$$\boldsymbol{\tau}_T + \mathbf{f}_T x \mathbf{l}_0 = \mathbf{I}\boldsymbol{\alpha} + \boldsymbol{\omega}\, x(I\boldsymbol{\omega}) \tag{10}$$

where $\mathbf{a}_T$ is the linear acceleration of the tail center of mass, m_T the mass of the tail, $\mathbf{g}$ the gravitational acceleration, $\mathbf{l}_0$ the radial vector from the tail joint to the tail center of mass, $\mathbf{I}$ the tail inertia matrix, $\boldsymbol{\alpha}$ the tail angular acceleration, and $\boldsymbol{\omega}$ the tail angular velocity.
Solving these equations in the Y direction (tail's rotation axis) and assuming a punctual mass, we obtain the mechanical torque that must be applied to the tail:

$$\tau_T = m_T l_0 a_{Tz'} \tag{11}$$

where z' is the Z component in the frame attached to the tail.
The linear acceleration of the tail center of mass (when no sliding) in the stance phase is

$$a_{Tz'} = l_0 \ddot{q}_0 + \ddot{P}_{z'} \tag{12}$$

where $\ddot{P}_{z'}$ is the linear acceleration of point P in the frame attached to the tail joint with respect to the inertial coordinate system projected in the z' axis.
In the stance phase and assuming there is not sliding, the tail joint has no Z component, and the X and Y components are because of the pendular movement of the legs around the ankles. As aforementioned, the biped without knees cannot avoid this sliding (see Figure 5) but it is kinematically possible for a biped with knees to do it.
The controller we propose applies a torque to the tail joint following Equation 13 (a feedforward term and a PD feedback controller):

$$\tau_T = m_T l_0 a_{Tr} + k_P e + k_D \dot{e} \tag{13}$$

where $e = q_{0r} - q_0$, $a_{Tr} = l_0 \ddot{q}_{0r}$, and k_P and k_D constants that must be tuned.
Taking into account the mechanical torque and ordering the terms we obtain the Laplace error equation,

$$(m_T l_0^2 s^2 + k_D s + k_P)E(s) = m_T l_0 \mathcal{L}(\ddot{P}) \tag{14}$$

If we select stable poles and $\lim_{s \to 0} s\mathcal{L}(\ddot{P}) = 0$, the steady-state error will be zero. The transient response will be very short when choosing the poles with high values. Figure 12 shows the error signal and the squared error integral.

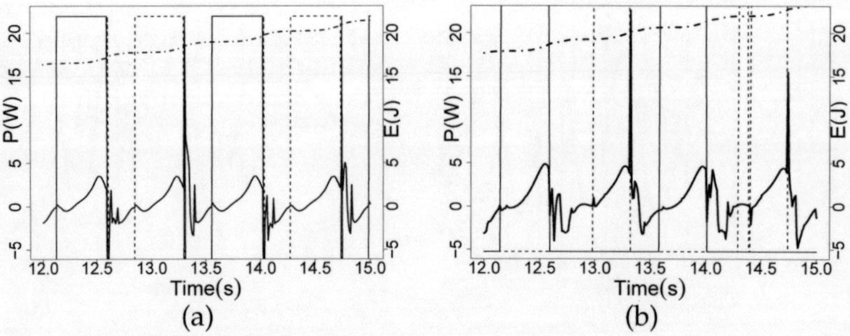

Fig. 13. Energy (dotdashed) and Power consumption (solid) for a robot (a) without knees and (b) with elastic knees in the time interval from 12 to 15 seconds.

Another important problem for biped robots is the mechanical power ($P(t)$) and energy ($E(t)$) consumption. Because the tail is the only actuated joint, we can define them using the following relations:

$$P(t) = \tau_{Tz}\dot{q}_0 \tag{15}$$

$$E = \int_0^t |P(t)|\, dt \tag{16}$$

Figure 13 shows the power and energy consumption.
As mentioned in Section 2.2 the limit cycle shown in Figure 10 could be used to analyze the cyclic stability of the biped. This study would proceed gait by gait searching a stable periodic motion, although it might not be a stable motion in every instant of time. In this work we do not analyze this important question, and we do not know if the system is stable in either one sense (cyclically) or the other (locally). Actually, we know that the robot with elastic knees does not fall and we consider it as a proof of its stability. This analysis remains to be done in a near future.

4. Simulation Studies

The robot has been simulated thanks to the *SimMechanics* toolbox included in the *Matlab* software. Robots with and without knees maintain the same weight. The femur and the tibia have been blocked in the robot with knees to simulate the robot without knees. The task performs a straight line walking for 15 simulated seconds where the tail's frequency oscillation is 0.7 Hz for both simulated robots.
The main parameters of the robot used in simulations are presented in Table 1 where the "K" corresponds to the robot with knees. (Note that $l_2 = l_{2K} + l_{3K}$). The constant "d" represents the distance between frontal and back bars of a leg.
Finally, the different values of the springs parameters are presented in Table 2.

5. Conclusions and future work

Comparing a biped with and without knees is a hard problem because of their structural differences. Nevertheless, in this work we have stated there are at least two measurements that

Name	Value	Name	Value	Name	Value
m_{tail}	*760 gr.*	l_0	*150 mm.*	d	*40 mm.*
m_{hip}	*210 gr.*	l_1	*100 mm.*	$lfoot_{C12}$	*320 mm.*
m_{legs}	*920 gr.*	l_2	*400 mm.*	$lfoot_{C13}$	*80 mm.*
m_{feet}	*280 gr.*	l_{2K}	*200 mm.*	$lfoot_{C24-AF}$	*120 mm.*
M_T	*2170 gr.*	l_{3K}	*200 mm.*		

Table 1. Simulation parameters: dimensions and masses

Spring	k	b	q^0 or x^0
Ankle	*10*	*0.4*	*0*
Femur/Knee	*8*	*0.1*	*0*
Knee	*750*	*100*	*-0.007*
Superior Hip	*200000*	*5000*	*0*

Table 2. Simulation parameters: springs

may serve as performance indexes: *(i)* the distance travelled considering the same experimental conditions and *(ii)* the capacity of the robot to walk with a higher tail frequency. Simulation results indicate that the robot with elastic knees is superior to the robot without knees because the former travels larger distances with the same oscillatory frequency (f=0.7 Hz). Moreover, the robot with elastic knees can walk with a higher frequency (f=0.8 Hz), at which the robot without knees falls down. The reason why the robot with knees travels a larger distance is not only because of this higher frequency capacity, but because the robot raises the feet higher in each stride. This is the result of leg spring combination.

As observed in the figures presented in this work, the performance of the two types of robots are very similar, in consumption (Figure 13), in the response of the tail controller (q_0) (Figure 12), in the reaction forces (Figure 8) and in the way the hip angle (q_1) oscillates to pass from the stance phase to the swing phase (Figure 9). Nevertheless, because of the kinematic differences between P_y,s it is difficult to draw definitive conclusions. We conjecture that it is possible for the robot with elastic knees to avoid the lateral sliding mentioned in this paper. This suggests a design of the robot in which the angle q_1 remains constant during the stance phase. Other possibility is a design in which the hip spring of the superior bar allows lateral balancing without sliding, but we have not yet addressed this question.

The way the robot with elastic knees walks is very different from the way the robot without knees walks, in a more compliant way. The limit cycles depicted in Figure 10 show clearly a big difference, but the problem remains in how to compare the two types of robots. One possibility is taking into account their skills. The foot raising height might be an useful criterion if the robot with knees could finally climb stairs. In Gutiérrez et al. (2008) the authors demonstrated that the robot without knees could go up and down different inclination slopes. This was achieved by tuning, in real time, the ankle spring parameter values. Therefore, it produced a modification on the robot equilibrium position, translated in different legs' inclinations. We have proved the same for the robot with knees, but it remains an open question if the robot can climb stairs or turn around. The relaxation of kinematic constraints, we have proposed in this work, points towards this line of research seeking the increase of its manoeuvrability.

6. References

Alexander, R. M. (2005). Walking made simple, *Science Magazine* **308**(5718): 58–59.

Berenguer, F. J. & Monasterio-Huelin, F. (2006). Easy design and construction of a biped walking mechanism with low power consumption, *Proc. of the 9th Int. Conf. Climbing and Walking Robots CLAWAR'06*, Springer-Verlag, Berlin, Germany, pp. 96–103.

Berenguer, F. J. & Monasterio-Huelin, F. (2007a). Stability and smoothness improvements for an underactuated biped with a tail, *Proc. of the 2007 IEEE Symp. on Industrial Electronics*, IEEE Press, Piscataway, NJ, pp. 2083–2088.

Berenguer, F. J. & Monasterio-Huelin, F. (2007b). Trajectory planning using oscillatory chirp functions applied to bipedal locomotion, *Proc. of the 4th Int. Conf. on Informatics in Control, Automation and Robotics*, IEEE Press, Piscataway, NJ, pp. 70–75.

Berenguer, F. J. & Monasterio-Huelin, F. (2008). Zappa, a quasi-passive biped walking robot with a tail. modeling, behavior and kinematic estimation using accelerometers, *IEEE Transactions on Industrial Electronics* **55**(9): 3281–3289.

Boeing, A. & Bräunl, T. (2004). Evolution of locomotion controllers for legged robot, *in* T. et al. (ed.), *Robotic Welding, Intelligence and Automation*, Vol. 299 of *Lecture Notes in Control and Information Sciences (LNCIS)*, Springer-Verlag, Berlin, Germany, pp. 228–240.

Collins, S. H. & Ruina, A. (2005). A bipedal walking robot with efficient and human-like gait, *Proc. of the 2005 IEEE Int. Conf. on Robotics and Automation*, IEEE Press, Piscataway, NJ, pp. 1983–1988.

Fumiya, I., Minekawa, Y., Rummel, J. & Seyfarth, A. (2009). Toward a humanlike biped robot with compliant legs, *Robotics and Automation Systems* **57**(2): 139–144.

Fumiya, I. & Pfeifer, R. (2006). Sensing through body dynamics, *Robotics and Autonomous Systems* **54**(8): 631–640.

Geyer, H. & Seyfarth, A. (2006). Walking and running dynamics explained by compliant legs: Consequences, general insights, and future directions, *Journal of Biomechanics* **39**(1): 361.

Gutiérrez, A., Berenguer, F. J. & Monasterio-Huelin, F. (2008). Evolution of neuro-controllers for trajectory planning applied to a bipedal walking robot with a tail, *in* A. Lazinika (ed.), *New Developments in Robotics, Automation and Control*, I-Tech Education and Publishing, Vienna, Austria.

Ham, R. V., Vanderborght, B., Damme, M. V., Verrelst, B. & D.Lefeber (2007). Maccepa, the mechanically adjustable compliance and controllable equilibrium position actuator: Design and implementation in a biped robot, *Robotics and Autonomous Systems* **55**(10): 761–768.

Hobbelen, D. G. E. & Wisse, M. (2007). Limit cycle walking, *in* M. Hackel (ed.), *Humanoid Robots, Human-like Machines*, I-Tech Education and Publishing, Vienna, Austria.

Hurmuzlu, Y., Génot, F. & Brogliato, B. (2004). Modeling, stability and control of biped robots A general framework, *Automatica* **40**(10): 1647–1664.

Kuo, A. (2007). The six determinants of gait and the inverted pendulum analogy: A dynamic walking perspective, *Human Movement Science* **26**(4): 617–656.

McGeer, T. (1990). Passive dynamic walking, *Int. Journal of Robotics Research* **9**(2): 62–82.

McMahon, T. (1984). *Muscles, Reflexes, and Locomotion*, Princeton Press, Princeton, NJ.

Pfeifer, R. & Bongard, J. C. (2007). *How the Body Shapes the Way We Think: A New View of Intelligence (Bradford Books)*, The MIT Press, Cambridge, MA.

Vukobratovic, M., Borovac, B., Surla, D. & Stokic, D. (1990). *Biped Locomotion: Dynamics, Stability, Control and Application*, Springer-Verlag, Berlin, Germany.

16

Quadrupedal Gait Generation Based on Human Feeling for Animal Type Robot

Hidekazu Suzuki and Hitoshi Nishi
Tokyo Polytechnic University & Fukui National College of Technology
Japan

1. Introduction

Animals have for long been recognized as being a positive force in healing processes (Baun et al., 1984). In recent years, animal-assisted therapy (AAT), which makes use of the healing effects of animals has attracted attention (Fine, 2006). Examples of the expected results of this type of therapy are buffering actions for stress, improvement of sociability and shortening of the medical treatment period through mental healing. Thus, the introduction of AAT is being considered in hospitals and health facilities. However, it is difficult to employ AAT in such facilities because of the risks of the spread of infection from animals to patients and the necessity of proper animal training.

Robot-assisted therapy (RAT), in which robots resembling animals are used instead of real animals, is important for patient safety (Shibata et al., 2005). Pet robots resembling various animals, such as the dog robot "AIBO", seal robot "Paro", etc., are used in this type of therapy. Banks et al. reported no difference between the effectiveness of a living dog and an AIBO robotic dog in reducing loneliness (Banks et al., 2008). Shibata et al. applied a mental commit robot, Paro, to RAT, and they verified that the interaction with Paro has psychological, physiological and social effects on people (Shibata et al., 2004; Wada et al., 2005). In these applications, it is important that the robot imitates the motions of living animal, especially essential motions, such as walking, running, etc.

However, it is difficult for the robot to walk and run like an animal because it is affected by various types of dynamic noise in the real world, in contrast to the ideal world. In recent years, many researchers have studied gait generation methods for various types of robots (Estremera & Santos, 2005; Kimura et al., 2005). A legged robot in the real world will have n-DOF (degrees of freedom) for movement, and it is difficult to solve the optimization problem in n-dimensional continuous state/action space to generate an adequate gait (Kimura et al., 2001). Therefore, evolutionary approaches, such as use of fuzzy logic, genetic algorithms, neural networks, or various hybrid systems, are employed for gait learning and parameter optimization (Inada & Ishii, 2003; Son et al., 2002). For example, Chernova et al. generated fast forward gaits using an evolutionary approach for quadruped robots (Chernova & Velosa, 2004). However, these gait generation methods for legged robots did not evaluate the degree to which the robot's movement approximated that of a living animal, because they were not designed for enhancement of the effects of RAT.

In the present study, therefore, we attempted to generate an animal gait for a quadrupedal robot using a genetic algorithm and gait patterns based on zoological characteristics (Suzuki et al., 2007). Moreover, a questionnaire study was performed to determine an adequate mix of several combinations of gait velocity and duty ratio for generated gait, and thus a more natural animal-like gait for the AIBO was chosen based on subjective human feelings from among the various gaits. Furthermore, parameters of each leg were adjusted again through an additional optimization on the ground.

2. Concept of Gait Generation

In this research, we used AIBO (ERS-7 M2, Sony), which is a well-known quadrupedal pet robot, as shown in Fig. 1. AIBO has 15 joints (head and legs), 3 DOF (degrees of freedom) at each leg, and 31 sensors. We can construct an application to control AIBO using OPEN-R SDK, a cross-development environment based on the C++ language provided by Sony.

Usually, when generating a gait for a robot, we often construct a robot model on the basis of dynamics. However, the dynamics of AIBO change in a complex manner depending on the situation in a real environment, and therefore strict modeling is difficult. Moreover, it may be even more difficult to define subjective human feelings for animals based on a model. Therefore, we generated a gait for AIBO on the basis of that of living animal and subjective human feelings.

Fig. 1. AIBO (ERS-7 M2 : Sony)

We attribute animal gait to that which achieves efficient propulsion. Moreover, both mono-leg propulsion and coordinated movement of each leg realize an efficient gait. Hence, we attempted to generate the orbit of a mono-leg, based on an animal's orbit, which can achieve efficient propulsion in the real world.

Figure 2 shows the normal gait of a walking dog. In this figure, (a) represents the dog's leg that is in contact with the ground and (d) represents the leg shape, which kicks out. Further, the start and end shapes of the leg are decided as shown in Fig. 3. However, the intermediary orbit in the real world is unknown. Therefore, we utilized a genetic algorithm (GA) (Michalewicz, 1994; Goldberg, 1989; Goldberg, 2002) to optimize the intermediary orbit of AIBO's leg.

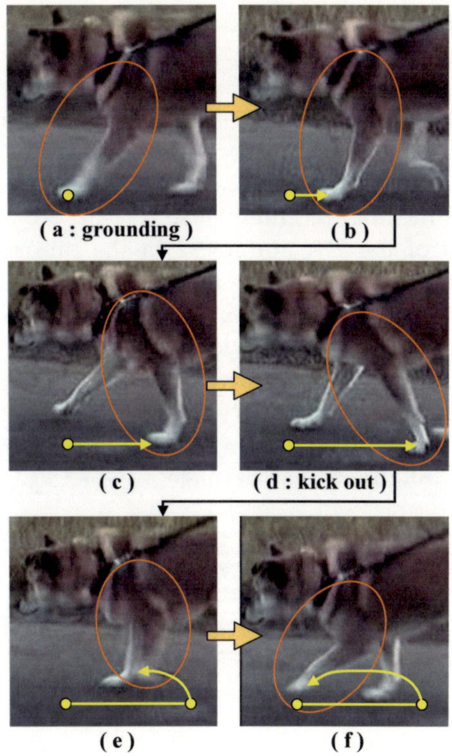

Fig. 2. Normal gait of a dog

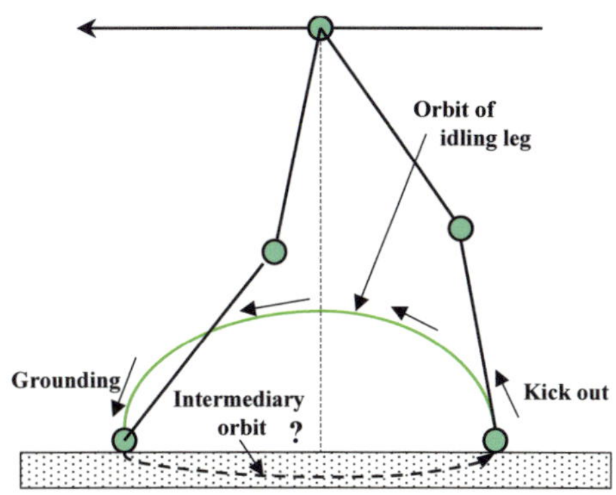

Fig. 3. Start and end shapes of the leg

3. Orbit Generation for AIBO's Leg

A genetic algorithm is an example of an AI program (Back, 1996) and is well known as a parallel search and optimization process that mimics natural selection and evolution. In the GA process, the search for a solution to a given problem is performed using a population of individuals as binary strings, which represent the potential solutions to that problem. The

GA is viewed as an optimization method as the iterative process of evolution toward better search solutions is equivalent to the process of optimizing the fitness function. The term "parallel," which is used in "parallel search" above, is related to the implicit parallelism of GA and has been explained previously by Goldberg (Goldberg, 1989; Goldberg, 2002). This concept means that even though the GA processes only *s* individuals in the population in each generation, we can obtain useful processing of around s^3 schemata in parallel without any special bookkeeping or memory requirements.
Figure 4 shows the genes of the GA employed in the present search, which has three parameters ($\theta 1$, $\theta 2$, r). Here, by studying the moving image of a dog's gait, we noted that there is a turning point that changes the velocity of the leg in front and behind. It appears that the function of the leg changes from providing support to driving. The three parameters ($\theta 1$, $\theta 2$, r) represent the leg shape at this turning point, as shown in Fig. 5, and T_g is the grounding time [ms]. Hence, the intermediary orbit is uniquely decided by the parameters ($\theta 1$, $\theta 2$, r). Briefly, the problem of generating a high propulsive orbit for AIBO's leg is changed to the problem of optimizing the parameters ($\theta 1$, $\theta 2$, r).

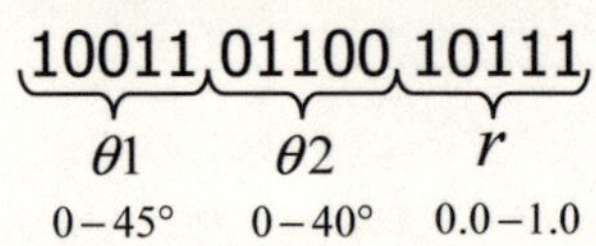

Fig. 4. Genes of GA

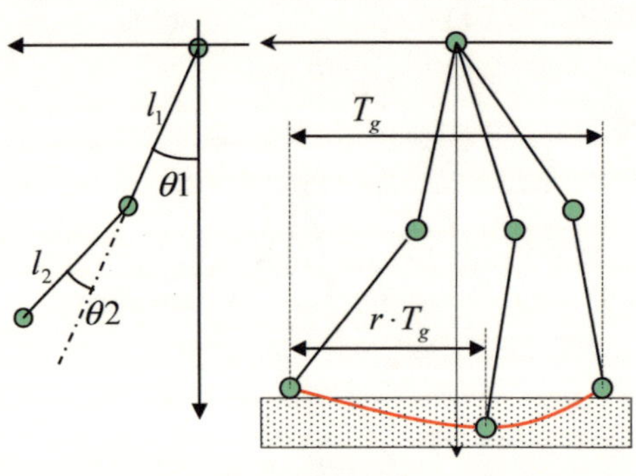

Fig. 5. $\theta 1$, $\theta 2$, r and T_g

The GA process is shown in Fig. 6. A population comprising a set of *s* individuals is used by the GA process to search for the target orbit in the real world. As the elitist model of the GA is adopted, the best sorted individual in the *N*-th population, designated as a vector $\boldsymbol{\phi}_1^N$ and possibly representing the leg's orbit, which can realize efficient propulsion in the real world is selected to survive. Let us denote the components of $\boldsymbol{\phi}_l^N$ expressing the orbit of the *l*-th individual in the *N*-th generation by $\theta 1_1^N$, $\theta 2_1^N$, and r_1^N.
In this study, we prepared a board attached with a free wheel, as shown in Fig. 7, to evaluate the propulsion caused by mono-leg motion in the real environment. Moreover, we adopted the measured advance distance of the evaluation board as the fitness value of the

GA search. In this evaluation system, AIBO moves the mono-leg only for one cycle based on the orbit represented by each individual of the population.

The obtained fitness values $E_1^N(\phi_1^N)$, $E_2^N(\phi_2^N)$, ..., $E_S^N(\phi_S^N)$ are sorted. Based on the ranking and a selection rate to die, the weakest individuals in terms of poor fitness values are replaced by newly created individuals. In creating the new individuals, random selection and random crossover are first performed. In this process, paired mates and two-point crossover are used. Next, a random bit-by-bit mutation (exchange of 1 by 0 or vice versa) is performed on the individuals obtained after the crossover. This ends the *N*-th generation and the population obtained after these operations constitute the population at the starting point of the (*N*+1)-th generation. The preceding steps are then repeated with the individuals in population *N*+1 to evolve the population toward the solution.

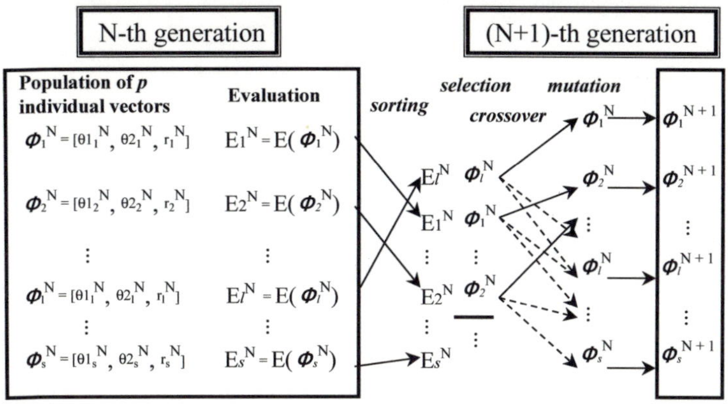

Fig. 6. Elitist model searching of a GA

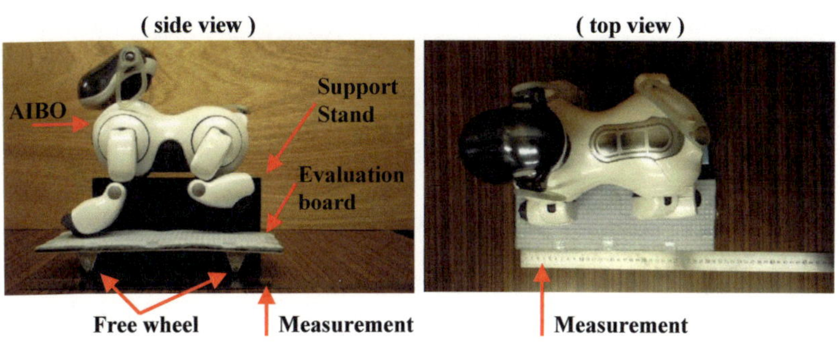

Fig. 7. Measurement method

In this experiment, we prepared three normal orbits, as shown in Fig. 8, using two-link inverse kinematics to compare the fitness value of the orbit optimized by the above GA process. Figure 9 shows the result of this experiment. Further, the orbit approximating an animal's gait and optimized by the GA shows a high evaluation value, i.e., a high propulsive force. In this GA process, the population size, selection rate, and mutation rate are 10 individuals, 0.5, and 0.3, respectively.

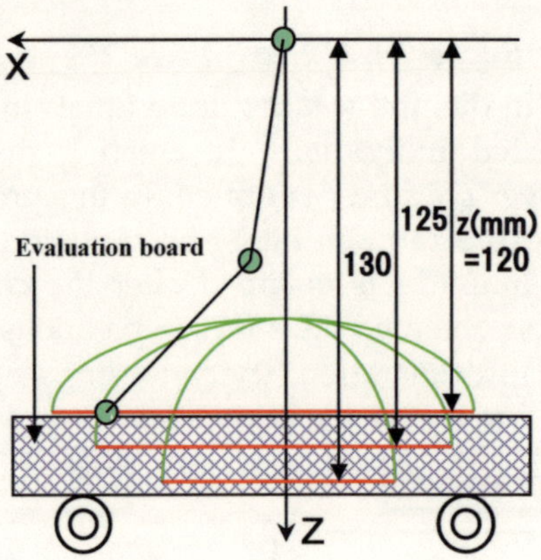

Fig. 8. Comparison orbits

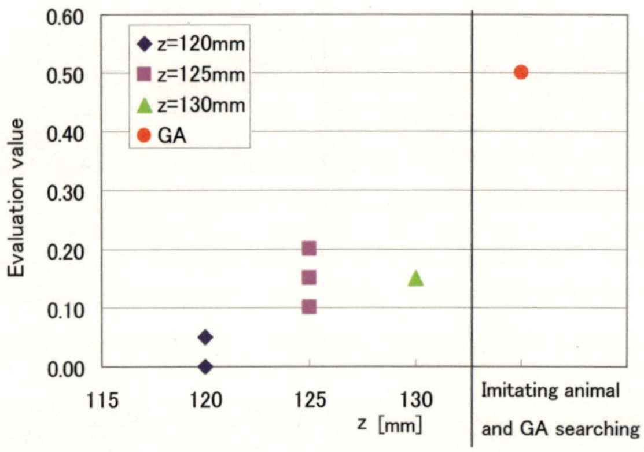

Fig. 9. Experimental result of GA

4. Quadrupedal Gait Based on Human Feeling

As described in the previous section, we constructed the orbit of the mono-leg that can provide efficient propulsive force by approximating an animal's gait and optimizing GA. Next, we addressed the coordination between each leg, which can realize an efficient gait. The gaits of various animals have already been studied and analyzed in the field of zoology. Moreover, Alexander et al. classified quadrupedal gait on the basis of energy cost, as shown in Fig. 10 (Alexander et al., 1980). In this figure, the numbers near each leg represent the phase difference based on the left forefoot; *d* is the duty ratio and it refers to the grounding ratio. In this classification, the phase difference between a dog's walking gait and running gait correspond to that between the "Walk" and "Trot" gaits shown in Figs. 10(a) and (b), respectively. We generated the quadrupedal gait of AIBO using both the above-mentioned optimum orbit of mono-leg and "Walk" gait to generate an animal-like walking gait.

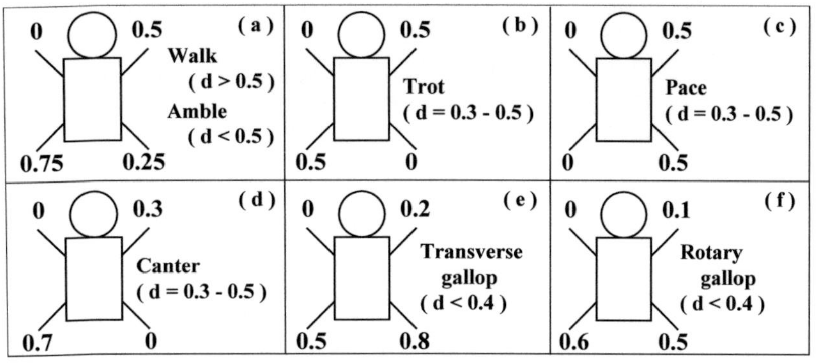

Fig. 10. Classification of quadrupedal gaits

		Duty ratio (grounding ratio)				
		0.51	0.56	0.63	0.69	0.75
T_{all} [ms]	200	58	54	38	67	36
	600	78	78	72	57	57
	1000	79	78	80	59	69
	1400	74	62	69	65	58
	1800	62	67	49	51	49

Table 1. Questionnaire related to subjective human feeling

In the "Walk" gait, the duty ratio generally decreases from 0.75 to 0.50 depending on the increment in the gait velocity. However, it is difficult to select an adequate mix of gait velocity and its duty ratio to cause a human observer to perceive an animal-like gait, because of the variable sensitivity of humans. Hence, we prepared a questionnaire study regarding several combinations of the gait velocity and duty ratio to determine an adequate mix. The results of the questionnaire study for 30 participants are shown in Table 1. In this table, T_{all} indicates the time period at motion cycle of mono-leg and includes the grounding time, which corresponds to T_g in Fig. 5 and is calculated as T_{all} x (*duty ratio*), and idling motion. This questionnaire study presented the participants with the moving image, the combined duty ratio of the 25 patterns, and T_{all}. Further, the participants assigned points from 1 (poor) to 5 (good, meaning the gait resembled that of a living animal) according to their subjective feelings regarding each moving image. Figures 11-13 show the results for duty ratios of 0.51, 0.63, and 0.75, respectively, for each value of T_{all}. Table 2 and Fig. 14 show the median of the polling number that seems to be the average subjective human feelings regarding the animal gaits.

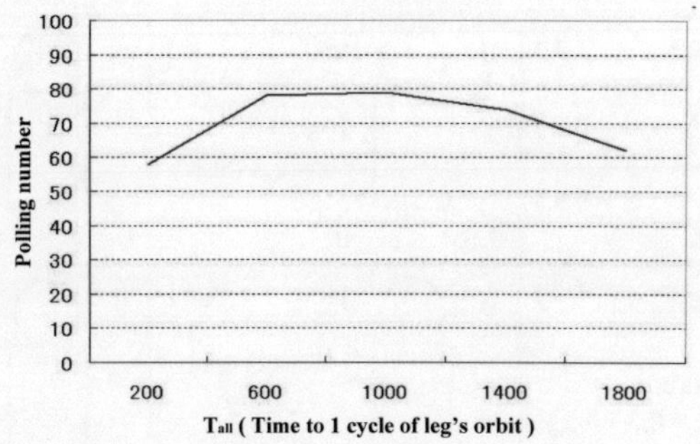

Fig. 11. Questionnaire data of duty ratio 0.51

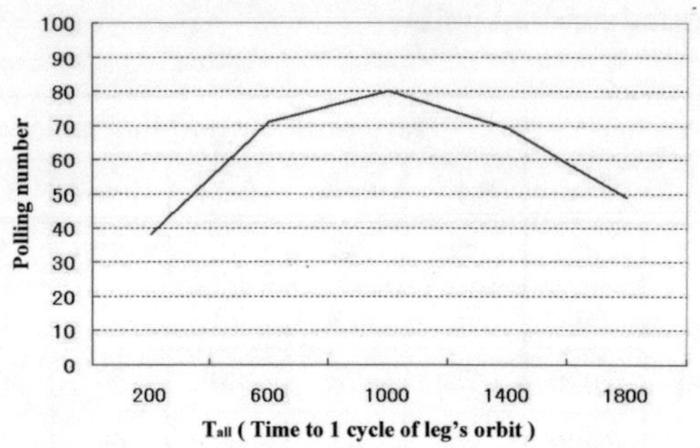

Fig. 12. Questionnaire data of duty ratio 0.63

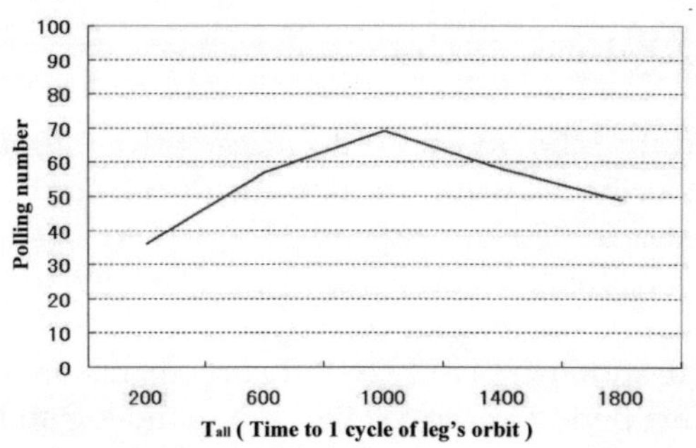

Fig. 13. Questionnaire data of duty ratio 0.75

	Duty ratio (grounding ratio)				
	0.51	0.56	0.63	0.69	0.75
Median of gait cycle [ms]	996	995	1020	1031	1038

Table 2. Median data of polling number

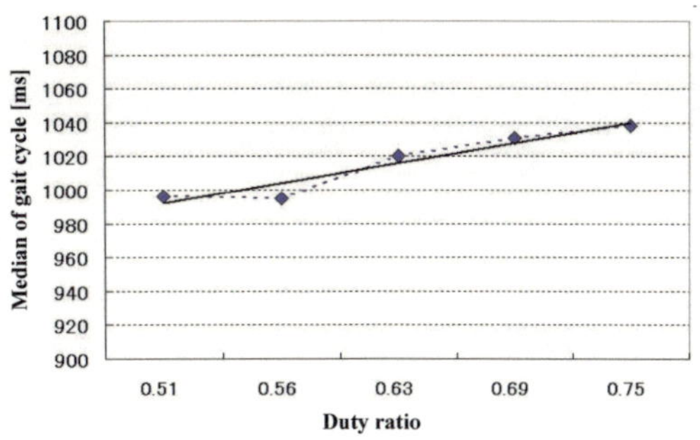

Fig. 14. Median of each duty ratio

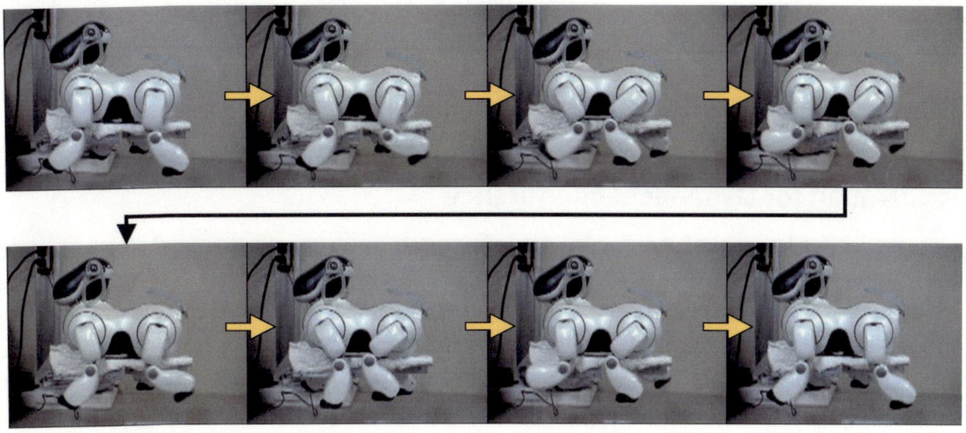

Fig. 15. Quadrupedal gait of AIBO based on animal gait

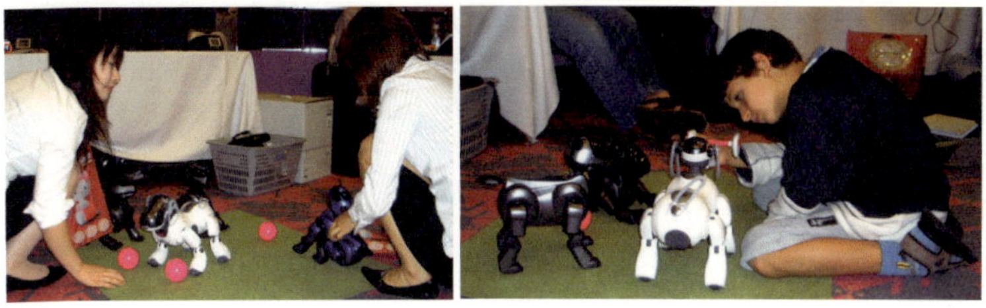

Fig. 16. Motion verification at conference

Figure 15 shows one of the animal gaits generated by the orbit of the mono-leg optimized by the GA, and an adequate mix of gait velocity and duty ratio based on subjective human feelings. In this gait, T_{all} is 1020 [ms] and the duty ratio is 0.63. Further, we have presented AIBO's gait generated by the above method at an international conference to verify the degree to which it approximates the natural gait of a living animal based on subjective human feelings (Fig. 16); we have confirmed that many viewers feel that this AIBO gait is fairly similar to that of a living animal.

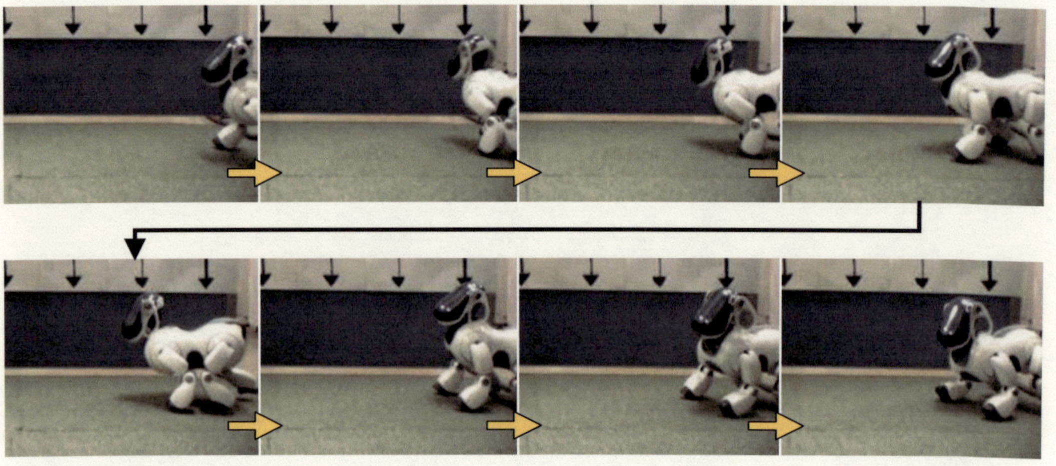

Fig. 17. Walking on ground

5. Modification for Dynamical Interference

We performed further experiment using the generated gait (Fig. 15) to check the adaptability to interaction with ground like that shown in Fig. 17. AIBO walked forward unsteadily, and the motion did not resemble the gait of a living animal. It seems that the generated gait cannot adapt to the dynamical interference by ground reaction.

So, we tried the additional optimization for generated gait to correct minor deviation of angle and timing for each joint. In this optimization experiment, the parameters ($\theta 1$, $\theta 2$) of each joint generated by above-mentioned process are modified slightly in the range of ±8[°] by GA. In the optimizing process, the walking distance at 5 cycles of gait is adopted as the evaluation value of the GA. Figure 18 shows the gaits of each generation in this experiment.

In the first half of the optimization, AIBO walked unsteadily and diagonally. However, at the 10-th generation, AIBO walked straight ahead stably and its gait resembled that of a living dog as well as the result of previous experiment (Fig. 15).

6. Conclusion

We proposed a method for generation of an animal gait for a quadrupedal robot. This method optimizes the orbit of the mono-leg using a GA based on the propulsive force and realizes the coordination of each leg on the basis of subjective human feelings. Moreover, we modified the generated gait by additional GA optimization to adapt to the dynamical interference by ground reaction.

We checked that AIBO walks straight ahead by proposed method, however, we also checked the centroid fluctuation at leg switching. It seems that the gait evaluation used for GA reproduction should include stability performance. In a future study, we intend to improve the evaluation method of optimization by including a body balance parameter.

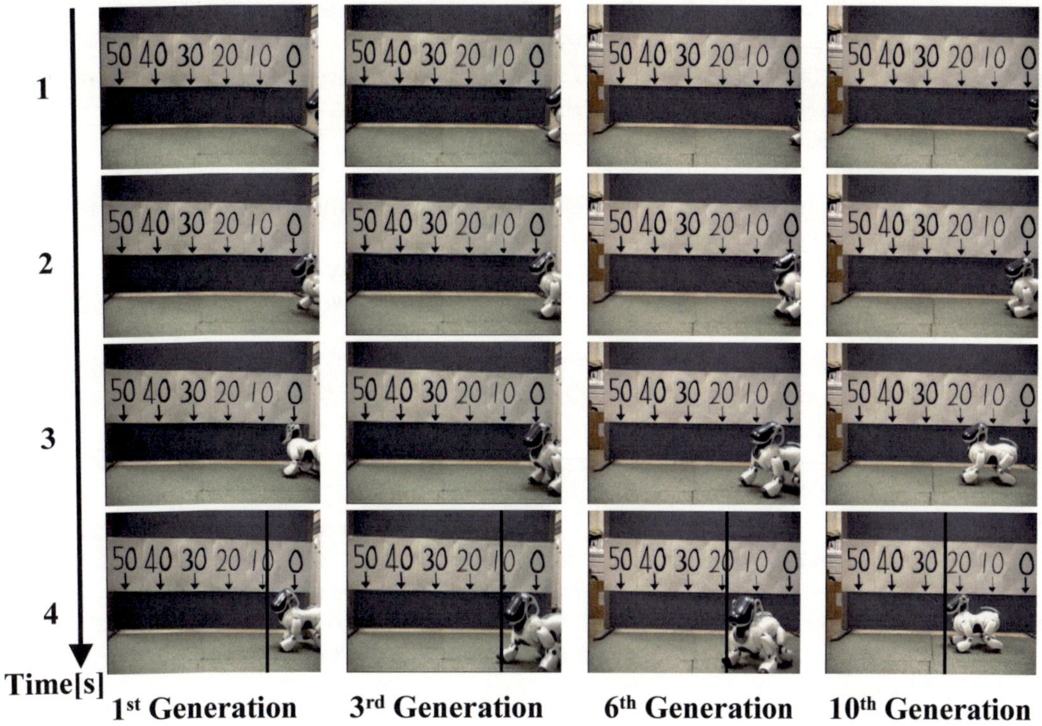

Fig. 18. Gait at each generations

7. References

Alexander, R. McN.; Jayes, A. S. & Ker, R. F. (1980). Estimation of energy cost for quadrupedal running gaits, *Journal of Zoology*, vol. 190, pp. 155-192.

Back, T. (1996). *Evolutionary Algorithms in Theory and Practice*, Oxford University Press, ISBN-9780195099713, Oxford.

Banks, M. R.; Willoughby, L. M & Banks, W. A. (2008). Animal-Assisted Therapy and Loneliness in Nursing Homes: Use of Robotic versus Living Dogs, *Journal of the American Medical Directors Association*, Vol. 9, No. 3, pp. 173-177.

Baun, M. M.; Bergstrom, N.; Langston, N. & Thoma, L. (1984). Physiological Effects of Human/Companion Animal Bonding, *Nursing Research*, Vol. 33, No. 3, pp. 126-129.

Chernova, S. & Velosa, M. (2004). An Evolutionary Approach to Gait Learning for Four-Legged Robots, *Proceedings of the 2004 IEEE/RSJ International Conference on Intelligent Robots and Systems*, pp. 2562-2567.

Estremera, J. & Santos, P. G. (2005). Generating Continuous Free Crab Gaits for Quadruped Robots on Irregular Terrain, *IEEE Transactions on Robotics*, Vol. 21, No. 6, pp. 1067-1076.

Fine, A. H. (2006). *Handbook on Animal-assisted Therapy: Theoretical Foundations and Guidelines for Practice*, Academic Press, ISBN-9780123694843, California.

Goldberg, D. E. (1989). *Genetic Algorithms in Search, Optimization, and Machine Learning*, Addison-Wesley, ISBN-9780201157673, Boston.

Goldberg, D. E. (2002). *The Design of Innovation: Lessons from and for Competent Genetic Algorithms*, Springer, ISBN-9781402070983, New York.

Inada, H. & Ishii, K. (2003). Behavior Generation of Bipedal Robot Using Central Pattern Generator (CPG) -1st Report: CPG Parameters Searching Method by Genetic Algorithm, *Proceedings of the 2003 IEEE/RSJ International Conference on Intelligent Robots and Systems*, pp. 2179-2184.

Kimura, H.; Yamashita, T. & Kobayashi, S. (2001). Reinforcement Learning of Walking Behavior for a Four-legged Robot, *40th IEEE Conference on Decision and Control*, pp. 411-416.

Kimura, H.; Fukuoka, Y. & Katabuti, H. (2005). Mechanical Design of a Quadruped "Tekken3&4" and Navigation System Using Laser Range Sensor, *Proceedings of International Symposium on Robotics*.

Michalewicz, Z. (1994). *Genetic Algorithms + Data Structures = Evolution Programs*, Spring-Verlag, ISBN-3540580905, New York.

Shibata, T.; Wada, K.; Saito, T & Tanie, K. (2005). Human Interactive Robot for Psychological Enrichment and Therapy, *Proceedings of the Symposium on Robot Companions: Hard Problems and Open Challenges in Robot-Human Interaction*, pp. 98-109.

Shibata, T.; Mitsui, T.; Wada, K. & Touda, A. (2001). Mental Commit Robot and its Application to Therapy of Children, *2001 IEEE/ASME International Conference on Advanced Intelligent Mechatronics Proceedings*, pp. 1053-1058.

Son, Y.; Kamano, T.; Yasuno, T.; Suzuki, T. & Harada, H. (2002). Target Tracking Control of Quadrupedal Robot Using CPG Network Tuned by Genetic Algorithm, *Proceedings of 2002 International Symposium on Advanced Control of Industrial Processes*, TA2D-4, pp. 661-666.

Suzuki, H.; Nishi, H.; Aburadani, A. & Inoue, S. (2007). Animal Gait Generation for Quadrupedal Robot, *Proceedings of Second International Conference on Innovative Computing, Information and Control*, CD-ROM, No. A03-02.

Wada, K.; Shibata, T.; Saito, T. & Tanie, K. (2004). Effects of Robot-Assisted Activity for Elderly People and Nurses at a Day Service Center, *Proceedings of the IEEE*, Vol. 92, No. 11, pp. 1780-1788.

17

Gait Based Directional Bias Detection of Four-Legged Walking Robots

Wei-Chung Teng and Ding-Jie Huang
National Taiwan University of Science and Technology
Taipei, Taiwan

1. Introduction

As an inevitable trend, more and more robots are designed to be sold as household products in recent years. Famous examples like AIBO, RoboSapien, and Kondo, though aiming on different functionality respectively, are all affordable by general family. Among the robots stated above, four-legged robots have great advantage on locomotion over stair, uneven or multilevel floor, and floor with scattered stuff. Four-legged robots can also be used as a mechanical mule and are capable of carrying significant payloads, such as BigDog manufactured by Boston Dynamics (Raibert et al., 2008). In this chapter, we discuss the directional bias problem in depth and introduce an approach to dynamically detect the direction bias utilizing gait pattern information and the feedback of accelerator sensor. To evaluate how effective this approach is, experiments are performed on two Sony's AIBO robots.

There are lots of research topics of four-legged robots such as balance control, gait generation, image recognition, walking bias detection, to name but a few. In this chapter, we focus on bias detection technique of four-legged robots. Comparing to mechanical mule, AIBO robot is designed to be light weighted and is equipped with plastic hemisphere on its feet such that it does not scratch the walking plane such as beech solid wood floor. This design is a reasonable result to a household robot, but it also makes the robot not able to step firmly, thus produces directional bias even when walking straight on flat and smooth plane. Since the directional bias of some AIBO robots is obvious and this kind of bias tends to accumulates as long as the robot is walking, it would be nice if there is an algorithm to automatically detect and correct the walking directional bias in real-time.

The most popular sensors used to detect heading direction of robots are video cameras, gyroscopes and accelerometers. Most image processing algorithms to detecting directional bias are time consuming and require more computing power than accelerometer based approaches. Since AIBO does not have gyroscope equipped, we choose to develop our algorithm according to accelerometer data. In theory, the distance of bias can be calculated by integrating the acceleration twice, but the acceleration data obtained from AIBO is not accurate enough to generate trustworthy data. Therefore, another data source is necessary to enhance our algorithm. After analyzed the characteristics of the three-axial acceleration sensors on AIBO, we confirmed the reliability of three-axial acceleration sensor on AIBO and found the relational model between the acceleration data and the walking gaits. As a result, it becomes obvious that the vibration and the axial bias being the main factors that cause the difference between integrated data and the real distance. We also observe that AIBO vibrates largely at some fixed time point during one gait cycle. It turns out that this observation becomes the basis of the bias detection algorithm.

The following sections are organized as follows. The hardware specification of AIBO robot is given at first, then a brief review to related work follows, after then a detailed description of developing bias detection algorithm is extended. The process to analyzing the acceleration data and how we utilize the statistical data are also explained in this part. Finally, we show results of experiments, including the accuracy rate, for the bias detection algorithm and give a conclusion.

2. Brief Description of the AIBO Robot

AIBO is considered a successful, though not profitable, product as a household four-legged robot. As mentioned above, our main experiment platform is AIBO ERS-220A made by SONY Corporation. The original design of AIBO is for entertainment purpose and is considered as a robot pet which can interact with human. Thus, it has lots of sensors for data feedback by which the built-in software utilizes to interact with environment and with its owner. The basic hardware specification of AIBO is shown in Table 1. As you may have noticed, the 64bit RISC processor and 32 MB main memory is not enough to execute complex computation in real-time. This is the main reason we exclude the CMOS camera when developing the bias detection algorithm.

Basic hardware	64bit RISC Processor. 32 MB Main memory.
Actuator	16 degrees of freedom with a continuous range of motion. Three for each of the legs, one from head light and three from the head.
Sensor	Temperature sensor. Infrared distance sensor. Acceleration sensor. Touch sensors(head,face,back,legs,tail). Vibration sensor. 100,000 CMOS image sensor

Table 1. The hardware specification of AIBO.

In addition to these various hardware components, SONY also provides a useful software framework, the OPEN-R SDK, for developer to develop customized programs. According to previous literatures (Wijbenga & van de Sanden, 2004; Hengst et al., 2000 ; Serra & Baillie, 2003; Rico et al., 2004), AIBO is easy to be controlled and can perform different actions by utilizing OPEN-R SDK. With this powerful framework, we can record data from three-axial acceleration sensors in AIBO. In the following sections, we discuss more about how to analyze these data.

According to the official document (Sony Corporation, 2004), the positive direction of X, Y and Z axis in the acceleration sensor are mapped to right, front and up of the robot respectively, as shown in Fig. 1. We use this definition in the following sections. Furthermore, by calculating the greatest common divisor of the collected data, we found that each axis has different unit length. The unit length is 0.086784 meter in X axis, 0.169080 meter in Y axis, and 0.172046 meter in Z axis. Since the unit length is different in three axes and it has great effect on the sensitivity of acceleration sensor, we must take this characteristic into consideration later.

For general gaits provided by OPEN-R SDK, the sampling rate of a frame is 125Hz, and AIBO takes 140 frames to complete one gait cycle. Thus, it costs 1.12 seconds in total for AIBO to execute a whole cycle. Besides, in order to gather more information about the walking status of AIBO, we also utilize the touch sensor under each foot. With the help of touch sensor's feedback, we can analyze and then can determine abnormal step timings according to AIBO's walking pattern.

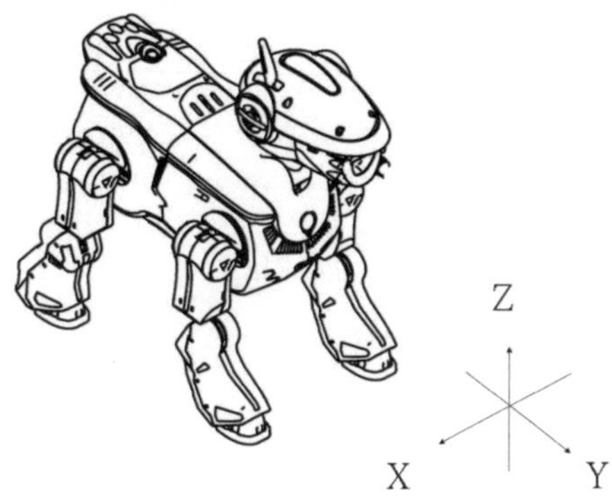

Fig. 1. The definition of axes of three-axial acceleration sensor in AIBO

3. Related Work and Previous Studies

According to previous studies, the position of a robot can be measured by using acceleration sensors (Liu & Pang, 1999). Since the data from acceleration sensors have accumulative error, other position information source like GPS can be useful to compensate the error. In 2004, one research (Vail & Veloso, 2004) uses acceleration data of AIBO to identify the type of surface under the robot as it walks. In addition, the acceleration data is also used to predict gait velocity by using a k-nearest neighbor approach. These researches are however focusing on measuring walking velocity and distance straight ahead, whereas we have interest only in bias of lateral direction.

As mentioned before, our goal is to find out the directional bias when AIBO is walking. In order to find out the directional bias, walking distance is a good starting point. It is proposed that low-cost acceleration sensors can be used to measure walking distance of human (Cho, 2002). It is observed that data feedback of acceleration sensors have errors due to sensor bias, input, axis misalignments, etc. Thus, an algorithm is developed and is applied in the measurement system to compensate these errors. Though this study aims on the walking pattern of human, it still serves as a good reference.

On the other hand, one previous study shows that the output of acceleration sensors in the AIBO robot is too shaky such that the calculated speed and walking distance do not converge in reasonable bounds (Westermark, 2005). Fortunately, it is still possible to determent the relation between actual walking distance and acceleration data by heuristic analysis, because acceleration data has some particular characteristics which would be affected by walking pattern and the type of walking surface.

4. The Algorithm to Detect Directional Bias

4.1 Basic concept before designing the detection algorithm

At the beginning, we use a naïve method, integrating the acceleration data twice, to generate distance. As stated by previous study, the integrated data are extremely different from the real distance. There are many possible reasons to this difference, and one is that the accelerator we use is a defective. Considering the characteristic of each acceleration sensor data from three axes, a series of preliminary experiments are designed to check the validity of the data, and the flowchart of the research method is illustrated in Fig. 2. For the first step, we need to confirm the reliability of three-axial acceleration sensors. We use Matlab software package to analyze the results of preliminary experiments. After gathered enough data, we then analyze acceleration data of each axis and find out the relation among three-axis acceleration sensors. Given the collected data is trustworthy, we may use the result and start to develop our method to detect directional bias.

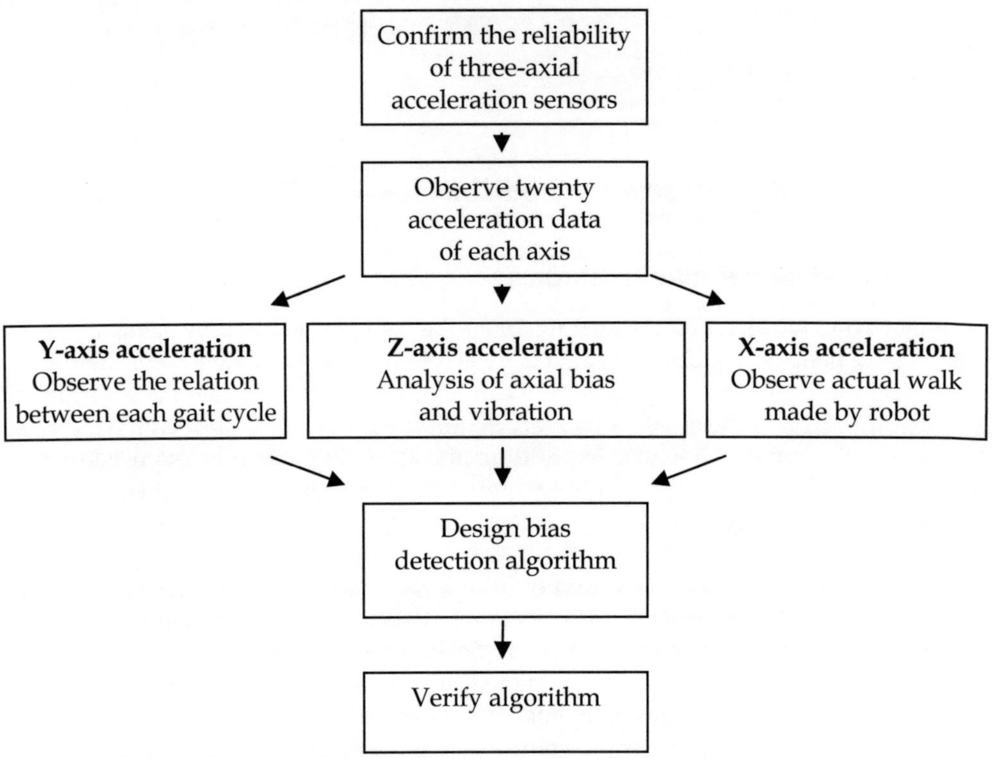

Fig. 2. Flowchart of the research method

4.2 Preliminary experiment

In order to examine the reliability of acceleration data gathered from AIBO, we setup a preliminary experiment. The designed experiment is to let AIBO walks for 5 gait cycles on top of pasteboard for 20 times. During this experiment, we record the acceleration data of each axis and drew all the data into one graph, as Fig. 3 shows. In other words, there are 20 overlapping record sets in the Fig. 3. We observe that acceleration in X axis is unstable in Fig. 3(a), but each gait cycle has similar pattern. For example, there are fixed 10 pulses during 5 gait cycles, and they almost happen in the same timing. The same phenomenon can be observed in Y-axis and Z-axis, they all have similar pattern during the 5 gait cycles.

Thus, it is reasonable to conclude that the three-axial acceleration sensor has property of specific repeatability. Differences between overlapped data are especially few in the first gait cycle, since the 20 lines are very similar to each other. However, after second gait cycle the vibrations are increasing caused by unstable walking gait, so each data are getting more and more different. Since the three-axial acceleration sensor on AIBO shows acceptable reliability, we may step further to analyze each axis data respectively.

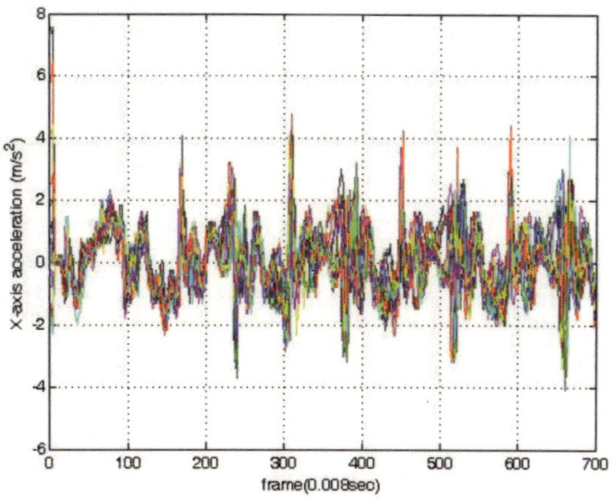

(a)

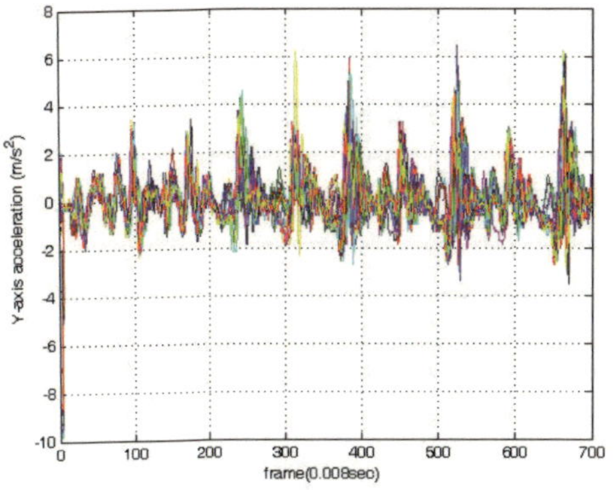

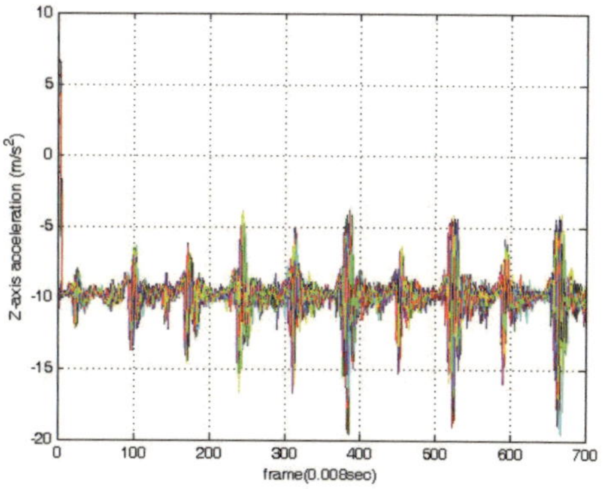

(b) (c)

Fig. 3. Overlapping the acceleration data in three axes during 5 gait cycles measured 20 times: (a) X-axis (b) Y-axis (c) Z-axis

4.3 The characteristics of acceleration sensor

After confirmed the reliability of three-axial acceleration sensors in AIBO, we start to analyze data of each axis. For the first step, we analyze Y-axis acceleration data by integrating the acceleration twice directly. As mentioned above, the integrated data are far from the real distance measured manually. Hence, we consider that there must be some kinds of factors that strongly affect the Y-axis acceleration data.

So far we can not compute the distance directly by using acceleration data, but the real distance has particular relation with the acceleration data. As the result of observation, we assume that the physical vibration of robot body during walking and the accompanied axial bias may be the main factors that cause the difference between integrated data and the real distance. It seems that the factors appear when AIBO is changing its gait step. This is because AIBO is on an unstable state when it is changing gait step. For example, the physical vibration, almost in vertical direction, occurs in the moment that the left front leg is raising up and the left back leg is putting down. This vibration causes the axial bias and a pulse. In order to find when the vibration occurs, we need to analyze Z-axis acceleration data.

In the second step, we analyze Z-axis acceleration data by simultaneously comparing the acceleration with output of touch sensor on the bottom of each leg. The reading from touch sensor is 0 or 1. If the reading of the touch sensor is 0, it means that AIBO is during swing phase; otherwise, it means that AIBO is during stance phase. In order to compare the touch sensor data with acceleration data in one picture, we add different fixed values on each reading of touch sensor. As illustrated in Fig. 4, "○" means the reading from right front touch sensor, "+" is from left front touch sensor, " ▷ " is from right back touch sensor, and " ◁ " means the reading from left back touch sensor. In addition, we find that the timings of pulses repeat regularly during the 5 gait cycles. It begins from 0.2 second (the 25th frame), and there are pulses every 0.56 second (70 frame). Comparing with each leg phase, the start time point 1 in Fig. 4 of these pulses is the time when right front leg is raising up and the right back leg is putting down. The time point 2 is the time when left front leg is raising up and the left back leg is putting down. Hence, the pulse is caused by the changes between two legs on the same side. This kind of vibrations would affect each axial data extremely. Thus, the data seem to be unreliable during this unstable period and should not be added to the integration.

Moreover, from Fig. 4 we can observe that AIBO starts its gait in the order of right front, left back, left front and right back, which is known as 1-4-2-3 footfall sequence. This gait, as shown in Fig. 5, has already applied to lots of research (Schmiedeler & Waldron, 1999; Toyomasu & Shinohara, 2003) and is the same with the notion that Hirose and Kunieda proposed in 1991 (Bekey, 2005). By observing walking AIBOs, we find that its walking direction is straight when it is raising right front leg. When it is putting right front leg down and raising left back leg up, it gets the most right position during a single gait. Then it raises left front leg and its walking direction is left forward. Eventually when it is putting left front leg down and raising right back leg up, it gets the most left position during a single gait.

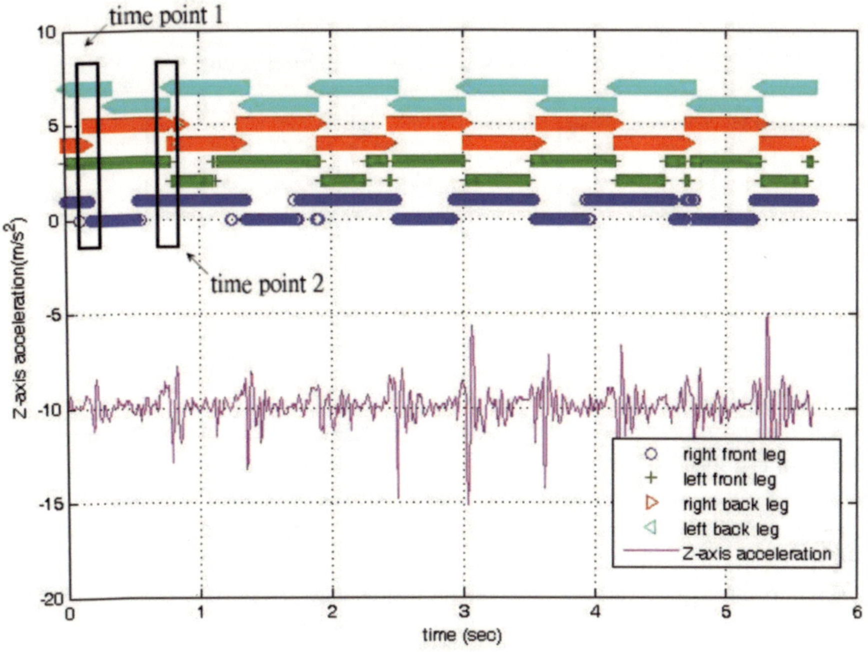

Fig. 4. Z-axis acceleration data during 5 gait cycles with stepping phases of 4 legs

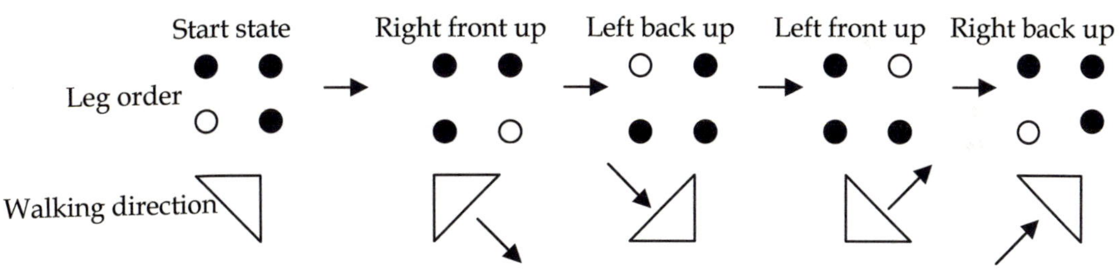

Fig. 5. Gait order of the AIBO robot with its walking direction in each footfall

In the third part, we want to know at what frame the vibrations start. We start to analyze X-axis acceleration comparing with the trajectory of AIBO. As Fig. 6 shows, the gap between dotted lines is the distance during one single gait cycle. The ending points of arrows stand for rightmost or leftmost positions during a gait cycle, and the two rectangles represent the space occupied by AIBO's body. The arrows point out the most right positions or the most left positions. We assume the velocity in X-axis is 0 when AIBO reaches the rightmost and the leftmost positions. This is because the instant velocity should be 0 when a robot is in the moment of changing its lateral direction. Therefore, we can use this characteristic to reset the velocity in X-axis every half gait cycle and integrate data from the next arrow partially. In this way we prevent the velocity error in X-axis to accumulate.

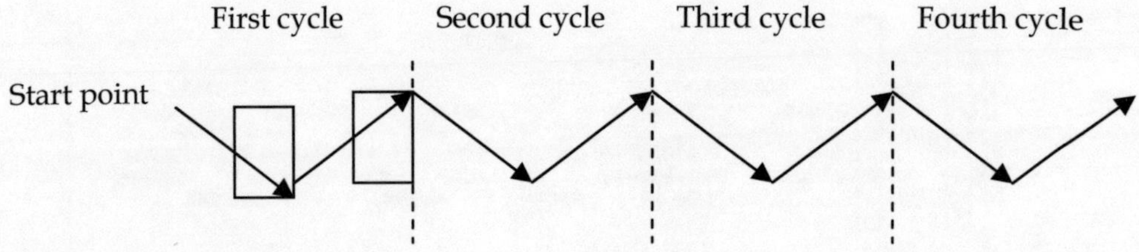

Fig. 6. The walking trajectory of AIBO robot with default gait

4.4 The bias detection algorithm

As we has discussed above, it seems reasonable to believe the acceleration sensors on AIBO are useful in some degree of accuracy. Nevertheless, we find that AIBO vibrates largely at some fixed time points, so it is important to take out the acceleration data while left leg or right leg in the swing phase individually. As shown in Fig. 7, according to the observation on walking pattern of each leg, data from the 23th to the 63th frame represent the acceleration when the right front leg is in the swing phases. And data from the 97th frame to the 137th frame represent the acceleration when the left front leg is in the swing phases. By integrating this two set of data twice individually and then summing these two values up, we can derive a value which we call partial integrating value (PIV). If the PIV is positive, the center position of AIBO is going to right position. If the PIV is negative, the center position of AIBO is going to left position. Thus, this method is our "single gaited bias detection algorithm".

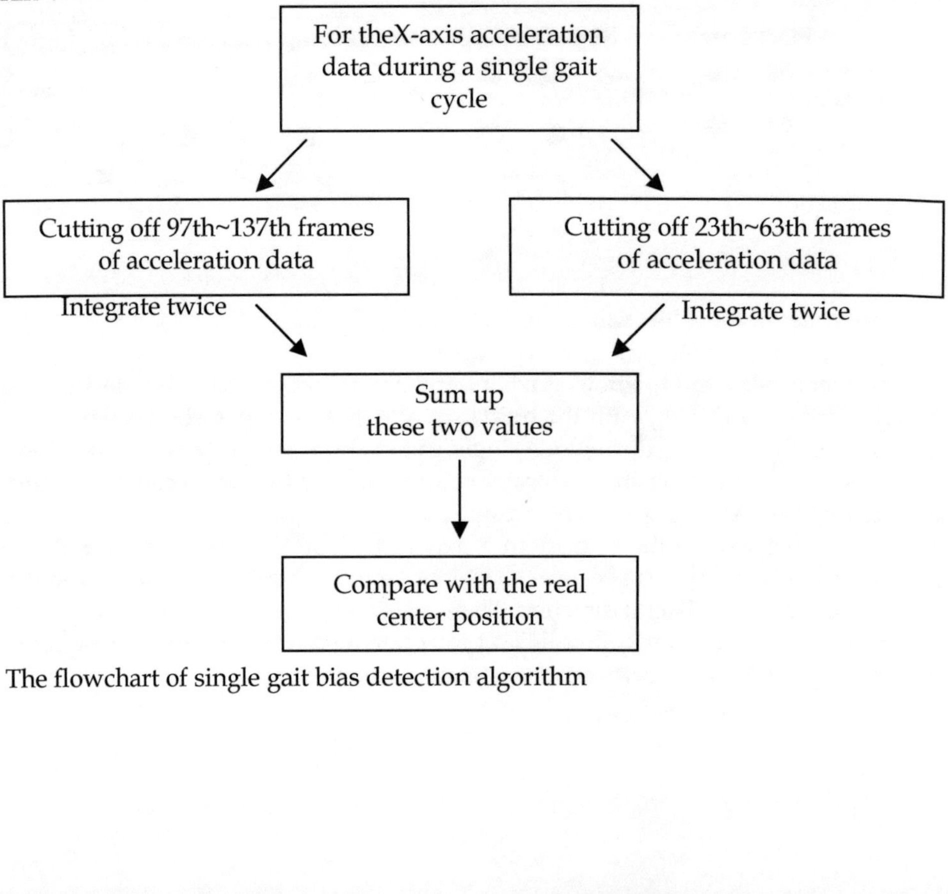

Fig. 7. The flowchart of single gait bias detection algorithm

Although we can compute the directional bias of first gait cycle by using this single gait bias detection algorithm, we cannot use this method directly to compute second gait cycle. It is because that the direction of Y-axis of AIBO might have changed at this time. The PIV value of the second gait cycle needs to be further computed with coordinate changes and then sums up with the PIV value of the first gait cycle. This concept is explained in Fig. 8 in which X_1 and Y_1 are X-axis bias and Y-axis displacement during the first gait cycle, X_2 and Y_2 are accumulated X-axis bias and Y-axis displacement during the second gait cycle. Furthermore, $X_2^{'}$ is X-axis bias generated in the second gait cycle by the local reference frame at the end of the first gait. The equations below therefore hold:

$$\tan\theta_1 = \frac{X_1}{Y_1}\ ,\ \tan\theta_2 = \frac{X_2}{Y_2} \text{ and } \tan\theta_3 = \frac{X_3}{Y_3}\ .$$

$$X_2^{'} = \sqrt{(X_2^2 + Y_2^2)}(\sin(\theta_1^{'} + \theta_2)) \qquad (1)$$

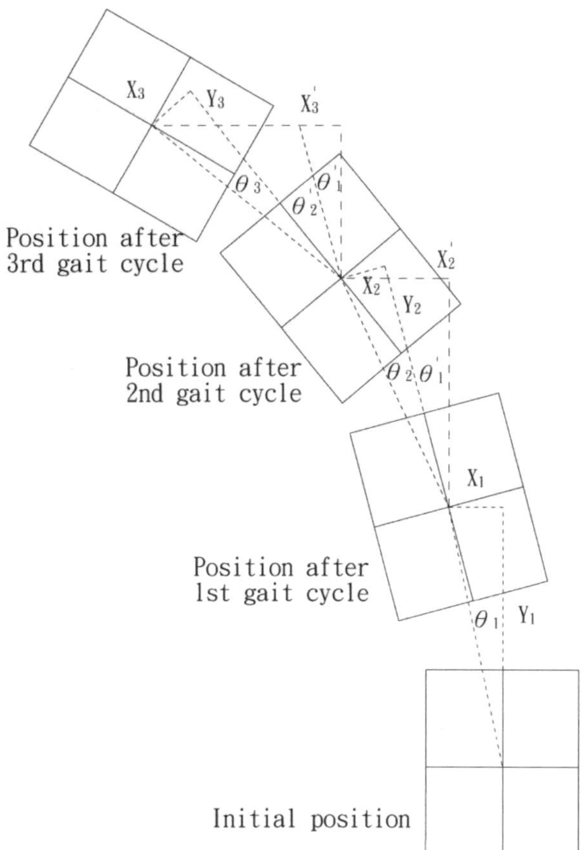

Fig. 8. The relative positions between each gait cycle

As we mentioned before, Y-axis displacement can not be trustworthy for walking distance over the first gait. Thus, we take average displacement (8.8cm) of Y-axis displacement from first gait cycle. Then, we let Y_1 =8.8 and Y_2 =8.8, and use single gait bias detection algorithm to find X_1 and X_2. If we know θ_1' and θ_2, we can compute the real bias X_2' by applying equation (1). θ_1' is the turning angle of AIBO between Y-axis of original position and Y-axis of first gait position. The real bias X_3' during third gait cycle is as followed:

$$X_3' = \sqrt{\left(X_3^2 + Y_3^2\right)}\left(\sin\left(\theta_1' + \theta_2' + \theta_3\right)\right) \quad (2)$$

θ_2' is the turning angle of AIBO between Y-axis of first gait position and Y-axis of first gait position. θ_1' and θ_2' are further discussed in the next section.

5. Experiment Results

To examine the effectiveness of the single gait bias detection algorithm, we gathered all sensor reading of a single gait cycle while AIBO walks on the pasteboard and record the center position of AIBO, and then repeat this experiment for 30 times. We list the first 10 records to explain the result of our experiments, as shown in Table 2. The X1 field is the X-axial position of left front leg, X2 is the X-axial position of left back leg, Xc is the center position of AIBO, or the average of X1 and X2, DIV is the value directly integrated twice from acceleration, and θ is the bias angle which is computed by X1 subtracting X2 and then divided by the constant distance (23cm) between front leg and back leg. If the PIV in Table 2 is positive, it means that the directional bias is right; otherwise, if the PIV in Table 2 is negative, it means that the directional bias is left.
Comparing Xc with DIV, they are different from each other from exp1 to exp10. The values of DIV are not only too large but also upside-down in positive-negative sign. Thus, the directional bias can not be simply computed by integrating X-axial acceleration. Comparing Xc with PIV, there are only two data (exp2 and exp7) in wrong judgment. If the value of Xc equal to 0, θ would help to explain the directional bias. Take exp3 for example, the value of θ is negative, so the directional bias is left, so the maximum error. The other data show that the proposed method can determine the directional bias. The correct rate of "single gaited bias detection algorithm" is 93.33%, because 28/30 data are correct.

NO.	X1	X2	Xc	DIV	PIV	θ
exp1	-1.3	0.7	-0.3	7.542884	-1.9959	-4.96974
exp2	-0.4	0.6	0.1	15.72617	-0.84257	-2.48955
exp3	-0.5	0.5	0	14.12934	-0.49655	-2.48955
exp4	-0.8	0.6	-0.1	13.84468	-0.64595	-3.48327
exp5	-1.2	0.6	-0.3	15.4393	-0.72316	-4.4749
exp6	-0.5	0.4	-0.05	8.680384	-0.54403	-2.24087
exp7	-0.6	0.8	0.1	8.871726	-1.71764	-3.48327
exp8	-0.5	0.4	-0.05	6.40316	-0.49071	-2.24087
exp9	-0.3	0.2	-0.05	17.49574	-0.83063	-1.24536
exp10	-1.2	0.6	-0.3	9.290792	-1.34884	-4.4749

Table 2. Experimental results (in unit of cm)

The scale factor of X-axis acceleration sensor is 0.086784 m/s^2, so the maximum error is ±0.043392 m/s^2. The integrated value can be represented as

$$\int f'(t)dt \approx \int f(t)dt + \Delta t \ n \ c \tag{3}$$

where $f'(t)$ is real value, $f(t)$ is the value from acceleration sensor, Δt is the time of each frame, n is a positive integer which is the number of frame, and c is a constant which is the constant scale factor. Hence, $\Delta t \cdot n \cdot c$ is the error after integration. If we let c = 0.043392, n = 140, Δt is 0.008, and use equation (3) twice, we can get the maximum error ±5.443cm after integrated twice. However, the X-axial center position of AIBO drifts from -0.3cm to 0.6cm during one single gait cycle. Therefore, it is a great improvement that our method can determine the directional bias under such a huge error.

In order to compute the directional bias of two gait cycles, the bias angle of first gait cycle is needed. The value of PIV and the bias angle of first gait cycle have been shown to be positively correlated with one another. In fact, the correlation coefficient is 0.855, which is a high correlation. Then, we compute the linear regression of them, and with regression function the bias angle can be easily transformed from PIV. Therefore, we can use (1) to detect the directional bias of two gait cycles. The same method is used to detect three gait cycles, four gait cycles, and five gait cycles.

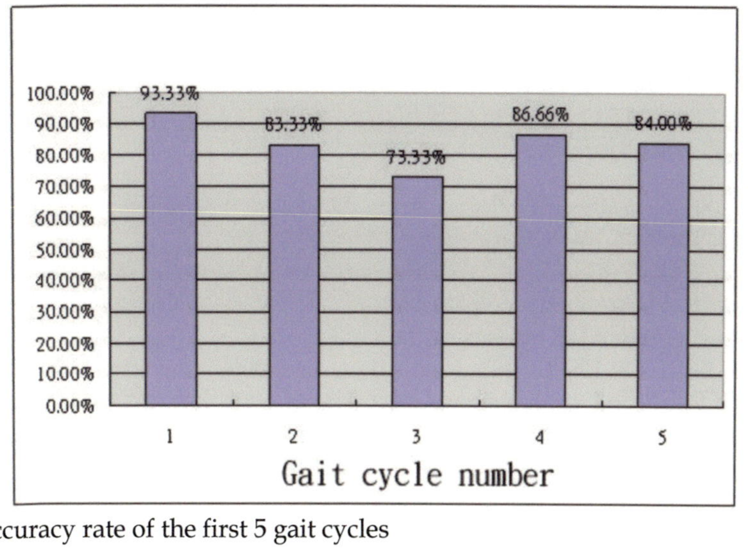

Fig. 9. The accuracy rate of the first 5 gait cycles

We plot the accuracy rate of each cycle from the first five gait cycles unto one graph, as Fig. 9 shows. The accuracy rate of the first gait cycle is up to 93.33%, so our algorithm can determine the correct directional bias generally. It is because that AIBO is in the most stable condition, so the robot is rarely affected by any kind of factor. The accuracy rate of second gait cycle is 83.33% on the downside, because the AIBO is getting unstable while walking. Under unstable condition, the value that our algorithm computes may be affected somehow.

The accuracy rate of third gait cycle is 73.33%; it means that the result of integration accumulates a lot of errors apparently and it causes the downward trend of the accuracy rate. However, the accuracy rate of fourth gait cycle goes up again to 86.66%, and the accuracy rate of fifth gait cycle is 84%. With Bias Detection Algorithm we can determine the directional bias of AIBO in first five gait cycles, and the accuracy rates are all over 73.33%. Although the directional bias cannot be determined absolutely, with the low sensitive sensors and limited computing power the performance of Bias Detection Algorithm is a good improvement.

6. Conclusion

We have demonstrated an effective approach to detect the directional bias of walking four-legged robots by using acceleration data and gait information. The precision of acceleration sensor equipped in AIBO is practical, but the difference between integrated walking distance and the actual one is not negligible. After analyzed the data from accelerometer, we found that AIBO vibrates largely at some fixed time point, so it is important to take out the acceleration data while left leg or right leg in the swing phase individually.

By observing the walking pattern of each leg, we decided to collect data from two fragments of the acceleration data and integrate these two data twice individually. Then, by summing these two values up, we get a value that we called it as partial integrating value (PIV). If the PIV is positive, the center position of AIBO is going to right position. If the PIV is negative, the center position of AIBO is going to left position. In order to make sure the correctness of our approach, several experiments were designed and executed. As a result, the accuracy rate of this single gaited bias detection algorithm is 93.33%, and the accuracy rate of this algorithm within five gaits can be up to 90%.

7. References

Bekey, G.A. (2005). *Autonomous Robots From Biological Inspiration for Implementation and Control.* The MIT Press, Cambridge, Massachusetts

Cho, S.Y.; Park, C.G. & Jee, G.I. (2002) Measurement System of Walking Distance Using Low-cost Accelerometers. *Proceedings of The 4th Asian Control Conference,* pp. 1799–1803, September 2002, Singapore

Hengst, B.; Ibbotson, D.; Pham, S.B. & Sammut, C. (2000). The UNSW United 2000 Sony Legged Robot Software System, University of New South Wales, Australia

Liu, H. & Pang, G. (1999) Accelerometer for Mobile Robot Positioning. *Proceedings Of the 1999 IEEE Industry Applications Society Conference,* vol.3, pp. 1735-1742, Phoenix, AZ, USA

Raibert, M.; Blankespoor, K.; Nelson, G. & Playter, R. (2008). Bigdog, the rough-terrain quaduped robot. *Proceedings of the 17th International Federation of Automation Control.* April 2008

Rico, F.M.; González-Careaga, R.; María, J.; Plaza, C. & Olivera, V.M. (2004). Programming Model Based on Concurrent Objects for the AIBO Robot, Universidad Rey Juan Carlos

Serra, F. & Baillie, J.-C. (2003). Aibo programming Using OPEN-R SDK. Tutorial, ENSTA, France

Schmiedeler, J.P. & Waldron, K.J. (1999). The Mechanics of Quadrupedal Galloping and the Future of Legged Vehicles. *The International Journal of Robotics Research,* Vol. 18, No. 12, pp. 1224-1234

Sony Corporation (2004), OPEN-R SDK

Toyomasu, M. & Shinohara, A. (2003). Developing Dynamic Gaits for Four Legged Robots. *Proceedings of The International Symposium on Information Science and Electrical Engineering,* pp. 577–580

Vail, D. & Veloso, M. (2004). Learning from Accelerometer Data on a Legged Robot, *Proceedings of the 5th IFAC/EURON Symposium on Intelligent Autonomous Vehicles,* Lisbon, Portugal.

Westermark, S. (2005). Programming tools for AIBO

Wijbenga, A. & van de Sanden, M. (2004). How To Make AIBO Do Tricks, University of Groningen, Netherlands

18

Locomotion analysis of hexapod robot

Xilun Ding, Zhiying Wang, Alberto Rovetta and J.M. Zhu
Beihang University, Politecnico di Milano
China, Italy

1. Introduction

Multi-legged robots display significant advantages with respect to wheeled ones for walking over rough terrain because they do not need continuous contact with the ground. In Multi-legged robots, hexapod robots, mechanical vehicles that walk on six legs, have attracted considerable attention in recent decades. There are several benefits for hexapods rover.

(a) Hexapod robot is easy to maintain static stability on three or more legs,
(b) It has a great deal of flexibility in how it can move.
(c) Hexapod robot is the most efficient one for statically stable walking. Preumount et al. 1991, observed that a larger number of legs more than six do not increase walking speed.
(d) Hexapod robots show robustness in case of leg faults
(e) Hexapods makes it possible for the robot to use one, two or three legs to work as hand and perform complex operations.

The most studied problem for multi-legged robots concerns how to determine the best sequence for lifting off and placing the feet (gait/locomotion planning). From the stability point of view, robot locomotion can be classified into dynamic locomotion, such as running and hopping, and statically stable locomotion as walking. Statically stable locomotion has the constraint that the moving body is stable at all times. The vertical projection of the centre of gravity of the robot must be within the convex of the supporting polygon linked positions of all supporting feet.

Statically stable gait is solely dependent on the design of bodies and legs. Hexapod gaits have been widely investigated as a function of shape and characteristics of the robot structure. In 1985, Kaneko et al. addressed the gait of a rectangular hexapod with decoupled freedoms where the propelling motion was generated by one degree of freedom (DOF). In 1988, Lee et al. realized an omnidirectional walking control system for a rectangular hexapod robot with adaptive suspension. A circular gait was studied for a layered hexapod robot (called Ambler) at the Carnegie Mellon University [Bares et al., 1989; Krotkov & Bares, 1991; Wettergreen, 1990] with rotating legs connected to the same vertical axis at six different heights. Hirose et al. in 1992 and 1998 and Gurocak in 1998 developed other two hexapods whose bodies were consisting of two different layers, each connected to three legs. The relative motion of the layers realized the omnidirectional robot gait in a simple way, but limiting the walking capability under leg faults. Two Lees in 2001 studied the gait of a special robot whose body was composed of three parts connected by revolute joints. Its flexible gait allowed it to overcome complex terrains, but its configuration was quite complicate for control system design. R Hex, introduced by Uluc et al. in 2001, is another hexapod robot with half-circle legs with a simple alternate tripod gait.

Most popular hexapods can be grouped into two categories, rectangular and hexagonal ones. Rectangular hexapods have a rectangular body with two groups of three legs distributed symmetrically on the two sides. Hexagonal hexapods have a round or hexagonal body with evenly distributed legs.

The gait of rectangular six-legged robots has motivated a number of theoretical researches and experiments which nowadays reached to some extent a state of maturity. In 1998 Lee et al. showed for rectangular hexapods the longitudinal stability margin, which is defined as the shortest distance from the vertical projection of center of gravity to the boundaries of the support pattern in the horizontal plane, of straight-line motion and crab walking. Song &Choi in 1990 defined the duty factor β as the fraction of cycle time in which a leg is in the supporting phase and they proved that the wave gait is optimally stable among all periodic and regular gaits for rectangular hexapods when $3/4 \leq\beta\leq 1$. Both the tripod gait and the problem of turning around a fixed point on an even terrain have been widely investigated and tested for a general rectangular hexapod with three DOF legs [Wang, 2005 and Su, 2004]. The so called 4+2 quadruped gaits [Huang and Nonami, 2003] have been demonstrated being able to tolerate faults [Yang & Kim, 1999]. A series of fault-tolerant gaits for hexapods were analyzed by Yang et al. [Yang & Kim, 1998a, 1998b, 2000 and 2003]. Their aim was to maintain the stability in case a fault event prevented a leg from supporting the robot. In 1975, Kugushev and Jaroshevskij proposed a terrain adaptive free gait that was non-periodic. McGhee et al. in and other researchers [Porta & Celaya, 2004; Erden & Leblebicioglŭ] went on studying free gaits of rectangular hexapod robots.

At the same time, the hexagonal hexapod robots were studied with inspiration from the insect family, demonstrate better performances for some aspects than rectangular robots. Kamikawa et al. in 2004 confirmed the ability to walk up and down a slope with the tripod gait by building a virtual smooth surface that approximates the exact ground. Yoneda et al. in 1997 enhanced the results of Song & Choi in 1990, developing a time-varying wave gait for hexagonal robots, in which velocity, duty factor and crab angle are changed according to terrain conditions. A. Preumon et al. in 1991 proved that hexagonal hexapods can easily steer in all directions and that they have longer stability margin, but he did not give a detailed theoretical analysis. Takahashi et al. in 2000 found that hexagonal robots rotate and move in all directions at the same time better than rectangular ones by comparing stability margin and stroke in wave gait, but no experimental results were presented. Chu and Pang in 2002 compared the fault tolerant gait and the 4+2 gait for two types of hexapods of the same size. They proved theoretically that hexagonal hexapod robots have superior stability margin, stride and turning ability compared to rectangular robots.

It is also worth to mention here a work carried out by Gonzale de Santos et al. [Gonzale de Santos et al., 2007a and Gonzale de Santos et al., 2007b]. They optimized the structure of rectangular hexapods and found that extending the length of middle legs of rectangular robots helps in saving energy. This outcome can be seen as a transition from rectangular six-legged robots to hexagonal ones.

2. Definitions

Several definitions are necessary to be introduced before locomotion planning.

1) Support/stance phase: a leg is said in its supporting/stance phase when it stands on the ground and its foot does not leave the ground.
2) Transfer/swing phase: a leg is said in its transferring/swing phase when it does not stand on the ground but move in the air.
3) Gait period/cycle time, *T*: a gait period/cycle time is a complete cycle of a leg including supporting phase and transferring phase.
4) Duty factor β: the duty factor β is the time fraction of stance phase of a leg to the cycle time *T*. $\beta = T_{si}/T_i$ where T_{si} denotes time of supporting phase of leg *i*; T_i denotes circle time of leg *i*.
5) Stroke length: the distance that the body moves thought the support phase of a leg.
6) Stride length: stride length is the distance the centre of gravity (COG) translates during one complete locomotion cycle.
7) Pitch length: the distance between the centers of the strokes of the isoceles legs.
8) Supporting polygon/pattern: the polygon the vertices of which are constructed on the horizontal plane by vertical projections of the foot-ground interaction points.
9) Statically stability margin (SSM): stability margin was defined for a given support polygon as the smallest of the distances from the COG. projection to the edges of the support polygon.
10) Longitudinal stability margin (LSM): the smallest of the distances from the COG. projection to the front and rear edges of the support polygon along the machine's longitudinal axis.
11) Crab Longitudinal Stability Margin (CLSM): The smallest of the distances from the COG. projection to the front and rear edges of the support polygon along the machine's motion axis.
12) Main walking direction stability margin (MDSM):the smallest of distance from projection of the C.G. to the front and rear edges of the support polygon along the main-walking direction
13) Kinematics margin: kinematics margin is defined as the distance from the current foothold of leg *i* to the boundary of the reachable area of leg *i*, measured in the opposite direction of body motion.
14) Periodic gait: a gait is periodic if similar states of the same leg during successive strokes occur at the same interval for all legs, that interval being the cycle time.
15) Symmetric gait: a gait is symmetric if the motion of legs of any right-left pair is exactly half a cycle out of phase.
16) Regular gait: A gait is said to be regular if all the legs have the same duty factor.
17) Body height: body height is the distance of the body center of mass from the support surface along the body vertical axis.
18) Protraction of leg: protraction is the forward movement of a leg relative to the body and ground.
19) Retraction of leg: retraction is the backward movement of a leg relative to the body with no movement of the leg relative to the ground.
20) Lateral offset: Lateral offset is the shortest distance between vertical projection of hip on the ground and the corresponding track.

21) Crab angle: it is defined as the angel from the longitudinal axis to the direction motion, which has the positive measure in the anti-clockwise direction.

3. Mechanism of Hexapods

Typical hexapod robots can be classified into rectangular and hexagonal ones (Fig.1). Rectangular hexapods inspired from insects have six legs distributed symmetrically along two sides, each side having three legs. Hexagonal hexapods have six legs distributed axisymmetrically around the body (that can be hexagonal or circular).

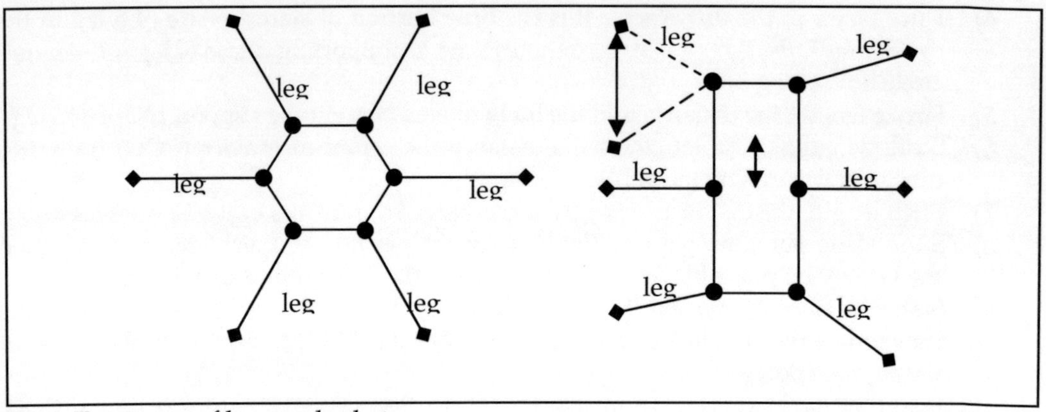

Fig. 1. Two types of hexapod robots

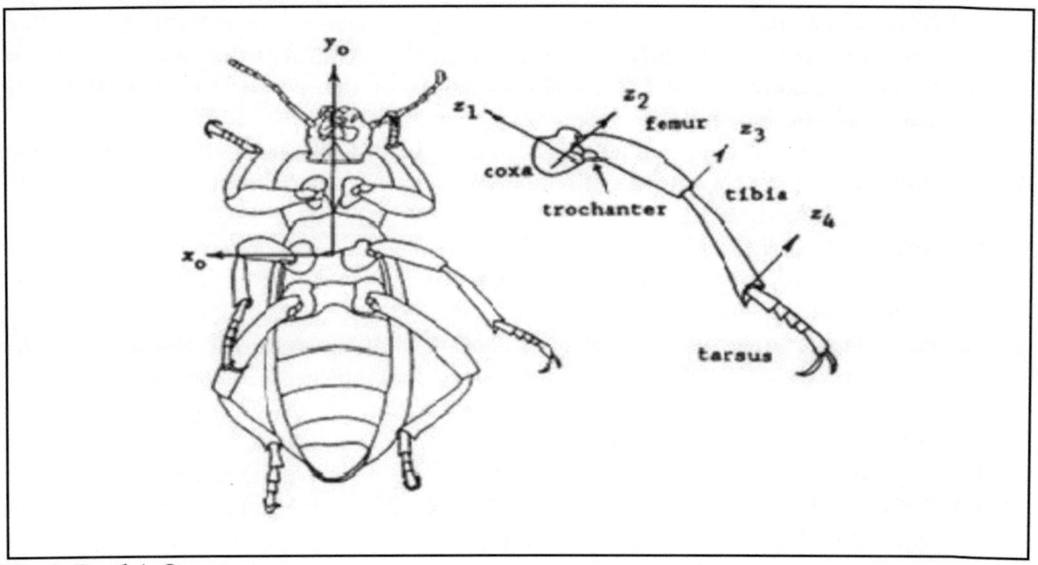

Fig. 2. Beetle's Structure

Typically, individual legs range from two to six degrees of freedom. Fichters [Fichter, E.F. & Fichter, B.L., 1988] have made a survey on insects' legs (Fig.2 as an example). They found that a general insect leg has four main segments: coxa, femur, tibia and tarsus. Most of the length of an insect leg is contributed by 2 long and nearly equally segments. The hinge joint between these segments allows one to fold back almost exactly along the other. Length of the tibia is highly correlated with that of the femur, the correlation coefficient ranges from 0.97 to 0.78. The coxa has no obvious correlation with femur or tibia. Similarly, the thigh and calf of mammals (dog for example in Fig.3) and human are almost equal. We will mainly talk about the locomotion of hexapod robots with leg structure in Fig.4. Three parts are connected together by two parallel revolute joints with rotating axes parallel to the ground, coax and knee. When all joints are at zero position, the link calf is perpendicular to the ground and the link thigh and calf are parallel to the ground. The hip is connected to the body by the waist joint that rotates around a vertical axis.

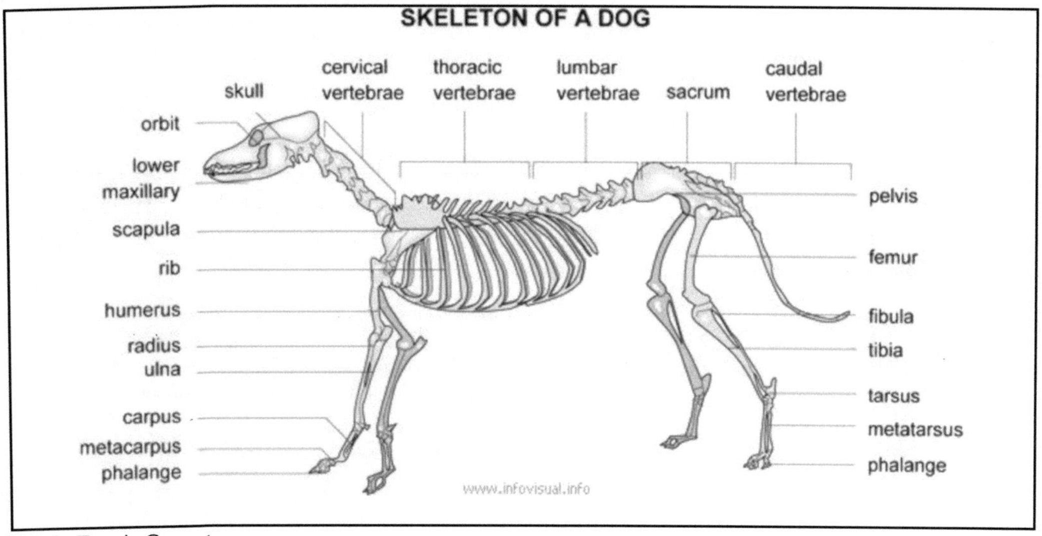

Fig. 3. Dog's Structure

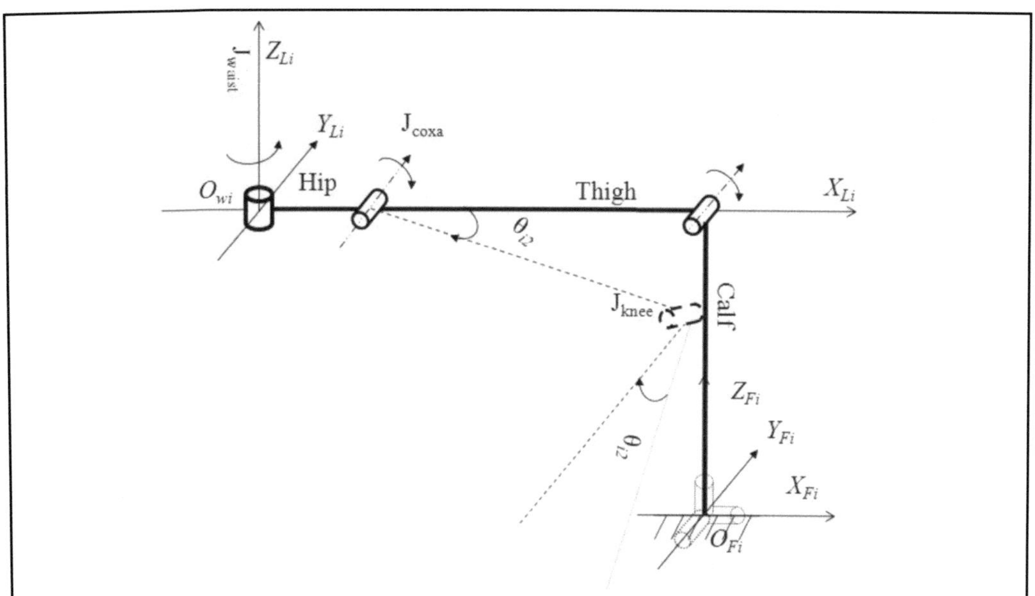

Fig. 4. General structure of a 3DOF (degrees of freedom) leg

4. Normal statically stable gaits

A hexapod has many types of statically stable gaits, such as regular gait, irregular gaits, periodic gaits and et al.
As for the regular periodic gaits, its gaits can be classified, according to the number of supporting legs during support phase, as 3 + 3 tripod gait with 3 supporting legs, 4 + 2 quadruped gait with four supporting legs and 5 +1 one by one gait with five supporting legs; according to the movement of legs, insect-wave gait which is the typical gait of rectangular six-legged robots, mammal-kick gait which is typical gait of rectangular quadruped robots and mixed gait which is typical multi-directional gait for hexagonal hexapod robots; the combination can be tripod insect-wave gait and so on. The typical irregular gait is so called free gait.

4.1 3+3 tripod gait

The tripod continuous gaits are characterized by having three legs standing on the ground for supporting and pushing the body forward, and the other three legs lifting off and swinging forward. In each gait period, the body moves two steps. The quickest tripod gait is when the duty factor β equals 1/2.

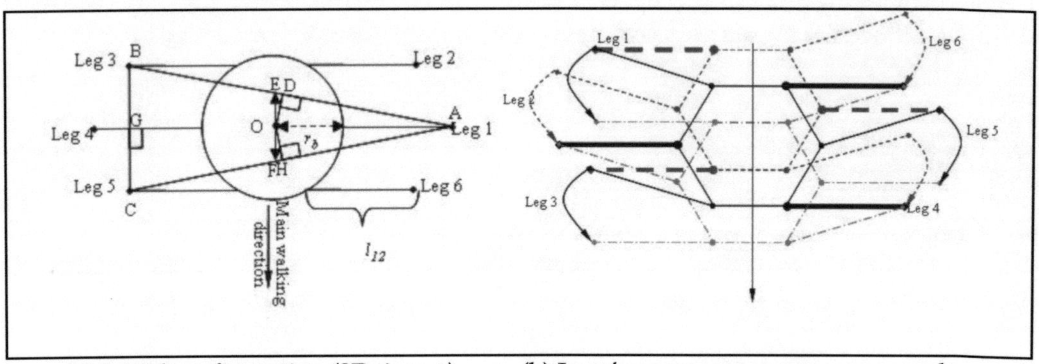

(a) Initial configuration (2D, insect) (b) Legs' movement sequence example

Fig. 5. Insect-wave tripod gait

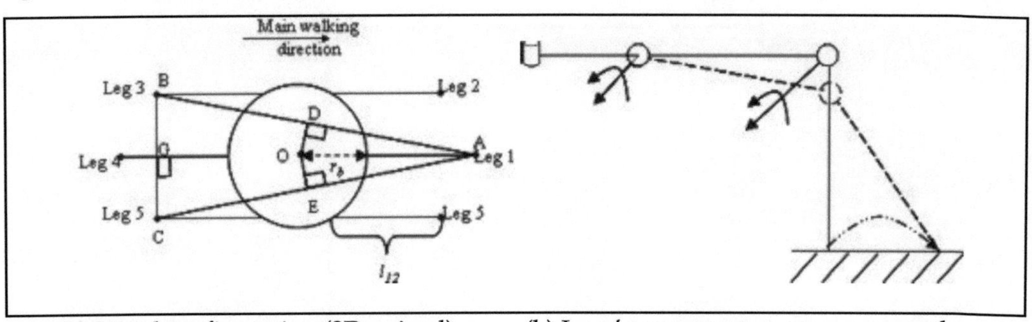

(a) Initial configuration (2D, mixed) (b) Legs' movement sequence example

Fig. 6. Mammal-kick tripod gait

In the initial configurations of insect-wave gait (see Fig.5) and mammal gait (see Fig.6), six legs of the robot are grouped into two and distributed along two sides as that of rectangular hexapod robots. Each group has three legs parallel. In Fig.5 (a) and Fig.6 (a). The positions of all waist joints are 0, -30, 30, 0, -30 and 30 degrees from leg 1 to leg 6, other joints angles are zeros.

The insect wave gait is characterized by a forward wave of stepping actions on each side of the body with a half-cycle phase shift between the two members of any right or left pair [63]. A scheme of the robot is sketched in Fig.5 (a), where the main direction of the movement, defined as main walking direction, is downwards, with legs swinging forward. Fig.4 (b) shows an example of legs sequence. In Fig.5 (b), the thick dashed or solid lines denote supporting legs. In the first step, leg 1, leg 3 and leg 5 are in stance phase and push the body forward; while leg 2, leg 4 and leg 6 swing ahead. In the second step, leg 2, leg 4 and leg 6 are in support phase and are responsible for pushing the body forward; leg 1, leg 3 and leg 5 then change to swing phase. After this, the procedure repeats again from the first step to the second step. The whole cycle includes two steps and the body is moved twice. In every step, the support polygon is an isosceles triangle ΔABC. The stroke length of supporting legs must make sure the gravity center of robot stays in side the support polygon, that means not surpass the stability margin.

In the mammal-kick gait legs generally move in a vertical plane like human's kicking out and trajectories of feet are along legs (Fig.6 (b)). The scheme of mammal-kick gait is depicted in Fig.6, and it walks mainly from left to right. The waist-joints do not work during mammal straight forward walking, but for turning. The support polygon is similar as with insect-wave gait and is an isosceles triangle ΔABC. During walking the front supporting legs retract and the rare supporting legs protrude so that the body is moving forward; on the contrary, the front swing legs are protrude and the rare swing legs retract. The legs' moving sequence is the same at that in insect-wave gait. The difference is just the configurations.

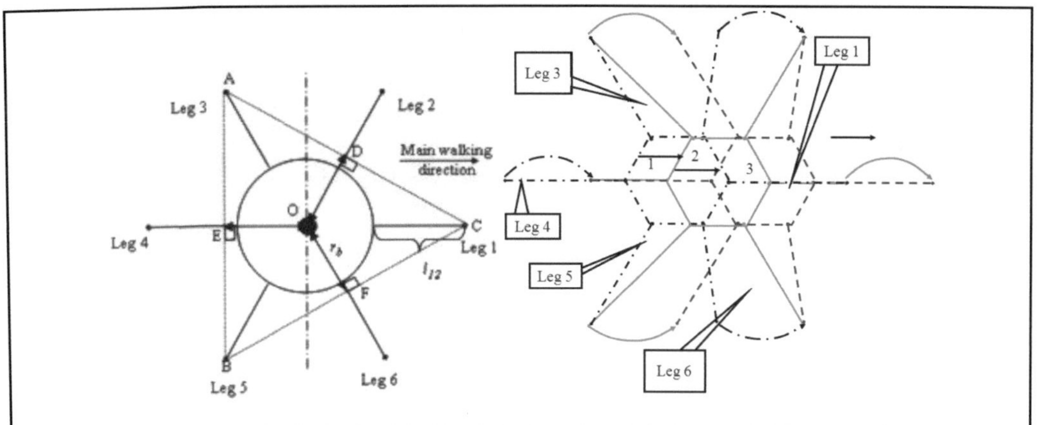

(a) Initial configuration (2D) (b) leg sequence

Fig. 7. Insect-mammal mixed tripod gait

In addition to the periodic tripod gaits mentioned above, we introduce here new type of mixed gait. In the initial configuration (see Fig.7) of insect-mammal mixed gait, all joint-angles are zeros. During walking, the mixed gait has a supporting area defined as a convex polygon connected all supporting legs, in the form of an equilateral triangle ΔABC or ΔDEF in Fig.7 (a)). The dark point in Fig.7 (b) is the gravity centre of the body. In every half period, one leg kicks off and two legs wave as insect-wave gait. Fig.7 (b) describes the walking sequence and 2D configuration of legs of the mixed gait. The legs' movement sequences are same as in other two gaits. The main walking direction is along the longitudinal axis of hip of leading leg, as shown in Fig.7.

From Fig.5 to Fig.7, it is can be seen that, for a given robot, the insect wave gait has the same size of supporting area ΔABC as the mammal gait; on the other hand, the mixed gait has the largest supporting area. In order to make a detail analysis, Song, Waldron and Choi in [Song and Choi, 1990] and [Song and Waldron, 1989] proved that wave gait has the optimum stability among all hexapod periodic and regular gaits in the range of $1/2 \leq \beta < 1$. While this is true for rectangular hexapod robots, it does not hold for hexagonal ones. The statically stability margin (*SSM*) and main-direction stability margin (*MDSM*) of three statically stable and continuous tripod gaits are compared based on one hexagonal hexapod robot whose parameters are listed in table 1. The stability results are reported in table 2 and table 3 respectively. In table 2 and table 3, the body heights, the distance from the bottom of the bodies to the ground, keep constant as length of calf (l_3); each link is assumed as a line and each joint is assumed as a point.

	Each leg			Body
	Hip	Thigh	Calf	
Mass, kg	m_1=0.80	m_2=2.00	m_3=2.00	m_b=10.9
Length, m	l_1=0.09	l_2=0.30	l_3=0.30	r_b=0.36

Table 1. Main physical parameters of hexapod robot example

	definition	formula	Hexxapod robot example	
			MDSM, m	Kinematic Limitation, m
Insect-wave gait	*min* $(\overline{OE}, \overline{OF})$ in Fig.5 (a)	$\frac{\sqrt{3}r_b(r_b+l_{12})}{4l_{12}+3r_b}$	<u>0.1471</u>	> 0.20
Mammal-kick gait	*min* $(\overline{OG}, \overline{OA})$ in Fig.6 (a)	$l_{12}+\frac{r_b}{2}$	0.5700	0.2196
Insect-mammal Mixed gait	*min* $(\overline{OE}, \overline{OC})$ in Fig.7 (a)	$\frac{1}{2}(r_b+l_{12})$	0.3750	0.2196

Table 2. MDSM of different tripod gaits (β = 1/2)

	definition	formula	hexapod robot example
Insect-wave gait	*min*$(\overline{OD}, \overline{OG}, \overline{OH})$ in Fig.5 (a)	$\frac{\sqrt{3}r_b(r_b-l_{12})}{2\sqrt{3r_b^2+4l_{12}^2-6l_{12}r_b}}$	0.1724m
Mammal-kick gait	*min*$(\overline{OD}, \overline{OE}, \overline{OG})$ in Fig.6 (a)	$\frac{\sqrt{3}r_b(r_b-l_{12})}{2\sqrt{3r_b^2+4l_{12}^2-6l_{12}r_b}}$	0.1724m
Insect-mammal Mixed gait	*min*$(\overline{OD}, \overline{OE}, \overline{OF})$ in Fig.7 (a)	$\frac{1}{2}(r_b+l_{12})$	<u>0.3750 m</u>

Table 3. SSM of different tripod gaits (β = 1/2)

As shown in table 2, the mammal-kick gait has the biggest MDSM but it loses this advantage because of kinematics limitation; the insect-wave gait has the smallest possible stride (14.71cm for the example) along main walking direction whereas the other two gaits have the same and much bigger possible stride. Synthetically, the insect-mammal mixed gait is optimally stable for hexagonal hexapod robots when $\beta=1/2$ and has stability advantage over the other two gaits while turning because of the biggest SSM.

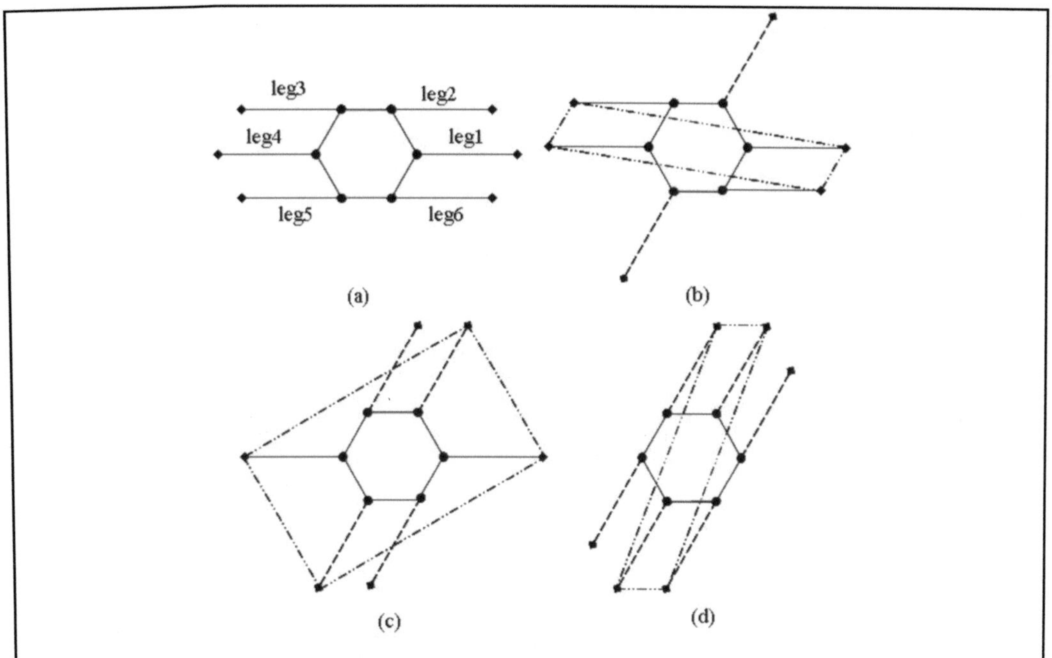

Fig. 8. 60 degree turning with inset-wave gait

Small angle turnings are easy for all three gaits. However, insect-wave gait needs special gaits to realize big-angle turning as stated in [Chu & Pang, 2002], [Wang, 2005] and [Zhang & Song, 1991], the same for mammal gait [Wang et al., 2007]. They have to stop and adjust legs at first for some big-angle turnings. Fig.8 shows examples of turning 60 degrees with insect or mammal gait. From the initial configuration in Fig.8 (a), the robot spends three steps to realize 60° turning. Quadrangles are supporting polygons. On the other hand, insect-mammal mixed gait can have big advantage on big-angle turning, especially at $\pm 60^{\circ}$, $\pm 120^{\circ}$ and 180°. With insect-mammal mixed gait, the robot just needs to reselect the leading leg for turning at $\pm 60^{\circ}$, $\pm 120^{\circ}$ and 180°, plus adjustment of crab angle it can realize any angle turning without stopping.

In the following Fig.9, Fig.10 and Fig.11, *R* and *S* denote revolute and spherical joint respectively; *f*, *k*, *c* and w denote foot, knee, coxa and waist, respectively. For instance, *Rc* specifies that the coxa is a revolute joint; *Sf* tells that between foot and ground, a virtual spherical joint is assumed.

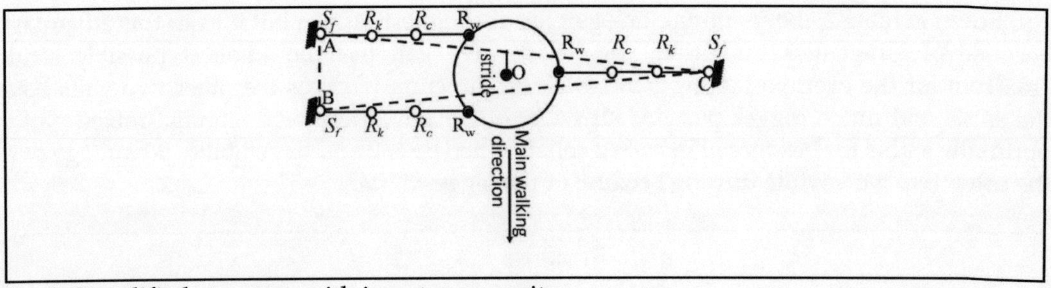

Fig. 9. Simplified structure with insect-wave gait

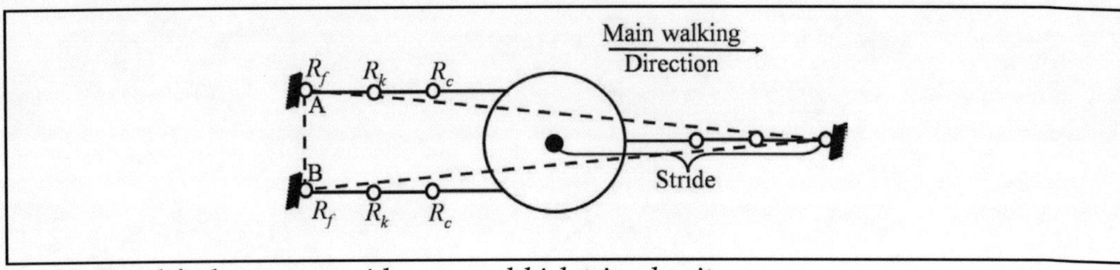

Fig. 10. Simplified structure with mammal-kick tripod gait

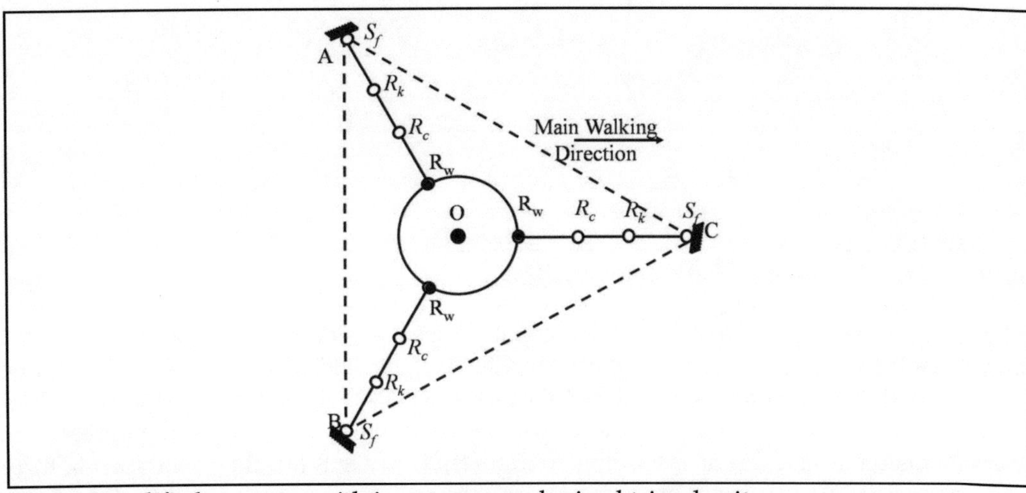

Fig. 11. Simplified structure with insect-mammal mixed tripod gait

In the insect wave gait, the waist joints are the most active joints during walking, and each foot needs three DOFs. The connection between each foot and the ground can be considered as a spherical joint (*Sf* in Fig.9.). The similar as insect wave gait, in mixed gait. The connection between each foot and the ground can be considered as a spherical joint (Sf in Fig.11). From the simplified structure, the mammal gait is easy to control. However, all legs with insect-wave gait have the same trajectories. It is therefore easiest to control. Just in insect-mammal gait, legs have different trajectory, but symmetric legs still have same trajectories.

3.2 4+2 quadruped gait

The rectangular hexapod robot has another type of gait, the "4+2" gait [Chu & Pang, 2002]. For this gait the legs are grouped into three groups. Every time there are four legs (two groups) standing on the ground to support the body, two other legs rise and walk ahead. In one gait period, there are three steps and the body moves only one step. The duty factor is 2/3. The hexagonal six-legged robot also has this gait with same leg sequences as that of a rectangular hexapod. One example can be:

1) Lifts leg 1 and leg 4, other legs support and push the body;
2) Leg 2 and leg 5 swing forward, all others support and push the body;
3) Leg 3 and leg 6 swing forward, the body is moved by others another step.
4) repeat procedure from 1) to 3).

This gait shows fault tolerant ability under certain conditions [Yang & Kim, 1998; Yang & Kim, 1999; Huang & Nonami, 2003; Chu & Pang, 2002], because three legs can support the body even if one supporting leg broken during walking. Chu and Pang had proved that the hexagonal robot by this gait has advantages compared with rectangular ones in stability, stride and turning ability, if the turning angle is within [-30 30] degrees.

3.3 5 + 1 one by one gait

The rectangular hexapod robot has another type of gait, the "4+2" gait [Chu & Pang, 2002]. For this gait the legs are grouped into three groups. Every time there are four legs (two groups) standing on the ground to support the body, two other legs rise and walk ahead. In one gait period, there are three steps and the body moves only one step. The duty factor is 2/3. The hexagonal six-legged robot also has three types' gaits with same leg sequences as that of a rectangular hexapod.

The leg-sequence of one by one gait can by any order, but generally legs move one after another following a clockwise or anti-clockwise order.

3.4 Free gait

Free gait proposed by Kugushev and Jaroshevskij in 1975 is characterized as non-periodic, non-regular, non-symetric and terrain adaptive. In a free gait, the leg sequence (i.e., the order in which leg transferences are executed), footholds, and body motions are planned in a nonfixed, but flexible way as a function of the trajectory, the ground features, and the machine's state. It is more flexible and adaptive than periodic and regular gaits on complicated terrain. A large number of free gaits for quadruped and hexapod robots have been developed to date. For more information, we can refer to [Pal & Jayaraian, 1990; Porta & Celaya, 2004; Estremera & Gonzalez de Santos, 2003 and 2005].

4. Fault tolerant gait

In arduous operating environments, robots may confront accidents and damage their legs; their legs may be dual-used as arms for some tasks, or some joints may suffer loss of control etc. In such cases, biped or quadruped robots would become statically unstable. However hexapods may still walk with static stable because their six legs provide redundancy. In this subsection we discuss these fault tolerant gaits.

4.1 Joint-lock

In this case, Yang [Yang, 2003] has already proposed a discontinuous tripod gait for rectangular hexapod robots.
However, with joint-lock a hexagonal hexapod may still maintain a continuous gait. The three possibilities for a single locked joint on one leg are discussed in the following.

1) Waist-joint-lock. In this case, the faulty leg cannot move in a horizontal plane, but it can swing in a vertical plane. The insect wave gait is difficult for this situation; whereas the mammal gait is still available by adjusting the other legs in parallel with the faulty leg. Also the mixed gait is possible if we chose the broken leg as the leading leg or the leg opposite as healing leg.
2) Knee or coax-joint-lock. For these two cases, the mammal gait and mixed gait are impossible to realize, but the insect gait is feasible, although not as efficient as before injury. If one whole leg is locked, the discontinuous tripod gait can be employed.

4.2 Loss of one leg

In the case of loss of one leg is due to fault or use for other tasks; two possibilities were considered in [Yang & Kim, 1998]. However, for symmetric hexagonal robot, there is only one case because the structure of every leg is the same and distributed evenly around the body. The 2+1+2 gait has same sequence as [Yang & Kim, 1998]. The difference is in the positions the leg. The legs of the gaits in [Yang & Kim, 1998] are overlapping. The symmetrical hexapod robot needs three steps to achieve this walk. During this procedure, the robot's body moves two steps.

4.3 Loss of two legs

There are three cases where two legs are either faulty or being used for other tasks. The positions of these two unavailable legs may be opposite, adjacent or separated-by-one (two damaged separated by one normal leg). Some studies [Takahashi et al., 2000] have been done in the first case, but there is a lack of study on the other two cases.

1) The opposite-legs case. Losing two opposite legs, for example, leg i and leg j the hexapod robot becomes a quadruped robot. It can walk with one of quadruped gaits, which have been widely studied. For example, the craw gait (Chen et al. 2006), the diagonal gait (Hirose & Matins, 1989), mammal-type "3+1"gait (Tsujita et al. 2001), "3+1"craw gait (Chen et al., 2006) which maintains static stability at each step, and the omni-directional updated quadruped free gait in [Estremara & Gonzalez de Santos, 2002; Estremara & Gonzalez de Santos, 2005].
2) The two-separated-by-one case and adjacent case. For these two cases the two unavailable legs are on the same side therefore it is almost impossible for a general rectangular hexapod robot to have statically stable locomotion. For a hexagonal robot the insect wave periodic gait is still available. The other four legs can be adjusted to suitable initial positions, as shown in Fig.12 for example. Fig.12 (a) is the case of losing leg 1 and leg 3. Fig.12 (b) shows the case where leg 1 and leg 2 are unavailable. Following the four-leg periodic gait sequence, robots can realize statically stable walking. The crab angle will be different. For example, if leg 1 and leg 2 or leg 1 and leg 3 are unusable, the crab angle will be $-\pi/6$. Fig. 13 lists the leg sequences for a separated-by-one fault tolerant gait. At each instant, there are three or four legs supporting the body. The mass centre is inside the supporting area.

For the adjacent case, the leg sequence is similar to the separated-by-one case after adjusting to suitable initial positions.

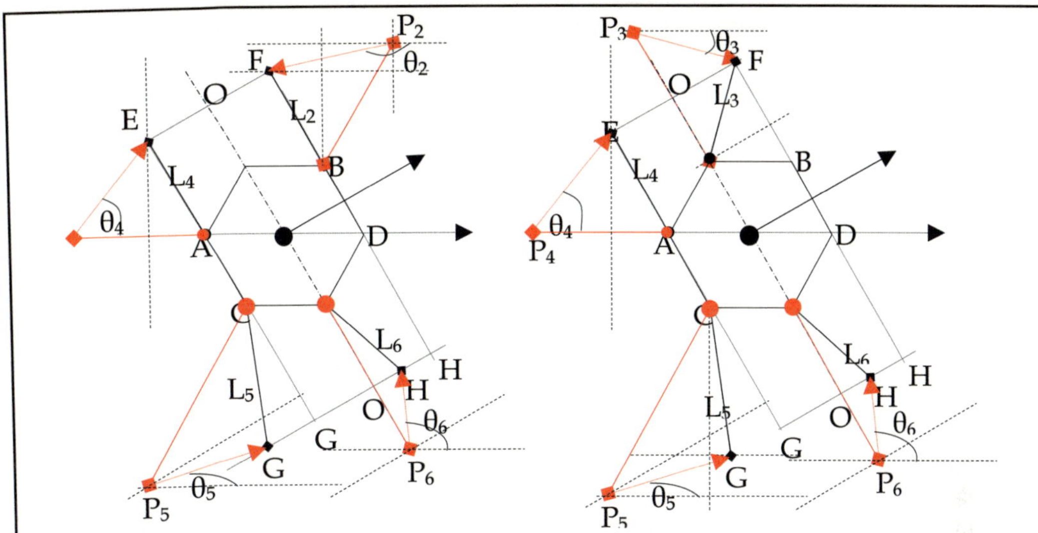

(a) Separated-by-one case (leg 1 and leg 3 are lost) (b) Adjacent case (leg 1 and leg 2 are lost)

Fig. 12 Initial state of four legs

To realize statically stable walking, there are several requirements in Fig.13:

1) $\overline{AE}=\overline{BF}=\overline{CG'}=\overline{DH'}=L$;

2) $L \geqslant R\cos(\pi/3)$;

3) $\overline{GG'}=\overline{HH'}=s$, the body stride;

4) $s<\overline{G'O_2}=\overline{O_2H'}$;

5) $\overline{EO_1}=\overline{O_1F}=\overline{O_2G'}=\overline{O_2H'}=R\sin\left(\frac{\pi}{3}\right)$

The rules for the quadruped insect wave gait are:

1) Rear legs (leg 4 and leg 5 in Fig.13) must not cross the central line (the point-dashed line in Fig.6-16) while moving ahead, so that the mass centre will also be in the subsequent supporting area.

2) Front legs (leg 1 and leg 2 in Fig.13) should not go back to the central line while the body (centre of mass) is moving ahead.

3) The stride of the swing legs is twice that of the body.

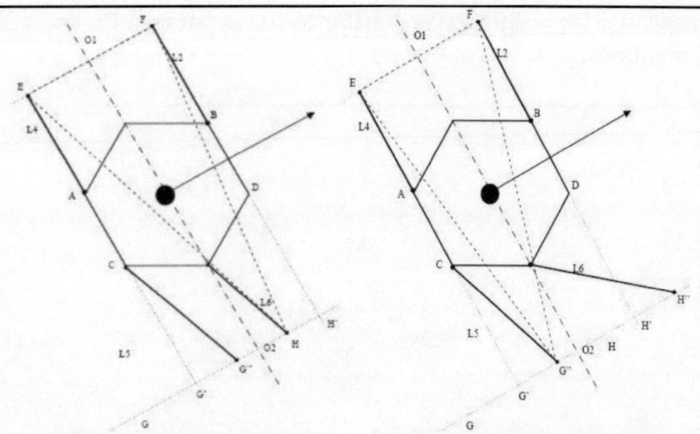

(a) Swing rear right leg, leg 5
Supporting area is △EFH

(b) Swing front right leg, leg6
Supporting area is △EFG″

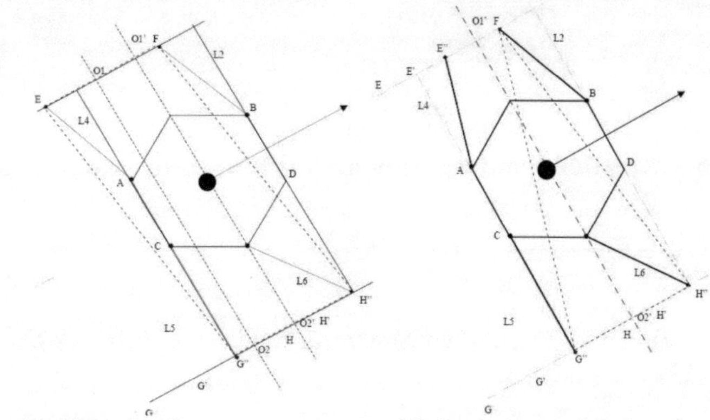

(c) Move body
Supporting area is □EFG″H″

(d) Swing rear left leg, leg4
Supporting area is △FG″H″

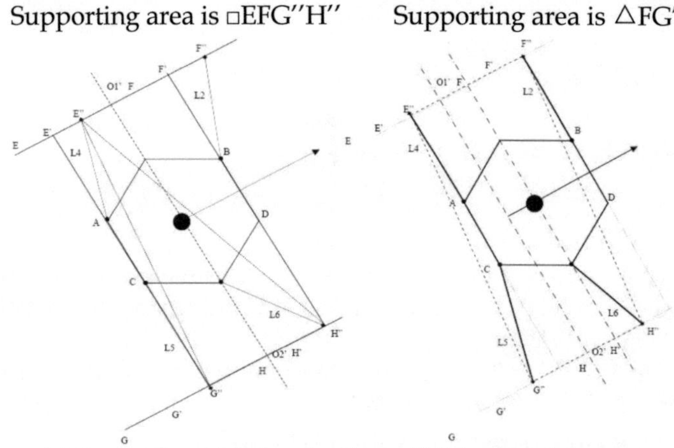

(e) Swing front left leg, leg2
Supporting area is △E″ G″H″

(f) Move body again
Supporting area is □E″ F″G″H″

Fig. 13. Leg sequences of separated-by-one case fault tolerant gait while two legs are broken

From the initial configuration of mixed gait to the fault tolerant initial state (Fig.12), we can adjust the legs according to following procedures (Equation 1 to Equation 8).
If two faults occur on two legs that separated by one, for example leg 1 and leg 3, the following procedure can be used to move the other legs from the original initial-positions to the fault tolerant initial-positions:

Leg 2 moves from P_2 to F with stride $\overline{P_2F}$ (Equation (1)) and rotates by angle θ_2 (Equation (2)) ; Leg 4 moves from P_4 to E with stride $\overline{P_4E}$ (Equation (3)) and rotates by angle θ_4 (Equation (4)) ; Leg 5 moves from P_5 to G with stride $\overline{P_5G}$ (Equation (5)) and rotates by angle θ_5 (Equation (6)); Leg 6 moves from P_6 to F with stride $\overline{P_6H}$ (Equation (7)) and rotates by angle θ_6 (Equation (8)).

$$\overline{P_2F} = \sqrt{(L\sin(\frac{\pi}{3}))^2 + (L-(l_1+l_2)\cos(\frac{\pi}{3}))^2} \tag{1}$$

$$\theta_2 = \mathrm{atan}(\frac{L-(l_1+l_2)\cos(\frac{\pi}{3})}{L\sin(\frac{\pi}{3})}) - \frac{\pi}{6} \tag{2}$$

$$\overline{P_4E} = \sqrt{((l_1+l_2) - L\cos(\frac{\pi}{3}))^2 + (L\sin(\frac{\pi}{3}))^2} \tag{3}$$

$$\theta_4 = \mathrm{atan}(\frac{L\sin(\frac{\pi}{3})}{(l_1+l_2) - L\cos(\frac{\pi}{3})}) \tag{4}$$

$$\overline{P_5G} = \sqrt{(L-(l_1+l_2)\cos(\frac{\pi}{3}))^2 + ((l_1+l_2)\sin(\frac{\pi}{3}) - s)^2} \tag{5}$$

$$\theta_5 = \frac{\pi}{6} - \mathrm{atan}(\frac{L-(l_1+l_2)\cos(\frac{\pi}{3})}{(l_1+l_2)\sin(\frac{\pi}{3}) - s}) \tag{6}$$

$$\theta_6 = (\frac{2\pi}{3} - \mathrm{atan}(\frac{R\sin(\frac{\pi}{3}) - s}{l_1+l_2-L+R\cos(\frac{\pi}{3})})) \tag{8}$$

$$\overline{P_6H} = \sqrt{(l_1+l_2-L+R\cos(\frac{\pi}{3}))^2 + (R\sin(\frac{\pi}{3}) - s)^2} \tag{7}$$

For the adjacent-legs case, the only difference is for the leg between the two faulty legs, leg 3 for example. The foot tip of leg 3 will move from P_3 to F with the following stride and rotation angle,

$$\overline{P_3F} = \sqrt{((l_1 + l_2 - L + R\cos(\frac{\pi}{3}))^2 + (R\sin(\frac{\pi}{3}))^2} \tag{9}$$

$$\theta_3 = -\frac{\pi}{3} + \operatorname{atan}\left(\frac{R\sin(\frac{\pi}{3})}{l_1 + l_2 - L + R\cos(\frac{\pi}{3})}\right) \tag{10}$$

In the above equations, $R\sin(\frac{\pi}{3}) \le L < l_1 + l_2$.

D. Loss of more than two legs

If more than two legs are lost, the robot is unable to maintain static stability while walking. Dynamic gaits may still be possible, such as the three-leg dynamics gait of Lee and Hirose, 2000. These will not be discussed further here.

5. Conclusion

In this chapter, the locomotion of symmetric hexapods has been studied in detail. We have presented a comprehensive study of hexagonal hexapod gaits including normal and fault tolerant ones. Gaits of rectangular and hexagonal six-legged robots have been compared from several aspects: stability, fault tolerance, terrain adaptability and walking ability. To facilitate simulations and experiments we have provided integrated kinematics of swinging and supporting legs for continuous gaits.

Hexagonal hexapod robots have been shown to be more flexible than rectangular ones. Moreover, hexagonal hexapods have many feasible gaits. In addition to the well-know insect gait and mammal gait, a new mixed gait for hexagonal six-legged robots has been proposed in this chapter which entails some features of both insect and mammal gaits. Except classified by legs movement as mentioned above, hexapod robots gaits are categorized according to the number of supporting legs during walking, as 3+3 tripod, 4+2 fault tolerant quadruped, and 5+1 one by one gaits. On account to the introduction of mixed gait, each numbered gait has one more form. Among three tripod-gait forms, the most stable is the mixed one. The mammal gait can reach the longest stride; whereas the continuous insect gait has the shortest maximum stride and poorest stability.

Thanks to their six legs, hexapod robots have redundancy and fault tolerance. Gaits where one leg is lost or two opposite legs are lost have been discussed in recent times. In this chapter we have tackled also the cases in which two adjacent legs or two separated by a normal leg are damaged. Algorithms for realizing these two fault-tolerant gaits have been detailed and validated with simulations.

6. References

Bares, J.; Hebert, M. et al. (1989). Ambler-An Autonomous Rover for Planetary Exploration. IEEE Computer, pp. 18-26(1989).

Chen, J. J.; Peattie, A. M. et al. (2006). Differential leg function in a sprawled-posture quadrupedal trotter. The Journal of Experimental Biology, Vol. 209, pp. 249-259(2006).

Chen, X.D.; Yi, S. & Jia, W.C. (2006). Motion Planning and Control of Multilegged Walking Robots. Publishing Press of Huazhong University of Technology and Science, Jun. 2006 (1).

Chu, S. K.-K. & Pang, G. K.-H. (2002). Comparison Between Different Model of Hexapod Robot in Fault-Tolerant Gait. IEEE Transactions on Systems, Man and Cybernetics, Part A, Vol. 32, Issue 6, Nov. 2002 pp. 752 -- 756.

Erden M.S. and Leblebicio ǔ lu K. (2008). Freegait generation with reinforcement learning for a six-legged robot. Robotics and Autonomous Systems, Vol. 56 pp. 199--212(2008).

Estremera, J. & Gonzalez de Santos, P. (2002). Free Gaits for Quadruped Robots Over Irregular Terrain. The International Journal of Robotics Research, Vol. 21, No. 2, Feb, 2002, pp. 115-130.

Estremera, J. & Gonzalez de Santos, P. (2005). Gnerating continuous free crab gaits for quadruped robots on irregular terrain. IEEE Transactions on robotics, Vol.21, No.6, pp.1067-1076(2005).

Fitcher, E.F. Fichter, B.L. (1988). A survey of legs of insects and spiders from a kinematic perspective. Proceedings., 1988 IEEE International Conference on Robotics and Automation, 1988., vol. 2, (24-29 Apr 1988), pp. 984-986.

Gonzalez de Santos, P.; Jcobano, A.; Garcia E. et al. (2007). A Six-legged Robot-based System for Humanitarian Demining Missions. Mechatronics, Vol. 17, pp. 417-430(2007).

Gonzalez de Santos, P.; Garcia, E. &Estremera J. (2007). Improving Walking-robot Performances by Optimizing Leg Distribution. Autonomous Robots, Vol.23, No.4, pp.247-258(2007).

Gurocak, H. B.; Peabody J. (1998). Design of a Robot that Walks in Any Direction. Journal of Robotic Systems, 15(2), pp. 75-83(1998).

Hirose, S.; Homma, K.; Matsuzawa S. & Hayakawa S. (1992). Parallel Link Walking Vehicle and Its Basic Experiments. 6th Symposium on intelligent Mobile Robots, pp. 7-8 (1992 in Japanese).

Hirose, S. A & Martins, F. (1989). Generalized standard leg trajectory for quadruped waking vehicle. Trans. Soc. Instrument Control Eng., 25(4), pp. 455-46(1989).

Huang Q.-J. and Nonami K. (2003). Humanitarian mine detecting six-legged walking robot and hybrid neuro walking control with position/force control, Mechatronics, Vol. 13 pp. 773--790(2003).

Kamikawa, K.; Arai, T. et al. (2004). Omni-Directional Gait of Multi-Legged Rescue Robot. Proceedings of the 2004 IEEE International Conference on Robotics and Automation, New Orleans, LA - 4111, pp. 2171-2176(2004).

Kaneko, M.; Abe M.and & Tanie K. (1985). A Hexapod Walking Machine with Decoupled Freedoms. IEEE Journal of Robotics and Automation, Vol. RA-1, No. 4, (December 1985), pp. 183-190.

Koyachi, N.; Ami,T. et al. (1995). Hexapod with Integrated Limb Mechanism of Leg and Arm. IEEE International Conference on Robotics and Automation Vol. 2, 21-27 May 1995 pp. 1952 - 1957.

Krotkov, E.; Bares J. and et al. (1991). Ambler: A six-legged planetary rover. Fifth international conference on advanced robotics-91 ICAR, pp. 717-722, (1991).

Kugushev E.I. and Jaroshevskij V.S. (1975). Problem of selecting a gait for an integrated locomotion robot. Process Fourth International conference. Artificial Intelligence, Tbilisi, Georgian SSR, USSR, Sept. 1975, pp. 789-793.

Lee B.-H. & Lee I.-K. (2001). The implementation of the gait and body structure for hexapod robot. IEEE International Symposium on Industrial Electronics ISIE 2001., Vol. 3, 12-16 (June 2001) pp.1959 - 1964.

Lee T.T.; Liao C.M. & Chen T.K. (1988). On the Stability Properties of Hexapod Tripod Gait. IEEE Journal of robotics and automation, Vol. 4. NO. 4. August, 1988, pp. 427-434.

Lee W.-J., Orin D.E. (1988). Omni-directional supervisory control of a multi-legged vehicle using periodic gait. IEEE Journal of Robotics and Automation, Vol. 4, Issue 6, (Dec. 1988) pp. 635 - 642.

Lee, Y.-J. & Hirose, S. (2000). Three-Legged Walking for Fault Tolerant Locomotion of a Quadruped Robot with Demining Mission. 2000 IEEE/RSJ International Conference on Intelligent Robots and Systems (IROS 2000), Vol. 2, 31 Oct.-5 Nov. 2000 pp. 973 - 978.

McGhee R.B. and Iswandhi G. I. (1979). Adaptive Locomotion of a Multilegged Robot over Rough Terrain. IEEE Tractions on system, man, and cybernetics, Vol., SMC-9, No. 4, April 1979, pp. 176-182.

Ota Y.; Inagaki Y. & et al. (1998). Research on a Six-Legged Walking Robot with Parallel Mechanism. Proceedings of the 1998 IEE WRSJ Intl. Conference on Intelligent Robots and Systems, Victoria, B.C., Canada (Oct. 1998), pp. 241-248.

Pal, P.K. & Jayaraian, K. (1990). A free gait for generalized motion. IEEE Transactions on robotics and automation, Vol. 6, No. 5, pp. 597-600(1990).

Porta J.M. and Celaya E. (2004). Reactive free gait generation to follow arbitrary trajectories with a hexapod robot. Robotics and Autonomous Systems, Vol. 47, pp. 187--201(2004).

Preumont, A; Alexadre, P & Ghuys, D. (1991). Gait analysis and implementation of a six leg walking machine. 91 ICAR., Fifth International Conference on Advanced Robotics. Robots in Unstructured Environments', 19-22 June, vol. 2, pp. 941 – 945(1991).

Saranli, U.; Buehler, M. & Koditschek D. (2001). RHex: A simple and highly mobile hexapod robot. International Journal of Robotics Research, v 20, n 7, July, 2001, pp. 616-631.

Song, S. M. & Choi B.S. (1990). The Optimally Stable Ranges of 2n-Legged Wave Gaits, IEEE Transactions on Systems, Man, and Cybernetics---part b: Cybernetics, Vol. 20, No. 4, July-August 1990, pp. 888-902.

Song S.M. and Waldron K.J. (1989). Machines that Walk: the Adaptive Suspension Vehicle. MIT Press, Cambridge (1989).

Su J. (2004). The Research of the Gait Planning and Control of the Multilegged Walking Robot. Master thesis (in Chinese). Huazhong University of Science & Technology, Wuhan 430074, P.R.China, Aprils, 2004.

Takahashi, Y. Arai,T., et al.(2000). Development of Multi-Limb Robot with Omnidirectional Manipulability and Mobility. Proceedings of the 2000 IEEE-RSJ International Conference on intelligent Robots and Systems, Vol. 3, 31 Oct.-5 Nov. 2000, pp.2012 - 2017.

Tsujita, K.; Tsuchiya, K.; & Onat A. (2001). Adaptive gait pattern control of a quadruped locomotion robot. 2001 IEEE/RSJ International Conference on Intelligent Robots and Systems, Vol. 4, 29 Oct.-3 Nov. 2001 pp. 2318 - 2325.

Wang X.-J. (2005). A Study of Locomotion and Force Planning for Multilegged Walking Robots. PHD thesis(in Chinese), Huazhong University of Science & Technology ,Wuhan 430074, P.R. China, October, 2005, pp. 95-100.

Wang, Z.-Y.; Ding, X.-L. & Rovetta A. (2007). Structure Design and Locomotion Analysis of a Novel Robot for Lunar Exploration. 12th IFToMM World Congress, Besançon (France), June 18-21, 2007.

Wettergreen, D.; Thomas, H & Thorpe C.(1990). Planning strategies for the ambler walking robot, IEEE International Conference on Systems Engineering, pp.9-11 (Aug.1990).

Yang,J.-M. & Kim J.-H. (1998). A Strategy of Optimal Fault Tolerant Gait for the Hexapod Robot in Crab Walking. IEEE International Conference on Robotics and Automation, Vol. 2, 16-20 May 1998 pp. 1695 - 1700.

Yang,J.-M. & Kim J.-H. (1998). Fault tolerant locomotion of the hexapod robot. IEEE Transactions on System, Man and Cybernetics.vol. 28, no. 1, Feb. 1998, pp. 109-116.

Yang,J.-M. & Kim J.-H. (1999). Optimal Fault Tolerant Gait Sequence of the Hexapod Robot with Overlapping Reachable Areas and Crab Walking. IEEE Transactions on System, Man and Cybernetics, Part A, Vol. 29, Issue 2, March 1999 pp.224 - 235.

Yang,J.-M. & Kim J.-H. (2000). A Fault Tolerant Gait for a Hexapod Robot over Uneven Terrain. IEEE Transactions on Systems, Man, and Cybernetics—part b: Cybernetics, vol. 30, no. 1, February 2000, pp. 172– 180.

Yang,J.-M. (2003). Fault-Tolerant Gait Generation for Locked Joint Failures. IEEE International Conference on Systems, Man and Cybernetics, Vol. 3, 5-8 Oct. 2003 pp. 2237 - 2242.

Yoneda K. and Suzuki K. (1997). Gait and Foot Trajectory Planning for Versatile Motions of a Six-Legged Robot. Journal of Robotic Systems, 14(2), pp. 121--133(1997).

Zhang, C.-d & Song, S.-M. (1991). Turning gait of a quadrupedal walking machine," IEEE International Conference on Robotics and Automation, Vol. 3, 9-11 April, 1991, pp. 703 - 708.

19

In situ self-reconfiguration of hexapod robot OSCAR using biologically inspired approaches

Bojan Jakimovski and Erik Maehle
Institute of Computer Engineering, University Lübeck
Germany

1. Introduction

A highly desirable feature for next generation robots operating on terrestrial or extraterrestrial environment (Chien et al., 2006) is that they possess the property of sustainable autonomic systems enveloping the self-management and self-x characteristics such: self-reconfiguration, self-optimization, and self-healing. Responses taken automatically by a system without real-time human intervention are called autonomic responses (Sterritt et al., 2006; Lewandowski et al., 2006). The self-x properties will enable the robot to continue with its mission tasks even in the cases when the robot has some faults within the system. The robot shall be able to reconfigure itself and continue with its mission tasks.

The autonomic concept was introduced with the IBM Manifesto for Autonomic Computing (IBM, 2001). This proposed several key elements important for the autonomic systems: self-configuration, self-healing, self-optimization, and self-protection which were inspired by the human body's autonomic nervous system. Complementary to the IBM's initiative, the Organic Computing initiative (DFG, 2004) on the other hand, proposes the means of achieving such self-x properties of the next generation of self-organizing embedded systems, inspired by information processing seen within the biological systems.

Transferring such biologically inspired paradigms (Hinchey & Sterritt, 2007) into computing systems and robots will enable the systems to perform in a more robust, safe, and flexible way. In that context we have been researching towards practically applying biologically inspired methodologies and developing novel procedures for next generations of self-reconfiguring and joint leg walking robots.

The rest of the paper is organized as follows: In the second chapter we give an overview of our hexapod robot prototype OSCAR on which we have conducted the experiments. We also introduce our innovative and patent pending mechanism for in-situ walking robot leg amputation and for robot reconfiguration.

In the third chapter we explain the biologically inspired fault detection method used for fault/anomaly detection. In the fourth chapter we describe the swarm intelligence concept for robot reconfiguration, which is used to perform a stable spatial reconfiguration of the hexapod walking robot. We also present the results from real experiments on self-reconfiguration performed on the experimental hexapod robot OSCAR using our robot leg amputation mechanism.

2. Organic Self-Configuring and Adapting Robot - OSCAR and innovative robot leg amputation mechanism

2.1 Self-Configuring and Adapting Robot - OSCAR and innovative robot leg amputation mechanism R-LEGAM

In order to perform real demonstrations on self-reconfiguration of a walking robot, we have built up a new prototype in our series of experimental hexapod robots - named OSCAR - Organic Self-Configuring and Adapting Robot (El Sayed Auf et al., 2006) (Jakimovski et al., 2006). The robot is constructed of a fiberglass framed body, six legs set up spatially in a circle with 60 degrees between each leg, three servos per leg, onboard control electronics, wireless camera, and additional sensors. The robot is shown in Fig. 1.

Fig. 1. OSCAR (Organic Self-Configuring and Adapting Robot)

2.2 Innovative robot leg amputation mechanism

In the newly built hexapod robot prototype, we have introduced an innovative mechanism called Robot Leg Amputation Mechanism (R-LEGAM), which is performing on demand robot leg amputation.

The mechanism is built out of electrical connectors which are used for providing the servos with power and also for the control signals for the bus on which the servos are connected. Additionally there are integrated springs in each amputation mechanism and those are used to eject the leg from the robot's body in the process of leg amputation. Every mechanism has an additional small servo that on demand moves small mechanical part that further releases the energy stored within the compressed springs and the robot's leg is ejected.

This mechanism is used in cases for example when the robot experiences some problems in some of its legs and in such circumstances can amputate the malfunctioned legs.

By this the robot does not have to carry on the malfunctioned legs during the rest of its mission. The advantages are that the weight is decreased, the energy from batteries is used effectively, and the operating range of the robot after performing the leg amputation is increased in comparison with having the robot still walking with the malfunctioned legs. On the other hand, the disadvantages are that the stability and the speed of the robot are decreasing with every amputated leg.

The robot's leg with the integrated leg ejection mechanism attached to the robot's body is presented in the Fig. 2.

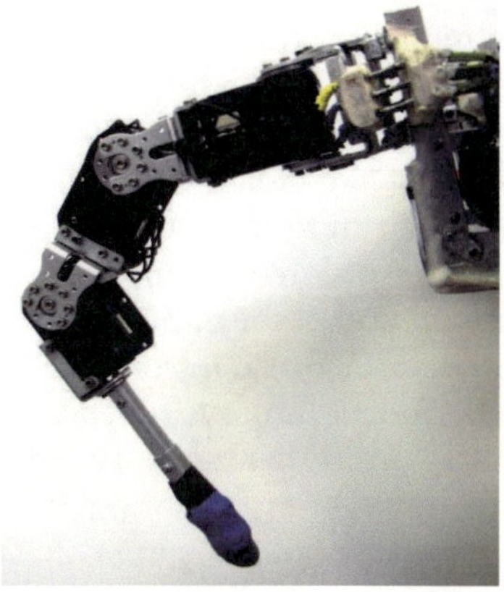

Fig. 2. OSCAR's leg with integrated leg ejection mechanism attached to the robot's body

When the robot's control architecture finds an anomaly within some leg, and it decides that the leg should be amputated, a signal is sent to the amputation mechanism which then detaches the leg from the robot's body.

In Fig. 3 the robot's leg is represented - detached from the robot's body.

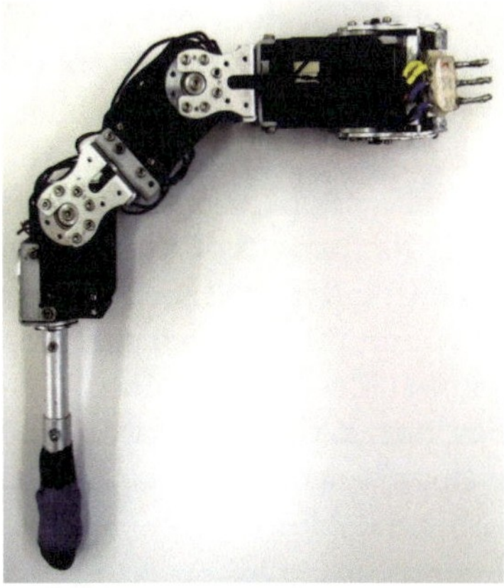

Fig. 3. OSCAR's leg detached from the robot's body after the amputation is performed.

3. Robot anomaly detection using artificial immune system based approach - RADE

Fault or anomaly detection is a scope of many projects, especially those related to reliable and safety-critical systems. For fault detection in general and for the domain of autonomous robots various approaches have been considered like: Time-Delay Neural Networks (TDNN) (Christensen et al., 2008), Recurrent Neural Networks (RNN) (Przystalka, 2006), particle filters (Verma & Simmons, 2006; Zhuo-hua et al., 2006). The mentioned approaches are mostly related to the procedure of synthesizing fault detection components based on the collected data in the training runs. After the fault detectors have been synthesized they are evaluated on fault scenarios in real time, or on the recorded data, and the fault positives or fault negatives are estimated.

However, building a full failure model for robotic systems and generating fault detectors for every possible fault scenario that may occur is often an exhaustive process (Haldar & Sarkar, 2006). To overcome this constraint the fault detection mechanism should possess self-adapting and learning properties, just like biological systems have.

In order to make our fault detection more adaptive to the environment and the current behavior of the robot, we have considered using the Artificial Immune System (AIS) paradigm as metaphoric implementation of the natural immune system.

Similar to the natural immune system, AIS also introduces a metric that allows the system to distinguish between self (correctly functioning system) and non-self (anomaly in the system) and also to memorize and detect specific patterns. Therefore main properties of AIS are: recognition, identification, adaptation, self-organization. It has been successfully applied for various domains: pattern recognition (Cao & Dasgupta, 2003), data mining (Nasraoui et al., 2005), network security (Pagnoni & Visconti, 2005), robotics (Michelan & Von Zuben, 2002), (Singh & Nair, 2005), (Sathyanath & Sahin, 2002), (Neal et al., 2006), (Canham et al., 2003), and others. There exist several different approaches (De Castro & Timmis, 2002) for AIS. However, the most commonly found in literature are: negative selection (Forrest et al., 1994), positive selection (Nino & Beltran, 2002), artificial immune networks (AIN) (Galeano et al., 2005) and clonal selection (De Castro & Von Zuben, 2002).

Negative selection mechanism is based on the ability of the immune system to learn to categorize between non-self and self by providing tolerance for the self. The negative selection consists of generating a set of detectors and evaluation of those detectors. Only the detectors that detect non-self, but do not react to self are considered for further detection. The positive selection, on the other hand, generates, evaluates and enables those detectors that can detect only the non-self. The clonal selection within AIS is based on the proliferation mechanism where self, upon recognizing non-self, starts to proliferate by cloning itself and also memorizing the pattern of the non-self (immune memory), so it has better responsiveness for the next encounter with such a particular non-self pattern.

For the detection of an anomaly within robot systems different immune system approaches have been considered, like inflammation (Sathyanath & Sahin, 2002), or usage of negative selection (Neal et al., 2006). For our RADE anomaly detection approach (Jakimovski & Maehle, 2008) we are using the clonal selection method in combination with fuzzy logic for representing the information within the AIS.

Within our robot's control architecture ORCA (Organic Robot Control Architecture) introduced in (Brockmann et al., 2005) and represented in Fig. 4. RADE is situated in monitoring and reasoning units called OCUs (Organic Control Units). OCUs are responsible for permanently monitoring the health status of the system and for taking suitable counteractions like reconfigurations in case malfunctions are detected. BCUs (Basic Control Units) implement the fault-free behaviors of the robotic system. Depending on its implementation, within the robot's control model several BCUs as behavioral units, several OCUs and a number of sensors and actuators can be present.

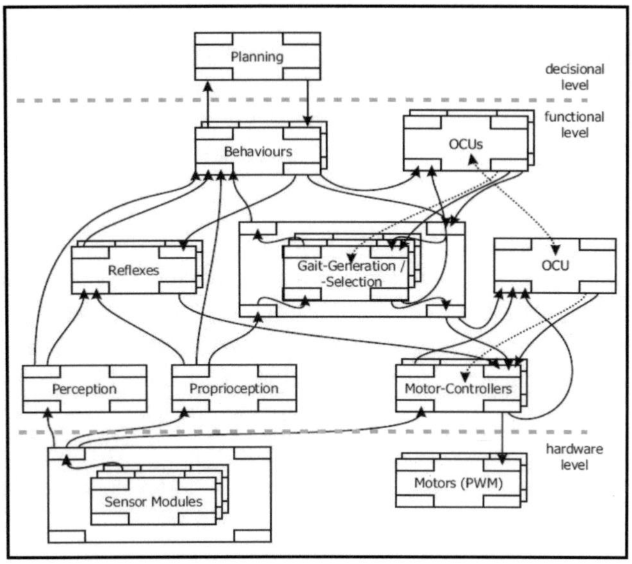

Fig. 4. ORCA (Organic Robot Control Architecture).

RADE uses fuzzy logic for representing information because an exact recognition is not necessary for triggering of an anomaly response. It is also similar to the way the association is made within the immune system, where given a similar (but not necessarily identical) stimulus, the response can be initiated. In a practical implementation, this would mean for example: if the behavior is walking then the acceleration level should not be low. Or in monitoring the servo's motor status, this can be interpreted as: by normal walking, the servo's current should not be high. In RADE such generalized self / non-self situations can be defined by fuzzy rules which are part of self / non-self sets. The rules in the non-self set detect when there is some anomaly present within the system, and the rules in the self set detect when the situation is not characterised as anomalous. The rules of both sets have the following structure:

IF X1&X2&...Xn THEN Anomaly is Y WITH weight_factor Z (1)

The "X1&X2&...Xn" in (1) represents the premise part which constitutes of monitored behaviours: walking, standing, etc.; and particular some characteristics like: current, acceleration, etc. The "Y" in (1) is the consequence part and can have two types of values: "anomaly is present" or "anomaly is absent". The weight factor "Z" represents the clonal proliferation within AIS, and is in a range from 0.0 to 1.0. The "Z" value in (1) will increase for some constant value (for example: 0.1) if the rule has "fired". In parallel to that the weights will decrease in all the rules belonging to the opposite set, just as the concentration of self/non-self drops being influenced by an increased concentration of non-self/self within the immune system. The firing level of each rule is therefore always adjusted, depending on the value of "Z". The weight factor "Z" has also another positive characteristic for the anomaly detection engine. Namely we want to reduce the factor of hand coded elements in RADE, and let the system dynamically adjusts itself to the current situation.

For example, in case we have coded fuzzy rules *without* using weights, depending on the manually pre-designed fuzzy membership sets for the premise parts of the rules, the rules can have an optimal response for some situation and perhaps a not satisfying response for other unforeseen situations. In case the fuzzy rules *have* weights, this would introduce two new features:

- The premise parts of the rules do not require any additional handcrafting and expert designing for their fuzzy membership sets. Therefore they can have some automatically generated "standard" triangular fuzzy membership sets, normally distributed within a valid range for the observed variable. For example such fuzzy membership sets for monitored variable "Current" having values in the range from min 0 to max 3 Amperes can be represented as:

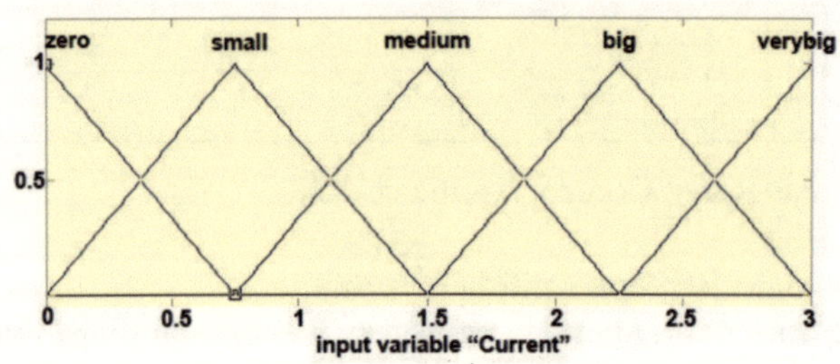

Fig. 5. Fuzzy membership set for monitored variable "Current".

  The other membership sets for other monitored variables (for example: acceleration) are going to have the same "standard" triangular fuzzy membership normally distributed for their input range. The nice thing for having such "standard" generated fuzzy membership sets is that they can be part of the learning process, which we plan to introduce in the next step of development of our anomaly detection engine. Having the "standard" fuzzy membership sets for each of the observed parameters, we may build up a rule-based learning system which incorporates only new situations since the fuzzy membership sets for the observed parameters are not going to be changed, and so it will be possible to distinguish between what has been already learned and what can be learned.

- The weight factors for such rules having the "standard" generated fuzzy membership sets allow the rule to adapt to the situation even without the rule being optimally pre-designed at start, i.e. having its membership sets optimally pre-designed. Therefore the changes of the weights depend on the particular situation and therefore contribute for the dynamics of overall system.

The previously discussed self and non-self rule sets and the dynamical change of their weights can be visually represented as in Figure 6.

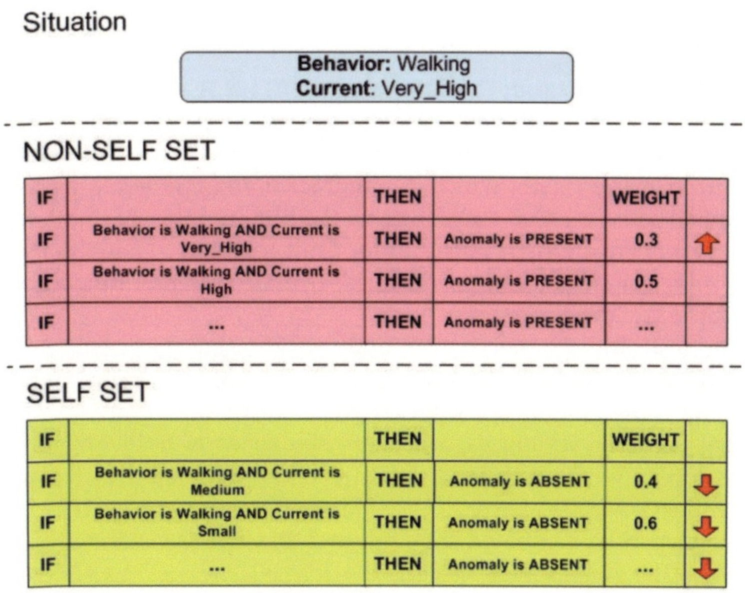

Fig. 6. Functioning principle of RADE and the dynamically changing weights within the self and non-self rule sets (for clearness, here only monitoring parameter "Current" is represented, although additional monitored parameter, like "Acceleration" can be considered as well).

As illustrated in the figure, when the premise within a rule belonging to the non-self set is satisfied, the rule "fires" and its weight is also increased by some constant value, e.g. by 0.1 from 0.3 to 0.4. At the same time, the weights of rules belonging to another set are decreased with the same constant value of 0.1. In such a way they lower their value from 0.4 on 0.3 or from 0.6 on 0.5 and so on.

In every computation step, a weighted output is calculated from such a fuzzy system which contains two membership functions: self and non-self. The value is in the range from 0.0 (no anomaly) to 1.0 (full anomaly) and represents the output of the RADE method. The output of RADE is computed in a defuzzification process as a centroid of fuzzy outputs of the "fired" rules. Therefore the output of RADE is influenced by the weight factor of each of the firing rules. The weight factor acts in the similar way as the secondary and subsequent responses within the immune system, i.e. the more the weight is associated to some rules, the more significant will be the response of those rules to the output of RADE in the next moment of their firing.

The change of weights therefore acts as some sort of short memory for events that occurred some moments ago. The anomaly output level of RADE depends on its short history and

also on actual system's state. Thus, the anomaly detection method RADE autonomously adapts to the situation and generates an appropriate health signal.
Within our experiments RADE method is used to detect a potential anomaly situation within the robot's system and robot's legs and to initiate a leg amputation and robot self-reconfiguration.

4 Self Reconfiguration of a joint-leg hexapod robot

4.1 Swarm Intelligence for Robot Reconfiguration (S.I.R.R.) method

Different approaches have been tried for overcoming the robot's malfunctioning. One of them is swarm robotics. Swarm robotics is a synonym for decentralized autonomous systems built out of many robots which communicate and cooperate with one another in order to accomplish some mission tasks. Due to the high number of entities in such swarm system, swarm robots have a high potential for fault tolerance. Namely, the failure within one or several robots may not compromise the overall system's functionality. (Winfield, 2006) (Hyun Yool et al., 2006).
However, we can also shift the macro view and consider that the robots themselves can be built out of several autonomous entities, where the failure of one or several such entities will not make the robot completely malfunctioned. Modular robots (Xuan et al., 2006) (Hancher & Hornby, 2006) are such an approach, where the robot is built out of many identical functional parts which when combined together and coordinated in the proper way can function like a robust robot. However, it is a still ongoing research on how those separately functional elements can be combined together and coordinated autonomously to function like one robot.
In our project, we have taken yet another direction with different types of granularity of the system. Namely for our walking hexapod robot, we have assumed that the robot legs themselves act like own entities. So we have considered that one robot leg is a functional entity built of several servos which cannot be considered as independent functional units, but only when combined together in one robot leg segment.
Therefore, when taking into account this type of granularity view, we have six legs acting like six independent entities which can be considered further for reconfiguration purposes.
Another part of the research is based on an idea on how to perform the reconfiguration on such functional entities within one hexapod robot with changeable morphology. The approach which we developed is based on the natural phenomena of flocks of birds and schools of fish and is named S.I.R.R. (Swarm Intelligence for Robot Reconfiguration) (Jakimovski et al., 2008). The flocking phenomenon was researched by Craig Reynolds who introduced boids as a distributed behavior model to simulate coordinated movements of the flocking of birds (Reynolds, 1987). Each member of a boid implements three rules and interacts only with its neighboring boid members.

These rules are defined as:

1. Separation rule: the member tries to avoid the crowding with other local boids;
2. Alignment rule: the member moves towards the average heading of the local boids;
3. Cohesion rule: the member moves towards the local average position of the other local boids.

In our experiments we have defined that each robot's leg represents one member of boid and since the robot has intrinsic symmetry with three legs on one side and three on another, the legs can be considered as members of two groups of boids - see Fig. 7 (a). Additionally for performing proper spatial robot reconfiguration we have considered that each boid besides the three above mentioned rules implements an additional two boid rules:

4. Boids are allowed to have n, n+1, or n-1 number of participants in their boid in comparison to the other boid, where n is the number of entities (legs) in the boid. In case there are n+2 more members in one boid in comparison to another one, the legs which belong to the "overcrowded" boid and that are situated on the edge of the boid, i.e. they have just one other neighbor member in the boid, will be moved to another boid until the "overcrowded" boid's size becomes n+1.

5. Members (legs) of one boid that join the other boid change their swinging and stance end-position parameters with respect to the parameters characteristic for that boid.

The fifth rule is used for example when boid (leg) changes from one to the other side of the robot's symmetry line.
In Fig. 7 we represent the usage of the above five boid rules used for a potential robot reconfiguration.

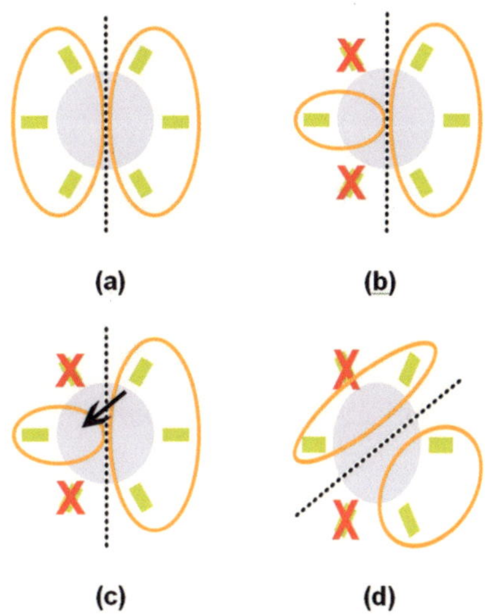

Fig. 7. Illustration of reconfiguration of a robot using the five boid rules within S.I.R.R. (Swarm Intelligence for Robot Reconfiguration) algorithm; (a) legs of the robot forming two boids; (b) two legs of the robot have malfunctioned; (c) member (leg) from "crowded" boid joins another less "crowded" boid; (d) situation after reconfiguration. The dotted line represents the robot's line of symmetry.

Fig. 7 (a) is the initial robot configuration with legs belonging to two boids. Fig. 7 (b) shows the situation when two legs have malfunctioned. Fig. 7 (c) shows that after the two legs have malfunctioned the number of members in the right boid does not equal the number of

members in the left boid. This matches the fourth rule used in our SIRR algorithm and therefore a member from the right boid transfers to the left boid and therefore reconfiguration is performed. Fig. 7 (d) shows the situation after the reconfiguration.
In Fig. 7 (d), we can also see that the robot's line of symmetry changes. This change of robot's symmetry line occurs in emergent way due to the reconfiguration of members in boids that takes place.

After the reconfiguration is performed using the SIRR approach, the robot has a more stable spatial configuration of its legs, since the angles and distance between the legs are more equally distributed. Such configurations enable the robot to have better static and dynamic weight distribution during walking.
In (Jakimovski et al., 2008) we have introduced and explained the Swarm Intelligence for Robot Reconfiguration approach in more detail.

4.2 Simulation - demonstration of self-reconfiguration of hexapod robot OSCAR

In order to validate the practical applicability of the S.I.R.R. approach, we have conducted an initial simulation test scenario. The test scenario considers several robot spatial leg posture reconfigurations, after some legs of the robot have been declared as malfunctioned. For each test scenario, we start with a fully functional configuration of the robot.
The leg numbering of the robot OSCAR is as following (Fig. 8):

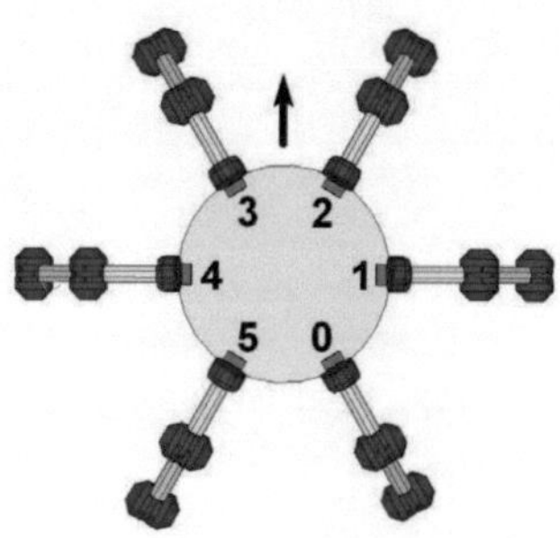

Fig. 8. Model of hexapod robot OSCAR with leg numbering. The arrow is showing the front direction of the robot.

We then assume that some randomly chosen legs have malfunctioned, and the reconfiguration of the robot is performed.
The scenario can be described as follows:

1. leg 0 has malfunctioned; reconfiguration;
2. leg 2 has malfunctioned; reconfiguration;
3. leg 3 has malfunctioned; reconfiguration;
4. leg 4 has malfunctioned; reconfiguration.

When a leg has malfunctioned, we assume that the leg is centered and it is lifted up. Then the reconfiguration takes place.

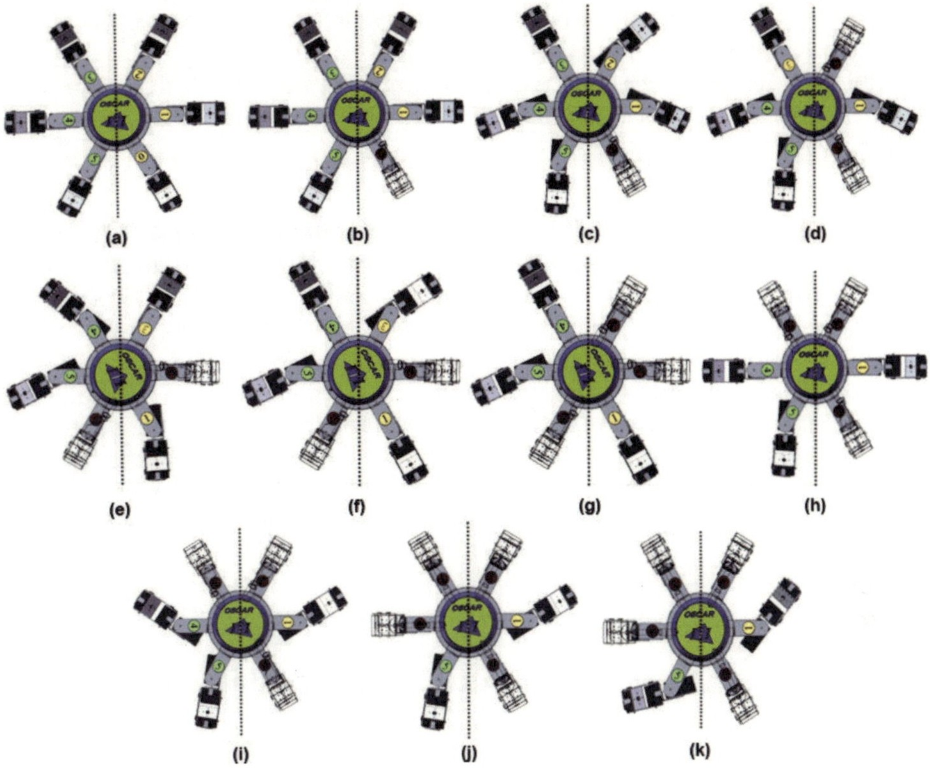

Fig. 8. Test scenario - reconfiguration of a robot using the S.I.R.R. approach; (a) Top model view of fully functional hexapod robot; (b) leg number 0 has malfunctioned; (c) after S.I.R.R. reconfiguration; (d) leg number 2 has malfunctioned; (e) S.I.R.R. reconfiguration with change of symmetry axis; (f) after S.I.R.R. reconfiguration; (g) leg number 3 has malfunctioned; (h) S.I.R.R. reconfiguration with a change of symmetry axis; (i) after S.I.R.R. reconfiguration; (j) leg number 4 has malfunctioned; (k) after S.I.R.R. reconfiguration; the dotted line represents the robot's line of symmetry.

The results of the simulated reconfiguration including the intermediate steps are represented in Fig. 8. The test scenario starts with a fully functional hexapod robot, Fig. 8(a). Then the leg 0 of the robot malfunctions, Fig. 8(b). After this, the S.I.R.R. reconfiguration goes into action and the spatial reconfiguration of the robot's legs is performed. Fig. 8(c) represents the robot's spatial configuration of the legs after the S.I.R.R. reconfiguration was done. After this, leg number 2 malfunctions, Fig. 8(d) S.I.R.R. reconfiguration is performed, and the direction of the axis of symmetry of the robot is adjusted, Fig. 8(e). The result of S.I.R.R. reconfiguration and new spatial arrangement of the legs is represented in Fig. 8(f). In Fig. 8(g), leg number 3 malfunctions; S.I.R.R. reconfiguration is performed, and the direction of the axis of symmetry of the robot is adjusted, Fig. 8(h). Reconfigured robot using S.I.R.R. is shown in Fig. 8(i). In Fig. 8(j), the robot's leg number 4 malfunctions. The final result after robot reconfiguration using S.I.R.R. is shown in Fig. 8(k). Although the algorithm has correctly distributed the legs, however the robot's stability is drastically impaired due to the only two legs configuration.

In comparison with the previous reconfiguration experiment of the robot, the results from the reconfiguration experiment using the S.I.R.R. approach show a better spatial reconfiguration of the robot's legs, in the sense of acquiring stability for the robot when a leg has malfunctioned, and in that way, enabling the robot to continue with its mission tasks, even in cases when it has mechanical failures in its legs.

4.3 Real case demonstration of self-reconfiguration of hexapod robot OSCAR

Initial positive simulation experimental results done with the SIRR method have motivated us to proceed with additional real robot experiments in which the goal is to perform in-situ real time hexapod robot reconfiguration with leg amputations and enable the hexapod robot to continue with its mission despite the malfunctioned legs. For achieving this requirement we have used the already introduced innovative robot leg amputation mechanism which enables the robot on demand to amputate the malfunctioned leg. When the monitoring unit in the robot's architecture detects that there is an anomaly present within the leg, it sends a control signal to ejection mechanism located on the robot's leg to initiate a leg ejection, i.e. to amputate the malfunctioning leg and then after to reconfigure the spatial positioning of the robots legs to

We have conducted the following demonstration scenario and simulation of leg defects:

1. First leg numbered 3 becomes malfunctioned and the robot performs SIRR reconfiguration;

2. Second leg number 1 becomes malfunctioned and the robot performs SIRR reconfiguration;

3. Third leg numbered 5 becomes malfunctioned and the robot performs SIRR reconfiguration;

This is represented in Fig. 9 (a) - (l).

As can be seen in the Fig. 9 (a), the robot starts with the initial six leg configuration. In the first fault case, leg number 3 becomes malfunctioned and the robot control architecture sends a signal to the leg amputation mechanism to amputate the leg number 3. This is shown in Fig. 9 (b)
After that the robot performs self-reconfiguration using the SIRR approach - Fig. 9 (c) and continues with its mission.
In the second fault case, leg number 1 becomes malfunctioned - Fig. 9 (d) and gets amputated - Fig. 9 (e) After that the robot performs self-reconfiguration - Fig. 9 (f) and continues with walking. In Fig. 9 (g) the third leg, number 5 becomes malfunctioned and gets amputated - Fig. 9 (h). After that the robot performs self-reconfiguration using the SIRR approach - Fig. 9 (i) and continues with walking - Fig. 9 (j) - (l).

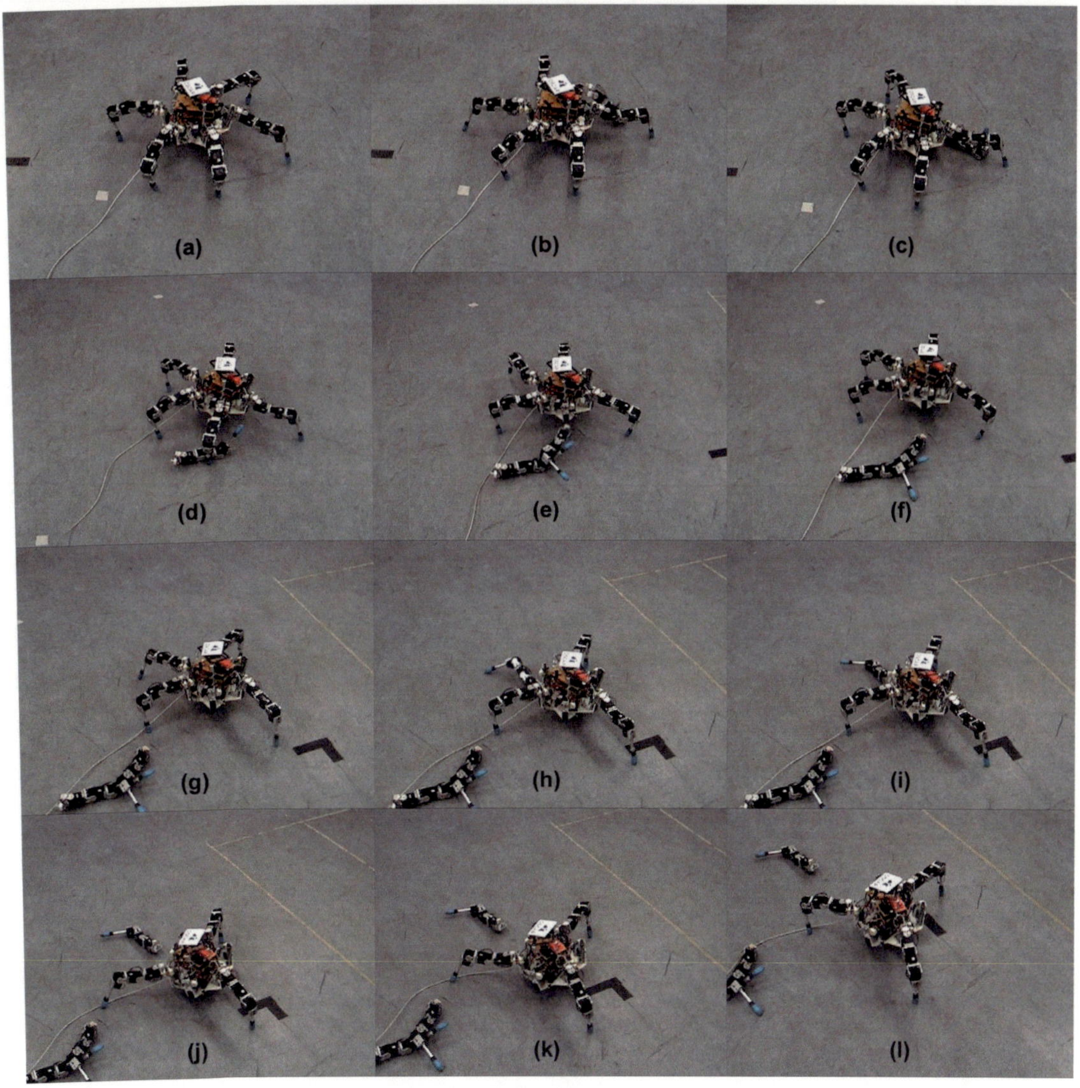

Fig. 9. Runtime reconfiguration of a hexapod robot OSCAR from 6 to 3 legs: (a) normal six legged configuration; (b) leg number 3 is malfunctioned and gets amputated; (c) robot performs reconfiguration using the SIRR approach and continues with walking; (d) leg number 1 becomes malfunctioned; (e) leg number 1 gets amputated; (f) robot performs reconfiguration using the SIRR approach and continues with walking; (g) leg number 5 becomes malfunctioned; (h) leg number 5 gets amputated; (i) robot performs reconfiguration using the SIRR approach and continues with walking; (j)-(l) robot OSCAR continues with its mission despite the loss of 3 legs.

We have made an analysis chart representing the ground contacts of legs by normal walking and by walking with leg amputations and robot self-reconfiguration. The results of these analyses can be seen in Fig. 10, Fig. 11.

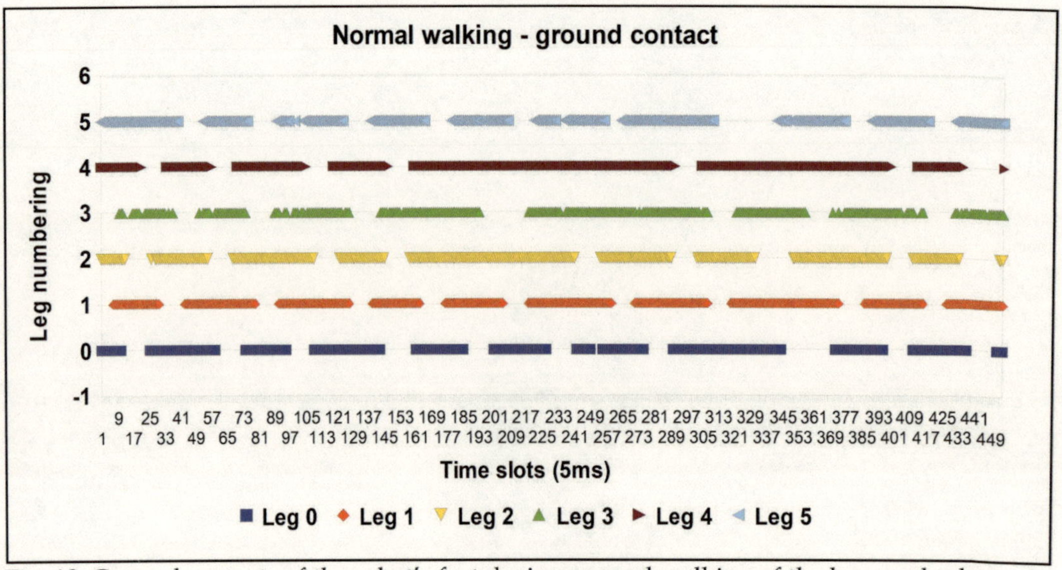

Fig. 10. Ground contacts of the robot's feet during normal walking of the hexapod robot

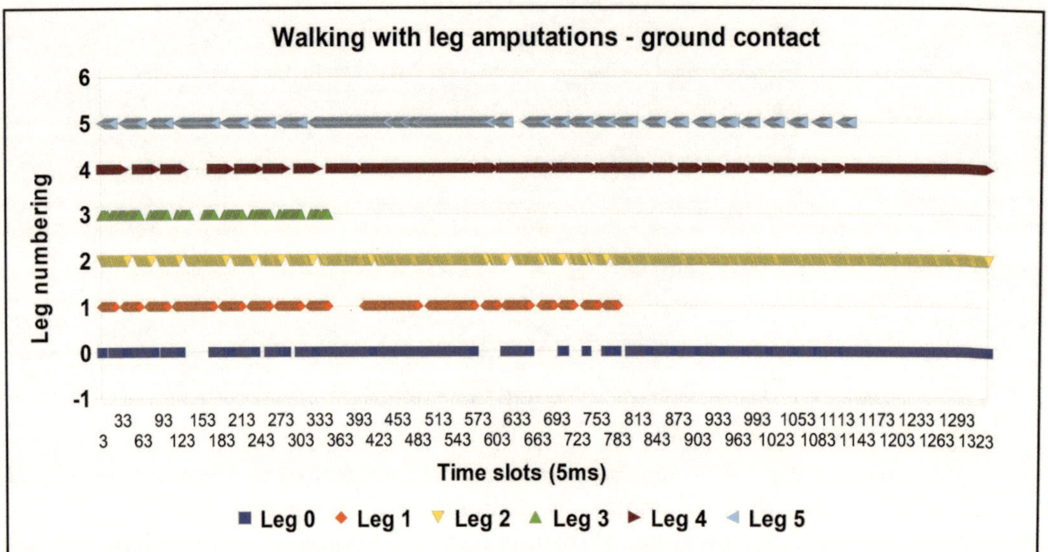

Fig. 11. Ground contacts of the robot's feet by walking of hexapod robot with leg amputations and self-reconfiguration

The robot in these experiments is walking with a biologically inspired emergent gait, which means that the gait is not "hard-wired" or by any means predefined. A simple rule is used which allows a leg to swing only if its two neighboring legs are on the ground (El Sayed Auf et al., 2006). By this, the gait pattern emerges from the local swing and stance phases of the robot's legs "joining" the "legs boid" at the particular robot's side after the reconfiguration has been performed. In Fig. 10 the chart represents the leg ground contacts for normal

walking - fully functional robot. In Fig. 11 the chart represents the leg ground contacts of the robot walking with leg amputations where we can see how the legs get amputated during the experiment, the leg ground contacts are lost and the robot still continues with walking. Leg number 3 gets amputated at time slot 335; Leg number 1 gets amputated at time slot 785; Leg number 5 gets amputated at time slot 1140. The swing phases are drastically shortened with each reconfiguration and after the time slot 1140 the robot still continues to walk although with very shortened swing phases comparing to relatively longer stance phases.
Additional measurements have been done on tracking the robot's heading while performing leg amputations and robot reconfigurations. With these measurements we wanted to test the straight walking and heading of the robot while it is performing leg amputations in different order of leg ejections and its influence on robot's walking.
The solid line in figures: Fig. 12. (a, b); Fig. 13 (a, b); Fig. 14 (a, b) represents the track of the robot during its walking. The arrow lines represent the heading of the robot. The initial heading angle is 270°.

Experiment 1:

- Fig. 12. (a) OSCAR performing leg amputations during its walking in the following order: 0, 1, 2 (in Fig. 12. a, from left to right and from up to down).

- Fig. 12. (b) Tracking of the robot's heading while the robot is amputating the legs during its walking in the following order: 0, 1, 2. The solid line represents the track of the robot during its walking. The arrow lines represent the heading of the robot during its walking.

Experiment 2:

- Fig. 13. (a) OSCAR performing leg amputations during its walking in the following order: 0, 2, 4 (in Fig. 13. a, from left to right and from up to down).

- Fig. 13. (b) Tracking of the robot's heading while the robot is amputating the legs during its walking in the following order: 0, 2, 4. The solid line represents the track of the robot during its walking. The arrow lines represent the heading of the robot during its walking.

Experiment 3:

- Fig. 14. (a) OSCAR performing leg amputations during its walking in the following order: 5, 1, 2 (in Fig. 14. a, from left to right and from up to down).

- Fig. 14. (b) Tracking of the robot's heading while the robot is amputating the legs during its walking in the following order: 5, 1, 2. The solid line represents the track of the robot during its walking. The arrow lines represent the heading of the robot during its walking.

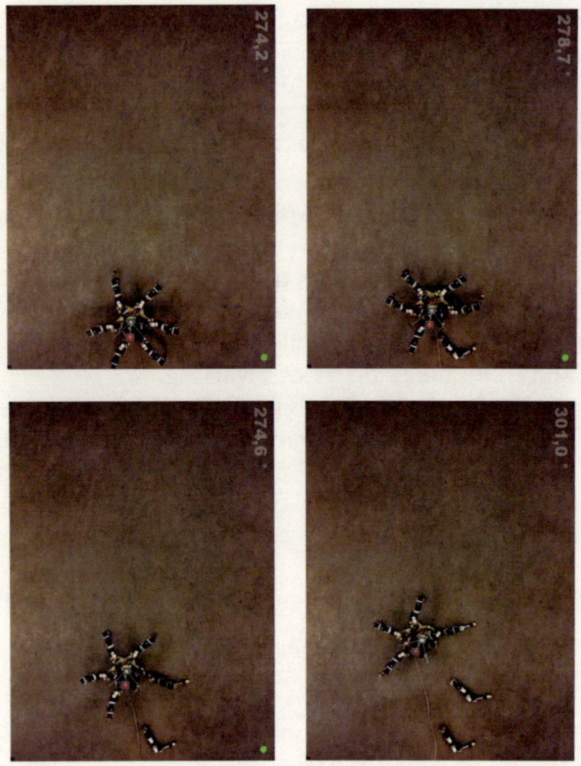

Fig. 12. (a). OSCAR performing leg amputations in the following order: fully functional, leg 0 amputated, leg 1 amputated, leg 2 amputated - from left to right and from up to down.

Fig. 12. (b) Tracking of robot's heading while the robot is ejecting the legs during its walking in the following order: 0, 1, 2.

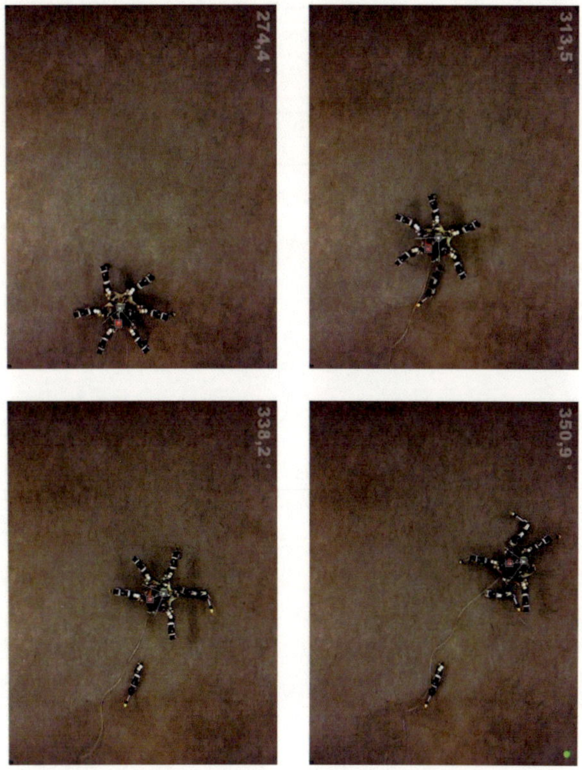

Fig. 13. (a) OSCAR performing leg amputations in the following order: fully functional, leg 0 amputated, leg 2 amputated, leg 4 amputated - from left to right and from up to down.

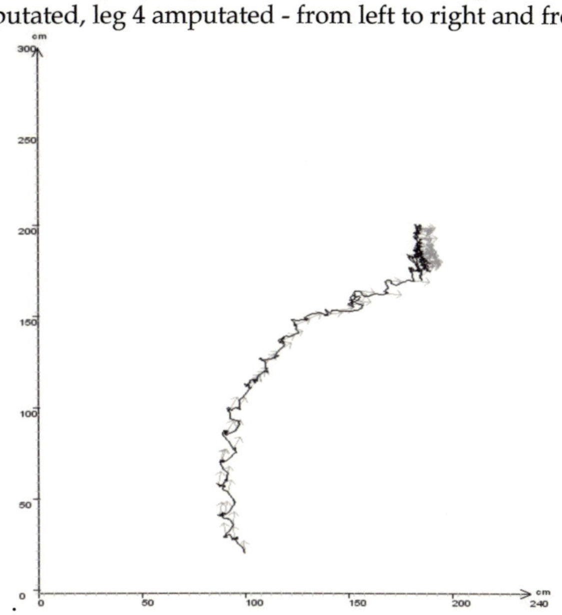

Fig. 13. (b) Tracking of robot's heading while the robot is ejecting the legs during its walking in the following order: 0, 2, 4.

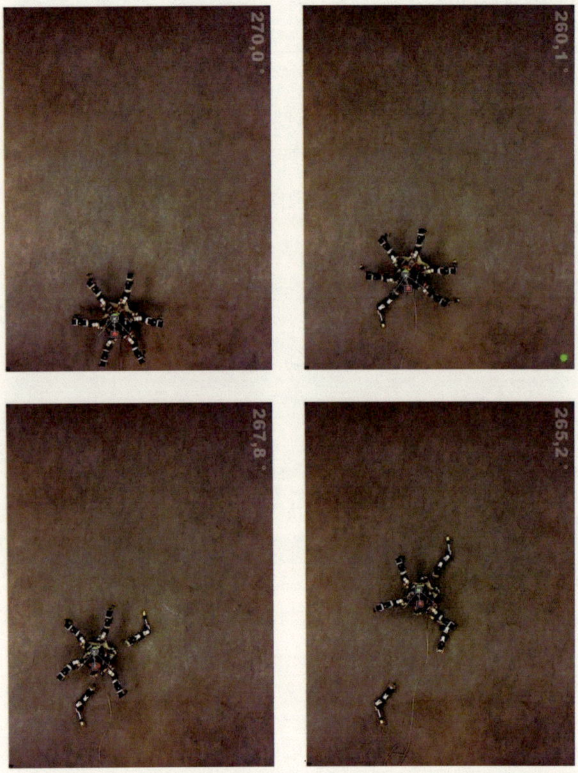

Fig. 14. (a) OSCAR performing leg amputations in the following order: fully functional, leg 5 amputated, leg 1 amputated, leg 2 amputated - from left to right and from up to down.

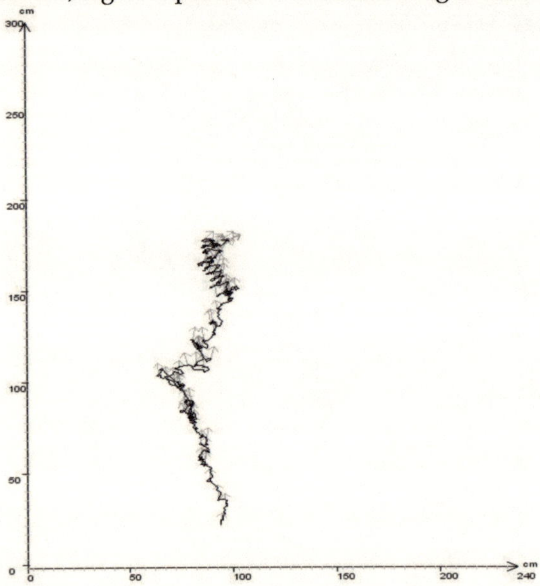

Fig. 14. (b) Tracking of robot's heading while the robot is ejecting the legs during its walking in the following order: 5, 1, 2.

On one hand it is "nice to have" a robotic system that exhibits emergent walking. However on the other hand, this kind of pure emergent walking has perhaps negative influence on how the robot is walking straight and its keeping the heading.
Despite this fact, we still wanted to measure how the robot deviates from the straight path (keeping the course to 270°) while performing the leg amputations and walking with emergent gait. The results show that even when the robot has malfunctions within its legs and performs legs amputations, it is still more or less capable to walk straight with slight turning in some cases (Fig.13). Although this deviation from course is present, we must take in account also that the robot has amputated legs and that the deviation is perhaps still not that radical - like for example: robot walking in circles immediately, or similar.
One additional idea that might be used to avoid or minimize such deviation from the main course is to couple the emergent behavior with some other behaviors like going right or left, which in that case will somehow intervene with the emergent walking gait in order to keep the robot on its course. This will be as extension to the research done on curve walking with robot OSCAR (El Sayed Auf et al., 2007). This idea will be analyzed further in future experiments done on self-reconfiguring walking robots.

5. Conclusion

In this section we have elaborated on biologically inspired methods and experiments done for real case hexapod robot self-reconfiguration. We have introduced a patent pending mechanism used for leg amputation by joint-leg walking robots which is practically used for reconfiguration cases by our hexapod robot OSCAR. Further, we have explained the artificial immune system based approach - RADE, used for monitoring the robot's health status and leg anomaly detection in joint-leg walking robot.
We have also introduced and explained the biologically inspired Swarm Intelligence for Robot Reconfiguration (S.I.R.R.) method which is used for performing in-situ robot self-reconfiguration. The S.I.R.R method is used for spatial distributing of the robot's legs when a reconfiguration is performed. So, the robot achieves a stable spatial configuration even when one or more legs are malfunctioned and get amputated from the robot's body.
Through experimental cases we have demonstrated how the hexapod robot OSCAR - despite the anomalies that occur within its legs - manages to amputate the malfunctioned legs, self-reconfigures and continues with walking. In these experiments also tracking measurements were done on tracking the robot's heading while it is performing leg amputations and self- reconfigurations.
The presented results from experiments on self-reconfiguration look promising, and therefore future work will consider an additional research on integrating self-reconfiguration with the walking robot's high-level behaviors aiming to improve the robot's heading after some reconfiguration is preformed. Additional work will be also done on improving the robustness and generic usefulness of the presented self-reconfiguration approach and its potential application for other types of robots.

6. Acknowledgment

This work is partly supported by German Research Foundation - DFG (associated to SPP 1183, MA 1412/8-1).

7. References

Brockmann W.; Maehle E. & Mösch F. (2005). Organic Fault-Tolerant Control Architecture for Robotic Applications. *4th IARP/IEEE-RAS/EURON Workshop on Dependable Robots in Human Environments*, Nagoya, Japan.

Canham R.; Jackson A. H. & Tyrrell A. (2003) Robot Error Detection Using an Artificial Immune System, *Proceedings of the 2003 NASA/DoD Conference on Evolvable Hardware*, 2003.

Cao Y. & Dasgupta D. (2003). An Immunogenetic Approach in Chemical Spectrum Recognition. *Edited volume Advances in Evolutionary Computing (Ghosh & Tsutsui, eds.), Springer-Verlag*.

Chien, S.; Doyle, R.; Davies, A.; Jonsson, A. & Lorenz, R. (2006), The Future of AI in Space, *IEEE Intelligent Systems*, pp. 64-69.

Christensen, A.L.; O'Grady, R.; Birattari, M. & Dorigo, M. (2008). Fault detection in autonomous robots based on fault injection and learning, *Journal Autonomous Robots*, Vol. 24, No. 1 (January, 2008), pp. 49-67.

De Castro L. N. & Timmis J. (2002). Artificial Immune Systems: A New Computational Intelligence Approach. *Springer-Verlag*, pp. 36-46.

De Castro, L.N. & Von Zuben, F.J. (2002). Learning and Optimization using the clonal selection principle, *IEEE Transaction on Evolutionary Computation*, pp. 239-251.

El Sayed Auf A.; Larionova S.; Litza M.; Mösch F.; Jakimovski B. & Maehle E. (2007). Ein Organic Computing Ansatz zur Steuerung einer sechsbeinigen Laufmaschine. *Autonome Mobile Systeme 2007*, pp. 233-239.

El Sayed Auf, A.; Mösch, F. & Litza, M. (2006). How the six-legged walking machine OSCAR handles leg amputations, *In Workshop on Bio-inspired Cooperative and Adaptive Behaviours in Robots*, pp. 115-124, Rome, Italy.

Forrest S.; Perelson A. S.; Allen L. & Cherukuri R. (1994). Self-Nonself Discrimination in a Computer, *In Proceedings of the 1994 IEEE Symposium on Research in Security and Privacy*, Los Alamitos, CA.

Galeano J.C.; Veloza-Suan A. & Gonzalez F.A. (2005) A comparative analysis of artificial immune network models, *Proceedings of the Conference on Genetic and Evolutionary Computation*, pp. 361-368, Washington DC, USA.

German Science Foundation (DFG) Priority Program SPP 1183 "Organic Computing", (http://www.organic-computing.de/spp), 2004.

Haldar, B. & Sarkar, N. (2006). Robust fault detection and isolation in mobile robot, *Proceedings of IFAC*, Beijing, China.

Hancher, M. D. & Hornby, G. S. (2006). A Modular Robotic System with Applications to Space Exploration, *2nd IEEE International Conference on Space Mission Challenges for Information Technology (SMC-IT'06)*, pp. 125-132.

Hinchey, M. G. & Sterritt, R. (2007). 99% (Biological) Inspiration...., *Fourth IEEE International Workshop on Engineering of Autonomic and Autonomous Systems (EASe'07)*, pp. 187-190.

Hyun Yool, K.; Lee, Y.; Choi', H.; Hyun Yool, B. & Hwan Kim, D. (2006). Swarm Robotics: Self Assembly, Physical Configuration, and Its Control, *SICE-ICASE International Joint Conference*, pp. 4276-4279.

IBM Research: Autonomic Computing, (http://www.ibm.com/research/autonomic), pp. 21- 30.

Jakimovski, B.; Litza, M.; Mösch, F. & El Sayed Auf, A. (2006). Development of an organic computing architecture for robot control, *In Informatik 2006 Workshop on Organic Computing - Status and Outlook*, pp. 145-152, Dresden.

Jakimovski, B.; Meyer B. & Maehle, E. (2008). Swarm Intelligence for Self-Reconfiguring Walking Robot, *IEEE Swarm Intelligence Symposium*, St. Louis, Missouri, USA, Sep. 21-23, 2008.

Jakimovski, B. & Maehle, E. (2008). Artificial Immune System Based Robot Anomaly Detection Engine for Fault Tolerant Robots, *5th International Conference on Autonomic and Trusted Computing (ATC-08)*, pp. 177-190, Oslo, Norway.

Lewandowski, S.M.; Van Hook, D.J.; O'Leary, G.C.; Haines, J.W. & Rossey, L.M. (2001), SARA: Survivable Autonomic Response Architecture, In *Proc. DARPA Information Survivability Conference and Exposition II*, Vol. 1, pp. 77-88.

Michelan R. & Von Zuben F.J. (2002). Decentralized control system for autonomous navigation based on an evolved artificial immune network, *Proceedings of the 2002 Congress on Evolutionary Computation*, Vol. 2., pp. 1021-1026.

Nasraoui O.; Cardona C.; & Rojas C. (2005). Using retrieval measures to asses similarity in mining dynamic web clickstreams, *Proceeding of the eleventh ACM SIGKDD International Conference on Knowledge Discovery in Data Mining KDD*, 2005.

Neal M.; Feyereisl J.; Rascuna R. & Wang X. (2006). Touch Me, I'm Fine: Robot Autonomy Using an Artificial Innate Immune System, *5th International Conference on Artificial Immune Systems*, Oeiras, Portugal.

Nino F. & Beltran O. (2002). A change detection software agent based on immune mixed selection, *Evolutionary Computation, CEC '02. Proceedings of the 2002 Congress on*, Vol. 1, pp. 693-698.

Pagnoni A. & Visconti A. (2005). An innate immune system for the protection of computer networks. *Proceedings of the 4th international symposium on information and communication technologies*, Vol.92, pp. 63-68.

Przystalka, P. (2006). Model-Based Fault Detection and Isolation Using Locally Recurrent Neural Networks, *Lecture Notes In Artificial Intelligence*, Vol. 5097, pp. 123 - 134.

Reynolds, C. (1987) Flocks, herds, and schools: A distributed behavioral model, *Comp. Graph, Vol. 21, No. 4*, 1987, pp. 25-34.

Sathyanath S. & Sahin F.(2002). AISIMAM – An Artificial Immune System Based Intelligent Multi Agent Model and its Application to a Mine Detection Problem, *1st International Conference on Artificial Immune Systems*, Canterbury, UK.

Singh C.T. & Nair S.B. (2005). An Artificial Immune System for a MultiAgent Robotics System, *Transactions of Engineering, Computing and Technology*, Vol.6, pp. 308-311.

Sterritt, R.; Hinchey, M.; Rouff, C.; Rash, J. & Truszkowski, W. (2006), Sustainable and Autonomic Space Exploration Missions, *2nd IEEE International Conference on Space Mission Challenges for Information Technology (SMC-IT'06)*, pp. 59-66.

Verma, V. & Simmons, R. (2006). Scalable robot fault detection and identification, *Robotics and Autonomous Systems*, Vol. 54, No. 2, pp. 184-191.

Winfield, A.F.T. (2006). Safety in numbers: fault-tolerance in robot swarms, *International Journal of Modelling, Identification and Control*, Vol. 1, No. 1, pp. 30-37.

Xuan, L.; Minglu, Z. & Wei, L. (2008). Methods to Modular Robot Design, *Second International Symposium on Intelligent Information Technology Application*, pp. 663-668.

Zhuo-hua, D.; Ming, F.; Zi-xing, C. & Jin-xia, Y. (2006). An adaptive particle filter for mobile robot fault diagnosis, *Journal of Central South University of Technology*, Vol. 13, No. 6 (December, 2006), pp. 689-693.

20

Softly Stable Walk Using Phased Compliance Control with Virtual Force for Multi-Legged Walking Robot

Qingjiu Huang
Tokyo Institute of Technology
Japan

1. Introduction

Recently, although the researches on the terrain adaptability of multi-legged walking robot have been widely performed (D. Wettergreen & C. Thorpe, 1996; T. Kubota, et al., 2000; Q. Huang, et al., 2000; Q. Huang & K. Nonami, 2000; Q. Huang & K. Nonami, 2003; K. Nonami & Q. Huang, 2003), it has not been put to be more widely practical use. This is because there are still some problems in the stable walking of multi-legged robot that need to be solved. For example, when the swing legs of robot moves, because the COG, supported weight, and moment of inertia of body change dynamically, the posture of robot body becomes unstable; furthermore, with the switch between the swing leg and the support leg, there occur the collisions and slippage between the foot and the ground. Because of the above uncertain disturbances, the tiny vibrations occur when the robot is walking. Until now, we proposed a robust control of posture and vibration based on a virtual suspension model for multi-legged walking robot to decrease the tiny vibrations when the robot walks (Q. Huang, et al., 2004; Q. Huang, et al., 2007). However, how to decrease the impact force between the foot and the terrain has not been solved yet. When the robot walks on irregular terrain or it bumps against the obstacle, due to the influence from the impact force between the foot and the ground, it is a possibility that the mechanical parts of robot are destroyed; moreover, the vibration in the robot body occurs and arouses the instability of posture. Therefore, it is necessary to decrease the impact force for the walking of the multi-legged robot.

Compliance control is one of the most effective control methods for the hand of manipulator to reduce impact force of contacting work (J. Huang, et al., 2002), because it can control relationship between the contact force and displacement of the hand. Recently, the compliance control was applied to biped walking robot (R. Quint, 1998). However, until now the compliance control is performed for decrease the vibration after impact force is generated, such as after the foot of the robot collides with the ground. It is impossible to reduce impact force perfectly as long as the compliance control is used after impact force is generated. So, counter measure which used the visual sensor to avoid object was proposed for manipulator in order to more effectively reduce the impact force (V. Mut, et al., 1998; X. Chen & H. Kano, 2005). And avoid action method that used virtual force to decelerate the motion speed of hand was proposed (T. Tsuji, et al., 1997). However, these methods don't be applied to the legged walking robot, the problem on the impact force between the foot and the terrain when the foot lifting and landing aren't solved yet. Moreover, the robot motion can only move uniformity using the current compliance control method because it always keeps constant stiffness gain and viscous gain. Therefore, the current compliance control method can realize the target motion, but it is hard to decrease impact force for multi-legged walking robot.

In this chapter, in order to realize the softly stable walk of multi-legged robot, we introduce a phased compliance control with virtual compliance force to reduce impact force between the foot and unexpected ground and obstacle (Q. Huang, et al., 2008). Moreover, we show a design of hierarchical control system for multi-legged walking robot, which is combined the proposed phased compliance control with a posture and vibration control based on a virtual suspension model, to realize the stable walking on unknown rough terrain. Finally, the effectiveness of the above introduced method is discussed using the walking experimental results of the developed six-legged walking robot.

2. A Six-Legged Walking Robot

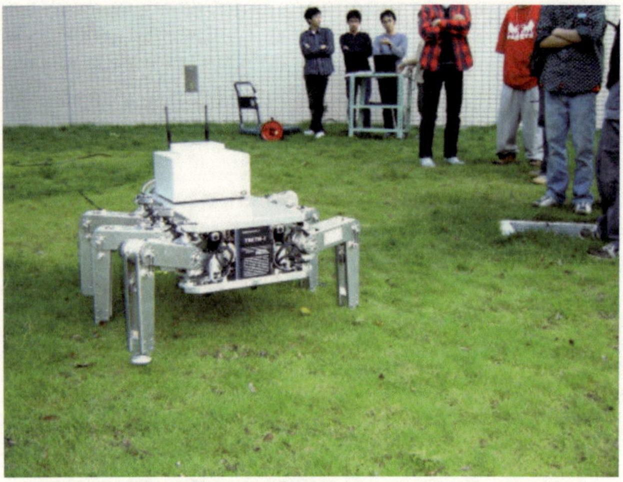

Fig. 1. The developed six-legged walking robot

Figure 1 shows our six-legged walking robot. The driving mechanism for each leg is a hybrid type mechanism composing a DC motor and a harmonic slowdown device through a rubber belt. Some accessories, such as computers, sensors, motor drive drivers and one AC power supply are mounted on the body of robot.

Figure 2 shows the schematic diagram of the robot with the detail measurements of its body. As shown, this robot is designed with three joints in each leg. By controlling the output torque of the motor for driving these three joints, the walking gait can be designed freely. θ_1~ θ_3 in Fig.1 show the rotation angles of the three joints, and each of them has a range of -90deg~ 270deg. This flexibility in foot is advantage for robot to walk on the irregular terrain. Each parameter encompassing the weight of each part of robot and the selected rated torque of the motor are shown in Table 1.

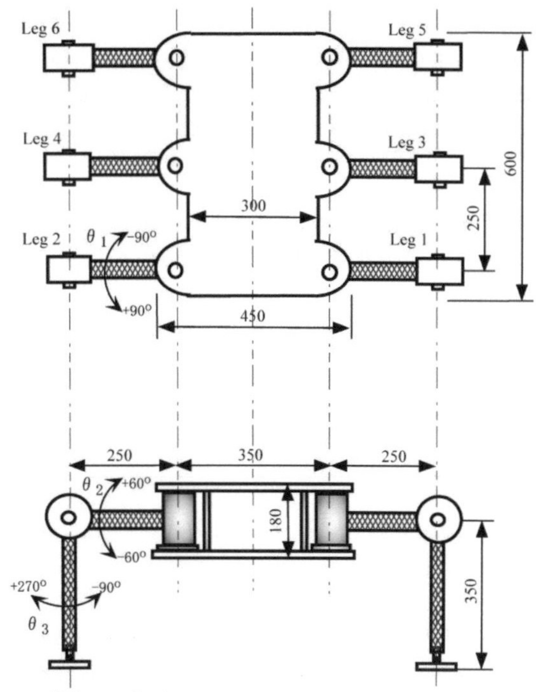

Fig. 2. Schematic diagram of the robot

Total weight [kg]	51.0
Loadable weight [kg]	40.0
Reduction ratio of harmonic drive β_1	80
Reduction ratio of harmonic drive β_2	120
Reduction ratio of harmonic drive β_3	80
Reduction ratio of timing belt γ_1	2.0
Reduction ratio of timing belt γ_1	4.0
Reduction ratio of timing belt γ_1	4.2
Rated torque of DC motor τ_{r1} [Nm]	0.14
Rated torque of DC motor τ_{r2} [Nm]	0.20
Rated torque of DC motor τ_{r3} [Nm]	0.14

Table 1. Specifications of the Robot

3. Virtual Compliance Control

3.1 Basic Controlling Expression

In this section, we introduce the conventional compliance control for the target trajectory tracking of the foot tip. The motion equation for the robot's legs with three joints can be expressed as follows.

$$\mathbf{M}(\theta)\ddot{\theta} + \mathbf{C}(\theta,\dot{\theta}) + \mathbf{D}\dot{\theta} + \mathbf{K}\theta + \mathbf{P}(\theta) = \boldsymbol{\tau} + \mathbf{J}^T(\theta)\mathbf{F} \tag{1}$$

where, θ=[θ_1 θ_2 θ_3]^T is a vector of three joints of a leg, $\mathbf{M}(\theta)$ is the inertia matrix, $\mathbf{C}(\theta,\dot{\theta})$ is the item considering centrifugal force and coriolis force, **D** is the coefficient matrix of viscous friction, **K** is the coefficient matrix of stiffness friction, **P** is the item of gravity, τ is the driving torque of the motor, $\mathbf{J}^T$ is the jacobian matrix, and **F** is the item representing the external force added to the foot tip.

According to Eq.(1), the driving input torque to the motors can be written as

$$\boldsymbol{\tau} = \mathbf{M}(\theta)\ddot{\theta} + \mathbf{h}(\theta,\dot{\theta}) - \mathbf{J}^T(\theta)\mathbf{F} \tag{2}$$

$$\mathbf{h}(\theta,\dot{\theta}) = \mathbf{C}(\theta,\dot{\theta}) + \mathbf{D}\dot{\theta} + \mathbf{K}\theta + \mathbf{P}(\theta)$$

Where, we define $\mathbf{p}_e$ as a error vector between the target trajectories and the real trajectories based on the foot's coordinates.

$$\mathbf{p}_e = \begin{bmatrix} X - X_r \\ Y - Y_r \\ Z - Z_r \end{bmatrix} \tag{3}$$

$\mathbf{p}_e$ obtained from above expression is used for compliance function between the force acting on foot. It is possible to control the foot trajectory tracking.

$$\mathbf{F} = \mathbf{C}_c\dot{\mathbf{p}}_e + \mathbf{K}_c\mathbf{p}_e \tag{4}$$

substituting Equations(3) and (4) into Eq. (2), we can obtain the control input torque as follows:

$$\boldsymbol{\tau} = \mathbf{M}(\theta)\ddot{\theta} + \mathbf{h}(\theta,\dot{\theta}) - \mathbf{J}^T(\theta)(\mathbf{C}_c\dot{\mathbf{p}}_e + \mathbf{K}_c\mathbf{p}_e) \tag{5}$$

This expression is a basic control equation for the conventional compliance control.

3.2 Virtual Compliance Force

In this chapter, we introduce to add a virtual compliance force to change the trajectory of the foot before the foot contacts the ground. The advantage of adding the virtual compliance force is that the relative velocity between the foot and the ground when landing ground becomes zero, and then the contact collision can be decreased.

Moreover, this virtual compliance is effective when leg takes off the ground. The conventional compliance control have a problem that the load of bottom of foot rapidly become zero when the foot taking off the ground, and the unbalance of the body happen rapidly too. So it need that our proposed virtual force can prevent unbalance to reduce foot load slowly. Where, we install a proximity sensor on the back side of the foot, and its effective range is set as 3 cm. Fig. 3 shows the outline model of the robot leg include proximity sensor and virtual force. Here, the proposed virtual compliance force is calculated using the **q** that is vector from root of leg to foot.

$$\mathbf{F}_v = \mathbf{C}_v\dot{\mathbf{q}} + \mathbf{K}_v\mathbf{q} \tag{7}$$

Where, $\mathbf{F}_v$ is a virtual compliance force matrix, $\mathbf{C}_v$ is a virtual viscosity gain matrix, $\mathbf{K}_v$ is a virtual stiffness gain matrix, Eq. (6) can work effectively if the foot orbit far from the ideal one, so it is able to decelerate rapidly when proximity sensor perceives the obstacle. Fig. 4 and Fig. 5 are showing a frame format of proximity sensor changing.

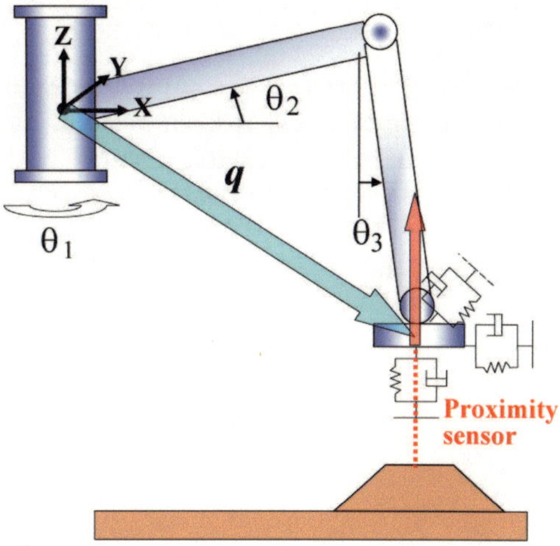

Fig. 3. Virtual compliance force

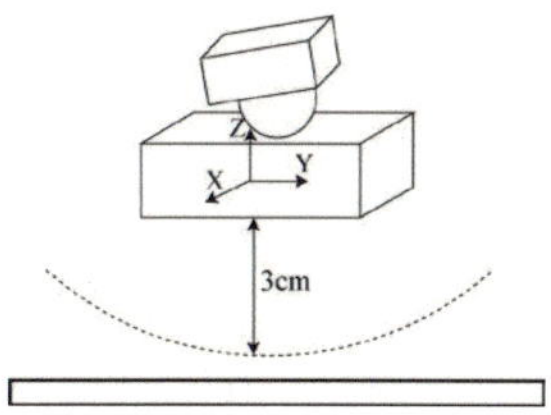

Fig. 4. Sensor off

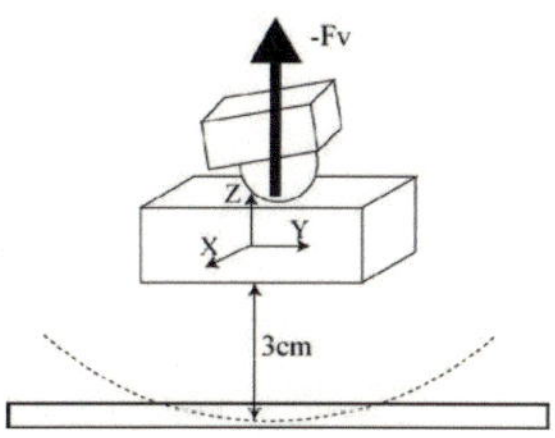

Fig. 5. Sensor on

However, this virtual force is needed while foot is near the objects and sensor is on, if the foot moves far away from the ground and obstacle, this virtual force is not only unnecessary but also disturbs a smooth track along the ideal orbit. So that, we rewrite $\mathbf{F}_v$ as follow.

$$\mathbf{F}_v = \begin{cases} 0 & \text{(sensor off)} \\ \mathbf{C}_v\dot{\mathbf{q}} + \mathbf{K}_v\mathbf{q} & \text{(sensor on)} \end{cases} \tag{6}$$

This virtual force divided in two cases to do flexible correspondence at the situation change. Concretely, when the sensor is off the virtual force $\mathbf{F}_v$ is zero, and when the sensor is on the virtual force $\mathbf{F}_v$ starts. Submitting the $\mathbf{F}_v$ into Eq. (5), we can get the control torques for the three joints of one leg of multi-legged robot as follows.

$$\boldsymbol{\tau} = \mathbf{M}(\theta)\ddot{\theta} + \mathbf{h}(\theta,\dot{\theta}) - \mathbf{J}^T(\theta)\left(\mathbf{C}_c\dot{\mathbf{p}}_e + \mathbf{K}_c\mathbf{p}_e - \mathbf{F}_v\right)$$

$$\boldsymbol{\tau} = \mathbf{M}(\theta)\ddot{\theta} + \mathbf{h}(\theta,\dot{\theta}) - \mathbf{J}^T(\theta)\left(\mathbf{C}_c\dot{\mathbf{p}}_e + \mathbf{K}_c\mathbf{p}_e - \mathbf{C}_v\dot{\mathbf{q}} - \mathbf{K}_v\mathbf{q}\right) \tag{8}$$

4. Phased Compliance Control

In the conventional compliance control, because the compliance gain matrix keep constant in all phases of the movement of the swing leg, it is impossible to response all unexpected situation walking on irregular terrain where needed stability and flexible correspondence. Therefore, in this paper, we divide the walking movement into six phases shown as in Fig. 6, the compliance gain matrix $\mathbf{C}_v$, $\mathbf{K}_v$, $\mathbf{C}_c$, $\mathbf{K}_c$ are variable in different phase.

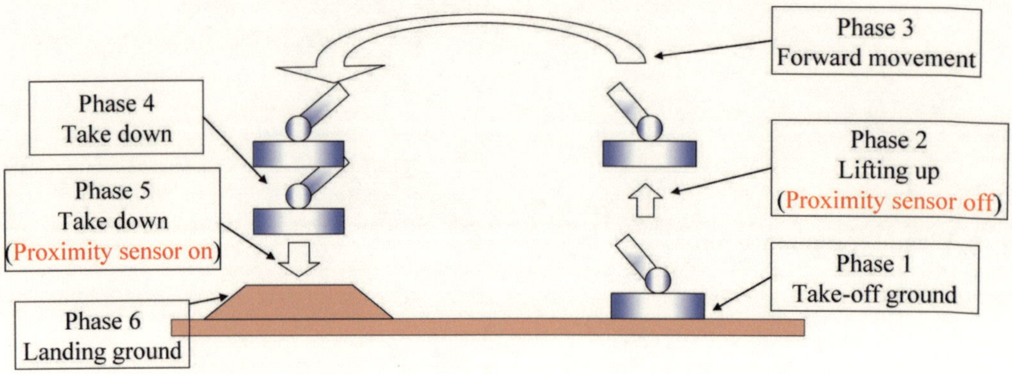

Fig. 8. Divided walking motion into six phases

Phase 1 is a period from the foot starts taking off ground until the proximity sensor becomes off. In this phase, the main purpose is to slowly decrease the force load of the foot using the virtual compliance force. By adjusting virtual compliance gain matrix $\mathbf{C}_v$, $\mathbf{K}_v$ the virtual force is generated to press the leg against ground and a smooth decrease in force load of the foot is achieved by slowly decreasing that virtual force. As a result, the generation of the impact power and a sudden balance-off of robot's body can be suppressed.

Phase 2 is a leg lifting phase that the control from the proximity sensor becomes off until end position of lifting leg. This phase is important to follow the ideal orbit to recover the phase 1's delay quickly. So the normal compliance gain $\mathbf{C}_c$, $\mathbf{K}_c$ are enlarged, and the virtual compliance gain $\mathbf{C}_v$, $\mathbf{K}_v$ make decrease to 0 by the time when the proximity sensor becomes off.

Phase 3 and 4 are the horizontal and descent movement of leg until the proximity sensor becomes on. These two phases are important to follow the ideal orbit, but horizontal movement have a danger of colliding with the obstacle toward x and y axially. As the measures, if the external force $\mathbf{F}$ is added to the foot, the normal compliance gain $\mathbf{C}_c$, $\mathbf{K}_c$ in the x and y directions becomes small to avoid the obstacle.

Phase 5 is a foot descend period from proximity sensor becomes on until the foot contacts with ground or obstacle. In this phase, deceleration of foot is put in preparation for contact with the object, and it has aimed to weaken the impact force when colliding. So that, the virtual compliance gain matrix $\mathbf{C}_v$, $\mathbf{K}_v$ are greatly taken and the sudden deceleration is enable.

Phase 6 is a control phase for the support leg that is not necessary to use virtual compliance. The normal compliance gain $\mathbf{C}_c$, $\mathbf{K}_c$ are set large because it is necessary not only to follow the orbit of the support leg but also to support the weight of the body. In this paper, to this phase 6, we use a posture and vibration control based on virtual suspension model

introduced in section 5 to generate torques for support legs to realize the stable posture of robot.
Smooth movement suitable for the situation is possible to use phased division of such phased compliance control.

5. A Hierarchical Control System Combined the Phased Compliance Control with a Posture and Vibration Control

However, because the phased compliance control is applicable only to swing legs of multi-legged walking robot, we designed a hierarchical control system for multi-legged walking robot, which is combined the proposed phased compliance control with a posture and vibration control based on virtual suspension model, to realize the stable walking on unknown rough terrain.
The control block diagram of the hierarchical control system is shown as in Fig. 9. According of this control block diagram, there are two parts in control system. The first part is a position control for swing legs using the phased compliance control designed in sections 3 and 4. The second part is posture and vibration control of robot body for support legs. In this part, the VSM and SMC are virtual suspension model and sliding mode control for constraining the robot's posture and damping the changes of its posture (Q. Huang, et al., 2004). And the CSM is Coordinate Shift Method proposed by our study, which can calculate the forces of each support leg to equilibrate the posture changes. The calculated forces are multiplied with the jacobian matrix, and are transported to the each joints of support leg. Where, we introduce the second part as follows.

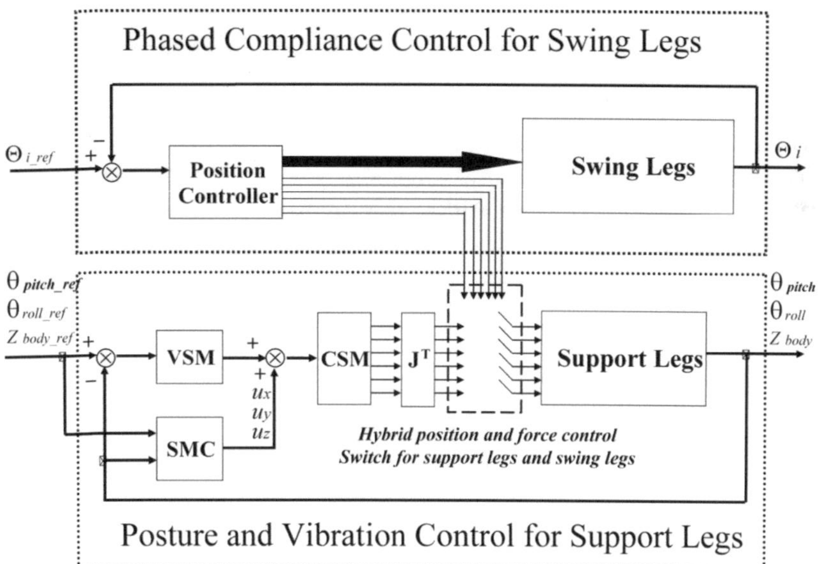

Fig. 9. A hierarchical control system combined phased compliance control with posture and vibration control

5.1 Suspension Control Using Sliding Mode Control Based on the Virtual Dynamic Model

To restrain the vibration of robot body, a suspension model is built in this study. As shown in Fig. 10, the ground is assumed to be rigid; a suspension model of one degree of freedom with virtual springs and dampers is designed in the vertical direction, the direction of the pitch angle, and the direction of the roll angle, respectively (Q. Huang, et al., 2004). In Fig. 10, the vertical direction of the robot body is defined to be the z coordinate axis, the advancing direction of the robot is defined to be the y coordinate axis, and the direction crosshatched to the above two directions is defined to be the x coordinate axis.

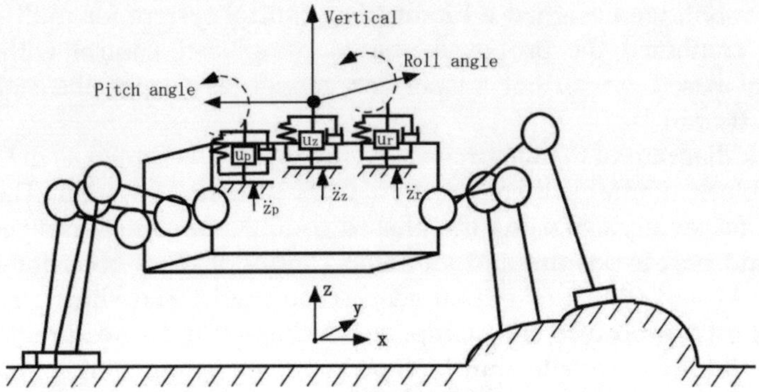

Fig. 10. The virtual suspension model considering of active control input and disturbances

Here, z, θ_p, θ_r are defined to be the changes in the vertical direction, the directions of the pitch angle and the roll angle from the balance place, respectively. In order to improve the excessive characteristic of the system response and to minify the overshot, the damping coefficient ζ is selected to be within 0.7~1.0. Besides, the natural angular frequency ω_n is enlarged as much as possible to avoid resonance because the vibrations in the vertical direction, the directions of the pitch angle and the roll angle, the disturbances from collisions and slippage between the ground and the robot leg, and dynamic changes of the supported weight and the centre of gravity are within low frequency. Furthermore, because of the enlarging of the natural angular frequency ω_n, good stability and excessive response characteristics within a wide frequency band can be realized in the virtual suspension dynamic system. Therefore, in this study, the natural angular frequency ω_n and the damping coefficient ζ are selected respectively as follows,

$$\omega_{nx} = 250 \qquad \omega_{ny} = 250 \qquad \omega_{nz} = 60$$

$$\varsigma_x = 0.800 \qquad \varsigma_x = 0.800 \qquad \varsigma_x = 0.800 \tag{8}$$

The body weight and rotary moment of inertia of the six-legged walking robot in this study are

$$I_x = 1.072[\mathrm{kgm}^2] \qquad I_y = 0.906[\mathrm{kgm}^2] \qquad \mathrm{M} = 16.753[\mathrm{kg}]$$

According of the natural angular frequency ω_n and the damping coefficient ζ, the parameters of the proposed virtual dynamic model are as follows,

$$K_Z = 167530.0[N/m] \qquad C_Z = 2680.5[Ns/m]$$

$$K_P = 67000.0[Nm/rad] \quad C_P = 428.8[Nms/rad]$$

$$K_r = 56643.8[Nm/rad] \quad C_r = 428.8[Nms/rad]$$

Where, the control input u_x, u_y, and u_z were designed by using sliding mode control (Q. Huang, et al., 2004; Q. Huang, et al., 2007; K. Nonami and H. Tian, 1994; X. Zhong, et al., 2007). So that, the motion equations for the vertical direction, the directions of the pitch angle of robot body and the roll angle of robot body are defined as

$$\left(M + \Delta M(t,\theta_{sw})\right)\ddot{z} = -K_z z - C_z \dot{z} + u_z + \ddot{z}_z \tag{9}$$

$$\left(I_x + \Delta I_x(t,\theta_{sw})\right)\ddot{\theta}_p = -K_p \theta_p - C_p \dot{\theta}_p + u_p + \ddot{z}_p \tag{10}$$

$$\left(I_y + \Delta I_y(t,\theta_{sw})\right)\ddot{\theta}_r = -K_r \theta_r - C_r \dot{\theta}_r + u_r + \ddot{z}_r \tag{11}$$

In this study, we designed the same control system for the vertical direction, the direction of the pitch angle, and the direction of the roll angle. Here, only the designed sliding mode control of one-type style servo system in the direction of the pitch angle is introduced. In the designed sliding mode control, the integral value of the difference between the target pitch angle and the actual pitch angle is defined to be a new state variable. Substituting the new state variable z_{en} into Eq. (9), an extended state equation can be derived. The extended state equation and the switching function σ are

$$\begin{cases} \dot{x}_e = A x_e + \boldsymbol{Bu} + \boldsymbol{Q} r + \boldsymbol{F} d \\ z_{en} = \int (r - \theta_p) dt \\ s = \boldsymbol{S x}_{en} \end{cases} \tag{12}$$

$$\boldsymbol{x}_e = [z_{en} \quad \theta_p \quad \dot{\theta}_p]^T \quad d = \ddot{z}_p$$

and, A, B, Q are expressed by

$$\boldsymbol{A} = \begin{bmatrix} 0 & -1 & 0 \\ 0 & 0 & 1 \\ 0 & -K_p / I_p & C_p / I_p \end{bmatrix} \quad \boldsymbol{B} = \begin{bmatrix} 0 \\ 0 \\ 1/I_p \end{bmatrix} \quad \boldsymbol{Q} = \begin{bmatrix} 1 \\ 0 \\ 0 \end{bmatrix}$$

Where, r is the target value of the pitch angle, and is zero here. d is the term denoting the disturbances. In sliding mode control, if the equivalent control input u_{eq} is without consideration of disturbances, from $\sigma = \dot{\sigma} = 0$, the linear input u_{lp} can be obtained from

$$u_{lp} = u_{eq} = -(\boldsymbol{SB})^{-1}(\boldsymbol{SAx}_e + \boldsymbol{SQ}r) \tag{13}$$

Switching matrix S is solved by using the solution of the Riccati equation. The real component of the eigenvalue of the equivalent control system is -50 here. The nonlinear input of the sliding mode control u_{nlp} is expressed by

$$u_{nlp} = -k(\boldsymbol{SB})^{-1} \frac{\sigma}{|\sigma| + \eta} \tag{14}$$

Nonlinear input u_{nlp} compensates for the uncertainty of the system, as the control input to constrain the system state variables within the switching plane. Here, the coefficient to repress the disturbances k = 8600, and the coefficient to avoid the chattering η = 0.1. Therefore, the input of the sliding mode control is u, which is composed of the linear control input u_{lp} and the nonlinear control input u_{nlp}.

6. Experiment and Discussion

6.1 Preparations for the Experiment

Because the purpose of this study is to restrain the vibration in the z direction and the directions of the pitch angle and the roll angle of the robot body when the robot walks, it is necessary to obtain the outputs in these three directions. As to the output angles in the directions of the pitch angle and the roll angle, they were measured by a slant sensor. In the vertical direction, the output can be calculated from the size of the robot body and the forces in the vertical direction of each leg. Here, in order to save the cost, we don't use the force sensor to observe the forces of each leg, rather than use the motor pseudo-torque.

6.1.1 The Observation by Using the Motor Pseudo-Torque

The motor torque is obtained by multiplying a torque coefficient to the motor electric current. However, because the vibration caused by noise is too big, instead of the motor torque, a pseudo-torque is used as the input torque. The pseudo-torque is the calculated torque of one sampling time before. In the servo electric circuit, the calculated pseudo-torque approximates to the actual consumed torque. Therefore, using the pseudo-torque, there is no the influence by the noise. Of course, a delay of one sampling time arises simultaneously. The influence caused by the delay can be ignored if the sampling time is small enough.

6.1.2 Conversion from Motor Torque to Force of the Tip of Each Leg

The force of the tip of leg, $f = [f_x \quad f_y \quad f_z]^T$, was calculated from the size of each link and the inverse of Jacobi matrix. The force of the tip of leg can be obtained as expressed as follows.

$$f = \left(J^T\right)^{-1} \tau \tag{15}$$

$$f_{asis} = \tau_2 + \frac{l_2 \sin\theta_2 + l_3 \cos(\theta_2 + \theta_3)\tau_3}{l_3 \cos(\theta_2 + \theta_3)} \tag{16}$$

$$f_z = \frac{1}{l_2 \tan(\theta_2 + \theta_3) \sin\theta_2 + \cos\theta_2} f_{asis} \tag{17}$$

$$\begin{aligned} f_y = {} & \frac{1}{\tan\theta_1}\left(\frac{\tau_1}{l_2 \cos\theta_2 + l_3 \sin(\theta_2 + \theta_3)}\right) \\ & + \frac{\tau_2}{l_3 \cos(\theta_2 + \theta_3)} - \tan(\theta_2 + \theta_3) f_z \end{aligned} \tag{18}$$

$$f_x = -\frac{\tau_1}{\sin\theta_1}\left(\frac{1}{l_2 \cos\theta_2 + l_3 \sin(\theta_2 + \theta_3)}\right) + \frac{1}{\tan\theta_1} f_y \tag{19}$$

6.2 Experimental Results

The experiments were performed by two kinds of gaits. The first is with one swing leg and five support legs called as pentapod gait; the second is with three swing legs and three support legs called as tripod gait. According of the walking experiments, the stable walking of the pentapod gait and the tripod gait were realized shown as in Fig. 11 and Fig. 12. Where, the experimental results of these two kinds of gait are introduced and discussed as follows.

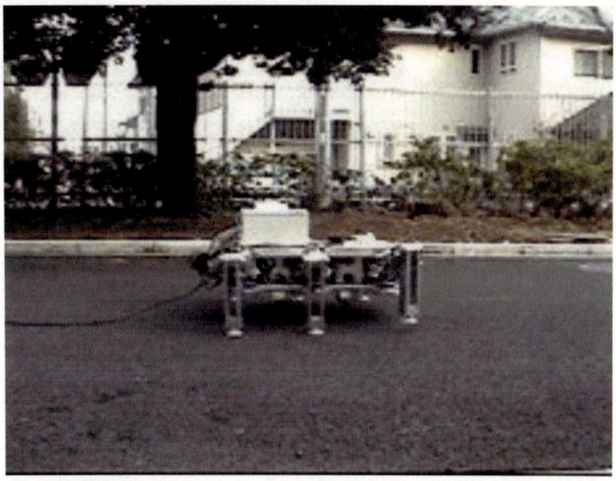

Fig. 11. Pentapod gaite walking

Fig. 12. Tripod gait walking

6.2.1 Pentapod gait

Pentapod gait is excellent in stability instead of slow the speed because always five legs are the support legs. Details of walking operation are shown Fig. 13. According to figure the legs are sequentially moved. Each parameter in pentapod gait shows as follows.

Length f tride = 180 [mm]
Max height of foot = 80 [mm]
Time of one step = 0.75 [s]

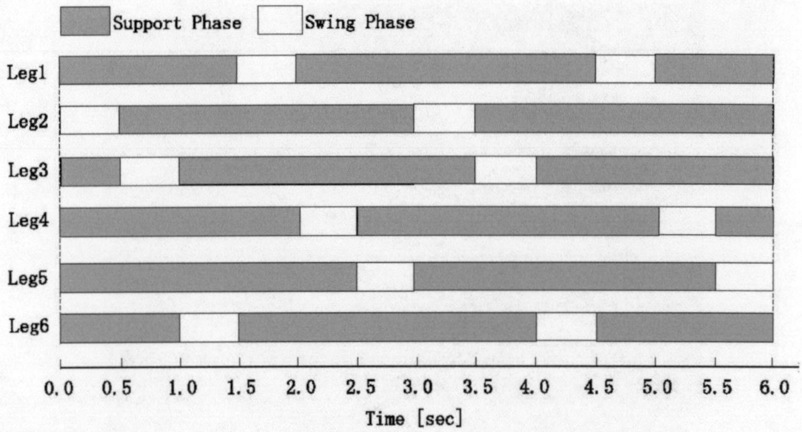

Fig. 13. Pentapod gait

The results of walking with pentapod gait are shown in Fig. 14~Fig. 19. In these figures, the dotted lines show the results of the conventional compliance control using the Eq. (5); the solid lines show the results of our proposed phased compliance control with virtual force.

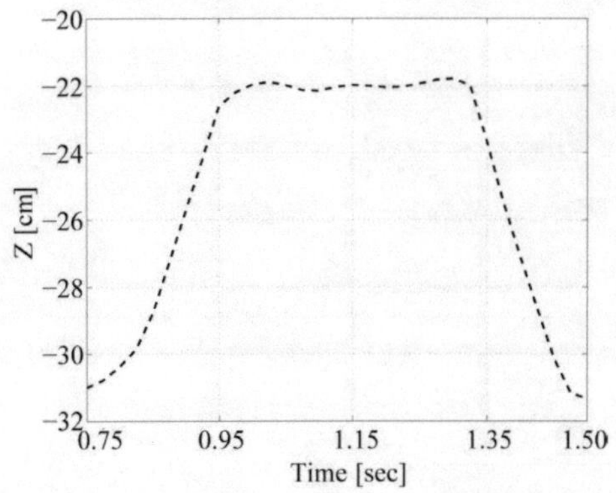

Fig. 14. Foot position change in Z direction using the conventional position control

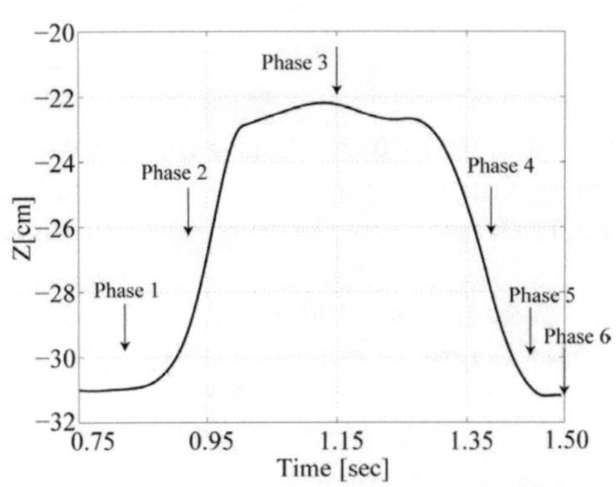

Fig. 15. Foot position change in Z direction using the proposed compliance control

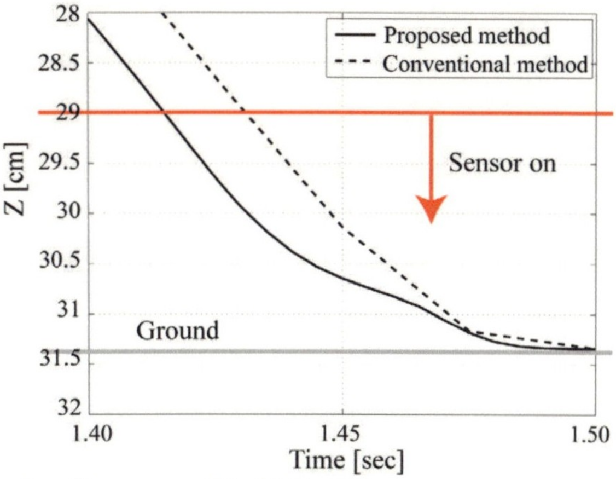

Fig. 16. Close-up of the landing ground in Z direction

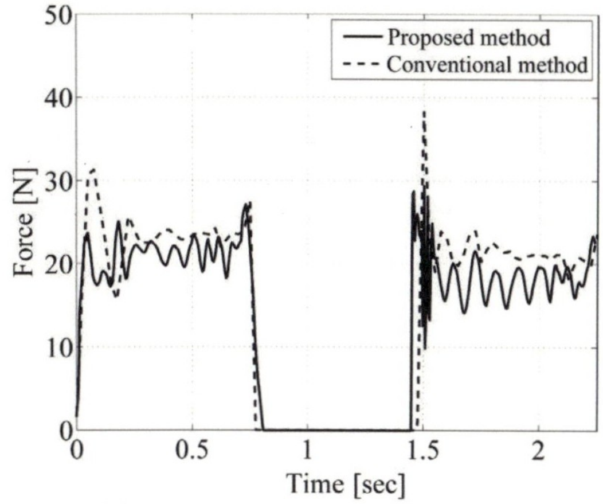

Fig. 17. Force load of the foot of leg 3

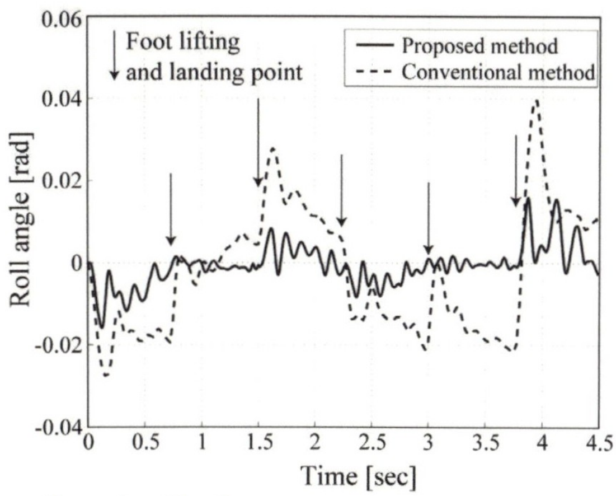

Fig. 18. Changes of the roll angle of body

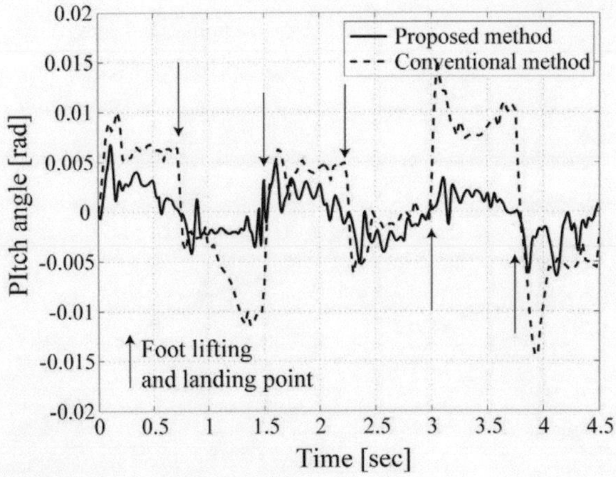

Fig. 19. Changes of the pitch angle of body

Firstly, according of Figures 14 and 15, about the changes in Z direction of the end position of the swing foot, both the proposed phased compliance control with virtual and the conventional compliance control are almost realized the ideal position finally. But in the time 0.75 sec~0.95 sec of lifting leg and 1.30 sec~1.50 sec of landing leg, the delay in the proposed phased compliance control with virtual force occurred that caused foot load slowly decreasing at phase 1 and slowing increasing at the phase 5 and phase 6 shown as in Fig. 17, so the changes of pitch angle and roll angle of the robot body become smaller than the results of the conventional compliance control shown as in Fig. 18 and Fig. 19.
The close-up of Z direction coordinates in the time 1.3 sec~1.5 sec of landing swing leg is shown Fig. 16. We can see that the proposed compliance control with virtual force generates a virtual force in the positive direction of Z direction coordinates at the time as the proximity sensor switched on.
And then, in Fig. 17, we can see the impact force between the foot and the terrain is reduced about 70% using our proposed method than the one using the conventional compliance control. And, according of Fig. 18 and Fig. 19, in the case of the proposed phased compliance control with virtual force, the changes of pitch angle and roll angle of the robot body are decreased about 50% than the results of the conventional compliance control. Especially, the biggest changes of pitch angle and roll angle of the robot body at the switch timing of lifting and landing the leg, have been decreased, because the control in phase 1, phase 5 and phase 6 work well using the virtual compliance force and impact force is decreased.

6.2.2 Tripod gait

Details of walking operation of tripod gait are shown Fig. 20. Each parameter in tripod gait shows as follows.

Length of stride = 180 [mm]
Max height of foot = 80 [mm]
Time of one step = 0.75 [s]

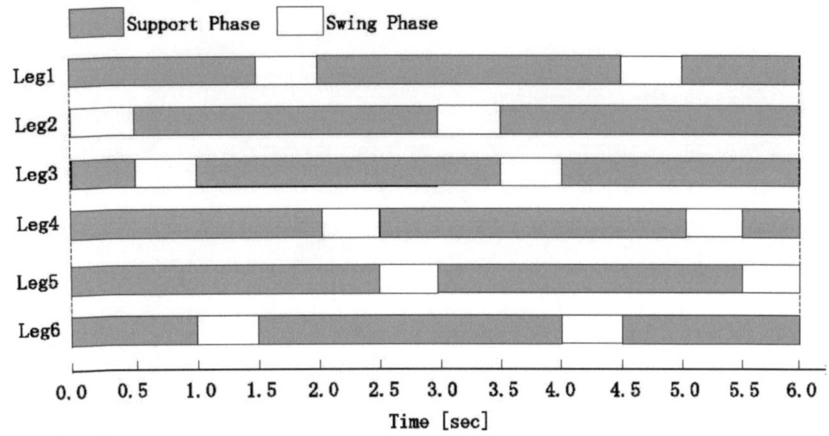

Fig. 20. Tripod gait

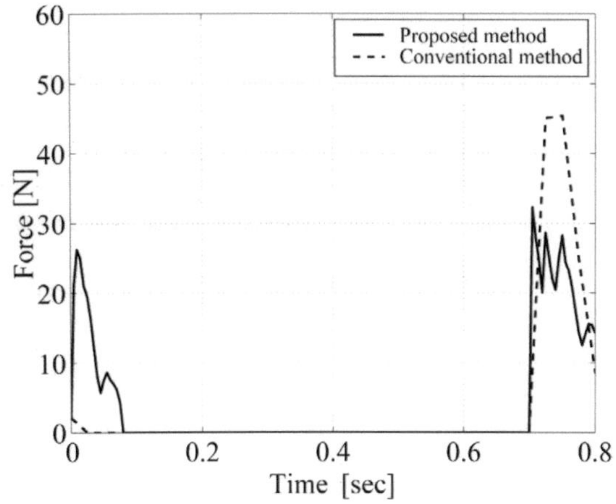

Fig. 21. Force load of the foot of leg 1

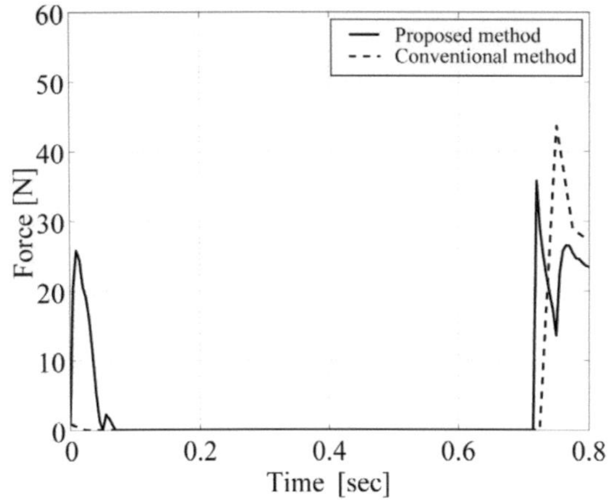

Fig. 22. Force load of the foot of leg 4

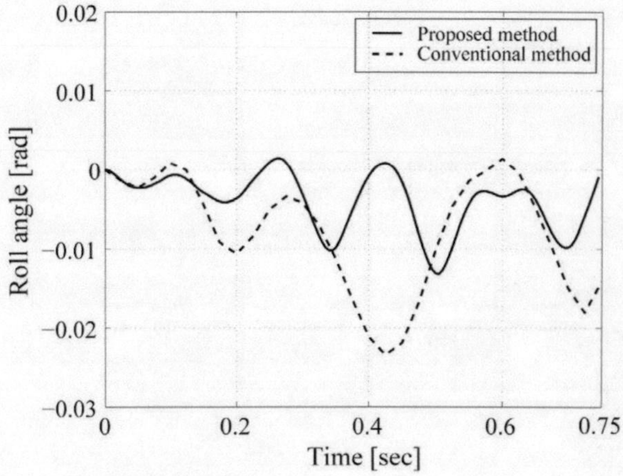

Fig. 23. Changes of the roll angle of body

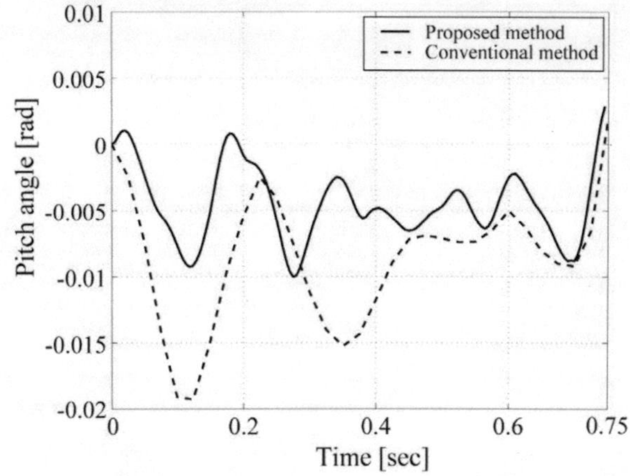

Fig. 24. Changes of the pitch angle of body

As the same results of the pentapod gait, in the time 0.0 sec~0.1 sec of lifting leg and 0.70 sec~0.78 sec of landing leg, the delay in the proposed phased compliance control with virtual force occurred that caused foot load slowly decreasing at phase 1 and slowing increasing at the phase 5 and phase 6 shown as in Fig. 21 and Fig. 22, so the changes of pitch angle and roll angle of the robot body become smaller than the results of the conventional compliance control shown as in Fig. 23 and Fig. 24. The impact force between the foot and the terrain is reduced about 70%, the changes of pitch angle and roll angle of the robot body are decreased about 50% using our proposed method comparing with the results using the conventional compliance control.

According of the above walking experimental results of pentapod gait and tripod gait, we can see the efficiency of our proposed compliance control with virtual force in aspect of decreasing the impact force between the foot and the terrain for multi-legged robot.

7. Conclusion

For the multi-legged walking robot, in order to decrease impact force between the foot and the terrain when the foot lifting and landing to realize its better stability, we did the following studies.

We introduced a phased compliance control using virtual compliance force for multi-legged walking robot, which can effectively reduce the impact force when the robot walks. Concretely, we add a virtual compliance force to change the trajectory of the foot before the foot contacts the ground based on the conventional compliance control. And then, we divide the walking movement into six phases to adjust the compliance gain to realize reducing the impact force when robot lifting and landing leg in the same walking speed as the conventional compliance control.

We designed a hierarchical control system for multi-legged walking robot, which is combined the proposed phased compliance control with a posture and vibration control based on virtual suspension model, to realize the stable walking on unknown rough terrain.

According of the walking experimental results of pentapod gait and tripod gait using a six-legged robot, the impact force between the foot and the terrain is reduced about 70%, the changes of pitch angle and roll angle of the robot body are decreased about 50% using our introduced method comparing with the results using the conventional compliance control, so the effectiveness of the proposed method for multi-legged walking robot was verified.

8. References

D. Wettergreen, C. Thorpe (1996). Developing Planning and Active Control for a Hexapod Robot, *Proc. 1996 IEEE Int. Conf. on Robotics and Automation*, pp. 2718-2723.

T. Kubota, H. Katoh, I. Nakatani (2000). Walking Rover with Multiple Legs for Planetary Exploration, *Proc. of the Third Int. Conf. on Climbing and Walking Robots*, pp. 795-788.

Q. Huang, K. Nonami, etc. (2000). CAD Model Based Autonomous Locomotion of Quadruped Robot by Using Correction of Trajectory Planning with RNN, Special Issue on Frontiers of Motion and Vibration Control, *JSME International Journal*, Series C, Vol. 43, No. 3, pp. 653-663.

Q. Huang, K. Nonami (2002). Neuro-Based Position and Force Hybrid Control of Six-Legged Walking Robot, Special Issue on Modern Trends on Mobile Robotics, *Journal of Robotics and Mechatronics*, Vol. 14, No. 4, pp. 534-543.

Q. Huang, K. Nonami (2003). Humanitarian Mine Detecting Six-Legged Walking Robot and Hybrid Neuro Walking Control with Position/Force Control, Special Issue on Computational Intelligence in Mechatronic Systems, *Mechatronics*, Elsevier, Vol. 13, No. 8-9, pp. 773-790.

K. Nonami, Q. Huang, etc. (2003). Development and Control of Mine Detection Robot COMET-II and COMET-III, *JSME International Journal*, Series C, Vol. 46, No. 3, pp. 881-890.

Q. Huang, M. Yanai, K. Oon, K. Nonami (2004). Robust Control of Posture and Vibration Based on Virtual Suspension Model for Six-Legged Walking Robot, *Proceedings of the 7th international conference of Motion and Vibration Control*, Washington University in St. Louis, America, CD-ROM, No. 41.

Q. Huang, Y. Fukuhara, X. Chen (2007). Posture and Vibration Control Based on Virtual Suspension Model Using Sliding Mode Control for Six-Legged Walking Robot, Special Issue on New Trends of Motion and Vibration Control, *Journal of System Design and Dynamics*, the JSME Dynamics, Mesurement and Control Division, Vol. 1, No. 2, pp. 180-191.

J. Huang, I. Toda, M. Matsuura (2002). Control a Redundant Robot Using Visual Information: Avoidance Control of a Moving Obstacle, *Transactions of the Japan Society of Mechanical Engineers*, Series C, Vol. 68, No. 674, pp. 2999-3006.

R. Quint (1998). Active leg compliance for passive walking, *Proceedings of the 1998 IEEE International Conference on Robotics and Automation.*

V. Mut, O. Nasisi, R. Carelli, B. Kuchen (1998). Tracking Adaptive Impedance Robot Control With Visual Feedback, *Proceedings of the 1998 IEEE International Conference on Robotics and Automation.*

X. Chen, H. Kano (2005). Dynamic Vision Based Motion Recovery, *International Journal of Innovative Computing, Information and Control*, Vol. 1, No. 3, pp. 509-525.

T. Tsuji, H. Akamatsu, M. Kaneko (1997). Non-Contact Impedance Control for Redundant Manipulators Using Visual Information, *Proceedings of the 1997 IEEE International Conference on Robotics and Automation.*

Q. Huang, K. Oka, X. Chen (2008). Phased Compliance Control with Virtual Force for Six-legged Walking Robot, Special Issue on New Trends in Advanced Control and Applications, *International Journal of Innovative Computing, Information and Control*, Vol. 4, No. 12, pp. 3359-3373.

K. Nonami, H. Tian (1994). Sliding Mode Control, CORONA Publishing Co., Ltd. (In Japanese)

X. Zhong, H. Xing and K. Fujimoto (2007). Sliding Mode Variable Structure Control for Uncertain Stochastic Systems, *International Journal of Innovative Computing, Information and Control*, Vol. 3, No. 2, pp. 397-406.

21

Biohybrid Walking Microrobot with Self-assembled Cardiomyocytes

Jinseok Kim[a], Eui-Sung Yoon[a] and Sukho Park[b]
[a]*Nano-Bio Research Center, Korea Institute of Science and Technology*
[b]*School of Mechanical Systems Engineering, Chonnam National University*
Republic of Korea

1. Introduction

1.1. Introduction to Microactuator

Since the late 1980s, due to the rapid progress in MEMS (Micro Electro Mechanical System) technology, various types of mechanical structures, micro size sensors and actuators, could be fabricated, and technological progress in VLSI and SMT (Surface Mounting Technology) has contributed to the miniaturization of many electronic parts, which has led to the active development of microrobots. With the development of microrobots, research on the integration technology of micro actuators and micro sensors for microrobots has been in strong demand. Previously developed electric motors have advantages of over 90% high conversion efficiency and handling convenience. Miniaturized conventional actuators can be used as actuators of microrobot, but miniaturization rapidly decreases the drive force. Therefore, to move very small size equipment, a small actuator device based on a new actuation principle is needed [1].

Recently, MEMS micromachining of SMA (Shape Memory Alloy), Piezoelectric Ceramic Material and functional polymer material has been used to develop a new actuator. This new actuator has drawn attention as a suitable actuator for a microsystem. This new actuator can be miniaturized and made lighter, and especially, its generating power per weight ratio is much higher than those of conventional electromotive and hydraulic actuators. Therefore, it is expected to be suitable for miniature robots and microrobots [2].

Many other small and light actuators can produce much power. Because their structure materials and drive principles are diverse, their suitability of miniaturization must be evaluated, and their characteristics must be fully understood with careful consideration of the applications.

Electrostatic and electromagnetic motors are representative micro actuators. Because their generated powers are proportioned to the area of the electrode, they are more advantageous in miniaturization than previous actuators, whose powers are proportioned to the volume.

Therefore, US-centered research has been actively developing micromotors, whose revolving part is smaller than the diameter of a hair strand by using silicon micromachining. Also, physical and chemical phenomena can be used as a micro actuator. For example, thermal expansion, piezoelectric effect, electric chemical reaction in addition to static electricity can be adopted as new actuation methods. Also, because each drive method has its own merits, the drive method that is suitable for meeting the actuation power requirements of the applied microsystem can be selected, so the most suitable design of the drive mechanism is needed. The basic requirements for micro actuators to be applied effectively to micro robots are as follows: compactness, light weight, high power density, high efficiency, good controllability, etc.

Polymers have many attractive characteristics; they are lightweight, inexpensive, fracture tolerant, and pliable. During the last twenty years, new polymers have emerged an additional important characteristic. This capability of the electroactive polymers (EAPs) has attracted much attention. Since they behave very similar to biological muscles, EAPs have acquired the moniker "artificial muscles." In the foreseeable future, robotic mechanisms actuated by EAPs will enable engineers to create devices previously imaginable only in science fiction [3].

Conventional microactuators use electrostatic, electromagnetic, pneumatic, piezoelectric, or thermal forces and require external power sources for operation. Although these actuators have been used in micro- and nano-scale systems, they have limitations in independent operation, such as in locomotion in a human digestive organ or migration in blood vessels. Unlike conventional actuators, cell-based actuators can use the glucose in physiological fluids as an energy source by converting glucose into ATP, and transforming this chemical energy into mechanical energy. Integrated systems like these cell-based actuators have been called hybrid systems, representing the material characteristics of micro- or nano-scale fabricated structures and biological components such as proteins and cells [4-5].

Actuator type	Max. Pressure (MPa)	Max. Efficiency (%)
Dielectric Elastomers	16.2	60~80
Electrostrict Polymer	15	-
Electromagnetic (Voice coil)	0.1	> 90
Piezoelectric	110~131	> 90
PVDF	4.8	-
SMA	> 200	< 10
Mechano-chemical polymer (Polyelectrolyte)	0.3	30
Natural Muscle (Human skeleton)	0.35	> 35
Natural Muscle (Mouse cardiac myocytes)	0.00905	-

Table 1. Comparison with the capabilities of various microactuators [3]

Natural muscles are considered highly optimized systems since they are fundamentally the same for all animals and the difference between species are small. The operation of muscles depends on chemically driven reversible hydrogen bonding between two polymers, actin and myosin. The reason why this paper chose the cardiac muscle of rat as an actuator for microrobot is that the cardiac muscle does not need external stimulation.

1.2. Muscle Powered Microrobot

Biomimetic refers to human-made processes, materials, devices, or systems that imitate nature. Billions of years of "natural R&D" have resulted in effective, optimized biological solutions that really work. By studying and mimicking nature's processes and structures, scientists and engineers can develop nature inspired solutions that are far more effective than solutions conceived and developed exclusively by man. Biomimetic artificial machines built with hybrid components (materials partly synthetic and partly biological in origin) offer the opportunity to combine enhanced sensitivity with robustness and the possibility to extend their application to diverse environmental conditions [6]. As examples of micro/nano machine using hybridization between organic and inorganic components, researches on adenosine triphosphate (ATP) biomolecular motors [7,8], a microorganism carrier in microchannel [9], a walking microdevice driven by micromuscle [10,11], and a pump actuated by cultured cardiomyocytes [12] were reported. Although, bimolecular motors are interesting, they can generate only between 5-60 pN forces [13] and are not robust to actuate microstructures. Micromuscles can be an alternative to microactuator for a micro-sized biomimetic system. Previous studies using cell based actuators [11-12] have shown very attractive results, but for advanced biomimetic systems, understanding and exploiting higher-order assemblies for micromuscles are key points of today's quest [6]. Structure and functional changes ensue in cardiac cell networks when cells are guided by three-dimensional scaffold topography, such as enhanced actin cytoskeleton organization, higher nuclear eccentricity [14] and altering gene expression, protein localization [15-16], cell signaling [17] and the intracellular calcium dynamics [18]. These topology-induced changes are expected to enhance the mechanical activity of cells [14]. However, until now, there has been no suitable tool to validate this enhancement.

We propose a biomimetic micromachine made of a silicone elastomer polydimethylsiloxane (PDMS) and self organized cardiomyocytes, which allows free motions in physiological liquids based on the increase of cell power in a 3D microenvironment. The fabrication method and the formation and function of the legs in the machines mimic the bottom-up process of nature, the patterns in jelly fish and the functions of a real heart. Fig.1 shows the microrobot structure mimicking the legs of a jelly fish.

To make it move like the synchronized motion of a jellyfish without external power supply, we primarily cultured the heart cells of a mouse on the microrobot device and used the beating motion of the heart cells as a power source. Also, to mimic the muscle bundle structure on the legs of a jellyfish, we made in vivo-like grooved surfaces. From this, we could obtain motions of the machine driven by the contractility of 3D controlled cardiomyocytes. We found the obvious increase of mechanical activity of cardiomyocytes in the grooved surfaces compared with that in flat surfaces by using culturing cardiomyocytes on flat and grooved PDMS microcantilevers. We can suggest the increase of force due to the arrangement and increment of the cells in the same area. The process and materials used in this study is noninvasive to cells, and have the potential of high throughput of the microdevice.

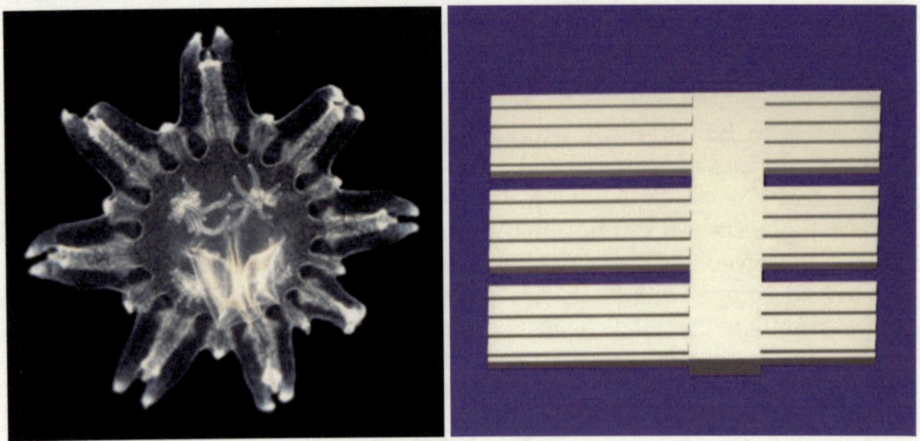

Fig. 1. A. Concept design to mimic the Ephyra of the moon jelly fish. Muscle bundles in middle of legs are used for movement. B. Mimicked polymer based backbone structure of a microrobot

2. Biopolymer Hybrid Devices

2.1. Design of biopolymer actuator & sensor

2.1.1. Flat surface cantilever

In this study, a microcantilever was designed and fabricated, using both synthetic and natural materials. Living cells were cultured on a cantilever made of flexible transparent PDMS elastomer. Therefore, the proposed cantilever can be considered as a hybrid biopolymer cantilever. PDMS has a low Young's modulus, which can be tuned from 360 to 868 kPa by controlling the ratios of its components [19], and thus is extremely sensitive to external stress and can be used to achieve large deflections. PDMS is also inexpensive, optically transparent and suitable for use with optical detection methods. Additionally, it is compatible with uses in biological studies because it is impermeable to water, nontoxic to cells, and permeable to gases. Finally, it can be easily fabricated and bonded to other surfaces [20]. Culturing primary cardiomyocytes on the microcantilever leads to self-organization of the cells on the structure, giving a parallel arrangement of cells in the hybrid system. The system can thus avoid problems arising from inefficient dissection and attachment of the muscle tissue to the microsystem by hand. Since the presented system was realized with simple micromolding a technique using PDMS, a large quantity of samples can be produced cheaply and easily. Our method is also safe and noninvasive, because the entire device is biocompatible and there are no hard contacts or structural constraints that can affect myocytes function or morphology. The important feature in our system over previous contractile force measurement techniques is that our system can quantitatively measure contractile forces on a specific microsized area in real time, and this capability should open opportunities for better understanding of the mechanisms of heart failure and promote further design of optimal microscale hybrid biopolymer actuators and microdevices.

2.1.2. Cell culture

The cardiomyocytes on the microcantilevers were obtained by aseptically isolating the heart of a neonatal Sprague–Dawley rat at day 1 and briefly washing it with Hank's balanced salt solution (Gibco Invitrogen Co., Grand Island, NY). After removing the ventricles, the remaining tissues were minced and incubated in a 0.3 mg/mL collagenase solution containing 0.6 mg/mL pancretin (Sigma Chemical Co.). The isolated cardiomyocytes were seeded directly onto the hybrid biopolymer cantilever at a cell density of 5_103 cells/mm2 and cultured in Dulbecco's modified Eagles' medium (Gibco Invitrogen) containing 10% fetal bovine serum (Sigma), 50 mg/mL streptomycin, and 50 mg/mL penicillin (Gibco Invitrogen) at 37 1C in air with 5% CO2. The cell activation level reached its maximum after 72–96 h of culture. The motion of the microcantilevers was measured at the 96th hour by using two microscopes,which monitored lateral and vertical motions, as described in Park et al. (2005) [21].

2.1.3. Molding master fabrication

By using a flexible and thin PDMS matrix layer, we intended to maximize the bending displacement of thin PDMS by generating a force through the arrangement of contractile cardiomyocytes. The flexible and thin PDMS matrix layer was fabricated by the sandwich molding process [22], as shown in Fig. 2. The sandwich molding process is a useful technique and was used to create a PDMS 3D microrobot structure with an ideal thickness of ,20 mm by pouring PDMS between a bottom Si wafer master and a top glass wafer master and then compressing the layers. The bottom master was fabricated on the Si wafer using a deep silicon etching process and a thick negative photoresist (PR) (SU-8). First, grooves were patterned onto a new silicon wafer using an AZ1512 PR. The Si wafer was then deeply etched to fabricate a grooved surface, as shown in Fig. 1(a). After removing the AZ1512 PR, the Si wafer was coated with SU-8 to develop the legs of the microrobot, as shown in Fig. 1(b). The microrobot legs were fabricated from the bottom Si master so that the grooved shape would be on the leg surface. The microrobot body was fabricated from the top glass master. After coating the glass wafer with a Cr/Au layer, we patterned the microrobot body using the AZ1512 PR, as shown in Fig. 1(c). Finally, the top master fabrication was completed by removing the Cr/Au layer from the etched glass wafer, as shown in Fig. 1(d).

2.1.4. Polymer device fabrication

Several previous reports have investigated the fabrication of 3D structures with polydimethylsiloxane (PDMS). The most common molding method is the sandwich molding technique using a single molding master [22, 23] In this method, the master, another flat wafer, and the PDMS are all stacked together to produce the 3D PDMS structures. Unfortunately, this method only allows specific geometries to be created on one side. Although a method for fabricating microstructures on both sides has been developed, it does not easily yield thin micromembranes and is difficult to use for wafer-level fabrication since the PDMS used as the top master is too flexible for high-pressure applications and may shrink during curing, particularly over areas greater than 1 cm^2. To overcome these problems, we propose a 3D molding aligner system that can align the two mold masters and stack them under high pressure to fabricate a 3D structure with micromembranes at the wafer level. To facilitate the detachment of the PDMS structures, the top and bottom masters were coated by plasma polymerization of C_4F_8 using inductively coupled plasma-reactive ion etcher (ICPRIE). The PDMS mixture was poured onto the bottom master, as shown in Fig. 1(e). After the top master was placed on the PDMS mixture, the stack, comprised of the bottom master, the PDMS, and the top master, was placed between two aluminum plates. To align the two masters using the 3D mold aligner, the geometries of the two masters were observed through the windows at the top of the wafer chuck. The two masters could be aligned to a precision of less than 2 mm using an optical microscope. Following the alignment, the stacked bottom master, PDMS, and top master were clamped, as shown in Fig. 1(f). The clamped stack was cured in an oven for 2 h at 100 °C. After curing, a thin PDMS replica was peeled from the master, as shown in Fig. 1(g).

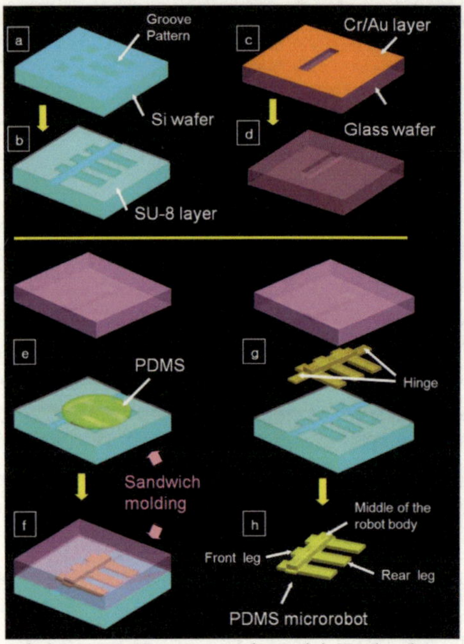

Fig. 2. Fabrication process of the PDMS microrobot. (a) Etched groove pattern on the bottom Si master. (b) Coated and patterned SU-8 layer on the bottom Si master. (c) Patterned Cr/Au layer on the top glass master. (d) Etched pattern on the top glass master. (e) Poured PDMS between the top and bottom masters. (f) Clamped top and bottom masters. (g) Detaching the molded PDMS

Figure 3 shows the concept of the process for making a hybrid device by primarily culturing neonatal rat heart cells on a micro device made with PDMS polymer. As reason for making a hybrid device and its working principle were previously explained, and the detailed fabrication method and process will be directly explained in this chapter.

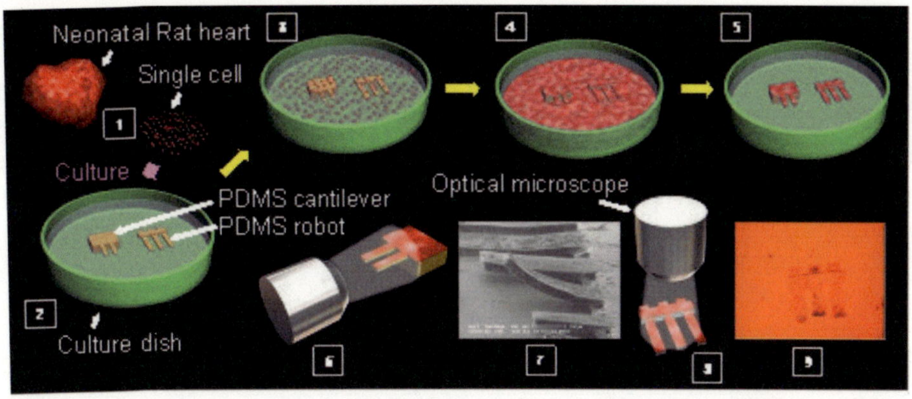

Fig. 3. Full preparation process of microrobot to observe the motion(1) Extract heart single cell from neonatal rat heart, (2) Prepared PDMS structure, whose surface will culture the cardiomyocytes , (3) Primary cardiomyocytes culture on the culture dish including PDMS structure, (4) Incubating cultured cardiomyocytes, (5) Detach PDMS structure from the original culture dish and move to a new culture dish to observe the movement of the PDMS structure, (6) Schematic image to observe the vertical movement, (7) Microscopic image of the vertical view, (8) Schematic image to observe the lateral movement, (9) Microscopic image of the lateral movement.

2.1.5. Engineered surface cantilever to concentrate the contraction force

A heart attack is the most dreaded heart disease. And chronic heart attacks, even slow-going, are the most common and fatal. Heart muscles weaken slowly and its cells simply can't contract. In this case, each heart muscle cell has longer pulsation and loses its force. Therefore, the information on the contractile force of the heart cell is expected to play an important role in finding the basic remedy and substitution for cells with heart disease [24].

Although force measurements at the cellular tissue and whole organ levels have been performed [25-26], influence from outside factors cannot be entirely excluded, giving rise to inaccurate measurements of the force of heart cells. Therefore, there have been several approaches proposed and methods developed to measure the contractile force of cardiomyocytes either directly or indirectly.

In the first approach, micro fabrication technology was used to make a force transducer that would reduce the size of the device for measuring the contractile force of cardiomyocytes [27]. In this system, two micro clamps hold each end of a cardiomyocyte cell to measure the contractile force of the cell. However, this approach involves cell manipulations, which may have unknown effects on the cells and their functions.

In the second approach, an array of micro-scale elastic posts was developed. The attached cells bend each post independently because the forces of a cell for adhesion were used locally. Poly-(dimethylsiloxane) (PDMS) pillar arrays were used as elastic posts [28]. However, the contact between the cell and the heads of pillars may have some effects on cell membranes and cardiomyocytes functions. Moreover, the spreading and morphology of the cell on the flat surface and on the pillar show different aspects, which seems to have a great influence on the contractile force of the cardiomyocytes cell.

The difference of the force of cardiomyocyte cells due to the surface change of the cantilevers was measured. Cantilevers are often used in MEMS devices, not only as actuators but also as sensors. The microcantilever can be used as a sensitive chemical/biological sensor through mechanical methods, by detecting the deflection of the cantilever due to surface stress. PDMS was used as the material of a biopolymer cantilever. PDMS is inexpensive, optically transparent, and suitable for use with optical detection methods. Also, it can be easily made and bonded to other surfaces [29].

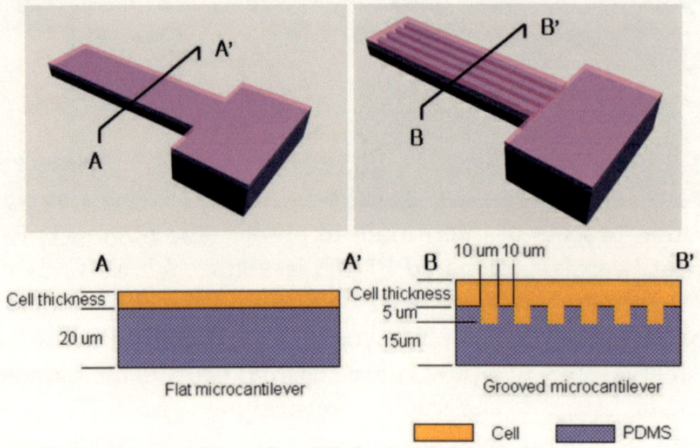

Fig. 4. Schematics of a cross-section of a 3D hybrid biopolymer microcantilever structure: left is a flat microcantilever and right is a grooved microcantilever.

The hybrid biopolymer microcantilever array consisted of five different sizes of microcantilevers, which were 50, 100, 150, 200, and 300μm wide and five times the widths in length, respectively. However, all cantilevers were 20μm thick. It was found that cantilevers longer than 1 mm frequently stuck onto the substrate or were bent, probably due to both the flexibility of the PDMS microcantilever and the cell mass formed by the self organizing cells. Therefore, as mentioned, all data obtained from cantilevers longer than 1 mm were eliminated. As the difference occurred in thickness in a grooved cantilever unlike the surface flat cantilever shown in Figure 4, the structural deflection change ratio was verified on the same force before culturing the cardiomyocytes cell.

	S1	S2	S3	S4	S5	S6	Ave.	S.D.
Displacement increment percentage of the grooved cantilever (%)	25.8	24.2	39.5	39.2	35.6	25.7	31.7	7.2
Contractile force increment percentage of the grooved cantilever (%) (analytical solution)	69.5	67.4	88.0	87.7	82.8	69.4	77.5	8.9
Contractile force increment percentage of the grooved cantilever (%) (FEM)	67.3	64.9	85.3	84.9	80.1	66.9	74.8	8.7

Table 2. Increments of the displacement and contractile force of the cardiomyocytes on the grooved microcantilever

A grooved cantilever was expected to bend much more than a surface flat cantilever upon a small force structurally, as shown in Figure 4. As expected, from the result of the simulation, it was confirmed that a grooved cantilever could bend more than a surface flat cantilever as shown in Table 2. Based on measured displacement and analytical calculation or FEM modeling, shear forces on the microcantilevers were calculated. Table 1 shows the increments of the displacement and contractile force of the cardiomyocytes on the grooved microcantilever compared with those of the flat microcantilever. To compare the contractile forces of cardiomyocytes on the grooved microcantilever with that on the flat one, the structural difference of two microcantilevers had to be considered.

Synthetically considering the above two factors, to yield the same amount of bending displacement by the two types of microcantilevers, the grooved microcantilever needed 34.8% higher contractile force than that of the flat microcantilever. Similarly, according to FEM analysis, the flat microcantilever yielded 32.8% more bending displacement to the same value of shear force than the grooved microcantilever. Nevertheless, the cardiomyocytes cultivated on the grooved microcantilever gave a 25–40% increase in bending displacement from the percentage increase in bending displacment on the flat microcantilever. The contractile force of the aligned cardiomyocytes on the grooved cantilever yielded a 67–88% increase in displacement, which is higher than the percentage increase of displacement of the cardiomyocyte on the flat cantilever. The result of FEM analysis, 65–85% increase in displacement, was similar to the analytical solution above. The contractility of the cardiomyocytes on the grooved microcantilever was much higher than that on the flat microcantilever. Previously, many researchers developed microtechnology methods to measure the contractile force of cardiomyocytes (Lin et al., 2000; Tan et al., 2003; Zhao and Zhang, 2005; Balaban et al., 2001). The stress was found to vary between 2 and 5 nN/mm2. The variation of stress from the cardiomyocytes, reported in the range from 2 to 5 nN/mm2, was verified using our flat microcantilever (Park et al., 2005). Also, in this paper, it was identified that the stress variation of cardiomyocytes concentrated on the grooved surface increased 4–10 nN/mm2 and the average contraction increased 65–85%.

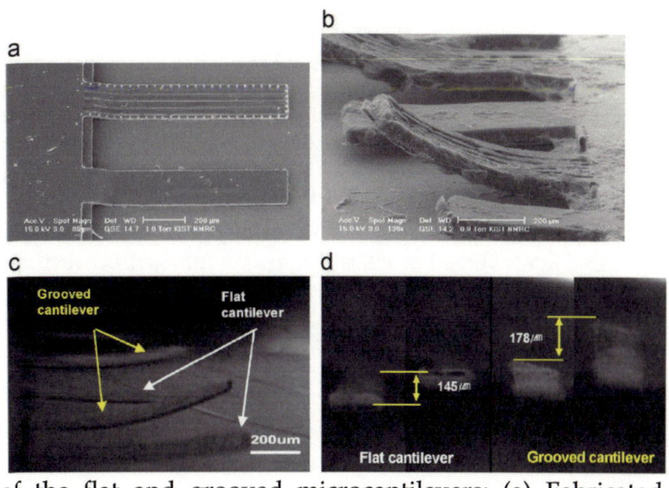

Fig. 5. Images of the flat and grooved microcantilevers: (a) Fabricated PDMS flat and grooved microcantilever, (b) ESEM image of the hybrid organic–inorganic flat and grooved microcantilevers, (c) and (d) still images from video recordings of the vertical motion of the 200x1000 m hybrid biopolymer microcantilevers.

Figure 5(a) and (b) show the fabricated PDMS microcantilevers and cultured cardiomyocytes on the flat and grooved microcantilevers. After the cardiomyocytes were cultured, the initial deflection of the grooved microcantilevers became larger than that of the flat microcantilevers. Compared with the actin filament of the cardiomyocytes on the flat microcantilever, that on the grooved microcantilever was well organized with a higher order. Figure 5 (c) and (d) show still images from video recordings of the vertical motion of the 200x1000 m hybrid biopolymer microcantilevers. The vertical motion of the hybrid biopolymer microcantilevers clearly showed the difference of the bending displacement between the flat and grooved microcantilevers. With respect to the deflections, the displacement of the grooved microcantilevers was larger than that of the flat microcantilevers. Analytical calculation and FEM analysis with ANSYS (ANSYS, Inc.) were performed to quantify the contractility of cardiomyocytes based on these experimental results. For a more realistic simulation, cardiomyocytes was modeled as a material contracting in the longitudinal direction in contact with the PDMS structure, and thus the contractile force was modeled as a shear force exerted at the interfacial area between the cardiomyocytes and PDMS due to the longitudinal contraction of the cardiomyocytes. Young's modulus and Poisson's ratio of the PDMS and cardiomyocytes were assumed to be 750 kPa and 0.49 [20], and 40 kPa [30] and 0.49 [31], respectively. Information on the height of the cells is essential for calibrating the focal pressure of the cardiomyocytes at an interface. In general, the thickness of the cells differs according to surface conditions. When cells are cultured on a flat surface, their height is approximately 5 mm [16,21]. However, the height of the cells on a grooved surface is approximately 10 mm [16].

3. Hybrid Biopolymer Microrobot

3.1. Introduction of biopolymer microrobot

A proposed micromachine was made of silicone elastomer polydimethylsiloxane (PDMS) and self organized cardiomyocytes, which allow free motions in physiological liquids based on the increase of cell power in a 3D microenvironment. In order to fabricate a 3D environment in a robot body, specially designed 3D molding aligner was invented [32].

3.2. Fabrication

3.2.1. Fabrication of molding master for 3D PDMS microrobot

As the same fabrication method of microcantilever, the sandwich molding process was used to fabricate a 3D PDMS microrobot. Figure 6 illustrates fabrication results from the precise alignment-based sandwich micromolding process. The silicon master with two level photoresist structures (Figure 6 (a)) and the glass master (Figure 6 (b)) were fabricated successfully. Figure 6 (c) shows the instant of alignment of the top and bottom masters carried out by our 3D micromolding aligner. A fabricated complex 3D microstructure with top and bottom geometries is shown in Figure 6 (d) as an example.

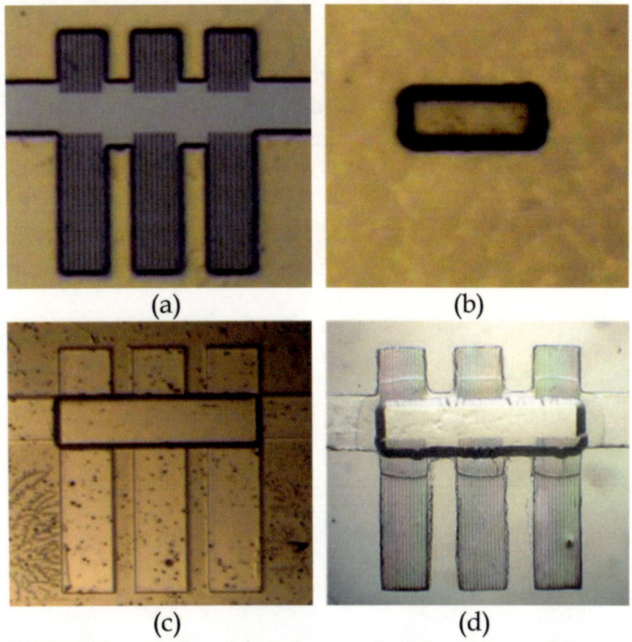

Fig. 6. Microrobot fabrication results using the sandwich micromolding process with precise alignment (a) Bottom silicon master. (b) Top glass master. (c) Alignment of top and bottom masters via a 3D-micromolding aligner. (d) the fabricated complex 3D PDMS structure.

To primary culture cardiomyocytes on the manufactured microrobot device through the PDMS molding process, a pre-process is needed. First, the PDMS device detached from the molding master is washed in 70% ethanol to remove impure particles. It is then immersed in 70% ethanol for an hour for sterilization. Then it is taken out, and dried for 30 minutes on the clean bench under UV light. The fresh PDMS surface is in hydrophobic condition and this prevents the adhesion of proteins and cells. Therefore, O_2 plasma treatment was applied to increase adhesion forces between the PDMS surface and extracellular matrix. PDMS consists of repeating -$OSi(CH_3)_2$- chains. The chain of CH_3 groups makes the surface of the PDMS hydrophobic. The surface can be changed to hydrophilic by exposure to O_2 plasma for 5 min. Plasma enhances cell adhesion onto the PDMS by oxidizing the surface of the PDMS to silanol (Si-OH) [20,69]. Then, the device is coated by immersing it in the ECM (Extra Cellular Matrix: fibronectin) solution overnight and taken out. Then, it is dried on the clean bench under UV light for an hour. Then the device to culture the cell primarily is ready, as shown in Figure 20 (2).

Then the cell is seeded on the device and 5 days later, the device is detached at the bottom and cut the hinge on the device. Then, the robot device can move freely. Just before observing through an optical microscope, turn over the device by pipetting and look at the walking motion of the microrobot.

3.3. Experiment

3.3.1. Monitoring of Walking Motion of Microrobot

To observe the vertical motion of the biomimetic microstructure, a CCD camera microscope (CCD CAM scope, SomeTech Vision) was used. Then, the motion of a biomimetic micromachine was measured using an inverted microscope (Olympus IX 81, Olympus) and captured by a digital camcorder (DCR-PC350, Sony Corp., Japan), as shown in Figure 20 (8). The captured movies were transferred into digital movie files at 30 frames/sec, and then we analyzed the walking motion of the biomimetic micromachine using sequential video frames.

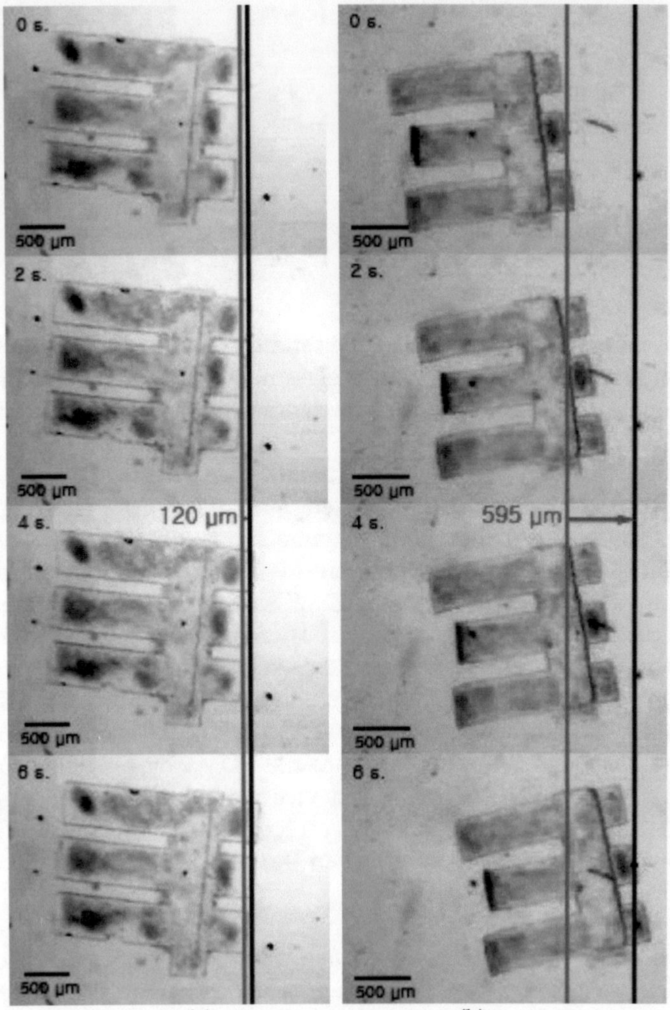

(a) (b)

Fig. 7. Sequential images for the movement of biomimetic micromachine
(a) flat surface cantilever type legs, (b) groove surface cantilever type legs

The biomimetic micromachine is designed and fabricated to achieve walking movement. The micromachine has asymmetric structure, which has 3 front and 3 rear legs of different length. To mimic the legs of the ephyra and to realize a stable and simple structure, multiple legs are considered and all the legs are connected to the middle of the robot body to synchronize with the cardiomyocytes. For the forward movement of the microrobot, the legs are aligned in the horizontal direction and the front legs are made to be shorter than the rear legs. To concentrate the cardiomyocyte cell force, groove patterns are engraved on the surface of the legs. After about 72 to 96 h of culture, cardiomyocytes began to beat synchronously and all legs shows vertical displacement owing to the contractile force of the cardiomyocytes. Finally, the biomimetic micromachine activated by the contractile force of the cardiomyocytes begins to move, as shown in in Figure 7. Figure 7 (a) and (b) show the movement of the micromachine with flat surface cantilever type legs and groove surface ones, respectively. In case of the micromachine with the flat surface cantilever type legs, the cardiomyocytes on the surface on legs show synchronous beating, but the microrobot appears to show only a small advancement. However, the micromachine with groove surface shows forward movement. The average beating frequency was 1.4 Hz and its average step stroke was 100 μm, and thus, the calculated speed was about 140 $\mu m \cdot s^{-1}$.

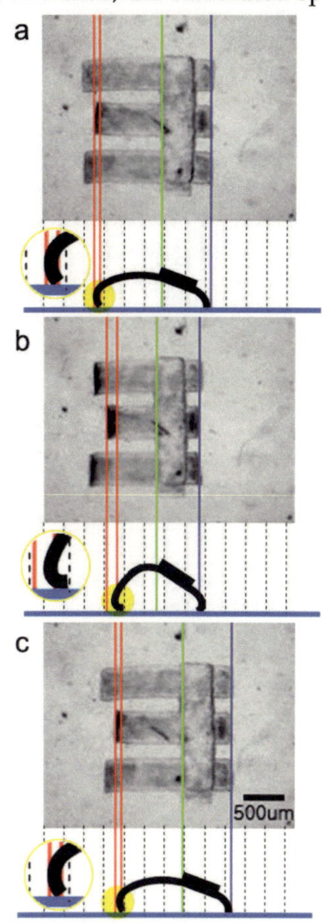

Fig. 8. Still images showing the sequential movement of the micromachine during one step (a) Before contraction of the leg (b) During contraction of the leg (c) After relaxation of the leg. Blue, green and red lines represent the edges of the vertical image of the micromachine at (a), (b) and (c) state, respectively.

However, the average speed of the biohybrid micromachine was measured to be about 125 μm·s-1. The difference between the calculated speed and the average speed will be explained with respect to the walking mechanism of the micromachine.
For forward movement, it is important that friction, related to the surface area of contact, be adjusted so that the microdevices can maximally use its muscle power [11]. The proposed micromachine has 3 front legs and 3 rear legs of different lengths. By the contractile force of the cardiomyocytes, cultured on the micromachine, the deflection of the rear legs of longer length is much larger than that of the front legs. At the end of the legs in contact with the substrate, the asymmetric design of the micromachine produces the difference of the surface area of contact and thus causes an imbalance of the friction force between the legs and the substrate. Therefore, the micromachine can move in the direction of the front legs of shorter length. The sequential video images of Figure 8 show the walking motion of the biomimetic micromachine during one step.

4. Measurement and Discussion

4.1 Long-term Monitoring of Microrobot

4.1.1. Experimental setup

In a hybrid device, the compatibility between the organic and inorganic materials is the most important parameter. Therefore, the cells on the structure were monitored for a long time and checked to determine if the proposed method was biocompatible and feasible. Although the cardiomyocytes started beating when they started to aggregate, but they was monitored from the 5th day when force was gathered by their synchronization.

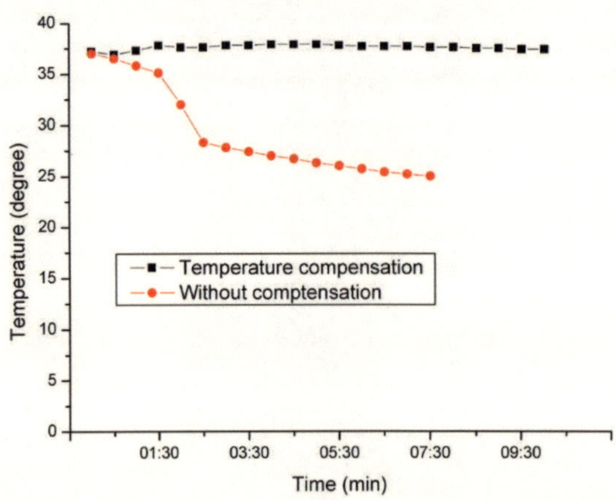

Fig. 9. Comparison of temperature variation between temperature compensation and temperature without compensation

To observe the hybrid device, the temperature change of the solution in the dish taken out from the incubator oven was checked. The graph in figure 9 shows the temperature change of the culture solution in the Petri dish filled with the cardiomyocytes when the cardiomyocyte cultured Petri dish was taken out from an incubator oven, and the temperature change was checked with an inverted microscope. When there is no temperature compensation, the inner temperature of the bioFAB is 25°. Because of the sudden drop in temperature, cells seemed to have been damaged. Therefore, temperature compensation was tried using a light bulb, and an experiment condition which could conserve the constant temperature was set up.

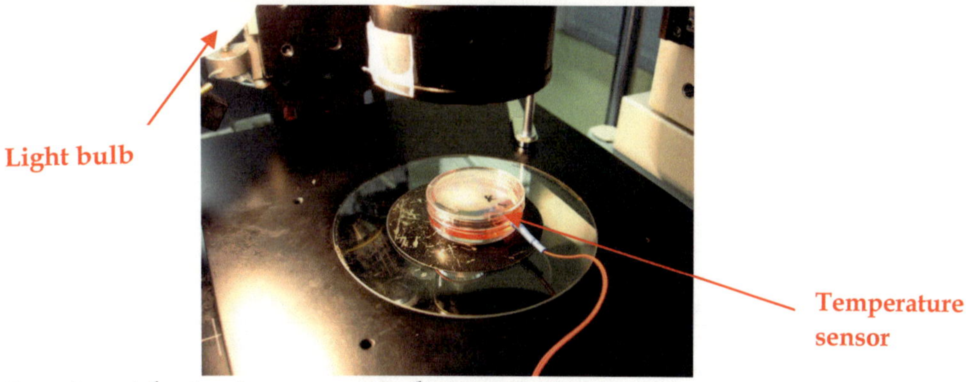

Fig. 10. Experimental setup to compensate the temperature

Figure 10 shows the experimental setup of temperature compensation using a light bulb and the setup measuring temperature by inserting a sensor in the Petri dish. The AD converted value of the voltage corresponding to the change of the resistance value in the thermostat was converted to temperature by interfacing.

4.2. Frequency and Displacement vs. Force

4.2.1. Frequency vs. Bending displacement

The beating frequency and the displacement of the structure due to the contractile force of the cardiomyocytes were measured from the 5th day. As shown in Figure 11 (a), frequency became slower with time. Bending displacement increased, and on the 10th day, it was biggest, as shown in Figure 11 (b).

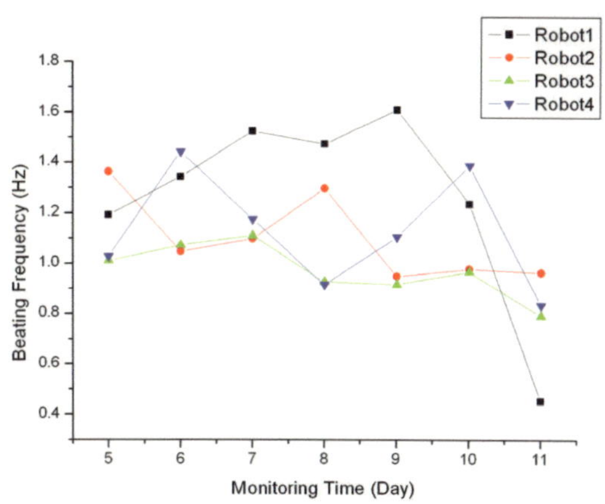

(a)

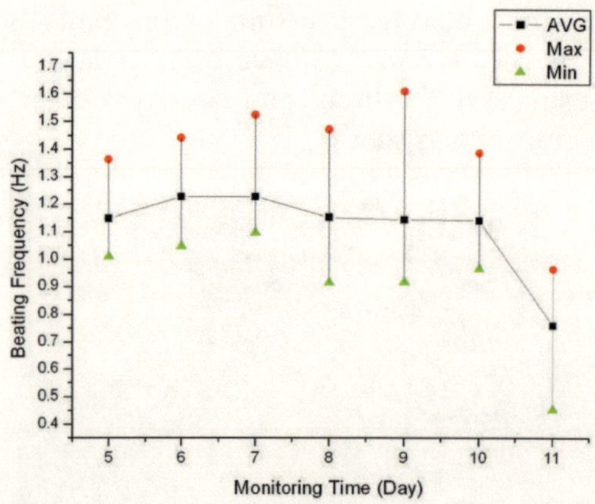

(b)

Fig. 11. Long-term monitoring results of the cardiomyocytes activity on the polymer structure (a) Beating frequency of the cardiomyocytes according to the number of days (b) Average frequency of the cardiomyocytes according to the number of days

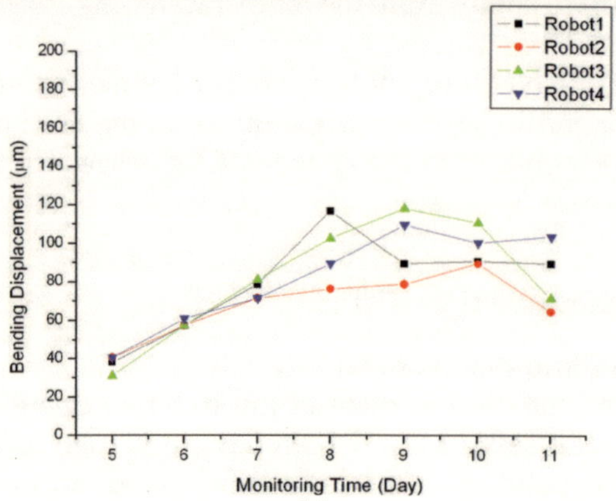

(a)

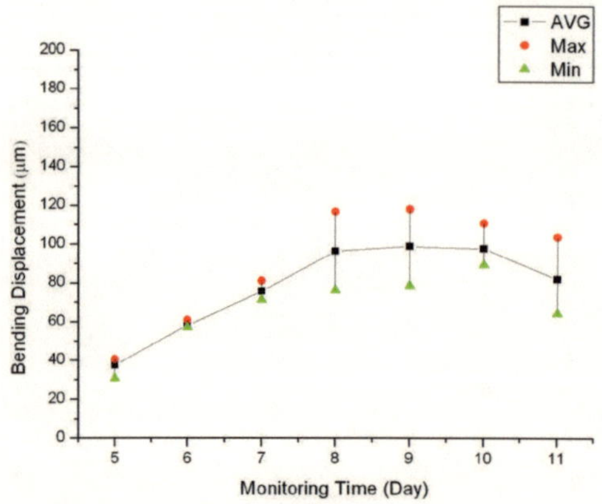

(b)

Fig. 12. Long-term monitoring results of the cardiomyocytes activity on the polymer structure (a) Step increment of the microrobot legs according to the number of days (b) Average step increment of the microrobot legs according to the number of days

Actually, the whole distance walked by the walking robot could not be measured, but when you multiply the frequency per day by the step increment, you can obtain the velocity per day. And when you multiply this result by time again, you can obtain the movement distance per day. So, by calculation, the total movement distance of the micro robot for 7 days is expected to be more than 50 meters.

As shown in the above experimental result, cardiomyocytes primary cultured on the polymer fabricated structure was checked to see if they were beating normally and operating well as an actuator. From this, it was found that PDMS used as the backbone of the hybrid device was biocompatible, and that the plasma treatment and extra cellular matrix coating used to increase the adhesion with cells made a stable and healthy environment for the cells to live. Therefore, a long-term actuator was found to be successful because it operated for over 10 days as an organic/inorganic hybrid actuator. Even though the beating frequency became slower, contractile force became stronger. Also, the microrobot constantly moved for 7 days and showed that it could move an incredibly long distance.

5. Conclusion

By growing rat muscle tissues on a polymer backbone, we demonstrated the movement of the cell powered microrobot. From the viewpoint of a walking microrobot driven by muscle, the cell powered microrobot has novel originality and great significance in the robotics field. The proposed microrobot has the following characteristics: (1) the self-assembled hybrid microrobot consists of biotic muscle cells and PDMS backbone, which is well-known as a biocompatible material. (2) The surface of PDMS backbone is engineered into a groove pattern, which promotes higher order cell concentration and realizes higher generative force of the muscle cells than a 2D culture surface. (3) PDMS is easily fabricated by the? a? micromolding procedure and thus high throughput and mass-production of the cell powered microrobot become possible. (4) The long term monitoring result of the primary cultured cardiomyocytes on engineered surfaces is firstly reported and analyzed. Consequently, the walking robot showed reliable and long-term actuation performances. During the organization of the cardiomyocyte-PDMS hybrid microrobot, the groove surface provided an adverse environment in the initial stages of cell attachment and at adaptation to the surfaces by the modifications in the specific gene expression of the isoforms of troponin and connexin. However, once the cells settled down, more vigorous and synchronous propelling forces were generated through the tighter cell-cell contacts, intracellular arrangements of vesicles and cytoskeltons and higher stimuli according to the microenvironment on the groove surfaces.

As quantitative results, the stress of 2–5 $nN/\mu m^2$ was calculated, which was induced on the flat surface of the microcantilever by the contractile force of the cardiomyocytes. Then, the force of the cardiomyocytes cultured on the groove surface microcantilever increased up to 60% higher than that on the flat surface cantilever. Finally, the forward movement of the micro-robot with the groove surface was monitored, and the measured speed was about 140 $\mu m \cdot s^{-1}$.

6. References

[1] Hiroyuki Fujita, (1998). Microactuators and micromachines, Proceedings of the IEEE, v.86 no. 8, pp. 1721-1732

[2] Friedrich, C.R.; Fang, J.; Warrington, R.O. (1997). Micromechatronics and the miniaturization of structures, devices, and systems, IEEE transactions on components, packaging, and manufacturing technology Part C, Manufacturing, v. 20 no. 1, pp. 31-38

[3] Bar-Cohen, Y., (2006). Biomimetics: biologically inspired technologies, Second Edition (SPIE Press Monograph) Vol. PM136

[4] Ho, D.; Chu, B.; Schmidt, J.J.; Brooks, E.K.; Montemagno, C.D. (2004). Hybrid protein-polymer biomimetic membranes , IEEE Trans. Nanotech. 3, pp. 256–263

[5] Neves, H.P.; Schmidt, J.J.; Soong, R.K.; Montemagno, C.D., (2002). Hybrid micro-and nanobiomechanical systems: a control perspective, World Automation Congr. Proc. 5th Biannual, pp 275–279

[6] Spatz, J. P. (2005). Bio-MEMS: Building up micromuscles, Nature Mater. 4, pp. 115-116

[7] Soong, R. K.; Bachand, G. D.; Neves, H. P.; Olkhovets, A.G.; Graighead, H. G.; and Montemagno, C. D. (2000). Powering an inorganic nanodevice with a biomolecular motor, Science 290, pp. 1555-1558

[8] Rondelez, Y.; Tresset, G.; Nakashima, T.; Kato-Yamada, Y.; Fujita, H.; Takeuchi, S.; and Noji, H. (2005). Highly coupled ATP synthesis by F1-ATPase single molecules, Nature, 433, pp. 773

[9] Weibel, D. B.; Garstecki, P.; Ryan, D.; DiLuzio, W. R.; Mayer, M.; Seto, J. E. and Whitesides, G. M. (2005). Microoxen: Microorganisms to move microscale loads, Proc. Natl. Acad. Sci. U. S. A., 102, pp. 11963-11967

[10] Neves, H.P., Schmidt, J.J., Soong, R.K., Montemagno, C.D., World Automation Congress, 2002. Proceedings of the 5th Biannual, v.14, pp.275-279 (2002).

[11] Xi, J.; Schmidt J. J. and Montemagno, C. D. (2005). Self-assembled microdevices driven by muscle, Nat. Mater., 4, 180

[12] Tanaka, Y.; Morishima, K.; Shimizu, T.; Kikuchi, A.; Yamato, M.; Okano, T.; and Kitamori, T. (2006). Demonstration of a PDMS-based bio-microactuator using cultured cardiomyocytes to drive, Lab on a chip. 6, pp. 362-368

[13] Fisher, M. E. & Kolomeisky, A. B. (1999). The force exerted by a molecular motor, Proc. Natl. Acad. Sci. U. S. A., 96, pp. 6597

[14] Entcheva, E. & H. Bien. (2003). Tension development and nuclear eccentricity in topographically controlled cardiac, Biomed. Microdevices. 5: pp. 163–168,

[15] Motlagh, D.; Hatman, T.; Desai, T & Russell, B. J (2003).Microfabricated grooves recapitulate neonatal myocyte connexin43 and N-cadherin, Biomed Mater Res A. 67, pp. 148-57

[16] Motlagh, D.; Senyo, S. E.; Desai, T and Russell, B. (2003).Microtextured substrata alter gene expression, protein localization and the shape of cardiac, Biomaterials, 24, pp. 2463-2476

[17] Dalby, M. J.; M. O. Riehle; S. J. Yarwood; C. D. W. Wilkinson and A. S. G. Curtis. (2003). Nucleus alignment and cell signaling in fibroblasts: response to a micro-grooved topography, Exper. Cell. Res. 284: pp. 274-282

[18] Yin, L.; Bien, H. and Entcheva, E. (2004). Scaffold topography alters intracellular calcium dynamics in cultured cardiomyocyte, Am J Physiol Heart Circ Physiol. 287, H pp. 1276-1285

[19] Armani, D.; Liu, C.; Aluru, N. (1999). Re-configurable fluid circuits by PDMS elastomer micromachining, IEEE Micro Electro Mechanical Systems; Orlando, FL, U.S.A. pp 222–227

[20] Park J Y; Ryu J; Choi S K; Seo E; Cha J M; Ryu S; Kim J; Kim B and Lee S H. (2005). Real-time measurement of the contractile forces of self-organized cardiomyocytes on hybrid, Anal. Chem. 77, pp. 6571–6580

[21] Sia, S.; Whitesides, G. (2003). Microfluidic devices fabricated in poly (dimethylsiloxane) for biological studies, Electrophoresis, 24, pp. 3563–3576

[22] B. H. Jo; L. M. Van Lerberghe; K. M. Motsegood and David J. Beebe (2000). Three-dimensional micro-channel fabrication in polydimethylsiloxane (PDMS) elastomer, J. Microelectromech. Syst., 9(1), 76–81.

[23] Y. Xia & G. Whitesides Replica molding with a polysiloxane mold provides this patterned microstructure, Angew. Chem., Int. Ed., 1998, 37, pp. 550–575.

[24] M. Zile, et. al., (1998). Constitutive properties of adult mammalian cardiac muscle cells, Circulation, vol. 98, pp. 567-579

[25] M. Kawai; Y. Saeki; Y. Zhao (1993).Crossbridge scheme and the kinetic constants of elementary steps deduced from chemically, Circulation research: J. American Heart Association, vol. 73, pp. 35-50

[26] K. P. Roos (1997). The Myocardium, 2nd ed. G.A. Langer, editor, Academic, New York. pp. 235-323

[27] G. Lin; R. Palmer; K. Pister, and P. Roos (2000). Surface micromachined polysilicon heart cell force transducer, IEEE Trans. J. Biomedical Engineering, vol. 48, pp. 996-1006

[28] Y. Zhao, X. Zhang, IEEE Micro Electro Mechanical Systems, pp. 834-837 (2005).

[29] S. Sia; G. Whitesides (2003). Microfluidic devices fabricated in poly (dimethylsiloxane) for biological studies, Electrophoresis, vol. 24, pp. 3563-3576

[30] Schmidt, J.J.; Montemagno, C.D. (2004). Bionanomechanical systems, Annual Review of Materials Research 34, pp. 315–337.

[31] Janz, R.F.; Grimm, A.F., (1972). Finite-element model for the mechanical behaviour of the left ventricle: prediction of deformation in the potassium-arrested rat heart, Circular Research 30, 244–252.

[32] Park, J.; Kim, J.; Roh, D.; Park, S.; Kim, B.; Chun, K. (2006). Fabrication of 3D thin polymer structures for hybrid sensors and actuators, Journal of Micromechanics and Microengineering 16, 1614–1619.

Theoretical and Experimental Study for Queueing System with Walking Distance

Daichi Yanagisawa[1,2], Yushi Suma[1], Akiyasu Tomoeda[3,4], Ayako Kimura[5], Kazumichi Ohtsuka[4], and Katsuhiro Nishinari[4,6]
[1]*Department of Aeronautics and Astronautics, School of Engineering, The University of Tokyo.*
[2]*Research Fellow of the Japan Society for the Promotion of Science.*
[3]*Meiji Institute for Advanced Study of Mathematical Sciences, Meiji University.*
[4]*Research Center for Advanced Science and Technology, The University of Tokyo.*
[5]*Mitsubishi Research Institute, Inc.*
[6]*PRESTO, Japan Science and Technology Agency.*
Japan

1. Introduction

Queueing theory has been considerably studied since Erlang started designing telephone exchanging system in 1909 Erlang (1909), and many important theories have been developed Burke (1956); Jackson (1957); Kendall (1953); Little (1961). Nowadays, it is applied to study many social systems such as the internet Kasahara (2002); Mukherjee & Manna (2005), a resource management system Barabasi (2005), a vehicular traffic system Helbing et al. (2006) and a pedestrian traffic system D. Helbing & Treiber (2005).

Traffic flow and pedestrian dynamics have been studied actively by using the theory of particle systems, fluid dynamics and cellular automaton Chowdhury et al. (2000); Helbing (2001). Especially, we simulate the dynamics of cars and pedestrians efficiently by using cellular automaton models since its time and space are both discrete. Pedestrians' movement is treated as a stochastic process in the floor field model Nishinari et al. (2004), which is a cellular automaton model for pedestrian dynamics. Then we have succeeded in calculating the total evacuation time analytically by using the floor field model, and studied the way of smooth evacuation Yanagisawa et al. (2009); Yanagisawa & Nishinari (2007).

Calculating the mean waiting time in the queueing system, whose bottleneck is a service window is similar to calculating the total evacuation time from a room, whose bottleneck is an exit. However, the former does not take into account of the effect of spatial structures, while the latter does.

According to the queueing theory, the waiting time of a fork-type queueing system is shorter than that of a parallel-type queueing system (Parallel) (Fig. 1 (a)). However, fork-type queueing system considered in the normal queueing theory (N-Fork) (Fig. 1 (b)) does not reflect the effect of the walking distances from the head of the queue to the service windows. The effect of the distances cannot be ignored in a system such as a large immigration inspection floor in an international airport since walking distances become very long.

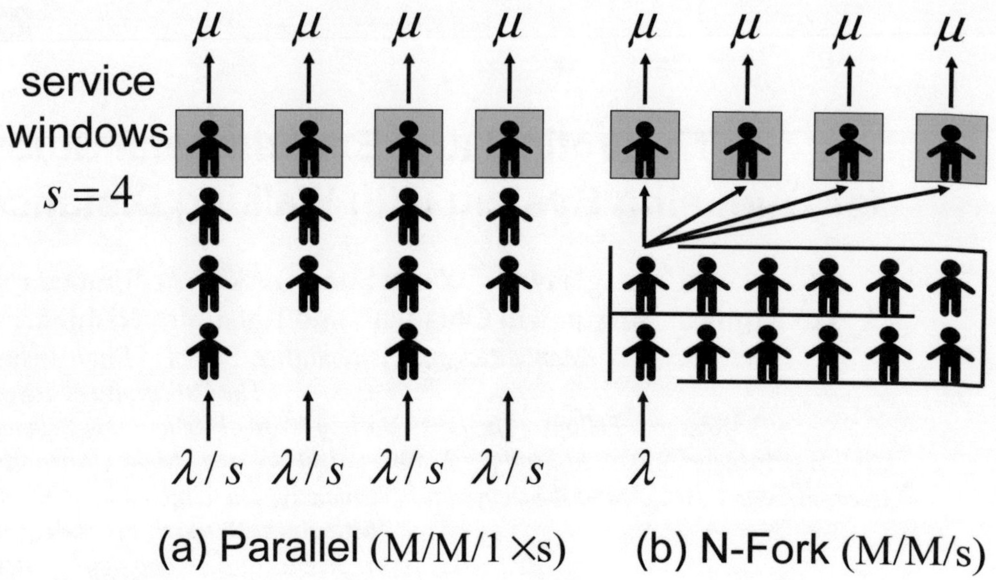

Fig. 1. Schematic views of queueing systems in the case $s = 4$. (a) Parallel $(M/M/1 \times 4)$. (b) N-Fork $(M/M/4)$. An ideal fork-type queueing system. A person at the head of the queue moves to the service window instantaneously when one of them becomes vacant.

Therefore, we have combined the queueing theory and the floor field model, and introduced the effect of delay in walking from the head of the queue to the service windows in the queueing theory. Suitable type of queueing system under various conditions is obtained by analyzing the new queueing theory. When there are plural service windows, the queueing theory indicates that a fork-type queueing system, which collects people into a single queue, is more efficient than a parallel-type queueing system, i.e., queues for each service windows. However, in our walking-distance introduced queueing theory, we find that the parallel-type queueing system is more efficient when sufficiently many people are waiting in the queues, and service time is shorter than walking time. Since the model is studied not only by simulation, but also by theoretical analysis, we have succeeded to obtain the diagram, which indicates the suitable type of queueing system according to the designing conditions.

In the fork-type queueing system, people at the head of the queue starts to move only when one of the service windows become vacant. Since this walking time reduces the efficiency, we consider two new methods for improving the efficiency of the queueing systems. First, we set the head of the queue at the center of the system to decrease the effect of walking distance. This transformation of the system decreases the walking distance to the service windows for pedestrians. Second, we have proposed to keep one person waiting at each service window when it is occupied by other person. Then the waiting person can instantaneously receive service at the window after the former person leave there.

In addition to the theoretical analysis, the experiments of the queueing system for people have been performed. We have verified that the mean waiting time of the parallel-type queueing system become smaller than that of fork-type in the congested situation, experimentally. When we have kept one person waiting at each window, the mean waiting time has dra-

matically decreased, i.e., the efficiency has improved, as we have expected from theoretical calculation.
Finally, we consider the situation where there are two kinds of people, whose service time is short and long. The analytical result says that we can decrease people's waiting time and their stress by setting up queues for each kind of people separately.

2. Walking-Distance introduced Queueing Theory

2.1 Distance introduced Fork-type Queueing System: D-Fork

In a parallel-type queueing system (Parallel) (Fig. 1 (a)), people wait just behind the former person, so that there is no delay in walking. While, in a fork-type queueing system, people take some time to walk from the head of the queue to the service windows. However, the walking time is not taken into account in a fork-type queueing system in the normal queueing theory (N-Fork) (Fig. 1 (b)), and people move from the head of the queue to the service windows instantaneously. Therefore, we consider D-Fork as in Fig. 2 by representing the walking distance using cellular automaton. The gray cells are window cells, and the numbers described in there are window numbers. The white cells are passage cells. Note that the letter "C" described in the passage cells represents the *common* passage cells. For example, both persons who are going to the window 3 and 4 pass the common passage cells. People sometimes cannot go forward in the common passage cells since there is a possibility that other people stand in front of them. The place that people are waiting, which is not divided into cells, is a queue. $s \in \mathbf{N}$, $\lambda \in [0, \infty)$, and $\mu \in [0, \infty)$ represent the number of service windows, the arrival rate, and the service rate, respectively. a and b represent the length of the passage, and k is the interval length between two service windows. The distance from the head of the queue to the service window $n \in [1, s]$ is described as $d_n = a + b + k(n-1)$. Fig. 2 represents the case $s = 4$, $a = 2$, $b = 2$, and $k = 2$. Service windows have two states: vacant and occupied. When a person at the head of the queue decides to move to the vacant service window n, it changes into occupied state. The person proceed to the service window by one cell with the rate $p \in [0, \infty)$ as the asymmetric simple exclusion process. A service starts when the person arrives at the service window, and after it finishes the state of the service window changes into vacant state.

2.2 Update Rules

The simulation of walking-distance introduced queueing systems consists of the following five steps per unit time step.

1. If there is at least one vacant service window and one person in the queue, and the first cell of the passage is vacant, then the person decide to proceed to a vacant service window which is the nearest to the head of the queue, and the state of the service window become occupied.
2. Add one person to the queue with the probability $\lambda \Delta t$, where Δt is the length of the unit time step.
3. Proceed each person in the passage cells to his/her service windows with the probability $p\Delta t$ if there is not other person at their proceeding cell.
4. Remove people at the service windows and change their states into vacant state with the probability $\mu \Delta t$.
5. If 1. takes into practice, proceed the person at the head of the queue to the first passage cell with the probability $p\Delta t$.

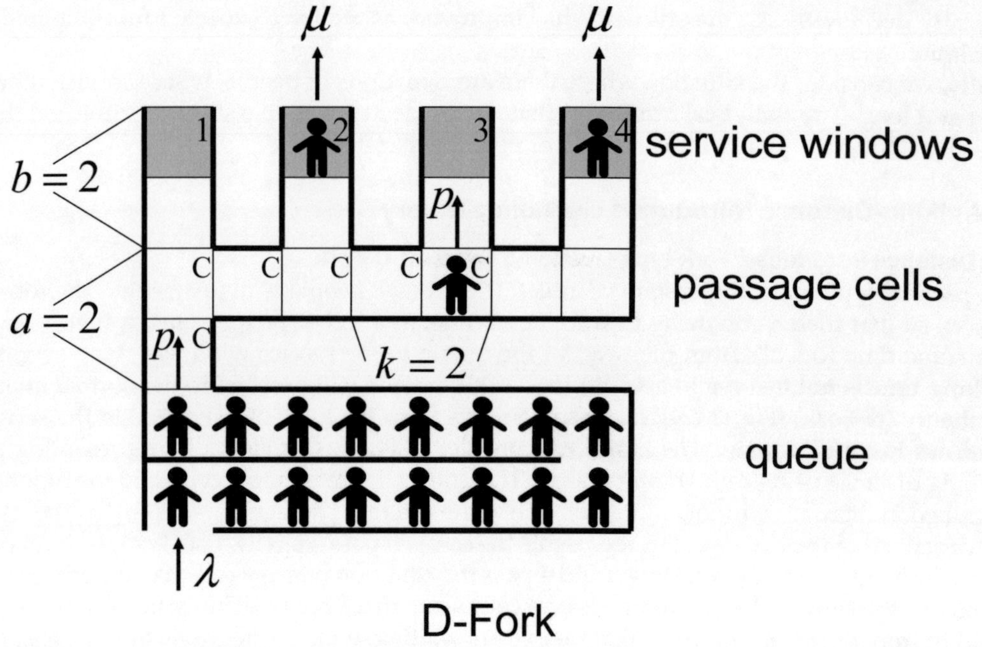

Fig. 2. Schematic view of D-Fork ($s = 4$).

2.3 Stationary Equations

We define the sum of the walking time and the service time at service window n as a throughput time τ_n and its reciprocal as a throughput rate μ_n. Here, we calculate the mean throughput rate $\hat{\mu}_n$ when n service windows are occupied, and obtain stationary equations of D-Fork. We suppose that all passage cells are vacant by mean field approximation. Then, the mean value of the throughput time $E(\tau_n)$ is described as follows.

$$E(\tau_n) = \frac{1}{\mu} + \frac{a + b + k(n-1)}{p}. \tag{1}$$

The throughput rate μ_n is obtained as

$$\mu_n = \frac{1}{E(\tau_n)} = \frac{1}{\frac{1}{\mu} + \frac{a+b}{p} + \frac{k(n-1)}{p}} = \frac{\mu}{1 + \alpha + 2\beta(n-1)}, \tag{2}$$

where

$$\alpha = \frac{\mu(a+b)}{p}, \qquad \beta = \frac{k\mu}{2p}. \tag{3}$$

In the case $2\beta(n-1)/(1+\alpha) \ll 1$, we calculate the mean throughput rate $\hat{\mu}_n$ as

$$\hat{\mu}_n = \frac{1}{n}\sum_{l=1}^{n} \mu_l \approx \frac{\mu}{1 + \alpha + \beta(n-1)}. \tag{4}$$

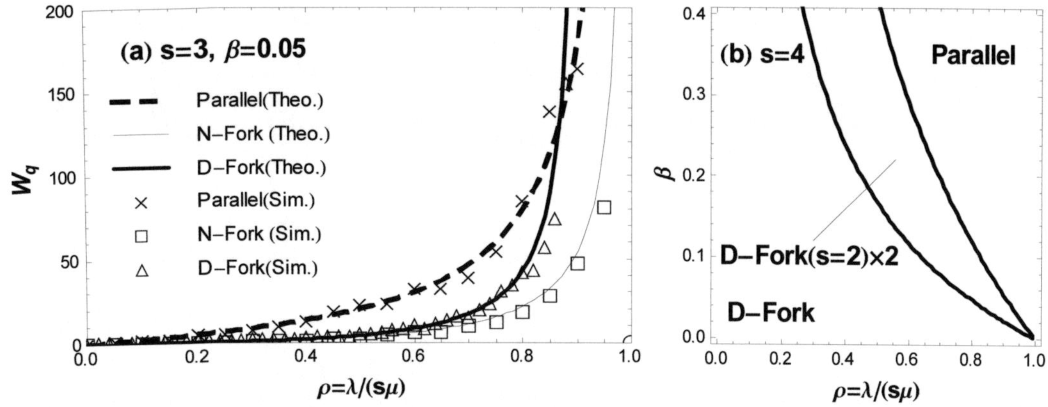

Fig. 3. (a) Comparison of W_q of Parallel, N-Fork, and D-Fork in the case $a = 1$, $b = 1$, $p = 1$, $\alpha = 0$, $\beta = 0.05$, $\mu = 0.05$. (b) A queueing system which makes the mean waiting time W_q minimum in the case $s = 4$.

By using (4) the stationary equations are described as follows:

$$\begin{aligned} \lambda P_0 &= \hat{\mu}_1 P_1 \\ \lambda P_{n-1} + (n+1)\hat{\mu}_{n+1} P_{n+1} &= (\lambda + n\hat{\mu}_n) P_n \quad (1 \le n \le s-1) \\ \lambda P_{n-1} + s\hat{\mu}_s P_{n+1} &= (\lambda + s\hat{\mu}_s) P_n \quad (n \ge s). \end{aligned} \tag{5}$$

We obtain the mean waiting time W_q by solving (5) analytically. In the case $\alpha = \beta = 0$, we have the stationary equations of $M/M/s$ Bolch et al. (1998) from (5), thus α and β represent the effect of walking time.

In our simulation the distribution of the throughput time is gamma distribution. We approximate it as exponential distribution in this calculation, however, when β is small the results from the exponential distribution approximated well to those from gamma distribution.

3. Comparison of a Parallel Queue and a Fork Queue

We compare the mean waiting time W_q of Parallel (Fig. 1 (a)), N-Fork (Fig. 1 (b)), and D-Fork (Fig. 2 (a)). Figure 3 (a) show W_q against the utilization $\rho(= \lambda/(s\mu))$. The results of analysis agree with those of the simulation very well. We see that W_q of N-Fork is smaller than that of Parallel and D-Fork in $0 \le \rho < 1$. There is a possibility that more than one person is waiting in one queue and no one is in the other queue in Parallel ($s \ge 2$), however there is no vacant service window in N-Fork when people are waiting in the system. This is the reason why W_q of N-Fork is always smaller than that of Parallel. Since N-Fork does not take into account of the effect of the walking distances, i.e. $\beta = 0$, it is obvious that W_q of N-Fork is smaller than that of D-Fork. The N-Fork is the most efficient of the three; however, it is an ideal system and does not exist in reality. By focusing on the curves of Parallel and D-Fork, we can clearly observe the crossing of them. This means that when the utilization ρ is small, i.e., there are not sufficiently many people in the system; we should form D-Fork to decrease the waiting time. On the contrary, when the utilization ρ is large, i.e., there are many people in the system, we should form Parallel. When β become large, the crossing point move to

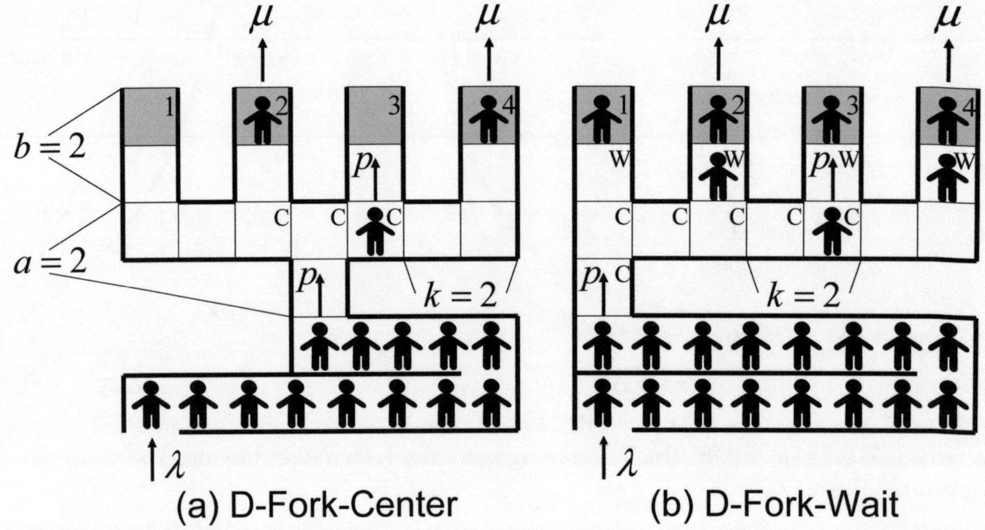

Fig. 4. Schematic views of walking-distance introduced queueing systems ($s = 4$). (a) D-Fork-Center. The head of the queue is at the center of the system. (b) D-Fork-Wait. People can wait at the cells, which are described as "W".

the left. The strong effect of the walking distances extend the suitable ρ region for Parallel. This agrees with our intuition, since D-Fork is influenced by the distances but Parallel does not. The reversal phenomenon of W_q is obtained for the first time by introducing the effect of distance.

Figure 3 (b) shows the type of queueing system, which minimize W_q against ρ and β in the case $s = 4$. This figure is useful for designing queueing systems. The curves divide the $\rho - \beta$ plane into three regions. In the lower left region W_q of D-Fork ($s = 4$) is the smallest, and in the upper right region W_q of Parallel is the smallest. Surprisingly, W_q of D-Fork ($s = 2$) $\times 2$ (Fig. 2 (b)) is the smallest in the middle region. This indicates that the choice of the type of queueing systems is not only Parallel and D-Fork, but also a combination of them. According to (3), β represents the ratio of walking time and service time. Therefore, D-Fork is suitable when service time is much longer than walking time. The value of β is small in most D-Fork in reality, however, in large queueing system such as an immigration inspection floor in the international airport, we should divide the large D-Fork into the several small D-Forks to decrease the effect of the walking distances.

4. Methods for Shortening Waiting Time in D-Fork

4.1 Set the Head of the Queue at the Center: D-Fork-Center

The head of the queue is usually set at the end of the system since we can efficiently use the space for the queue. However, people have to walk a long distance to the farthest window as in Fig. 2. Thus, we propose to set the head of the queue at the center (D-Fork-Center) as in Fig. 4 (a). Then, the mean throughput rate is described as follows:

$$\hat{\mu}_n \approx \frac{\mu}{1 + \alpha + \frac{\beta}{2}\left(n + \frac{1+(-1)^{n+1}}{2n}\right)} \tag{6}$$

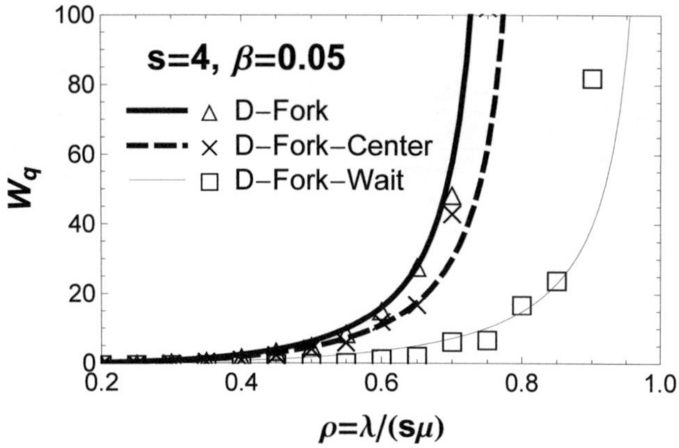

Fig. 5. Comparison of the mean waiting time W_q among D-Fork, D-Fork-Center, and D-Fork-Wait in the case $s = 4$, $\alpha = 0.15$, $\beta = 0.05$, $\mu = 0.05$.

Comparing (4) and (6), we see that the coefficient of n in (6) is half of that in (4). Therefore, the effect of the walking distance approximately becomes half if we set the head of the queue at the center. The mean waiting time is approximately calculated by replacing β of D-Fork with $\beta/2$ in Sec. 4.3. Note, that the (6) represents the $\hat{\mu}_n$ in the case that both s and k are even number. The mathematical formulation of the other cases are described in Ref. Yanagisawa et al. (2008).

4.2 Keep One Person Waiting at the Window: D-Fork-Wait

The walking distance in D-Fork is essentially problematic, since it delays the start of services, i.e., people have to walk the passage before they start to receive the service. Thus, we propose to keep one person waiting at the service window. We call a queueing system which this method is applied to as D-Fork-Wait (Fig. 4 (b)). Since people are waiting just next to the service windows, they can receive service instantaneously when their former people leave there. The delay in walking is almost removed by this method, i.e., the effect of walking distance does not need to be considered. Therefore, the mean waiting time is approximately calculated by the expression for N-Fork.

4.3 Approximated Calculation and Simulation

Figure 5 show the mean waiting time W_q in D-Fork, D-Fork-Center, and D-Fork-Wait. We see that W_q in D-Fork-Center is always smaller than that in D-Fork in $0 \leq \rho < 1$. This result verifies that we can decrease W_q by setting the head of the queue at the center. We also find that W_q in D-Fork-Wait is the smallest of the three since it almost completely removes the effect of walking time by keep one person waiting at the window. The well correspondence between the results of the theoretical analysis and simulation in $\rho \leq 0.85$ also verifies our assumption that D-Fork-Wait becomes close to N-Fork. However, in the large-ρ region, the result of the simulation becomes larger than that of the theoretical analysis. In D-Fork-Wait, there is a possibility that $2s$ people are in the window cells or passage cells, while s people is the maximum in the other cases. Thus, when ρ is large, people in the common passage cells cannot often proceed since other people are in front of them. This jam in the common

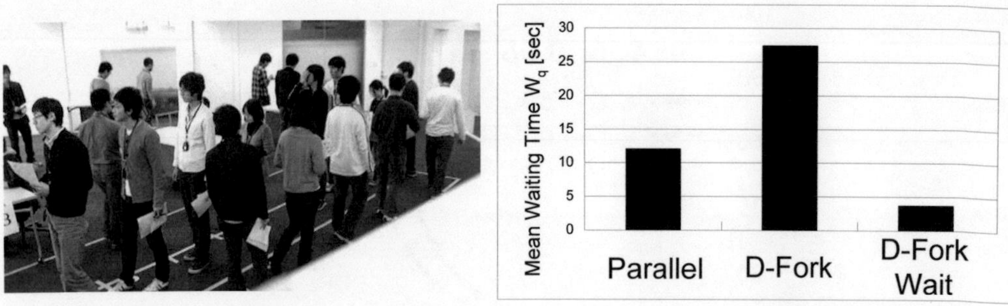

(a) (b)

Fig. 6. (a) Schematic view of the experiment. (b) The experimental mean waiting time for Parallel, D-Fork, and D-Fork-Wait.

passage cells, which is not considered in the theoretical calculation, increase the waiting time in D-Fork-Wait when ρ is large.

5. Experiments

We have performed the experiments to examine the two results in the former section. 1. There is a case that W_q in Parallel becomes smaller than that in D-Fork. 2. We can decrease W_q by keeping one person waiting at the window. We made the queueing system as in Fig. 1 (a), 2, and 4 (b), whose parameters are $s = 4$, $a = 1$ [m], $b = 0.5$ [m], $k = 3$ [m], $\lambda = 188/600$ [persons/sec], and $\mu = 1/8$ [persons/sec]. Figure 6 (a) is the snap shot of the experiment. Participants of the experiments enter the system and line up in the queue when the staff says to do so. They proceed to the windows and receive service. After that they wait at the starting position until the staff let him/her enter the system again. We put 188 people in 600 [sec] in one experiment. Note, that the arrival was random while the service was deterministic for simplicity. According to the Pollaczek-Khintchine formula Bolch et al. (1998), W_q becomes small when the service is deterministic. Since this effect acts on the all kinds of the queueing systems in the same way, the results are not critically influenced by deterministic service. Therefore, we can examine the result of the theoretical analysis and simulations by these experiments.

Figure 6 (b) shows the result of the experiments. We see that W_q in Parallel is smaller than that in D-Fork. This result verifies our theoretical analysis and simulation by using the walking-distance introduced queueing theory. The reversal of W_q between Parallel and D-Fork is observed experimentally for the first time in this paper. We also find that W_q becomes dramatically small in D-Fork-Wait. This new result indicates that the method "Keep one person waiting" is an effective way to shorten the waiting time empirically.

6. Simultaneous Arrival of Many People

In the former sections, we discuss the queueing systems, whose distribution of interarrival time is exponential distribution, i.e. random arrival. Here, we consider the situation that many people arrive at the same time, such as arrivals of people who alight from trains and airplanes.

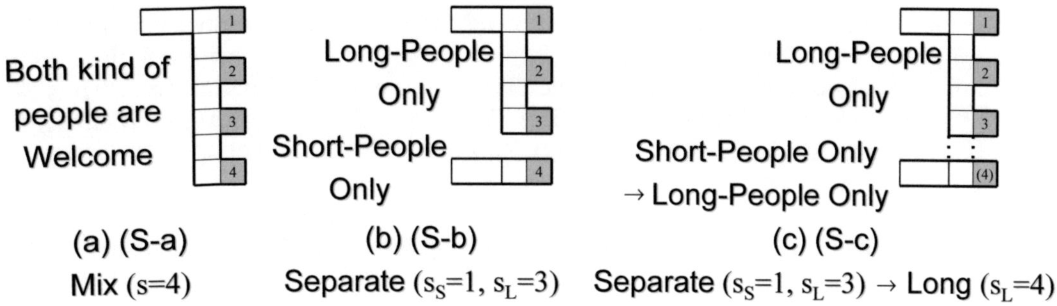

Fig. 7. Schematic views of queueing systems. (a) (S-a) Mix queueing system ($s = 4$). (b) (S-b) Separate queueing system ($s_S = 1$, $s_L = 3$). (c) (S-c) Separate queueing system ($s_S = 1$, $s_L = 3$) $\rightarrow$ Queueing system for LP ($s_L = 4$).

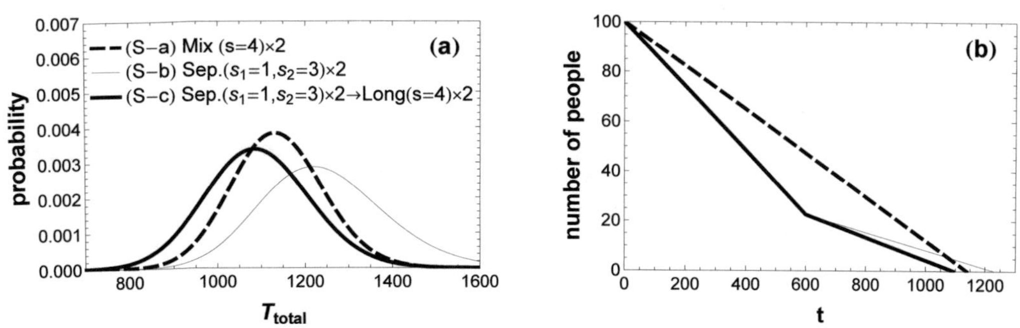

Fig. 8. (a) The probability distribution of the total throughput time T_{total}. (b) Number of people in the queueing system against time step t. The value of plot S-b and S-c decrease dramatically, when people start to be given services. This is because SPs get out from the system quickly by avoiding the disturbance by LPs.

The total throughput time T_{total} that all people finish leaving the system is calculated and studied to decrease it. We consider the system, which has four service windows. There are two kinds of people whose service time are short (SP) and long (LP). The service rate of SP and the number of service windows only for SP are denoted as μ_S and s_S, respectively. Similarly, those of LP are described as μ_L and s_L. We have three strategies as follows:

(S-a) Mix ($s = 4$) (Fig. 7 (a))
Both SPs and LPs use the same windows.

(S-b) Separate ($s_S = 1$, $s_L = 3$) (Fig. 7 (b))
SPs use the window only for SP and LPs use the windows only for LP.

(S-c) Separate ($s_S = 1$, $s_L = 3$) $\rightarrow$ Long ($s_L = 4$) (Fig. 7 (c))
First same as (b), but after all SPs have left, the window, which was only for SP, is open for LP.

The distribution of the T_{total} when 50 SPs and 50 LPs arrive at the same time is described in Fig. 8 (a). We see that the mean of T_{total} of S-c is the smallest and that of S-b is the largest. When S-b or S-c is adopted, the queue of the SP is not affected by the distances. Thus SPs leave the system quickly. After all SPs have left, all four service windows are used efficiently in S-c, however, one window is not used in S-b. This makes S-c the best and S-b the worst.
We also discuss the stress of waiting people by Fig. 8 (b), which describes the number of people waiting in the queue against the time step t. People suffer from a stress when they are waiting in the queue and do not when they leave the system. Therefore, an area surrounded by the axes and the curves represents the sum of all people's stress until they left. Clearly, we observe that (Area of S-c) $<$ (Area of S-b) $<$ (Area of S-a). It is interesting that when we compare T_{total}, S-a is better than S-b, however, comparing the stress, the result is opposite. We also find that S-c is the best strategy from the both point of views: T_{total} and the stress.
If the arrival of people is random, it is difficult to adopt S-c since we cannot find out when to change the type of the queueing system. However, if we have information about people's arrival, we can decrease both waiting time and stress of people by adapting the type of the queueing system into proper one.
When we apply this study to the real system, the service rates are estimated by the data from the measurement.

7. Conclusion

We have introduced the effect of walking distance from the head of the queue to the service windows and shown that the performance of a parallel-type queueing system is better than that of a fork-type queueing system when there are sufficiently many people in the system. The effectiveness of the two new methods is also studied. The mean waiting time becomes small when we set the head of the queue at the center or keep one person waiting at each service window in a fork-type queueing system since the effect of walking distance decreases. It also turns out that dynamical change of a queueing system decrease both waiting time and stress of people when two kinds of people, whose service time are short and long, come into the system at the same time.
We would like to emphasize that our study is based on the theoretical analysis, simulations, and experiments. Therefore, the results in this paper are reliable enough to apply to the queueing systems in the real world. It is an important future work to study the effect of the costs raised by the two methods. When we set the head of the queue at the center, the space for waiting people decreases, and at least one clerk is needed at the head of the queue to keep one person waiting at each window.

Acknowledgement

We thank Dai Nippon Printing Co., Ltd. in Japan for the assistance of the experiment, which is described in Sec. 5. This work is financially supported by Japan Society for the Promotion of Science and Japan Science and Technology Agency.

8. References

Barabasi, A.-L. (2005). The origin of bursts and heavy tails in human dynamics, *Nature* **435**: 207–211.

Bolch, G., Greiner, S., de Meer, H. & Trivedi, K. (1998). *Queueing Networks and Markov Chains*, A Wiley-Interscience Publication, U.S.A.

Burke, P. J. (1956). The output of a queueing system, *Operations Research* **4**(6): 699–704.

Chowdhury, D., Santen, L. & Schadschneider, A. (2000). *Phys. Rep.* **329**: 199.

D. Helbing, R. J. & Treiber, M. (2005). Analytical investigation of oscillations in intersecting flows of pedestrian and vehicle traffic, *Phys. Rev. E* **72**: 046130.

Erlang, A. K. (1909). The theory of probabilities and telephone conversations, *Nyt. Tidsskr. Mat. Ser. B* **20**: 33–39.

Helbing, D. (2001). Traffic and related self-driven many-particle systems, *Rev. Mod. Phys.* **73**: 1067–1141.

Helbing, D., Treiber, M. & Kesting, A. (2006). Understanding interarrival and interdeparture time statistics from interactions in queueing systems, *Physica A* **363**: 62–72.

Jackson, J. K. (1957). Networks of waiting lines, *Operations Research* **5**(4): 518–521.

Kasahara, S. (2002). Towards queueing theory for the internet design, *TECHNICAL REPORT OF IEICE* **101**(649): 25–30. (in Japanese).

Kendall, D. G. (1953). Stochastic processes occurring in the theory of queues and their analysis by the method of the imbedded markov chain, *Ann. Math. Statist.* **24**(3): 338–354.

Little, J. C. D. (1961). A proof for the queueing formula $l = \lambda w$, *Operations Research* **9**(3): 383–387.

Mukherjee, G. & Manna, S. S. (2005). Phase transition in a directed traffic flow network, *Phys. Rev. E* **71**: 066108.

Nishinari, K., Kirchner, A., Namazi, A., Schadschneider, A. & Nonmembers (2004). *IEICE Trans. Inf. Syst.* **E87-D**: 726.

Yanagisawa, D., Kimura, A., Tomoeda, A., Nishi, R., Suma, Y., Ohtsuka, K. & Nishinari, K. (2009). Introduction of frictional and turning function for pedestrian outflow with an obstacle, *Phys. Rev. E* **80**: 036110.

Yanagisawa, D. & Nishinari, K. (2007). Mean-field theory for pedestrian outflow through an exit, *Phys. Rev. E* **76**: 061117.

Yanagisawa, D., Tomoeda, A., Kimura, A. & Nishinari, K. (2008). Analysis on queueing systems by walking-distance introduced queueing theory, *Journal of JSIAM* **18**(4): 507–534. (in Japanese).

23

Intention-Based Walking Support for Paraplegia Patients with Robot Suit HAL

Kenta Suzuki[1], Gouji Mito[1], Hiroaki Kawamoto[2],
Yasuhisa Hasegawa[1] and Yoshiyuki Sankai[1]
[1]*Graduate School of Systems and Information Engineering, University of Tsukuba*
[2]*Japan Association for the Advancement of Medical Equipment*
Tennodai 1-1-1, Tsukuba, 305-8573, Japan
Email: cybernoid@golem.kz.tsukuba.ac.jp

Abstract

This paper proposes an algorithm to estimate human intentions related with walking in order to comfortably and safely support a paraplegia patient's walk. A robot suit "HAL" has been developed for an enhancement of healthy person's activities and for support of physically challenged person's daily life. Assisting method based on bioelectrical signals such as myoelectricity successfully supports healthy person's walking. These bioelectrical signals, however, cannot be measured properly from a paraplegia patient. Therefore another interface that can estimate patients' intentions without any manual controller are desired for robot control since a manual controller deprives a patient of his/her hands' freedom. Estimation of patients' intentions contributes to support not only comfortably but also safely, because an inconformity between the robot suit motion and the patient motion results in his/her stumbling or falling. The proposed algorithm, therefore, estimates patient's intentions from a floor reaction force reflecting patient's weight shift during walking and standing. The effectiveness of this algorithm is investigated through experiments on a paraplegia patient who has a sensory paralysis on both legs, especially his left leg. We show that HAL supports patient's walk properly, estimating his intentions based on floor reaction force.

Keywords: robot suit, paraplegia, walking support, intention estimation, floor reaction force

1. INTRODUCTION

People may have muscle rigidity, relaxation, involuntary contraction of muscle, and sensory paralysis due to cerebral paralysis, stroke, spinal cord injury, muscular dystrophy and post-polio syndrome. Even if people do not suffer from these physical problems, aging brings various troubles on his/her motility. Most people who have problems on the lower limbs due to these symptoms or aging are unable to walk and are bedridden all day long at worst. Moreover, this situation depresses the patients' feelings, for instance bedridden patients lose his/her life worth living. Caregivers including the patient's family also receive hard works to look after him/her, once a person has a trouble in the motility. To relieve these problems and to support the patient's independent life, it is quite important to provide a safe and convenient transportation device. A wheelchair is now used in most cases as a transportation device for patients with gait disorder. It is convenient for the patients because they can move easily as long as an enough muscular power is left in their upper body. Even if a patient has weakness of the arms, a motorized wheelchair could be used. However, wheelchairs have some problems in its using environment and the user's posture. Especially, wheelchair users are apt to keep sitting posture for a long time and have less opportunities to exercise their own lower bodies. That may cause a decrease in not only muscular power of lower body with paralysis but also residual physical functions. This problem could be solved if a patient with paraplegia could walk on his/her legs as a healthy person does. Therefore, a device which helps a patient walk in his/her standing posture would be one of the solutions since he/she can locomote with his/her leg receiving a physical support. Several devices for walking support have been developed. In our study, a wearable type robot "Robot suit HAL (Hybrid Assistive Limb)" has been developed in order to physically support wearer's daily activities and heavy works. HAL-1 utilizing DC motors and ball screws shown in Fig. 1(a) was developed as the first prototype of HAL [1], and it enhanced wearer's walking by amplifying wearer's own joint torque. After developing some prototypes, HAL-3 shown in Fig. 1(b) was developed toward a more suitable system to be used in actual daily life [2, 3]. These robot suits have a power unit on each hip and knee joint, and they support functional motions of lower limbs with multiple joints simultaneously. After that, HAL-5 (see Fig. 1(c)), that is demonstrated at the 2005 World Exposition in Aichi, has been developed for whole body support. It assists human motions involving wearer's upper-body activities such as carrying heavy loads. Meanwhile, "RoboKnee" [4] and "Wearable Walking Helper" [5] have been developed to support the knee motion by using linear actuators. However, it is difficult for these two devices to support a patient with paraplegia since these devices cannot support their multiple joints in lower limbs simultaneously. As an exoskeleton to assist soldiers, disaster relief workers and other emergency personnel who needs to move long distance on foot on their fields, Kazerooni et al., [6, 7] has developed "BLEEX" that supports human's walking while carrying heavy loads on his/her back. This exoskeleton is not designed for welfare purposes, and it is too large and heavy (75 kg including exoskeleton weight and maximum payload) for patients to handle as their own supporting devices in actual daily life. To provide effective physical support according to each wearer's condition, it is necessary to strongly focus on control algorithm as well as mechanism of supporting devices. The robot suit HAL has a cybernic control system that is a hybrid control algorithm consisted of "Cybernic voluntary control (Bio-cybernic control)" and "Cybernic autonomous control (Cybernic robot control)". The cybernic control system can provide suitable physical support to wearers in various conditions such as a healthy person, a physically challenged person and so on by using two algorithms as complementary controls.

The features of each control algorithm are described below. The cybernic voluntary control provides physical support according to his/her voluntary muscle activity. Power units of HAL generate power assist torque by amplifying wearer's own joint torque estimated from his/her bioelectrical signals, and the support motions are consequently controlled by wear's signal adjustment. This control was used for power assist of healthy person's activities [8], for example walking and standing up from sitting posture, and we confirmed the cybernic voluntary control successfully supported a wearer's motion. The bioelectrical signals including myoelectricity are useful and reliable information to estimate human's motion intentions because the signals are measured just before corresponding muscle activities. Thus, the wearers receive the physical support directly by unconscious interface using the bioelectrical signals, which realize much more easily operation than manual controllers such as a joystick. HAL

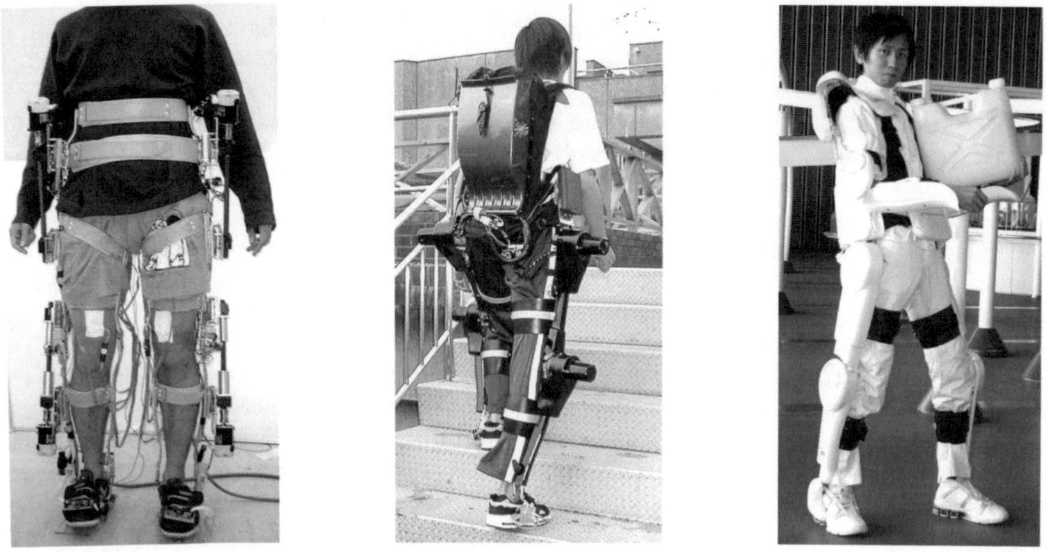

(a) HAL-1 (1999) (b) HAL-3 (2001) (c) HAL-5 (2005)

Fig. 1. Representative conventional robot suits we have developed. (a) and (b) HAL supports wearer's lower body motion. (c) HAL supports their whole body motion. A twenty kilogram load is carried on wearer's single arm.

can physically support patients with some handicaps on their lower limbs as well as healthy people because HAL supports functional motions with multiple joints simultaneously, covering whole of lower limbs. However, as a whole, a patient with gait disorder is not able to receive walking support by the cybernic voluntary control because the signals that induce a broken walking pattern are not used for the power assist, and no signal is observed in the severest case. In that case, the cybernic autonomous control can provide an effective physical support.

The cybernic autonomous control autonomously provides a desired functional motion generated according to wearer's body constitutions, conditions and purposes of motion support. While the bioelectrical signals are mainly used in the cybernic voluntary control, various kinds of information except for the bioelectrical signals, such as reaction force and joint angle can be used to provide comfortable physical supports. It can be applied to rehabilitation and walking support for the patients as well as power assist for healthy people and it enables HAL to be used as alternate body functions for their handicaps or weakness of muscular power. In that case, HAL needs to observe wearer's conditions and motion intentions from any motion information instead of his/her bioelectrical signals in order to provide a suitable support with a suitable moment. HAL-3 with the cybernic autonomous control successfully enhances healthy person's walking, stair-climbing, standing up from sitting posture and cycling, synchronizing with his/her body conditions [9]. In that work, floor reaction forces and joint angles are used as motion information to detect wearer's conditions. Posture control as well as sensing and recognition for environment including a wearer is essential technologies for an entirely autonomous physical support, but they remain to be solved. In this paper, the cybernic autonomous control among the cybernic control system is applied to the robot suit HAL in order to support a paraplegia patient's walk. Our conventional cybernic autonomous control algorithm [9] cannot be applied to them directly due to variety of patients' body constitutions and handicaps. Generally, the human intentions in his/her mind are essentially independent from the physical interactions between a body and an environment. As far as we know, no current technologies can directly measure and extract the human intentions. However, we can sometimes guess the human intentions in his/her mind from his/her appearances or motions. Besides, we can estimate his/her corresponding intentions if we observe a motion or an appearance that is closely connected with his/her intentions. According to conventional works on human transient walking [10, 11], a COG shift to one leg is prior motion to a walk. That motion is an indispensable to swing a leg and can be observed earlier than a bioelectrical signal such as myoelectricity, because it is observed before a human starts swinging a leg, while a bioelectrical signal is observed when corresponding muscles start contracting. The COG shift can be used for an early and smart trigger to start walking supports, because the shift is involved into preliminary motions for a walk and human does not have to operate any manual switch to start the walking supports. On the other hand, gait stopping is similar to the time-reverse motion of the gait initiation, and the COG stops at around the center of both supporting legs. Therefore this paper proposes an intention estimator that can estimate his/her walking intentions from the COG shift that is closely connected with his/her intention. We define that intention-based support (including the walking support) is to provide a physical support for the next wearer's desired motion that can be predicted based on the current state or motion induced by his/her intention. In a case of walking, a human shifts the COG to a supporting leg side before he/she starts swinging a leg. If the robot suit HAL can sense the COG shift induced by his/her intention, it can predict his/her walking start and then start walking support. Our project aims to realize the comfortable walking supports for paraplegia patients that reflect the patients' intentions on the start and stop of walking, cycle and stride of walking motion, walking direction and so on. We call the walking support conforming to these various intentions of walking "Intention-based walking support". It is hoped that the intention-based walking support improves the usability, safety and reliability of the robot suit HAL. As the first step, this paper focuses on three kinds of intentions: start and stop of walking and the beginning to swing a leg, and proposes a control algorithm that uses patient's residual physical functions effectively. We need to observe not only the COG shift in a lateral plane but also the forward COG shift and bending of the upper body in order to distinguish the gait initiation from other similar motions such as just stepping or changing a supporting leg for a leg relaxation. However, the robot suit HAL can understand his/her intention if we instruct the wearer to shift the COG to either of his/her legs in order to receive the physical support for swinging a leg. Therefore, floor reaction force can be one of reliable information that reflects his/her intentions without any manual interfaces if a patient can control his/her weight balance in lateral plane by holding a walking frame with own hands. The purpose of this study is that HAL helps a patient with paraplegia walk in a standing posture. Based on our conventional works, two additional functions should be developed for this purpose. First, HAL should generate a suitable bipedal walk according to patient's body constitutions. Reference trajectories for each joint support should be designed in another way because the bioelectrical signals are not observed from a patient with paraplegia. The reference motions consist of swinging wearer's leg, supporting his/her weight and shifting his/her weight from one leg to the other. Second, HAL should provide walking support according to patient's intentions that are estimated from wearer's COG shift. To achieve two functions mentioned above, this paper takes the following approaches. They are:

1. To achieve the bipedal locomotion partially based on walking patterns of a healthy person,
2. To estimate wearer's intentions from his/her COG shift that is observed by the floor reaction force and
3. To synchronize support motions with estimated wearer's intentions: the walk start, stop and the beginning to swing a leg.

The following section explains assumptions and approach of this study. Section 3 introduces the robot suit "HAL-5 Type-C" used in this experiment. Section 4 describes the proposed algorithm for walking support and intention estimation. Section 5 shows experimental results and verifies the performance of the proposed algorithm in HAL-5 Type-C. Finally, section 6 is the conclusion.

2. ASSUMPTIONS AND APPROACH

In this paper, a proposed algorithm is applied to the walking support for a paraplegia patient called "subject A" in this paper. He has sensory paralysis on both legs, especially left leg because of spinal cord injury by traffic accident. He can keep standing posture and slowly walk by himself with two canes. In this case, we cannot measure proper bioelectrical signals to estimate his intention during walking because of disorder of neural transmission. We, therefore, use floor reaction force instead of the bioelectrical signals in this experiment. Floor reaction force (FRF) reflects his weight shift during walking and standing. It should be noted that he can control his balance holding a walking frame and that our algorithm can estimate his intentions from his FRF. That is our algorithm synchronizes the physical support with his intentions through his controlled weight balance by using not any manual controllers such as a joystick but FRF during walking and standing. The reference patterns to the patient are extracted from healthy person's walk. The healthy person's walking motion could be suitable to the patient if he/she has the same body constitution as the healthy person. The extracted walking motion, however, should be adjusted according to the patient's body constitution and handicap conditions, for example a walking cycle and amplitude of each joint trajectory in swinging a leg.

3. ROBOT SUIT HAL

In the experiment, the robot suit HAL-5 clinical type (HAL-5 Type-C) which is made for the subject A is used. Figure 2 shows the overview of HAL-5 Type-C and Fig. 3 is its system configurations. As in the case of the conventional type of HAL (HAL-3), HAL-5 Type-C consists of power units, exoskeletal frames, sensors and a controller. Power units are attached on each hip and knee joints and actuate each joint by their torques. On ankle joints, springs are attached so that wearer's ankle joints could come back to a normal angle even if any external forces do not affect the joints. The spring action contributes to avoiding collisions between a toe of a swing leg and a floor. The exoskeletal frames are fixed to wearer's legs with molded plastic bands, and transmit torques of the power units to his/her legs. There are angular sensors and FRF sensors to measure motion information of HAL-5 Type-C and a wearer for wearer's intention estimation. Potentiometers as angular sensors are attached to the each joint to measure the joint angles. FRF sensors utilizing the semiconductor-type pressure sensor are implemented in shoes. Figure 4 shows the appearance of the shoes of HAL-5 Type-C with built-in FRF sensors. The weight of a wearer including HAL-5 Type-C is transferred onto the sensor unit and measured by the pressure sensors. These sensors can also measure the distribution of load between a toe part and a heel part during walking and standing because two sensors are built in the front and rear of the shoe sole inside. In addition, a computer and batteries are attached on a wearer's waist, and motor drivers and other electrical circuits for the signal processing are allocated on each power unit. Compared with the robot suit HAL-3 (see Fig. 1(b)), HAL-5 Type-C is improved for patients' daily use since there is no large backpack on his/her back and a width of the power units in the back view becomes thin enough to pass through narrow spaces as shown in Fig. 2. Figure 5 shows angles and rotation directions of each joint described in this paper.

4. CONTROLLER DESIGN

In this section, we explain a controller for walking support system. Walking motion in this work shall be consist of three functions including swinging a leg, landing and supporting a body as shown in Fig. 6. In this paper, we call each span of three functions "swing phase", "landing phase" and "support phase". In the swing phase, the patterns extracted from healthy person's walk are applied as the reference patterns of the proportional and derivative (PD) control for the corresponding joints of a wearer. The reference patterns are used for the corresponding leg's control synchronizing with wearer's intention estimated by our proposed algorithm. In the landing phase, we realize the leg function for a foot landing by not tracking reference patterns but applying constant-value control. Based on our conventional work [12], we found that the knee joint of a wearer at landing instance is apt to be flexed by his/her own weight and much torque beyond the torque tolerance is needed to compensate for the knee bend. Therefore the knee joint has to be extended earlier than the reference pattern by constant-value control. In the support phase as well as the landing phase, the leg is supported by constant-value control in order to support his weight by one leg. The following sub-sections explain the details of the controller algorithm.

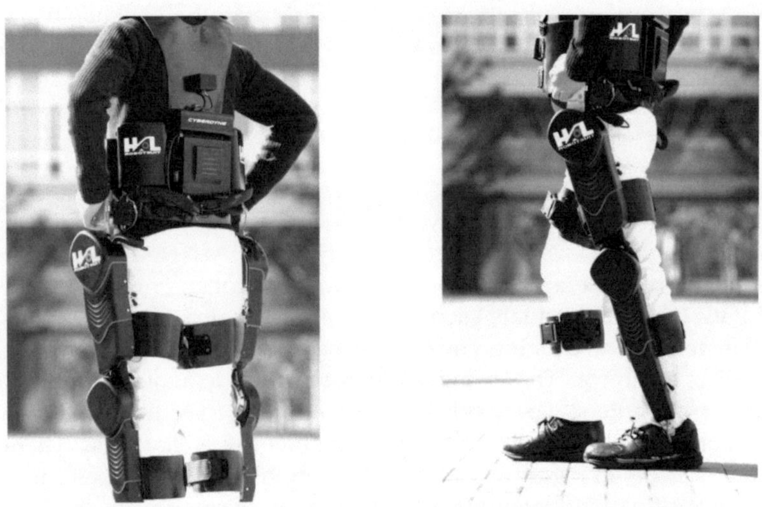

(a) Back view. (b) Side view.

Fig. 2. HAL-5 Type-C developed for walking support of a paraplegia patient. Total weight is 15 kg.

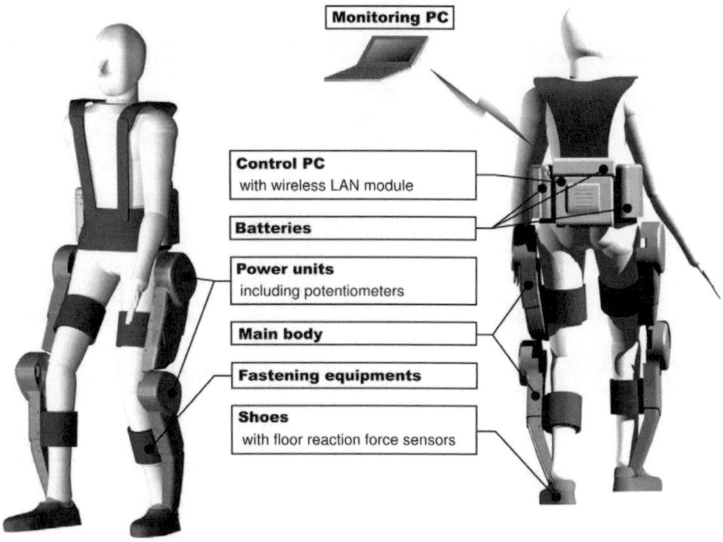

Fig. 3. System configurations of HAL-5 Type-C.

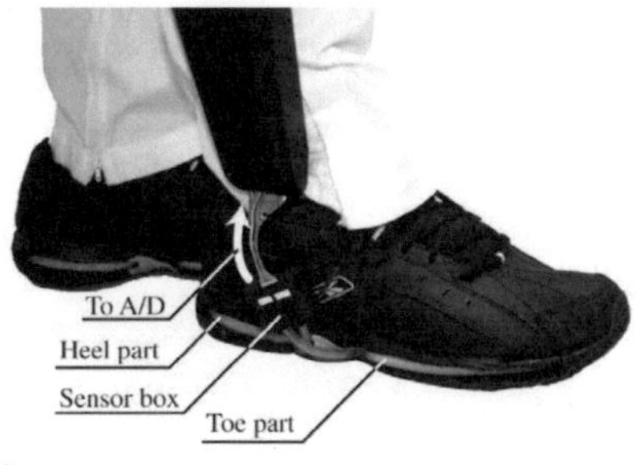

Fig. 4. Built-in floor reaction force sensors.

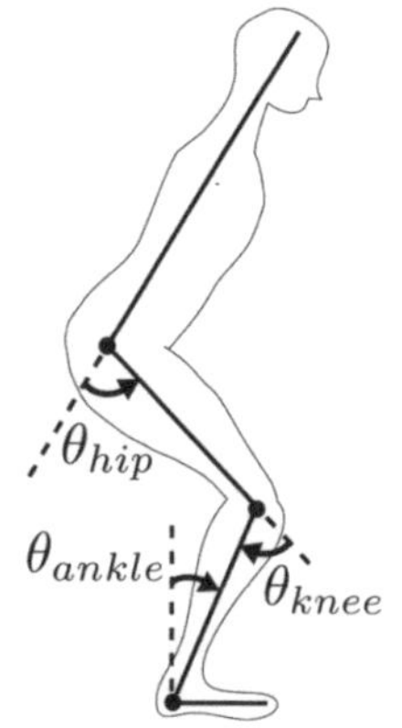

Fig. 5. Rotation directions of each joint.

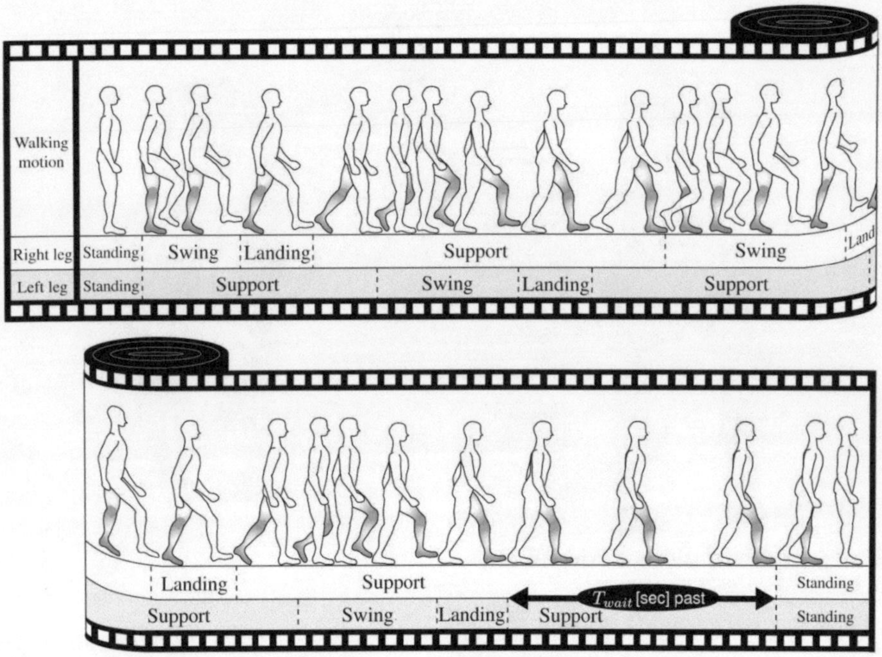

Fig. 6. Three functions in walking motion.

4.1 Reference pattern generation

As mentioned above, a swing leg in the swing phase is supported by applying reference walking patterns measured in healthy person's walk. The reference patterns are generated in the following process.

1. To measure angle data of hip and knee joints in healthy person's walk.
2. To divide a sequence of the measured walk pattern into patterns of each step and then average the walk patterns.
3. To divide the averaged pattern into three phases and extract a pattern in the swing phase.

At first, we measure a healthy person's walk to acquire the angle data of hip and knee joints during walk. In this experiment, we measure a normal walk of a man in his twenties, who has the similar body constitutions including height, weight and length of legs to the subject A. Second, a sequence of the measured walk pattern is divided into patterns in each step and then they are averaged. At this stage, we should pay attention that habits of walking and asymmetry between right and left leg step are not reflected in the extracted patterns strongly. Figure 7(a) shows walking patterns in one step averaged in this experiment.

Finally, the averaged walking patterns are divided into patterns in the swing, landing and support phase. The swing phase is between a moment when a foot leaves a floor and a moment when a thigh is full flexed. The landing phase continues until a moment when a foot of the swing leg contacts a ground, and the support phase continues until a moment when one step finishes. The walking patterns extracted from a healthy person's walk are shown in Fig. 7(b), (c) and (d). Namely, Fig. 7(b) shows the reference angle patterns in the swing phase

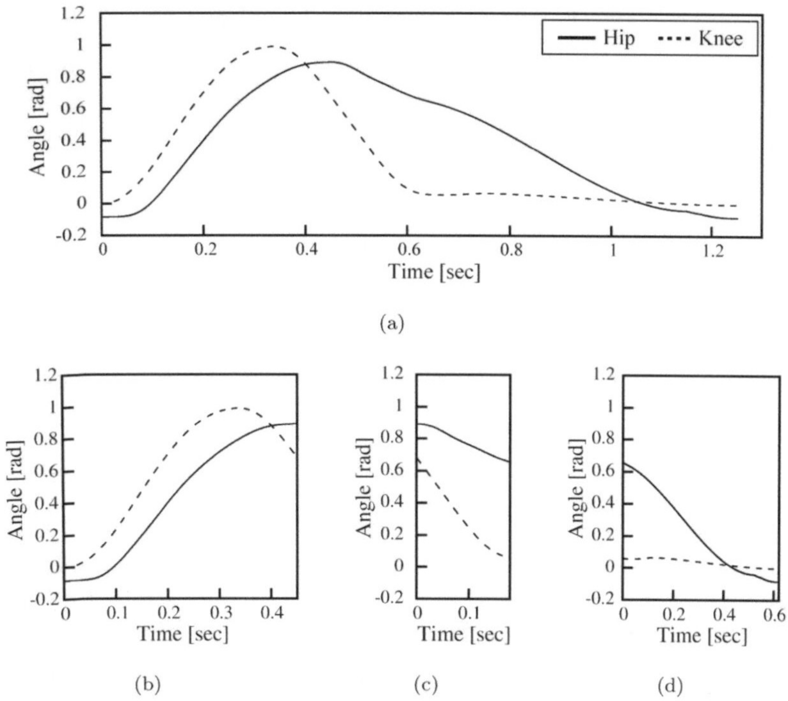

Fig. 7. Reference walking patterns of joint angle. (a) Patterns in one cycle of walk. (b) Patterns in the swing phase. (c) Patterns in the landing phase. (d) Patterns in the support phase.

used in this walking support. The PD controller to drive a leg swing needs reference angular velocity patterns as well as the angle patterns, and the angular velocity patterns are generated by differentiating the angle patterns with respect to time. In addition, the time scales of the reference patterns are linearly shorten or lengthen so that the walking cycle could be adjusted to a wearer's intentions or a wearer's body constitutions.

On the other hand, a swing leg in the landing phase is supported by constant-value control for the preparation of patient's weight support. The reference angle and angular velocity in the landing and support phase are empirically set. Table 1 shows reference values in each phase of walking support. In this table, θ_{href} and $\dot{\theta}_{href}$ show the reference angle and angular velocity of a hip joint respectively, and θ_{kref} and $\dot{\theta}_{kref}$ show the reference angle and angular velocity of a knee joint respectively. In addition, the hip and knee joints should be straightened through the landing and support phase in order to support a wearer's weight by one leg. Therefore, the reference angle of hip joint in the landing phase is 0 rad. Table 1, however, shows the reference angle of knee joint is not 0 rad but -0.052 rad. This over extension of the knee joint can prevent the knee joint from bending due to an impact of landing a foot and gravity. In general, it is quite harmful for human to extend the knee joint excessively, but HAL does not extend wearer's joints beyond the range of motion since fastening equipments of HAL made of rigid plastic has a little flexibility and mechanical limiters at knee joints prevent the joints from extending more than that angle. HAL controls the joint angle to keep the reference values in the support phase until the end of the single leg support phase when a foot of an opposite side swing leg touches on a floor. After the foot of the swing leg makes a contact with a floor,

	Swing phase	Landing phase	Support phase
θ_{href}[rad]	Fig. 7(b)	0.0	0.0 (-0.7)
$\dot{\theta}_{href}$[rad/sec]	Time derivative of Fig. 7(b)	0.0	0.0
θ_{kref}[rad]	Fig. 7(b)	-0.052	-0.052
$\dot{\theta}_{kref}$[rad/sec]	Time derivative of Fig. 7(b)	0.0	0.0

Table 1. Reference values in one cycle of walking support.

the reference hip joint angle of the supporting leg switches from 0.0 rad to -0.7 rad shown in parentheses of Table 1. This hip extension contributes to the smooth weight shift from a current supporting leg to a following one. Reference angular velocity of both joints in the landing and support phase consistently maintains 0.0 rad/sec through one cycle of walking support.

4.2 Intention estimator

Estimation of patients' intentions contributes to support not only comfortably but also safely, because an inconformity between the robot suit motion and the patient motion results in his stumbling or falling. Instead of the bioelectrical signals used for the control of the conventional HAL, the floor reaction force is used for an intention estimation of the subject A who can control his weight balance using two canes with his hands. The floor reaction force reflects the position of center of gravity (COG) and COG could be the reliable information for the intention estimation. For example, a leg could leave a floor and work as a swing leg safely if it does not support his/her weight. A support system "HAL" estimates which leg supports a wearer's weight, when a wearer begins to swing a right or left leg and when he/she wants to stop walking. At first, for example, a right leg is considered to be a support leg when a foot contact condition:

$$f_{rh} > \alpha_{rh} \quad \text{or} \tag{1}$$

$$f_{rt} > \alpha_{rt} \tag{2}$$

is satisfied, where f_{rh} and f_{rt} are FRF of a right foot heel side and toe side, respectively. In addition, α_{rh} and α_{rt} are thresholds to detect a landing of a right foot. In general, the condition (1) is applied in advance of (2) since a healthy person puts a heel of a swing leg on a floor in advance of a toe. Patients with paralysis on legs such as the subject A, however, have a foot weighed down and may put a toe of a swing leg on a floor in advance of a heel. The condition (2) is effective in detecting the landing in cases of paraplegia patients. On the other hand, a left leg is considered to be a support leg when a foot contact condition:

$$f_{lh} > \alpha_{lh} \quad \text{or} \tag{3}$$

$$f_{lt} > \alpha_{lt} \tag{4}$$

is satisfied, where f_{lh} and f_{lt} are FRF of a left foot heel side and toe side, respectively. In addition, α_{lh} and α_{lt} are thresholds to detect a landing of a left foot.
Second, for example, HAL estimates the intention that a wearer wants to swing a right leg when swing start conditions:

$$f_{rh} < \beta_{rh} \quad \text{and} \tag{5}$$

$$f_{rt} < \beta_{rt} \quad (6)$$

are satisfied, where β_{rh} and β_{rt} are thresholds to detect a moment when each part of a right foot leaves a floor. On the other hand, HAL estimates the intention that a wearer wants to swing a left leg when swing start conditions:

$$f_{lh} < \beta_{lh} \quad \text{and} \quad (7)$$

$$f_{lt} < \beta_{lt} \quad (8)$$

are satisfied, where β_{lh} and β_{lt} are thresholds to detect a moment when each part of a left foot leaves a floor. In this study, the following two constraint conditions are added to the above conditions for more stable estimation of wearer's intentions.

1. Do not start to swing a leg unless a foot of the opposite side leg is on a floor.
2. Do not swing the same leg sequentially.

HAL estimates the intention that a wearer wants to stop in his/her tracks if it pasts a certain time before the swing start conditions (5) and (6), or (7) and (8) are satisfied. In the walking support, HAL stops the sequential walking supports and helps a wearer come back to the standing posture when a condition:

$$t_{cur} - t_r > T_{wait} \quad \text{or} \quad (9)$$

$$t_{cur} - t_l > T_{wait} \quad (10)$$

is satisfied, where t_{cur}, t_r and t_l are the current time and the time when the last right or left foot touches on a floor. In addition, T_{wait} is a temporal threshold to switch the walking support to the standing posture support. In this moment, the reference angles of all joints are almost zero, therefore a backward leg is replaced around a forward leg if a load on the backward leg becomes almost zero by his/her weight shift. We set $T_{wait} = 5.0$ sec in this experiment.

4.3 Control Architecture

Bipedal locomotion using patient's legs is achieved by the tracking control and by phase synchronization of motion support with patient's intention. This control consists of the PD control using reference walking patterns based on healthy person's walk as shown in Fig. 7(a) in the swing phase and the constant-value control in the landing and support phase. Figure 8 shows a block diagram for this tracking control and phase synchronization. The human intention estimator (HIE) located in the upper-left part in the figure has the FRF as inputs for the estimation algorithms described in the section 4.2. Three blocks under the HIE are a library of the reference patterns in the swing phase and the reference values in the landing and support phase. The HIE allocates these references to two legs during walking. There are six ordinary PD control blocks on the right side of the HIE and the library. The upper three blocks are controllers for the right leg and the lower ones are for the left leg. The command voltages τ_r and τ_l to the power units on both legs are calculated by:

$$\boldsymbol{\tau}_r = \boldsymbol{K}_r(\boldsymbol{C}_r\boldsymbol{\theta}_{ref} - \boldsymbol{\theta}_r) + \hat{\boldsymbol{K}}_r(\boldsymbol{C}_r\dot{\boldsymbol{\theta}}_{ref} - \dot{\boldsymbol{\theta}}_r) \quad \text{and} \quad (11)$$

$$\boldsymbol{\tau}_l = \boldsymbol{K}_l(\boldsymbol{C}_l\boldsymbol{\theta}_{ref} - \boldsymbol{\theta}_l) + \hat{\boldsymbol{K}}_l(\boldsymbol{C}_l\dot{\boldsymbol{\theta}}_{ref} - \dot{\boldsymbol{\theta}}_l), \tag{12}$$

where $\boldsymbol{\theta}_r$ and $\boldsymbol{\theta}_l$ are the actual wearer's leg joint angles, $\dot{\boldsymbol{\theta}}_r$ and $\dot{\boldsymbol{\theta}}_l$ are angular velocities and subscripts r and l mean right and left, respectively. In addition, $\boldsymbol{\theta}_{ref}$ and $\dot{\boldsymbol{\theta}}_{ref}$ are the reference joint angles and the reference angular velocities, respectively. These variables including $\boldsymbol{\tau}_r$ and $\boldsymbol{\tau}_l$ have two elements that correspond to two joints: hip and knee joint. $\boldsymbol{\tau}_r, \boldsymbol{\tau}_l, \boldsymbol{\theta}_r, \boldsymbol{\theta}_l, \dot{\boldsymbol{\theta}}_r, \dot{\boldsymbol{\theta}}_l, \boldsymbol{\theta}_{ref}$ and $\dot{\boldsymbol{\theta}}_{ref}$ are given as follows:

$$\boldsymbol{\tau}_r = \begin{bmatrix} \tau_{rh} \\ \tau_{rk} \end{bmatrix}, \quad \boldsymbol{\tau}_l = \begin{bmatrix} \tau_{lh} \\ \tau_{lk} \end{bmatrix}, \tag{13}$$

$$\boldsymbol{\theta}_r = \begin{bmatrix} \theta_{rh} \\ \theta_{rk} \end{bmatrix}, \quad \boldsymbol{\theta}_l = \begin{bmatrix} \theta_{lh} \\ \theta_{lk} \end{bmatrix}, \quad \dot{\boldsymbol{\theta}}_r = \begin{bmatrix} \dot{\theta}_{rh} \\ \dot{\theta}_{rk} \end{bmatrix}, \quad \dot{\boldsymbol{\theta}}_l = \begin{bmatrix} \dot{\theta}_{lh} \\ \dot{\theta}_{lk} \end{bmatrix}, \tag{14}$$

$$\boldsymbol{\theta}_{ref} = \begin{bmatrix} \theta_{href} \\ \theta_{kref} \end{bmatrix}, \quad \dot{\boldsymbol{\theta}}_{ref} = \begin{bmatrix} \dot{\theta}_{href} \\ \dot{\theta}_{kref} \end{bmatrix}, \tag{15}$$

where subscripts rh, rk, lh and lk mean right hip joint, right knee joint, left hip joint and left knee joint, respectively. On the other hand, $\boldsymbol{K}_r$ and $\boldsymbol{K}_l$ are feedback gains of the joint angle errors, and $\hat{\boldsymbol{K}}_r$ and $\hat{\boldsymbol{K}}_l$ are feedback gains of the joint angular velocity errors. The different feedback gains are used in the swing, landing or support phase independently by adopting this control architecture. In addition, $\boldsymbol{C}_r$ and $\boldsymbol{C}_l$ are gains to the reference joint angles and angular velocities. These gains can adjust a joint flexion and a stride length in a wearer's supported walk. In this experiment, we set $\boldsymbol{C}_l$ larger than $\boldsymbol{C}_r$ in order to avoid collisions of a left leg which has a more severe paralysis with a floor in the swing phase. $\boldsymbol{K}_r, \boldsymbol{K}_l, \hat{\boldsymbol{K}}_r, \hat{\boldsymbol{K}}_l, \boldsymbol{C}_r$ and $\boldsymbol{C}_l$ are diagonal matrixes which are given as follows:

$$\boldsymbol{K}_r = \begin{bmatrix} k_{rh} & 0 \\ 0 & k_{rk} \end{bmatrix}, \quad \boldsymbol{K}_l = \begin{bmatrix} k_{lh} & 0 \\ 0 & k_{lk} \end{bmatrix}, \quad \hat{\boldsymbol{K}}_r = \begin{bmatrix} \hat{k}_{rh} & 0 \\ 0 & \hat{k}_{rk} \end{bmatrix}, \quad \hat{\boldsymbol{K}}_l = \begin{bmatrix} \hat{k}_{lh} & 0 \\ 0 & \hat{k}_{lk} \end{bmatrix}, \tag{16}$$

$$\boldsymbol{C}_r = \begin{bmatrix} c_{rh} & 0 \\ 0 & c_{rk} \end{bmatrix}, \quad \boldsymbol{C}_l = \begin{bmatrix} c_{lh} & 0 \\ 0 & c_{lk} \end{bmatrix}. \tag{17}$$

Moreover, the PD gains of swing leg control k_{rh}, k_{lh}, $\hat{k}_{rh}$, $\hat{k}_{lh}$, k_{rk}, k_{lk}, $\hat{k}_{rk}$ and $\hat{k}_{lk}$ were determined based on frequency responses and step responses of hip and knee joints. The concrete procedure is described in Appendix A.

The control flow for the walking support is as follows. At first, HAL supports a wearer's standing posture. Once the conditions shown in the equations (5) and (6) are satisfied, HAL starts the PD control for the swing phase in a right leg and for the support phase in a left leg. On the other hand, HAL starts the PD control for the swing phase in a left leg and the support phase in a right leg once the conditions shown in the equations (7) and (8) are satisfied. The PD control for a swing leg continues until HAL finishes the reference swing patterns. After that, HAL runs the constant-value control for the landing phase until the condition shown in the equation (1) or (2) is satisfied in a case of a right leg and until the condition shown in the equation (3) or (4) is satisfied in a case of a left leg. The other leg continues the control for the support phase. After HAL detects a contact between a foot of a swing leg and a floor, HAL runs the constant-value control for the support phase on both legs and continues the control until the next swing start conditions are satisfied. If the conditions are not satisfied,

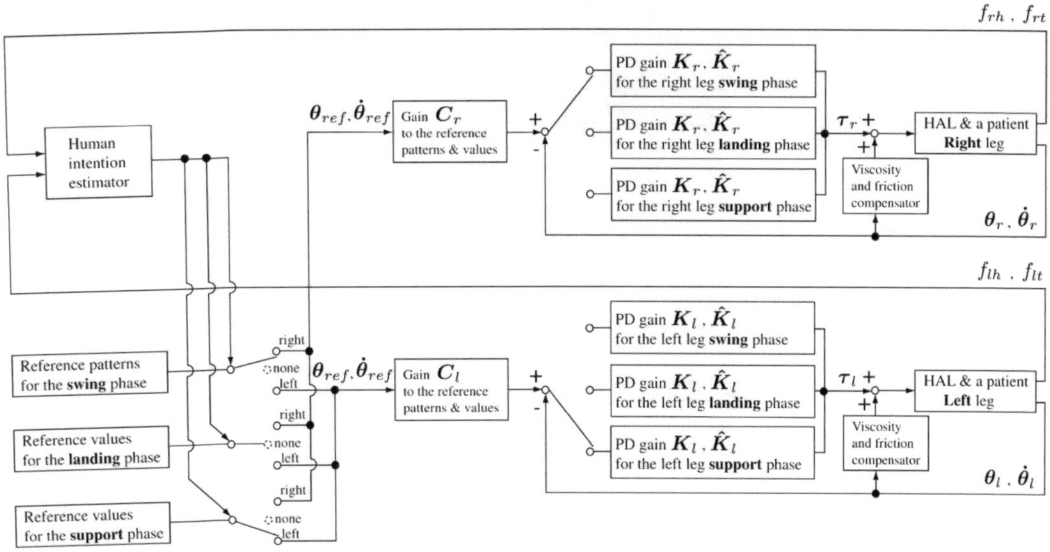

Fig. 8. Block diagram for tracking control.

two legs are kept at the final posture of the step. However, the reference angles of all joints are almost zero in this phase, therefore a backward leg is replaced around a forward leg if a load on the backward leg becomes almost zero by his/her weight shift. Thus, a wearer can come back to the standing posture. This algorithm can synchronize walking support with human intentions at a walk start instance, a walk stop instance as well as the beginning of leg swing during walking. In addition to those walking support, HAL compensates viscosity and static friction of the power units [3].

5. EXPERIMENT

The subject A is the patient who has a strong sensory paralysis especially on the left leg and can walk slowly using two canes with his both hands. Since he can stand by himself, the support aim with HAL is to help his leg swinging forward and sustaining his weight (65 kg). This support contributes to stabilize his walk by pushing a swing leg forward and by avoiding collisions of a swing leg with a floor. In this experiment, the patient is supposed to keep his own stability by holding a walking frame with his arms and a staff supports the walking frame for the sake of the patient's safety as shown in Fig. 9.

5.1 Experimental setup

In this experiment, the whole thresholds to detect a moment when a foot leaves a floor or contacts on a floor expressed as α_{rh}, α_{rt}, α_{lh}, α_{lt}, β_{rh}, β_{rt}, β_{lh} and β_{lt} are finally set to 50 N based on the subject's weight and his impression after some trials. On the other hand, the feedback gains for the joint control k_{rh}, k_{lh}, $\hat{k}_{rh}$, $\hat{k}_{lh}$, k_{rk}, k_{lk}, $\hat{k}_{rk}$ and $\hat{k}_{lk}$, the gains to the reference joint angle and velocity errors c_{rh}, c_{rk}, c_{lh} and c_{lk} and a time span for swinging a leg are adjusted through some trials reflecting the subject's impression. The time span for swinging a leg is finally set to 0.9 sec at the time.

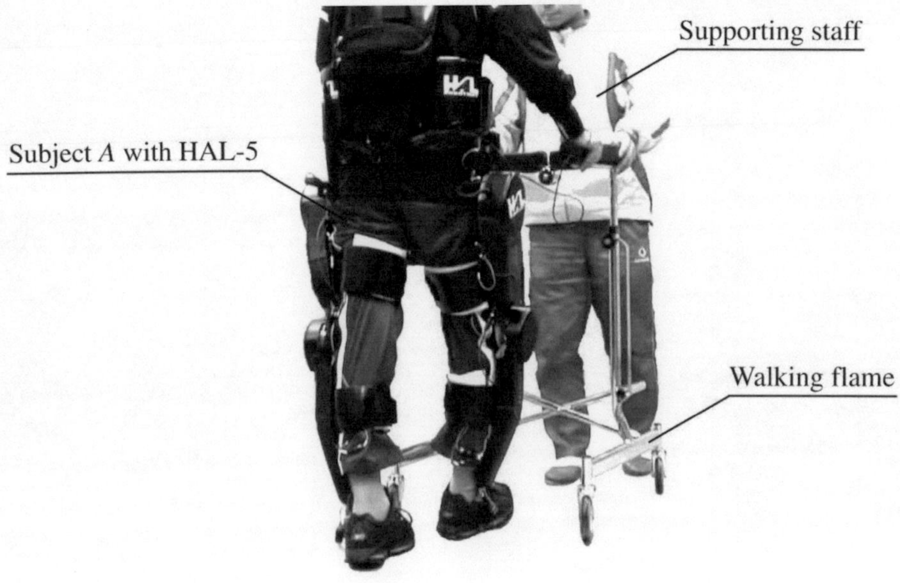

Fig. 9. Experimental setting.

5.2 Results

Figures 10 and 11 show the FRF data and phase transitions on each leg during a walking support. In both figures, one leg performs as the support leg up to a toe-off moment when the equation (6) or (8) is satisfied and then the leg performs as the swing leg for 0.9 sec and the leg begins to support his weight as the support leg from a heel-on moment when the equation (1) or (3) is satisfied shortly after the start of the landing phase. In addition, Fig. 12 shows the FRF data on both legs and the phase transitions at the start of walking support. The FRF of the heel part is almost zero since the subject A leans on the walking frame for the sake of safety. On the other hand, the FRF of the toe part reflects the shift of his COG. At first, he stands on his legs with a load distribution which the right leg supports about 250 N and the left leg supports about 350 N. After that, he shifts his COG in a direction toward his left side, and finally the right and left leg begins to perform as a swing leg and support leg, respectively when the equations (5) and (6) are satisfied. HAL starts supporting the walk of the subject A synchronizing his intentions. Figures 13 and 14 show his each joint angles, their references and torques of the power units during walking support. From the results of joint angles in these figures, his hip and knee joints follow the reference angles in a almost part of time in one cycle of the supported walk. HAL supports his walk based on a healthy person's walk as shown in Fig. 7. On the other hand, the results in a latter part of the swing phases show his joints do not follow the references, especially knee joint on his left leg which has a severe sensory trouble. The knee joint of the subject A resists the actuator of HAL since he does not get used to receiving the physical support. The tracking error will be small after enough training for relaxation of the knee joint in the swing phase.

6. CONCLUSIONS

In this chapter, we have proposed the algorithm to estimate patients' intentions so that the HAL-5 Type-C could support a patient with paraplegia to walk. The estimation algorithm based on the floor reaction force was investigated through the walking support experiments for a patient with a sensory paralysis on both legs. The cycle of reference walking patterns was adjusted for the patient and the walking support based on the reference walking was achieved, synchronizing with a patient's intentions estimated by the algorithm. We confirmed that the algorithm successfully estimated corresponding to a patient's intentions. However, it did not stabilize a patient's body posture and he had to maintain his balance using a walking frame with his hands. One of our future works is to develop a stabilizing algorithm and mechanism so that his hand regains its own functions.

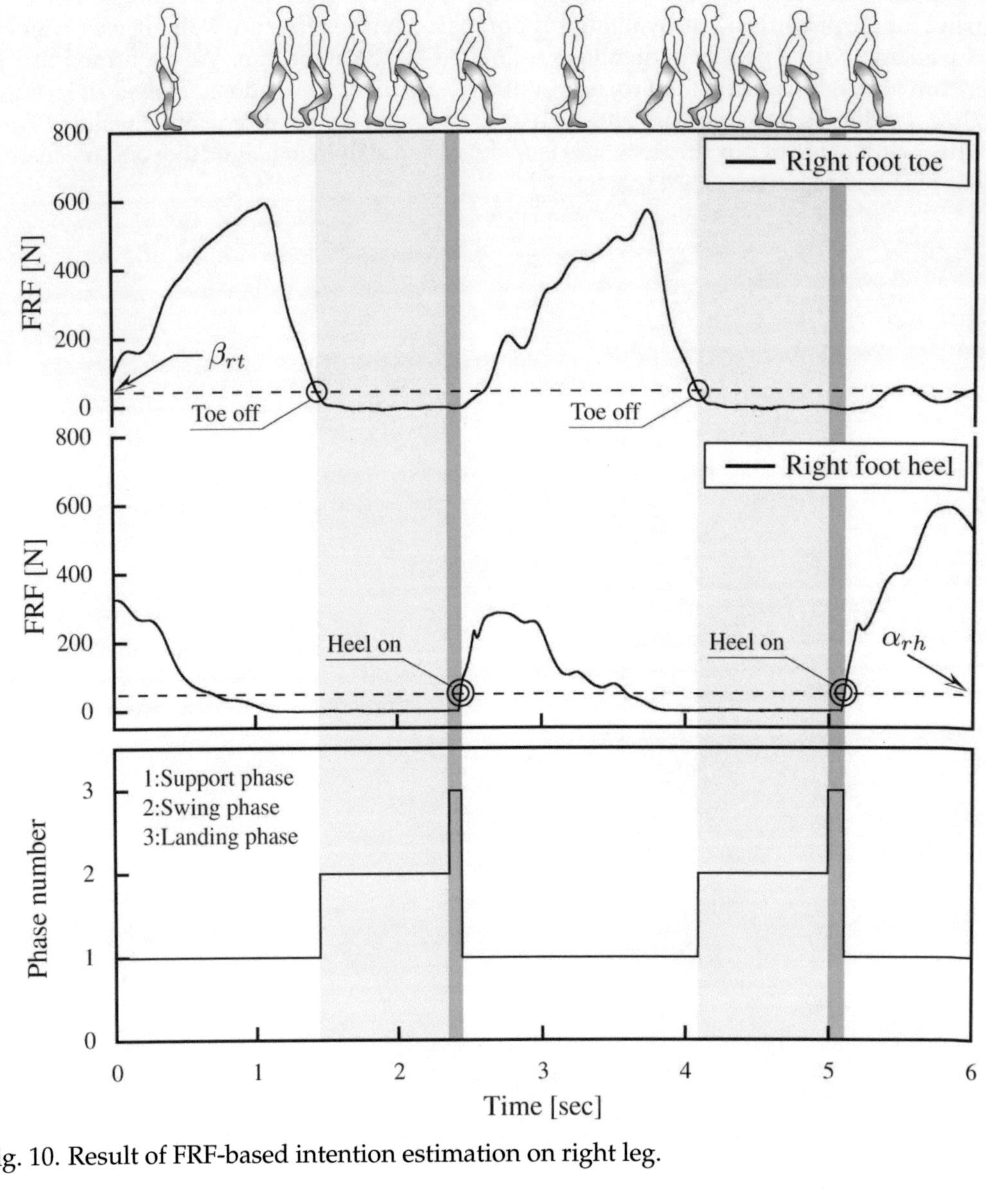

Fig. 10. Result of FRF-based intention estimation on right leg.

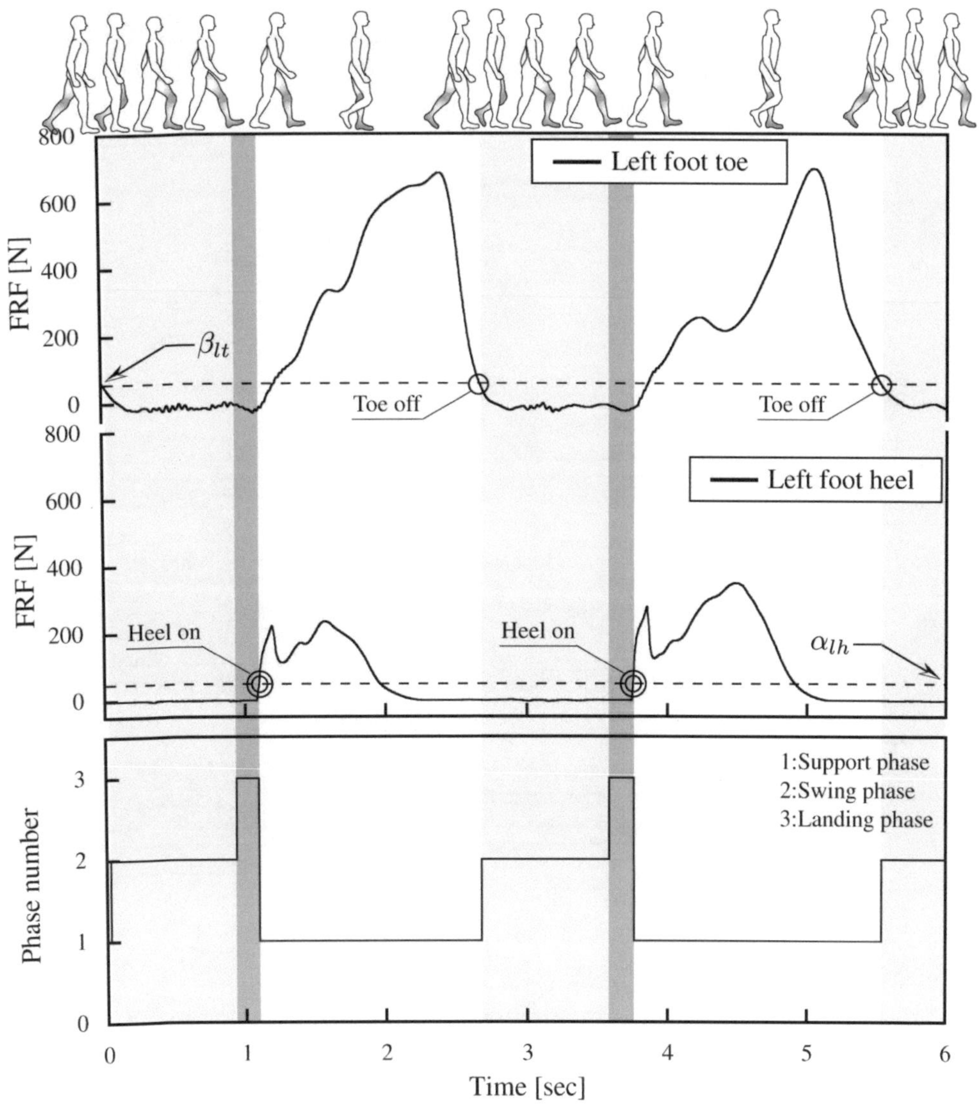

Fig. 11. Result of FRF-based intention estimation on left leg.

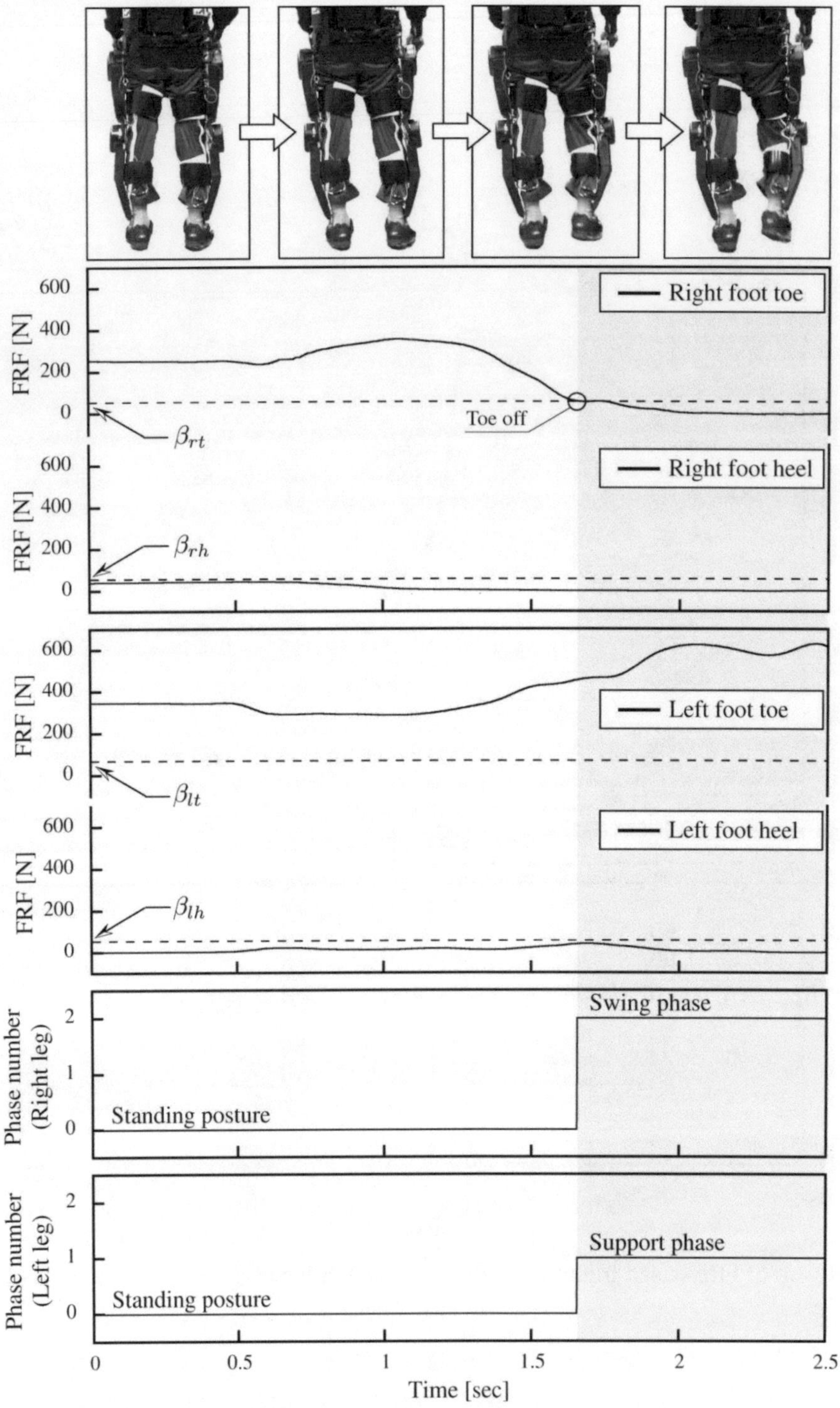

Fig. 12. Start of walking support by intention estimation.

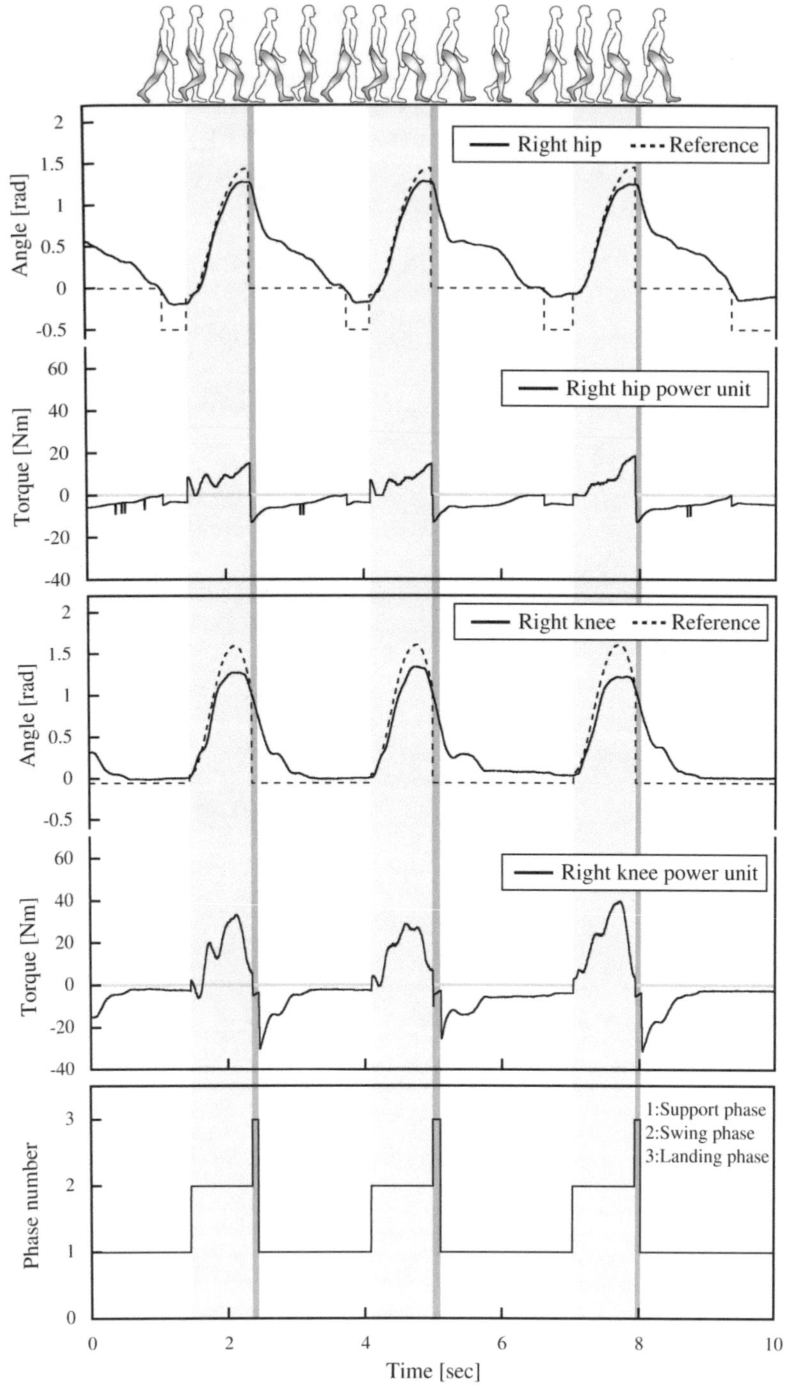

Fig. 13. Right leg joint angles with reference angles and power units' torques in each phase.

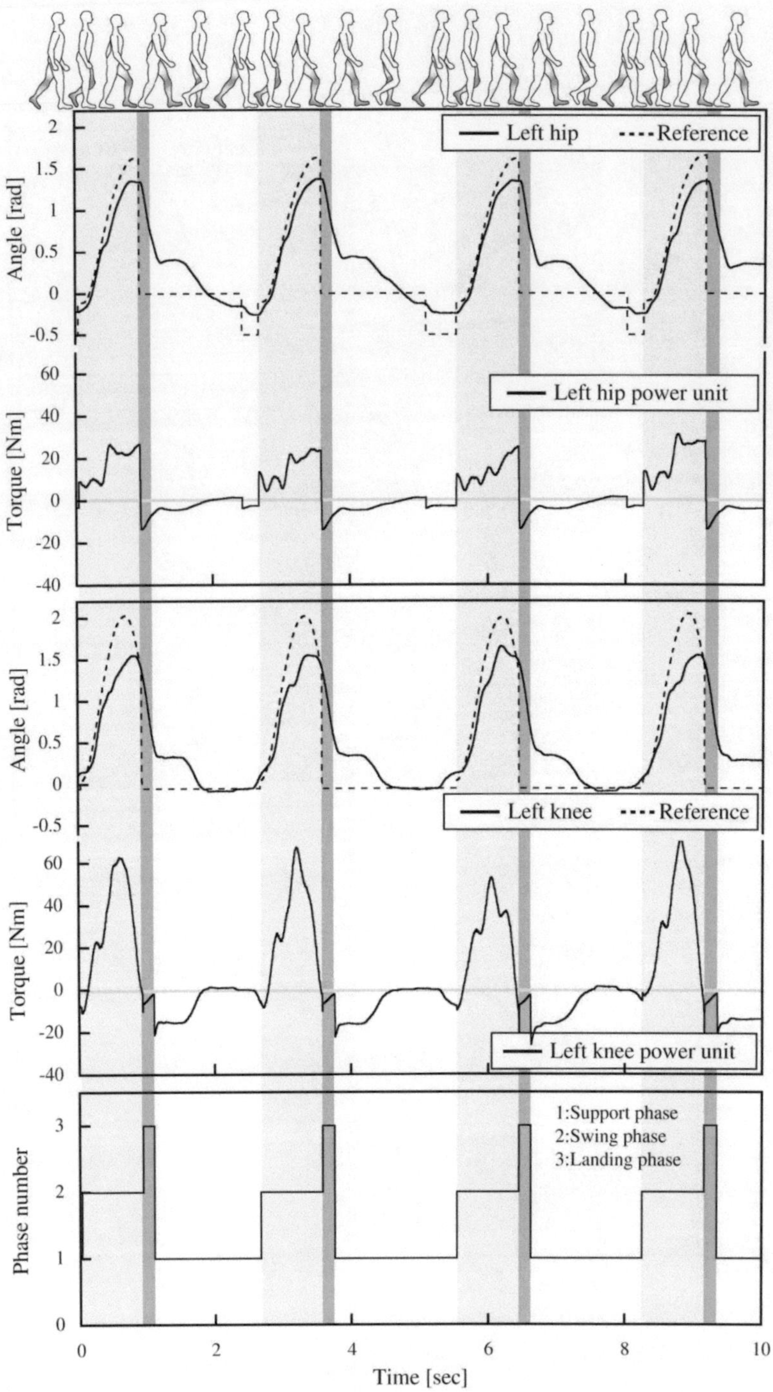

Fig. 14. Left leg joint angles with reference angles and power units' torques in each phase.

Acknowledgments

This research was partially supported by Grant-in-Aid for Scientific Research (General Research (A), No. 17206045, 2005-2008) from the Ministry of Education, Culture, Sports, Science and Technology of Japan and by Health and Labour Sciences Research Grants (No. H17-Physi-005, 2005-2007) from the Ministry of Health, Labour and Welfare of Japan.

7. REFERENCES

[1] Junpei Okamura, Hiroshi Tanaka and Yoshiyuki Sankai, EMG-based Prototype Powered Assistive System for Walking Aid, in *Proc. of Asian Symposium on Industrial Automation and Robotics (ASIAR'99)*, Bangkok, pp.229-234 (1999).

[2] Takao Nakai, Suwoong Lee, Hiroaki Kawamoto and Yoshiyuki Sankai, Development of Powered Assistive Leg for Walking Aid using EMG and Linux, in *Proc. of Asian Symposium on Industrial Automation and Robotics (ASIAR'01)*, Bangkok, pp.295-299 (2001).

[3] Lee S. and Sankai Y., Power Assist Control for Walking Aid with HAL-3 Based on EMG and Impedance Adjustment around Knee Joint, *Proc. of IEEE/RSJ International Conference on Intelligent Robots and Systems (IROS 2002)*, EPFL, Switzerland, pp.1499-1504 (2002).

[4] Jerry E. Pratt, Benjamin T. Krupp, Christopher J. Morse and Steven H. Collins, The RoboKnee: An Exoskeleton for Enhancing Strength and Endurance During Walking, in *Proc. of the 2004 IEEE International Conference on Robotics and Automation (ICRA '04)*, New Orleans, pp.2430-2435 (2004).

[5] Takahiko Nakamura, Kazunari Saito and Kazuhiro Kosuge, Control of Wearable Walking Support System Based on Human-Model and GRF, *Proc. of the 2005 IEEE International Conference on Robotics and Automation (ICRA '05)*, Barcelona, pp. 4405-4410 (2005).

[6] H. Kazerooni, Ryan Steger and Lihua Huang, Hybrid Control of the Berkeley Lower Extremity Exoskeleton (BLEEX), *The International Journal of Robotics Research*, vol. 25, pp. 561-573 (2006).

[7] Ryan Steger, Sung Hoon Kim and H. Kazerooni, Control Scheme and Networked Control Architecture for the Berkeley Lower Extremity Exoskeleton (BLEEX), in *Proc. of the 2006 IEEE International Conference on Robotics and Automation (ICRA '06)*, Orlando, Florida, pp. 3469-3476 (2006).

[8] Hiroaki Kawamoto and Yoshiyuki Sankai, Power Assist System HAL-3 for Gait Disorder Person, *Proc. of the 2002 International Conference on Computers Helping People with Special Needs (IC-CHP 2002)*, Linz, Austria, pp. 196-203 (2002).

[9] Hiroaki Kawamoto and Yoshiyuki Sankai, Power assist method based on Phase Sequence and muscle force condition for HAL, *Advanced Robotics*, vol. 19, No. 7, pp. 717-734 (2005).

[10] Masashi SATO, Hidetaka IKEUCHI, Ryozo KATOH and Tadashi YAMASHITA, Experimental Analysis of Reaction Force and Motion of Center of Gravity during Human Gait Initiation (Characteristics of Transferring from Transient Phase to Steady-State Phase), *Trans. JSME (Series C)*, Vol.59, No.566, pp.3101-3107 (1993). (in Japanese)

[11] H. Ikeuchi, K. Shinkoda, R. Katoh, M.Sato, T. Yamashita, Analysis of Human Transient Walking by Wavelet Transform, *Proc. of The Third International Symposium on Artificial Life and Robotics*, Beppu, Japan, pp.695-698 (1998).

[12] Suzuki K., Kawamura Y., Hayashi T., Sakurai T., Hasegawa Y. and Sankai Y., Intention-Based Walking Support for Paraplegia Patient, *Proc. of the 2005 IEEE International Conference on Systems, Man and Cybernetics (SMC2005)*, Hawaii, pp.2707-2713 (2005).

Appendix A

The PD gains of swing leg control k_{rh}, k_{lh}, $\hat{k}_{rh}$, $\hat{k}_{lh}$, k_{rk}, k_{lk}, $\hat{k}_{rk}$ and $\hat{k}_{lk}$ (see equations (16)) were designed in preliminary experiments. In this appendix section, proportional gains for the hip joints k_{rh} and k_{lh} are expressed by k_{*h}, and proportional gains for the knee joints k_{rk} and k_{lk} are expressed by k_{*k} for convenience of explanation. On the other hand, derivative gains for the hip joints $\hat{k}_{rh}$ and $\hat{k}_{lh}$ are expressed by $\hat{k}_{*h}$, and derivative gains for the knee joints $\hat{k}_{rk}$ and $\hat{k}_{lk}$ are expressed by $\hat{k}_{*k}$. Figure 15 shows experimental environments, where one leg can swing freely and the other leg supports a subject's weight. At first, hip or knee joint is controlled by the PD feedback control to examine the frequency response. The reference joint angle patterns are expressed by sine function with seven different frequencies, ranging from 0.1 to 3.0 Hz, and with 1.0 rad peak-to-peak amplitude. Five different PD gains, ranging that the proportional gain is from 20.0 to 200.0 and the derivative gain is from 0.02 to 0.20, are tested. Figures 16 and 18 show the frequency responses of hip and knee joints on Bode plot.

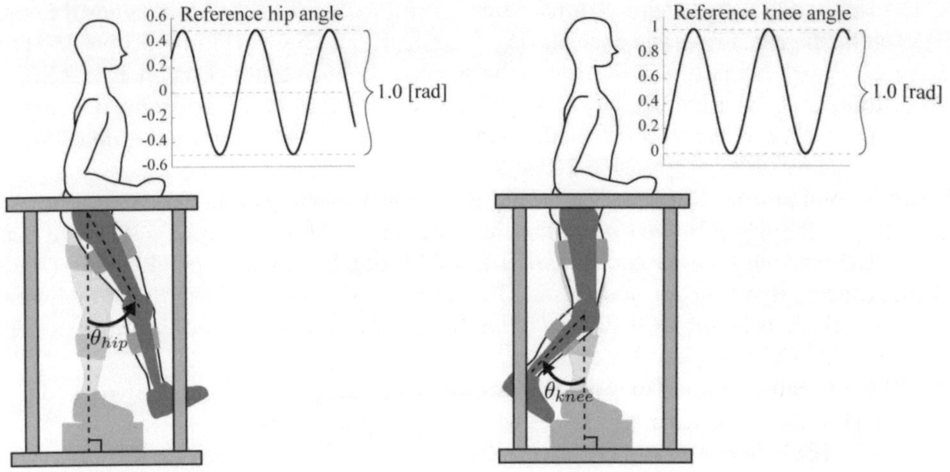

(a) Experimental motion for hip joint. (b) Experimental motion for knee joint.

Fig. 15. Experimental settings for each joint frequency response. A subject with the robot suit HAL stands inside a frame with one leg on a raised block so that the other leg can swing freely. The subject are asked to keep the upper body upright and completely relax the leg to follow reference joint motions produced by the HAL's power units. Two sine curves in the figure show the reference joint angle patterns on each joint.

Then, a unit step response of each joint is examined with six different PD gains. Figures 17 and 19 show the responses of hip and knee joints. From the viewpoint of amplitude characteristic, resonance frequency and phase shift on the hip joint shown in Fig. 16, the sufficient response could be obtained when the range that k_{*h} was from 100.0 to 200.0 and $\hat{k}_{*h}$ was from 0.10 to 0.20 at less than 0.5 Hz, which equaled leg swing frequency in this walking support. Besides, Figure 17 shows little overshoot and sufficient convergence at the proportional gain of k_{*h} = 100.0 and the derivative gain of $\hat{k}_{*h}$ = 0.10 for the unit step response. In consideration of those results, we set the hip joint feedback gains k_{*h} and $\hat{k}_{*h}$ equaled 100.0 and 0.10 in the actual walking supports. In the same way, the sufficient response was observed when the range that k_{*k} was from 100.0 to 200.0 and $\hat{k}_{*k}$ was from 0.10 to 0.20 at less than 1.0 Hz on the knee joint,

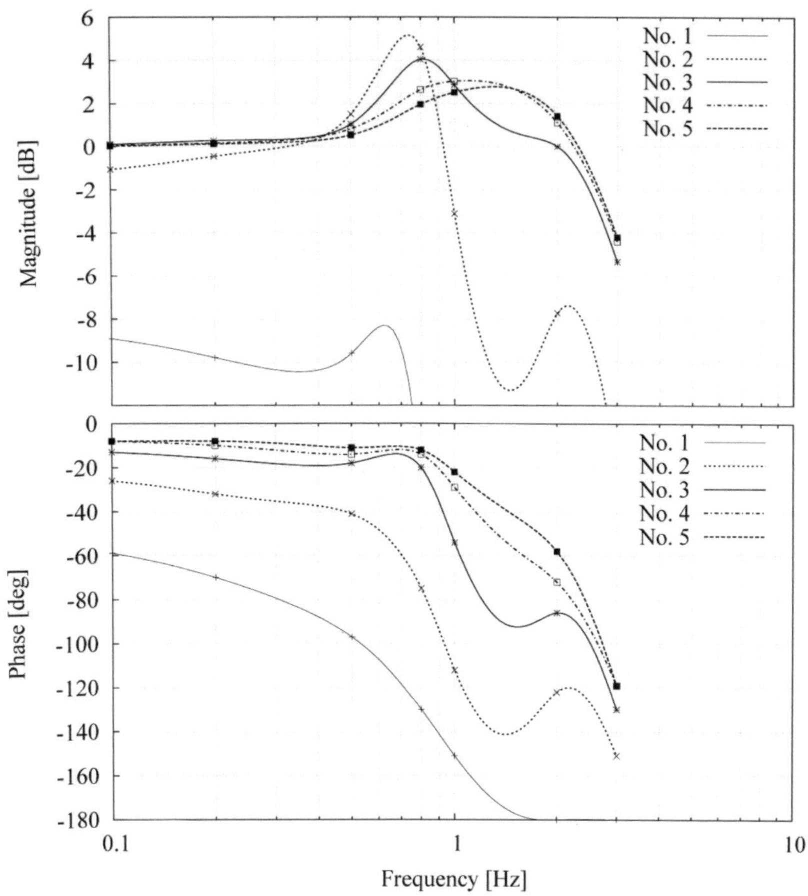

Fig. 16. Frequency response of hip joint shown in Bode plot. Upper graph and lower one show amplitude characteristic and phase characteristic, respectively. Line No. 1 is a characteristic curve in a case of "$k_{*h} = 20.0$ and $\hat{k}_{*h} = 0.02$". Similarly, line No. 2, No. 3, No. 4 and No. 5 show that of "$k_{*h} = 50.0$ and $\hat{k}_{*h} = 0.05$", "$k_{*h} = 100.0$ and $\hat{k}_{*h} = 0.10$", "$k_{*h} = 150.0$ and $\hat{k}_{*h} = 0.15$" and "$k_{*h} = 200.0$ and $\hat{k}_{*h} = 0.20$", respectively. The all five lines are drawn using a cubic spline curve in order to express the correspondence with seven points in the same set of feedback gains. They therefore interpolate the experimental data and would not precisely express the real values.

as shown in Fig. 18. We also set the knee joint feedback gains of 100.0 and 0.10 from the viewpoint of overshoot and oscillation on the step response, as shown in Fig. 19.

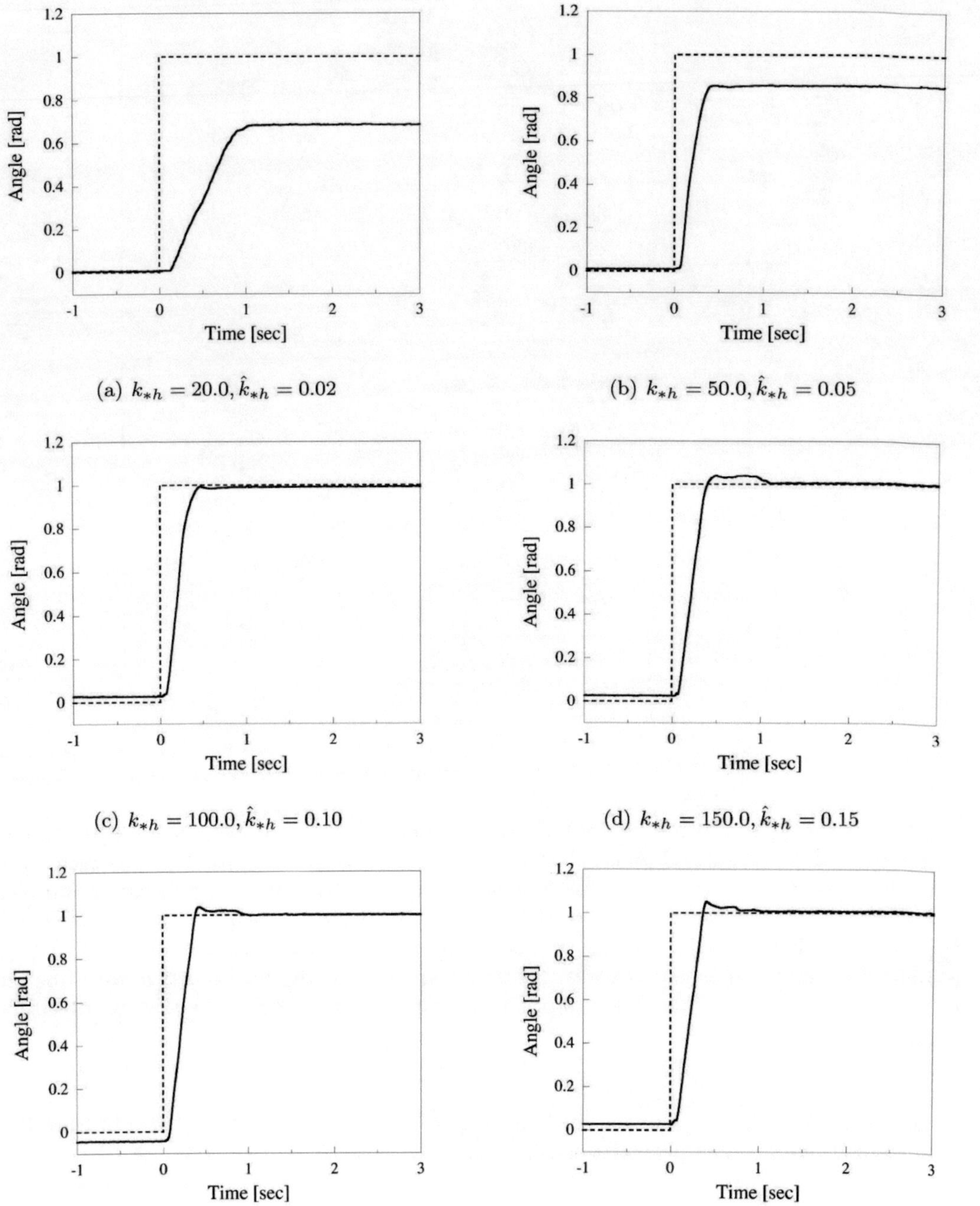

(a) $k_{*h} = 20.0, \hat{k}_{*h} = 0.02$

(b) $k_{*h} = 50.0, \hat{k}_{*h} = 0.05$

(c) $k_{*h} = 100.0, \hat{k}_{*h} = 0.10$

(d) $k_{*h} = 150.0, \hat{k}_{*h} = 0.15$

(e) $k_{*h} = 200.0, \hat{k}_{*h} = 0.20$

(f) $k_{*h} = 300.0, \hat{k}_{*h} = 0.30$

Fig. 17. Step responses of hip joint on six kinds of feedback gains (PD gains). Dashed lines and solid lines in the graphs show step inputs and actual angular variations, respectively.

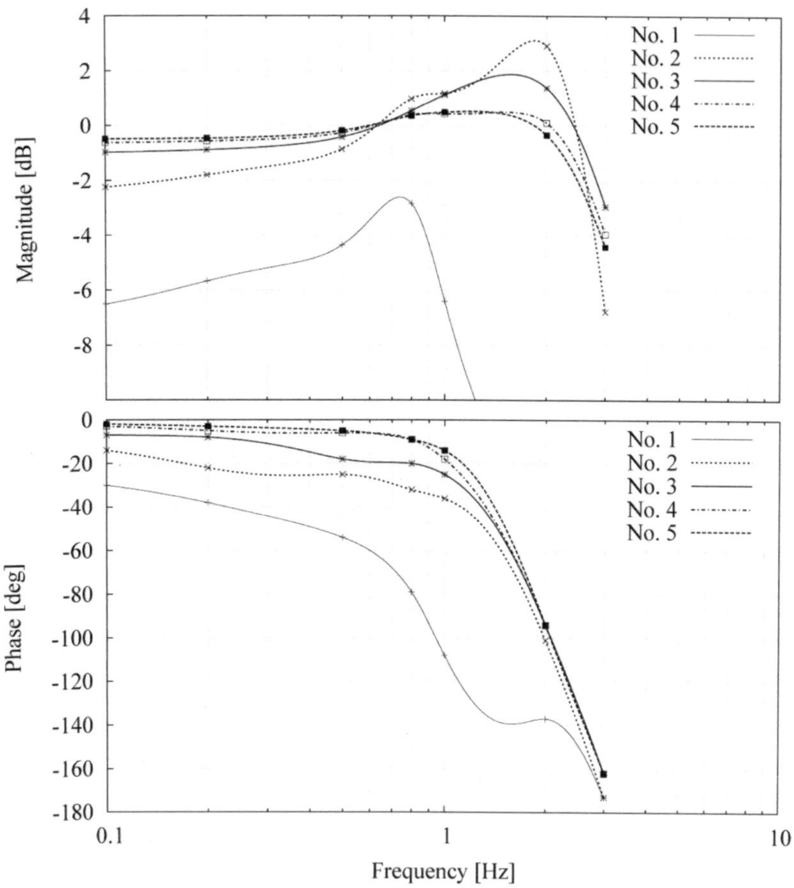

Fig. 18. Frequency response of knee joint shown in Bode plot. Upper graph and lower one show amplitude characteristic and phase characteristic, respectively. Line No. 1 is a characteristic curve in a case of "$k_{*k} = 20.0$ and $\hat{k}_{*k} = 0.02$". Similarly, line No. 2, No. 3, No. 4 and No. 5 show that of "$k_{*k} = 50.0$ and $\hat{k}_{*k} = 0.05$", "$k_{*k} = 100.0$ and $\hat{k}_{*k} = 0.10$", "$k_{*k} = 150.0$ and $\hat{k}_{*k} = 0.15$" and "$k_{*k} = 200.0$ and $\hat{k}_{*k} = 0.20$", respectively. The all five lines are drawn using a cubic spline curve in order to express the correspondence with seven points in the same set of feedback gains. They therefore interpolate the experimental data and would not precisely express the real values.

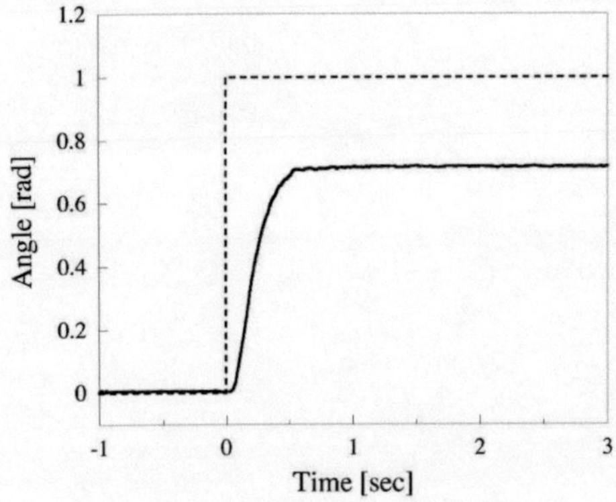

(a) $k_{*k} = 20.0, \hat{k}_{*k} = 0.02$

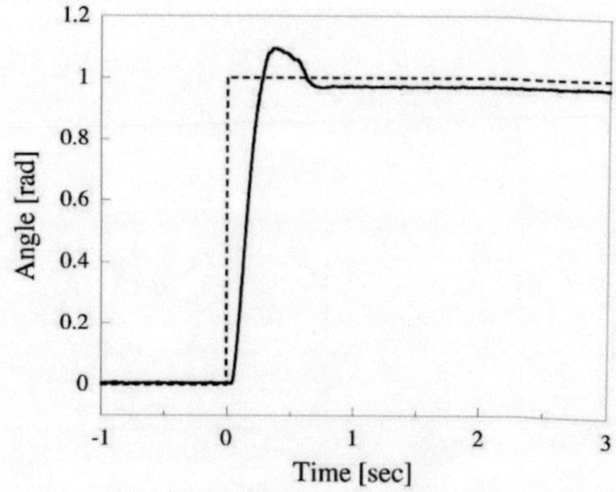

(b) $k_{*k} = 50.0, \hat{k}_{*k} = 0.05$

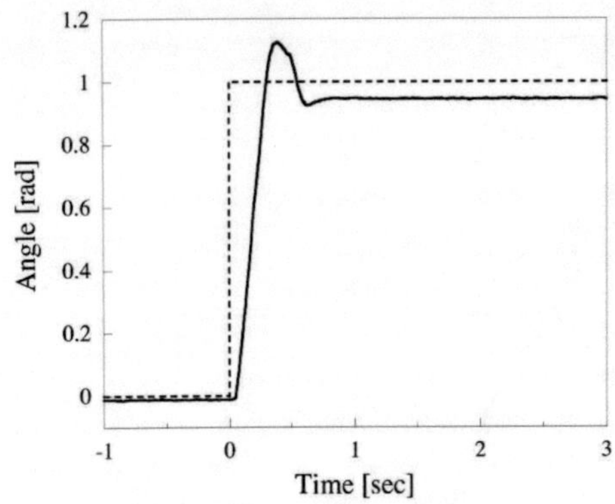

(c) $k_{*k} = 100.0, \hat{k}_{*k} = 0.10$

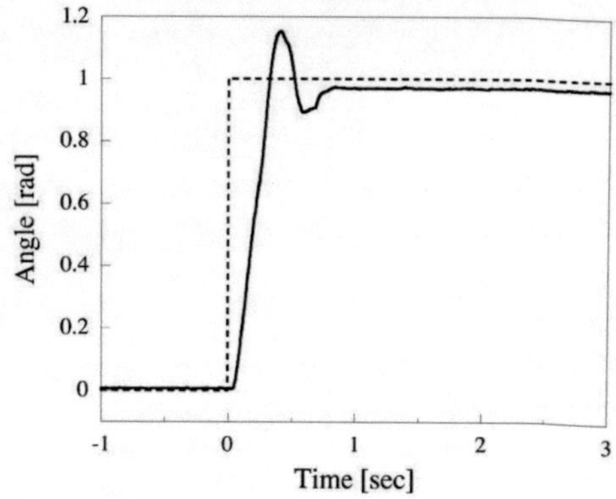

(d) $k_{*k} = 150.0, \hat{k}_{*k} = 0.15$

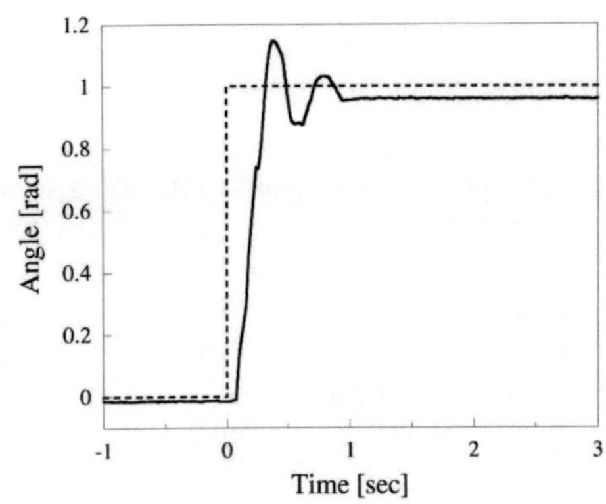

(e) $k_{*k} = 200.0, \hat{k}_{*k} = 0.20$

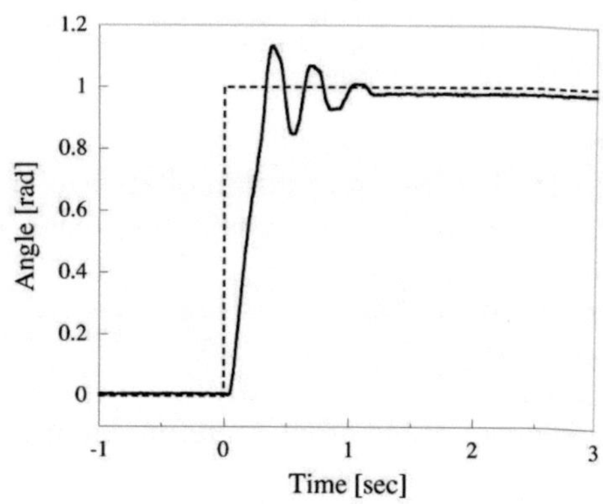

(f) $k_{*k} = 300.0, \hat{k}_{*k} = 0.30$

Fig. 19. Step responses of knee joints on six kinds of feedback gains (PD gains). Dashed lines and solid lines in the graphs shows step inputs and actual angular variations, respectively.

24

Development of Vision Based Person Following Module for Mobile Robots in RT-Middleware

Hiroshi Takemura, Zentaro Nemoto, Keita Ito and Hiroshi Mizoguchi
Tokyo University of Science
Japan

1. Introduction

Recently, the role a robot can play in a person's daily life has increased drastically over the years. There are many researches that aim to communicate and interact with human in a real environment. In order to achieve such purpose in a real environment, robots need the ability to see the person and adjust to the person's behavior. Such ability is an indispensable function for safety and security as well as carrying out tasks. The ability to automatically follow a person is a key enabling technology for mobile robots to effectively interact with the surrounding world. If a robot combines a communication function with speech-recognition, a grasping function with arms and a person following function, we believe that the robot can serve the person well (Fig.1). The robot must follow the person intelligently not only in a limited environment such as an indoor office but also in real world such as a crowded area outdoors. In order to continue following the person intelligently in an indoor/outdoor environment, the robot needs a robust adaptive control system under varying illuminations.

In this chapter, we describe the robust person following method based on disparity images and HSV color spaces from a color stereo camera and distance information from the laser range sensor. The proposed person following method applies for an inverted pendulum type robot, Segway RMP200 (Segway Japan, Ltd.) and a humanoid type service robot, enon (FUJITSU Ltd.). The validity of the proposed methods is confirmed though the person following experiments in an indoor/outdoor environment under varying illuminations.

Fig. 1. Overview of Person Following Mobile Robot Service

2. Person Following Mobile Robot

In order to follow a person, the first thing the robot must do is find the target person. Many researches using fixed camera images find the target person by using such characteristics as the face, human body shape, background subtraction, etc.(2006, Kwon et al., 2005, Negishi et al., 2004, Ohya & Munekata, 2002, Schlegel et al., 1998, Schulz et al., 2003, Sidenbladh et al., 1999). (The following problems, however, occur when the robot follows the target person by using a camera located on the robot, because both of the camera and the target person are moving.)
1) Background changing problem. The background is always changing because the camera is moving.
2) Occlusion problem. Camera images change according to distance between the robot and the target person.
3) Position problem. The image size of the target person is changeable.
4) Direction problem. The robot cannot see the target person's face at all times. Face tracking is not the best method in this case.
5) Multi person problem. Robot has to find the target person even if several people exit in the frame.
6) Brightness problem. The light condition, sunshine and shadow become a major problem when the robot moves between indoor/outdoor environments.
7) Real-time problem. Calculation cost is important for person following.

We proposed a robust person following method to solve such problems. The proposed method uses three information; the disparity images, HSV color spaces images from a color stereo camera, and the distance information from the laser range sensor. The proposed method does not depend on the image size, the target person's shape and face, but depends on the color histogram of the target person's clothes and the distance between target person and background. We also proposed the multi HS histogram sequential update method for a color extraction method to realize the person following in real surrounding.

3. Person's Area Detection Image Processing Method

3.1 HS Histogram Based Color Extraction Method

This section describes the target person detection method based on target color extraction image processing method. First, the target person's color images are registered by captured multi view point around the target person. Usually, clothing color is not the same throughout, print and color are not same on the front and back, or on the left and right side. If only one color of the person's clothes is registered, the robot cannot find the target person when the target person changes his/her walking direction. Fig.2 shows the template histogram created by a front image of the target person. The x axis indicates H value of HSV color space, y axis is S value of HSV color space, z axis is a normalized number of HS pixels. Fig.3 also shows the template histogram created by the back image of the target person. In this example, the front template has two peaks indicating blue and purple, and the back side temple has one peak indicating purple. These templates show that there are differences found in the colors found on the front and back side of the same person's clothes, even if same person.

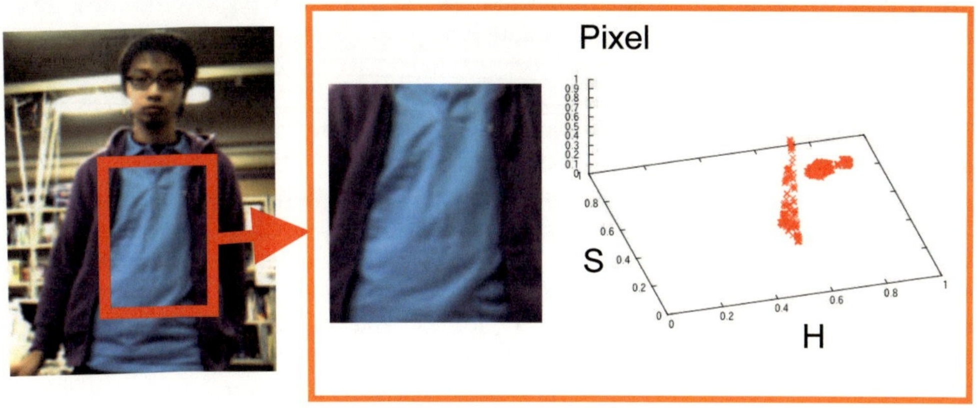

Fig. 2. Template Histogram (front side)

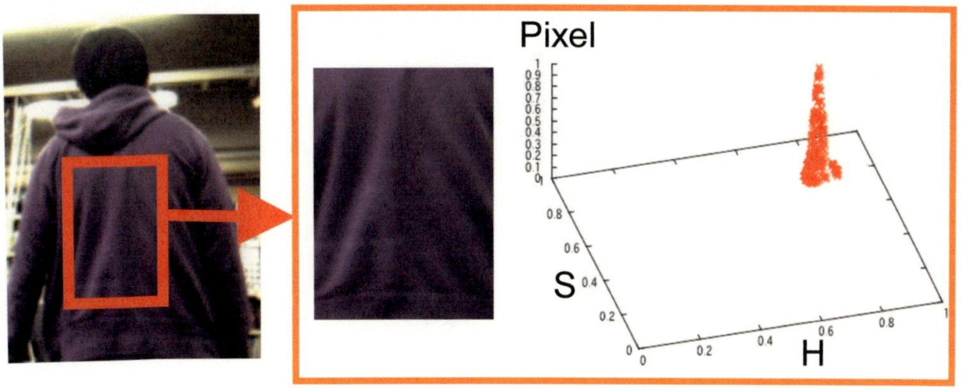

Fig. 3. Template Histogram (back side)

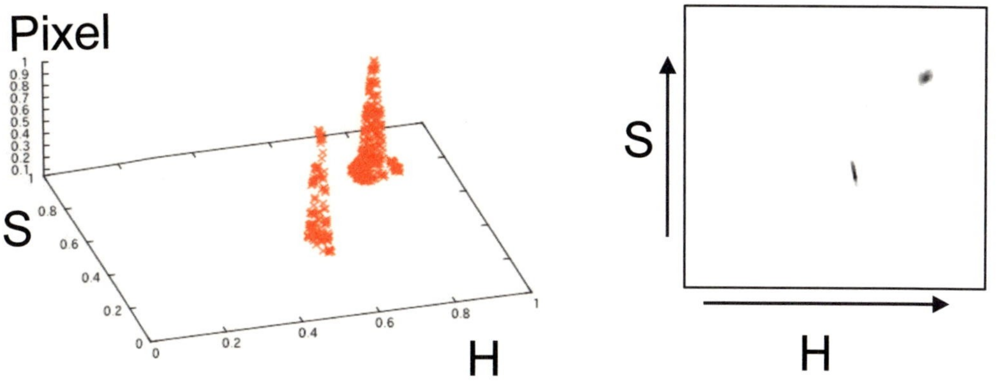

Fig. 4. Base Template Histogram

Fig.4 is a base template histogram connected by the front and back templates. The base template is also normalized by multi template histogram. Fig.4 (right) indicates the number of pixels as the brightness (0 to 1).

The next step is to cluster the base template histogram by using the Nearest Neighbor algorithm. The NN method is a method for classifying objects based on closest training examples in the feature space. The NN method was applied to the base template histogram, and the result is shown in Fig.5. There is no guarantee that a target person wears unique color clothes. The proposed method can extract multi colors by using the clustered histogram. In this example, the base template histogram divided into clusters. One of them indicates a blue color on target person's clothes, and the other a purple color. The advantages of using clustering of the base template histogram are that multi color templates are available and that some small color space and noise can be neglect.

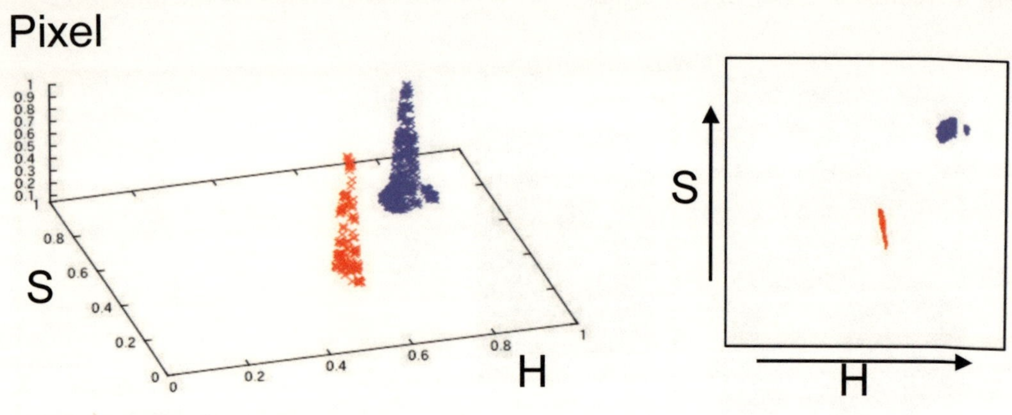

Fig. 5. Clustered Histogram by NN Method

3.2 Updating HS Histogram by Anisotropic Asymmetric Gaussian Distribution Model

The clustered histogram is not a Gaussian distribution and the HS histogram has an anisotropic characteristic. The S value is more changeable than the H value under varying illumination. The weight of each color cluster is calculated by using an anisotropic asymmetric Gaussian kernel function. The center cluster position (u,v) and the distribution value $\sigma^2(\theta)$ are needed to define the anisotropic asymmetric Gaussian distribution in the H-S plane. The definition of the anisotropic asymmetric Gaussian distribution model (Umeda et al., 2006) is given by

$$W(h,s) \equiv W(r,\theta) = e^{-\frac{r^2}{2\sigma^2(\theta)}}, \tag{1}$$

where r is a distance from the center of a cluster to the outer perimeter of the cluster and θ is an angle $(0<\theta<2\pi)$ of orientation around the center. The H-S rectangular coordinate system is transformed to r- θ polar coordinate system. Fig.6 shows the calculation result of the anisotropic asymmetric Gaussian distribution filter for weighting the HS histogram.

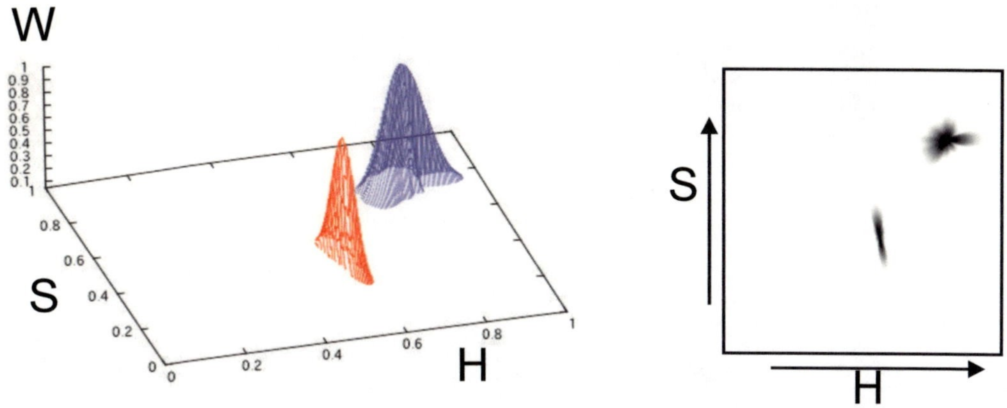

Fig. 6. Anisotropic Asymmetric Gaussian Distribution Filter for Weighting Histogram

In order to extract the target colors under varying illumination and cut noise, the weighted histogram has to update according to the lighting condition. The weighted histogram (H_w) updates based on the target person's area. $HS_{new}(t)$ is a current histogram acquired by the detected target person's area. $H_w(t\text{-}1)$ is old histogram. W is weight calculated by equation 1. $H_w(t)$ is calculated as follows:

$$Hw(t) = \frac{Hw(t-1) + W \times HS_{new}(t)}{\max(Hw(t-1) + W \times HS_{new}(t))} \tag{2}$$

3.3 Target Person Detection Processing Flow

The processing flow of the target person area determination method by using color, distance information and weighted histogram is shown in Fig.7.

1) Background Elimination: The background is eliminated by using the distance information from the stereo camera to reduce the color matching calculation cost. The area, which has the distance far from threshold value, is defined as the background area.

2) Color Extraction: The target person's color is extracted based on the weighted HS histogram.

3) Target Area Detection: The target person's area is detected based on the extracted pixels and a profiling process (Itoh et al., 2006). The profiling process is the position search method used maximum sum of x axis and y axis of the extracted pixels.

4) Histogram Update: The weighted histogram is updated based on the target person's area. In order to adjust to varying lighting condition, the weighted histogram is continuously to update using Equation (2).

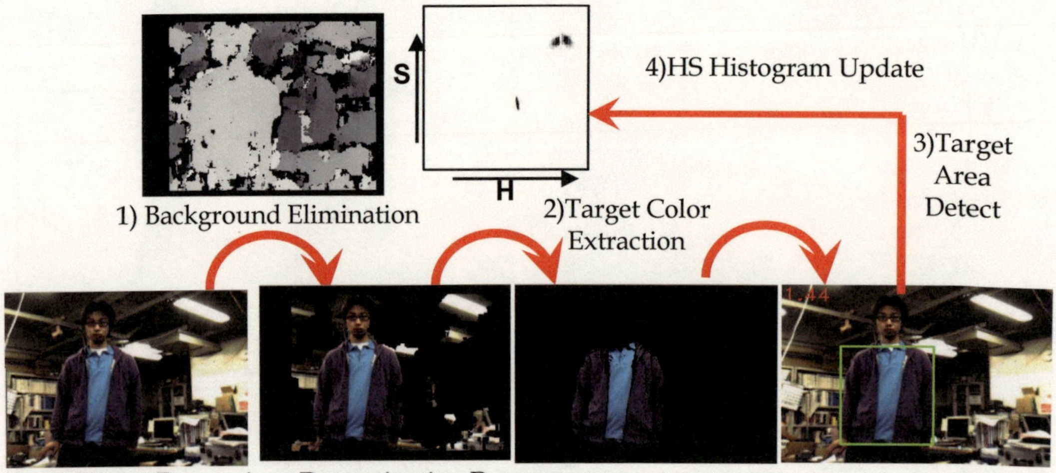

Fig. 7. Target Person Area Determination Process

3.4 Target Person's Area Detection Experiment Results

Fig. 8 shows the target person's area detection experiment results. In Fig.8, (a) are the target person's area detection results, (b) the target color extraction results, (c) HS_{new} histogram of the detected target person's area, and (d) are the updated weighted $H_w(t)$ histogram. The bottom numbers in the results indicate processing time [second]. The green rectangles indicate the target person's detected area. The lighting condition is changing; normal condition, becoming darker, becoming brighter and normal condition. As the results of time-series of (a), the target person could be tracked successfully in this experiment regardless of the size, shape, direction of target person and the varying illumination. It is understood that the histogram could be update according to the varying illumination. The background color, however, was not extracted, and only the target person could be tracked. This is the effect of the introduced anisotropic asymmetric Gaussian distribution filter. It estimated the changeable color range of the target person and the update process was actually performed only within the range.

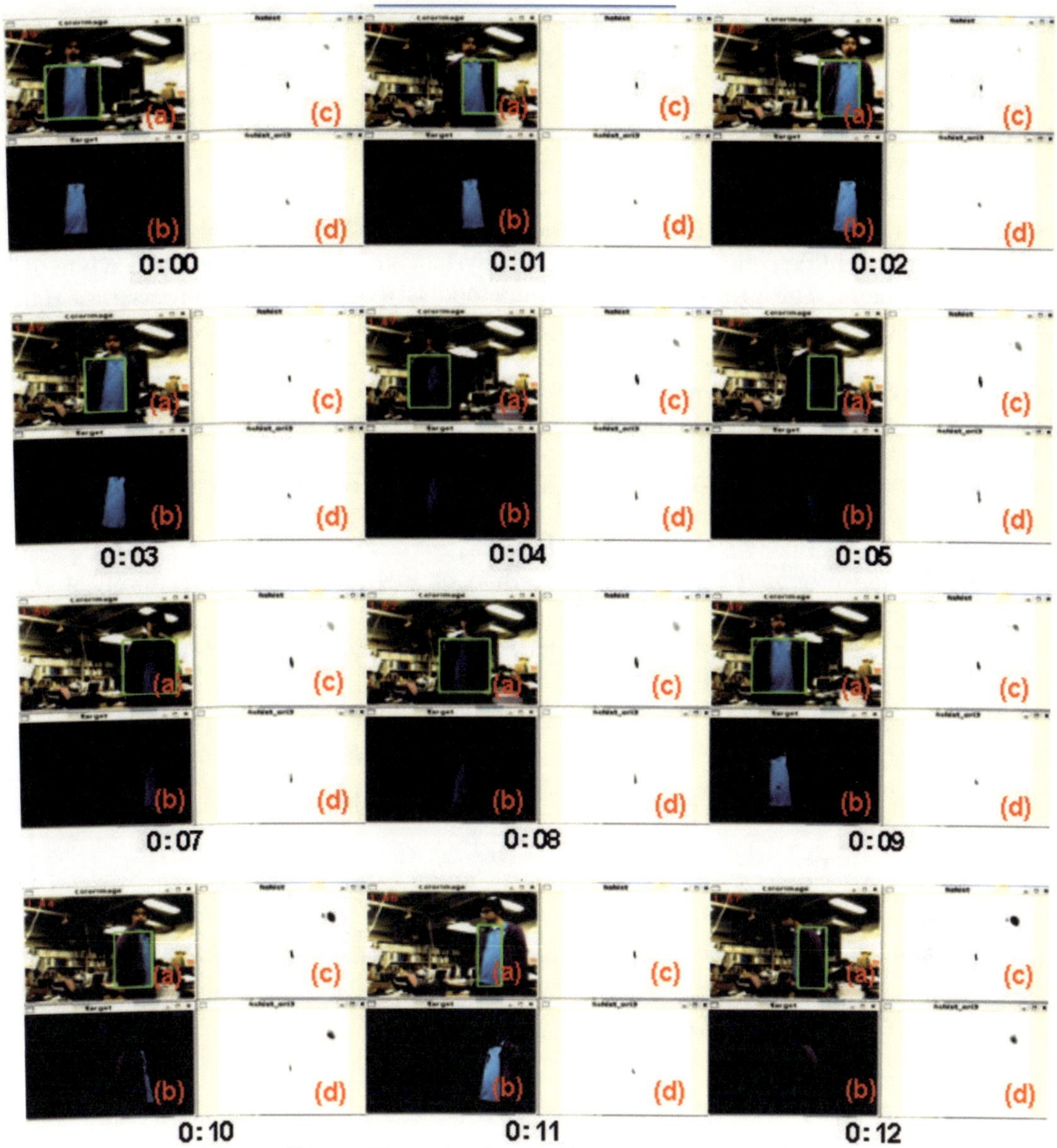

Fig. 8. Results of Target Person Detection Image Processing

4. Target Person Tracking by Laser Range Finder

4.1 Robust Target Person Tracking by LRF

A normal camera has a narrow dynamic range, the black out effect or the white out effect are caused when the robot moves from indoor to outdoor and vice versa, or moves against the sun, even if the proposed HS histogram sequential updating color extraction method is applied. The proposed method does not adjust to such a major illumination change. Thus, we also proposed the robust target tracking method by using the Laser Range Finder (LRF: UTM-30LX, HOKUYO CO., LTD.). In this method, the target person is detected by the proposed stereo camera based method. The LRF is a laser range sensor, which can be used in an outdoor environment. LRF is not greatly influenced by the lighting condition.

4.2 LRF Target Tracking Method

The Nearest Neighbor Clustering method based on the input data from LRF (Hashimoto et al., 2006, Ishihara et al., 2002, Kawata et al., 2005, Okusako et al., 2006, Lindstrom & Eklundh, 2001) applies to the LRF target tracking method. The LRF can be used stably and can adjust to such varying lighting conditions of the camera system such as a black out or a white out caused by sunlight. The LRF target tracking method is that the cluster of distance data around the robot is calculated by the Nearest Neighbor Clustering method and the target person's cluster is searched and tracked based on a shape of cluster and a Euclidean distance between each cluster. The process flow is described as follows:

(1) Converting distance date from the LRF to 2-dimensinal coordinate data (Fig.9 (a)).

(2) Clustering by NN method (Fig.9 (b)).

(3) Numbering only human size clusters and calculating center of each target person candidate's clusters (Fig.9 (c)).

(4) Tracking the target person's cluster that is the nearest position of the target person's area calculated by the proposed camera based method or that is the nearest position of the pervious target person's cluster.

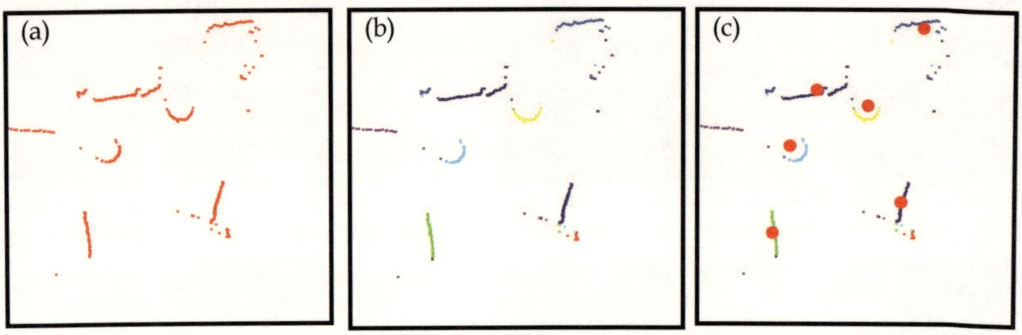

Fig. 9. Results of Target Person Detection Image Processing: (a) Input Data (b) Results of Clustering (c) Center of Each Class

5. Target Person's Position Determination

The target person's position is determined based on the result of the proposed target person's area detection method and the proposed LRF target tracking method. In order to adapt to the changing image size and shape of the target person, the proposed target person's area detection method uses the color information as the target person's identification. The proposed LRF target tracking method calculates the target cluster position based on the size of 2-dimentional cross section shape and the time-series data. Although the camera based method has some trouble adjusting to rapid lighting condition change, the camera based method can re-detect when some occlusions occurred and the target is lost. Although the LRF target tracking method was trouble tracking the target person when some occlusions occurred and cannot determine the target person by using only LRF data, the LRF target tracking method can adjust to rapid changing lighting condition. The target person's position is determined by integrated the result of the target person's area and the result of the LRF target tracking method to adjust to occlusion and changing lighting condition.

5.1 Integration of Camera Information and LRF Distance Information

The target person's area and the target person's cluster position should be the same when the robot moves in a stable environment. Fig.10 shows the integrated camera and LRF results. The outputs of both methods are the same. However, in real environment, the lighting condition is unstable and some occlusions occur by obstacles and other people. Sometimes the robot cannot detect the target person's area and the robot cannot track the target person's cluster. Although the camera system can track the target person as well as detect the target person, the camera system has a major problem (white/black out, against the sunlight) when the robot moves from indoors to outdoors and vice verse. The LRF target tracking method is not effect by sunlight and can be used in an outdoor environment. However, the target person cannot be detected by only using the LRF method. Another method is needed to detect the target person's initial position when the robot has lost the target person. Thus, the integration of the two difference method can compensate the potential weakness. At first, the target person's area is detected by using the camera based method. Then, the target person's position is tracked by the LRF based method based on the camera based method result. Although the output from the LRF based method is used as the robot command input, the HS histogram continues to update according to the lighting condition. When the LRF based method has lost the target person's position by occlusion or other reason, the camera based method re-detects the target person's area by using the updating HS histogram method.

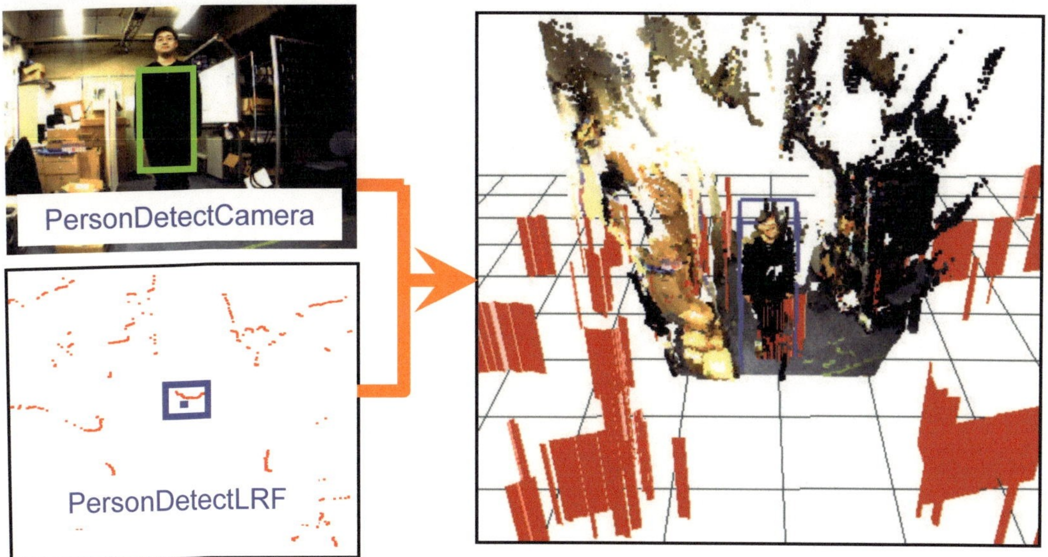

Fig. 10. Integration of Target Person's Position Calculated by the Image Method and the LRF Method

6. Indoor/Outdoor Target Person Following Experiment

6.1 Evaluation of Target Person Following Method

The snapshots of the indoor target person following experiment are shown in Fig. 11. The robot can follow the target person in an indoor environment. The inverted pendulum type robot, Segway RMP200 (Segway Japan, Ltd.), is used as the robot platform. The robot has a color stereo camera (STOC: Videre Co. Ltd.) located on a two-axis (pan/tilt) motion control device (Biclops PT: TRACLabs Inc.). A LRF (UTM-30LX: HOKYOU Co. Ltd.) is also located on the robot. All devices are controlled by one laptop computer. Fig.12 shows the trajectories of the target person and the robot calculated by the robot's odometer. The robot keeps a distance from the target person within 3 meter during the target person following. The target person following ability of the robot in simple environment is verified through this experiment.

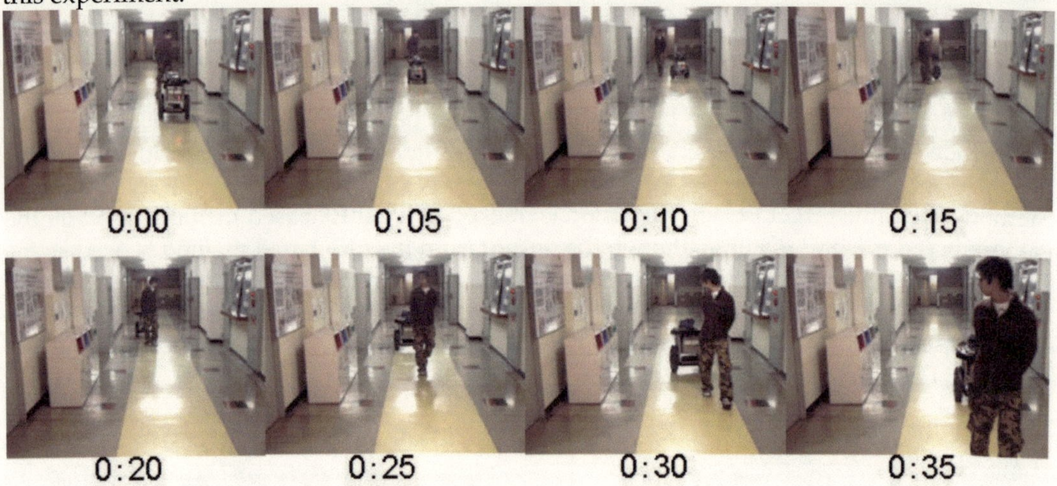

Fig. 11. Snapshots of Target Following Test

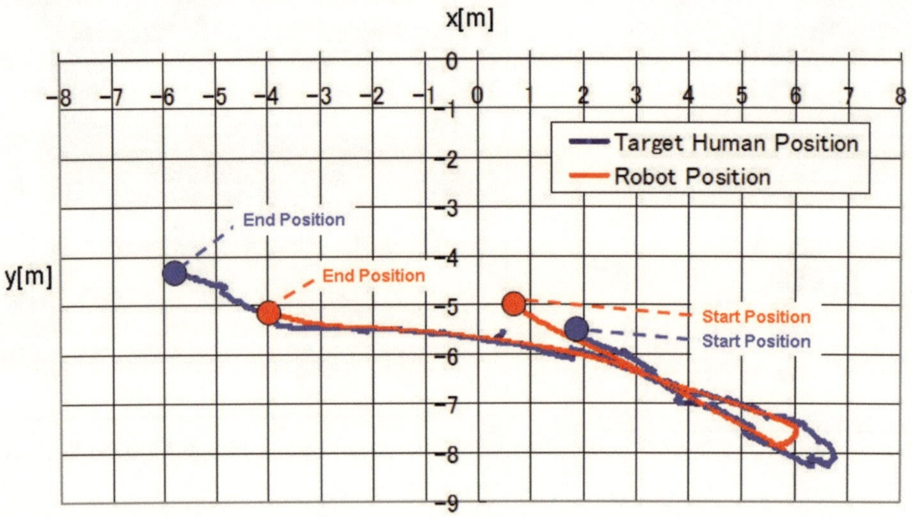

Fig. 12. The Target and the Robot Trajectory during the Target Following Test

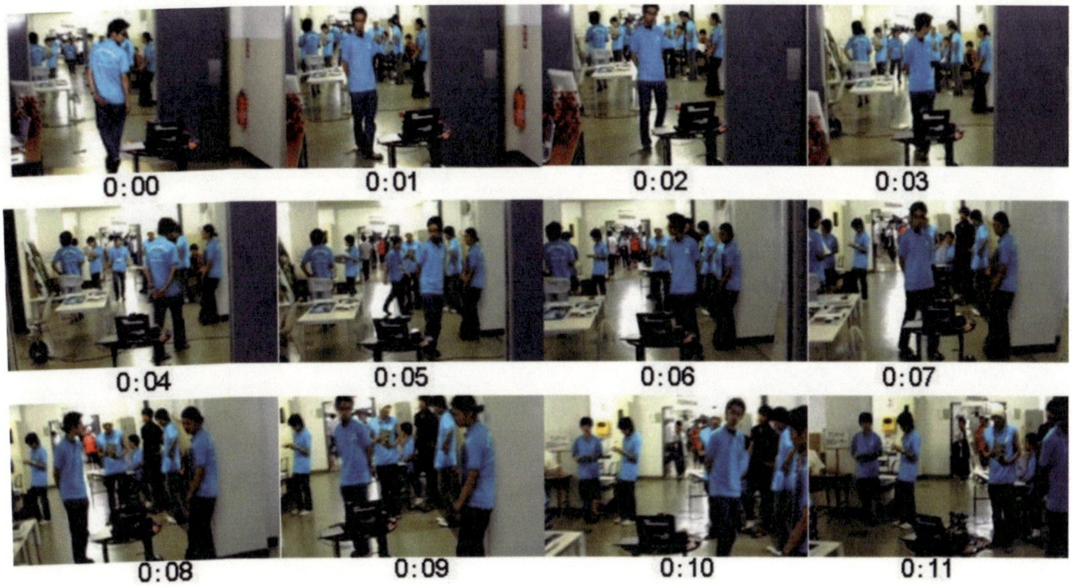

Fig. 13. Person Following Experiment (Indoor Environment)

6.2. Indoor to Outdoor Target Person Following Experiments

The pervious experiment was performed in an indoor environment with a stable lighting condition. In this section, the experiments of target following is conducted in a more complicated environment, and the robustness of the proposed method are evaluated. Fig.13 shows the experiment result of an indoor environment with multi persons wearing the same color shirt. The space that the robot can move is limited and crowded with people. In such environment, the robot must not make a false recognition and the robot must adjust to the target person's movement. The proposed method uses the background subtraction based on the distance information of the stereo camera and also uses the time-series data. Therefore the robot can detect and track the target person in such a narrow space with other person wearing the same color shirt. The robot has the function to adjust to the target person's velocity. When the target person walks slowly in a crowded space, the robot tries to reduce the distance from the target person, and moves slowly according to the target person's walking velocity. Thus the robot achieves the target person following in such environment.

Fig. 14 shows the experiment results when the robot moves from indoor to outdoor. The major illumination change occurs when the robot moves from indoor to outdoor, even human is dazzled by the bright sunlight. The image device usually has a narrow dynamic range, and the stereo camera is hard to adjust to varying illumination compared to human eyes. However, the LRF is little effected by the sunlight because the LRF uses a laser to measure the distance around the sensor. Therefore the robot can track and follow the target person in such lighting conditions.

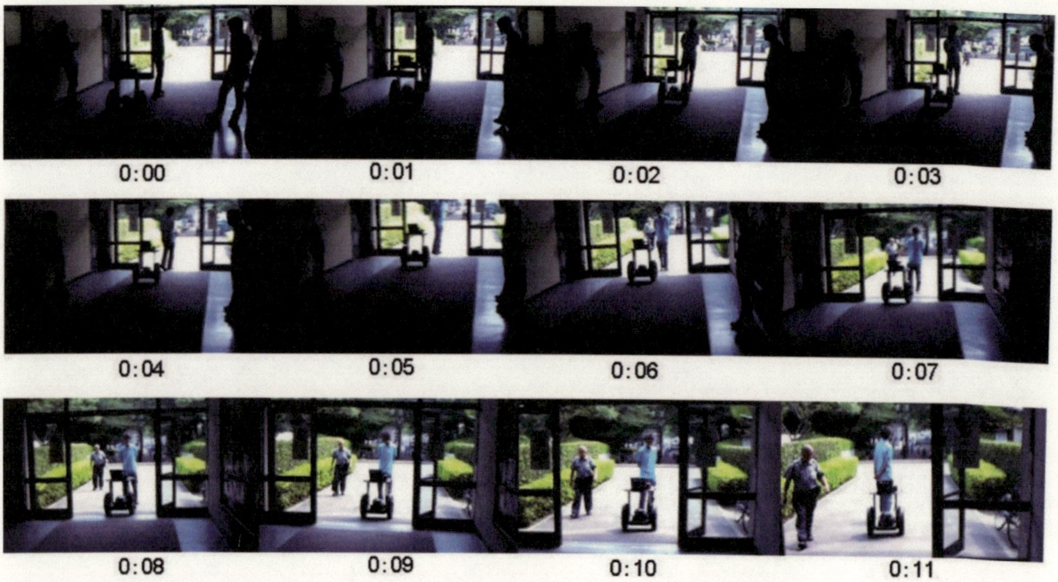

Fig. 14 Person Following Experiment (From Indoor to Outdoor)

Fig. 15. Person Following Experiment (Outdoor Environment)

Fig. 15 shows the person following experiment result when the robot moves in an outdoor environment. In comparison to the indoor environment, uneven surface, sunshine and shadow become serious problems in an outdoor environment. However, the robot also performs the target person following in an outdoor environment in where several people wearing the same color shirt exist. The robot is used the two wheels inverted pendulum type robot and has the ability to move on uneven surfaces. The robot adjusts color changes according to varying illumination greater than indoor environments by updating the color histogram. As in the previous experiment, when the robot moves against the sun, the robot cannot detect the target person by using only the camera system. However, the robot can continuously perform the target following by using the LRF tracking system in conjunction with the camera system. The LRF system is effected only slightly by sunshine. There are several people wearing the same color shirt around the target person. The robot can separate the target person and the others by the distance information. The robot also uses the time-series location information to separate the target person from the crowd.

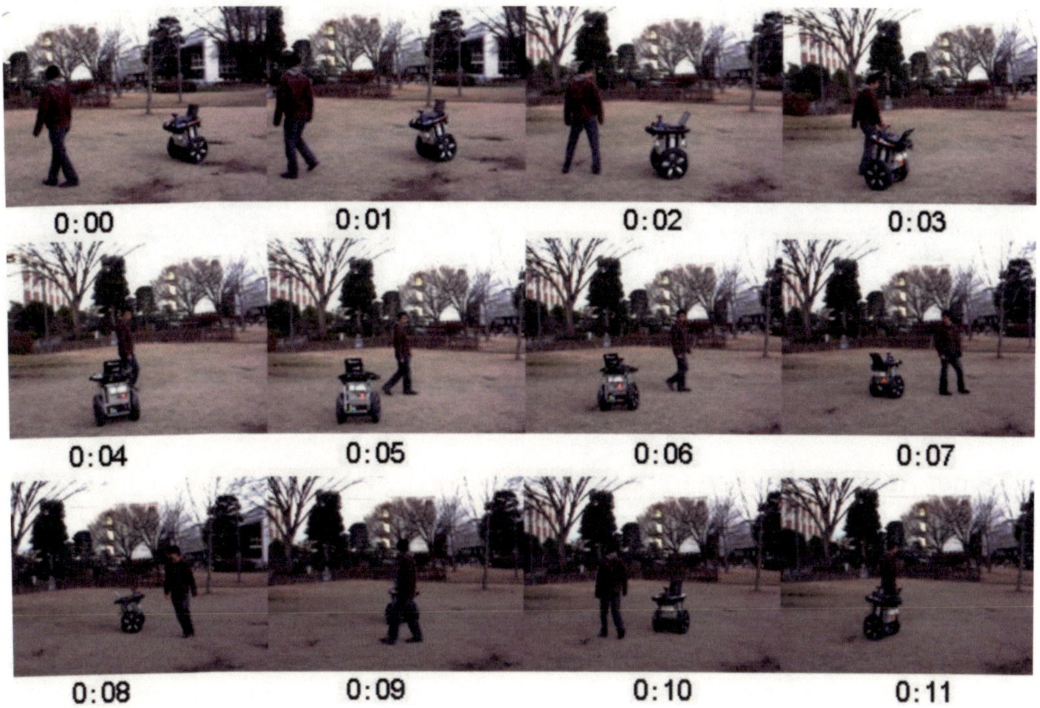

Fig.16. Person Following Experiment (Outdoor and Irregular Terrain Environment)

Fig.16 shows the experiment result when the robot moves on an outdoor irregular terrain. When the robot moves on a slope, the robot tries to keep a balance by tilting against slope because the robot is an inverted pendulum type robot. The tilt angle of the robot on an uneven terrain is larger than on an even terrain. In such case, the LRF located on the top of the robot rapidly tilts and the target tracking by the LRF will lose the target person. However, when the LRF target tracking method has lost the target person, the camera based method can re-detect the target person by the target person's color information. Therefore the target person following ability of the robot in real environment is verified through the experiment.

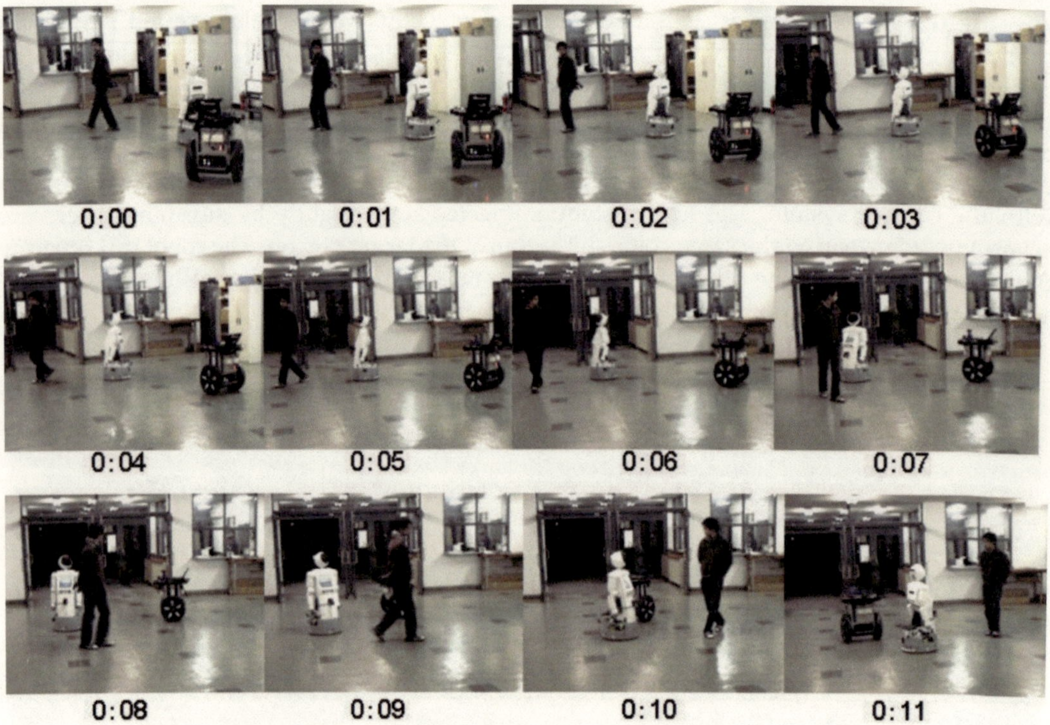

Fig. 17. Snapshots of Person–Robot Following Experiment

7. Target Robot Following Experiment

The proposed target person following method detects, tracks and follows the target person by using the target person's clothes color histogram and the target person's position history information. The method can be applied not only to human but also to other object, because the method does not use the temperature distribution of the target person by thermography or the shape of the target person for image processing. Additionally, the proposed methods can easily change hardware, because all the proposed methods are implemented using RT-Middleware (Ando, N. et al., 2005, Ikezoe, A. et al., 2006). Fig.17 shows the experiment result of a robot following another robot. Two different type of robots are used in this experiment. One is the Segway RMP 200, and the other enon (Kanade et al., 2006) a humanoid robot. Enon has two active wheels, two degree of freedom of his neck, a stereo camera (STOC) located on his face, and a LRF (UTM-30LX) located on his front. The proposed method is applied to these two robots. Enon attempted to follow the target person and Segway RMP attempted to follow enon. Thus there experimental result support that the proposed method can also be applied to cooperating operations between two or more robots.

8. Conclusion

In this chapter, we proposed the robust target person following method based on a color stereo camera and a laser range finder in a real environment. All proposed methods are implemented using RT-Middleware and therefore the hardware can be easily changed. The proposed method is applied to two different types of robots, one is an inverted pendulum robot (Segway RMP), and the other is a humanoid robot (enon). The target person following experiments is conducted in an indoor/outdoor environment and on regular/irregular terrain. The robot can perform the target person following in such environment. The validity and utility of the proposed method is verified through these experiments. The target robot following experiment is also conducted to confirm the robustness of the proposed method. The experimental results support that the proposed method can also be applied to cooperating operation between two or more robots.

9. Acknowledgment

A part of this work was performed as a project of the New Energy and Industrial Technology Development Organization (NEDO).

10. References

Ando, N.; Suehiro, T.; Kitagaki, k.; Kotoku, T. & Yoon, W.K. (2005). RT-middleware: distributed component middleware for RT (robot technology), *Proceedings of IEEE/RSJ International Conference on Intelligent Robots and Systems*, pp. 3933- 3938.

Hashimoto, M.; Ogata, S.; Oba, F. & OKada, S. (2006). Multi-Moving Target Tracking with In-Vehicle Laser Range Sensor, *Transactions of the Japan Society of Mechanical Engineers, C72 (717)*, pp. 1566-1573.

Ikezoe, A.; Nakamoto, H. & Nagase, M. (2006). Development of RT-Middleware for Image Recognition Module, *Proceedings of SICE-ICASE International Joint Conference*, pp. 2036-2041.

Ishihara, N.; Zhao, H. & Shibasaki, R. (2002). Tracking passenger movement with ground-based laser scanner, *Proceedings of the 8th Symposium on Sensing via Image Information*, pp. 13-16.

Itoh, K.; Kikuchi, T.; Takemura, H. & Mizoguchi, H. (2006). Development of a Person Following Mobile Robot in Complicated Background by Using Distance and Color Information, *Proceedings of the 32nd Annual Conference of the IEEE Industrial Electronics Society*, pp. 3839-3844 .

Kanade, S.; Murase, Y.; Sawasaki, N. & Asada, T. (2006). Development of the Service Robot "enon", *Journal of the Robotics Society of Japan, 24 (3)*, pp. 288-291.

Kawata, H.; Mori, T.; & Yuta, S. (2005). Design and Realization of 2-Dimensional Optical Range Sensor for Environment Recognition in Mobile Robots, *Jounal of Robotics and Mechatronics*, pp. 116-120.

Kwon, H.; Yoon, Y.; Park, J. B. & Kak, A. C. (2005). Person Tracking with a Mobile Robot using Two Uncalibrated Independently Moving Cameras, *Proceedings of the 2005 IEEE International Conference on Robotics and Automation*, pp. 2888-2894.

Lindstrom, M. & Eklundh, J.O. (2001). Detecting and Tracking Moving Objects from a Mobile Platform using a Laser Range Scanner, *Proceedings of 2001 IEEE/RSJ International Conference on Intelligent Robots and Systems*, pp. 1364-1369.

Negishi, Y.; Miura, J. & Shirai, Y. (2004). Mobile Robot Navigation in Unknown Environments Using Omni directional Stereo and Laser Range Finder, *Proceedings of 2004 IEEE/RSJ International Conference on Intelligent Robots and Systems*, pp. 2737-2742.

Ohya, A. & Munekata, T. (2002). Intelligent Escort Robot Moving together with Human - Interaction in Accompanying Behavior-, *Proceedings 2002 FIRA Robot World Congress*, pp.31-35.

Okusako, S. & Sakane, S. (2006). Human Tracking with a Mobile Robot using a Laser Range-Finder, *Journal of the Robotics Society of Japan 24 (5)*, pp. 605-613.

Schlegel, C.; Illmann, J. & Jaberg, H. (1998). Vision Based Person Tracking with a Mobile Robot, *Proceedings of the 9th British Machine Vision Conference, Southampton*, pp. 418-427.

Schulz, D.; Burgard, W.; Fox, D. & Cremers, A.B. (2003). People Tracking with a Mobile Robot Using Sample-based Joint Probabilistic Data Association Filters, *International Journal of Robotics Research, vol. 22, no. 2*, pp. 99-116.

Sidenbladh, H.; Kragic, D. & Christensen, H. I. (1999). A Person Following Behavior for a Mobile Robot, *Proceedings of the 1999 IEEE International Conference on Robotics and Automation*, pp. 670-675.

Uemura, H.; Tan, J.K. & Ishikawa, S. (2006). A Color Tracker Employing a Two-dimensional Color Histogram Under Changeable Illumination, *Proceedings of 2006 IECON*, pp. 3273-3278.

A-B Autonomy of A Shape-shifting Robot "AMOEBA-I" for USAR

Yuechao Wang, Jinguo Liu and Bin Li
State Key Laboratory of Robotics
Shenyang Institute of Automation
Chinese Academy of Sciences, P.R. China

1. Introduction

The frequent nature disasters and man-made catastrophes during the last decades such as earthquake, typhoon, hurricane, radiation and terrorist attack have aroused people's attention on the importance of Urban Search and Rescue (USAR). Although people have more watchfulness than before, a large number of people still have died in unprofessional rescue due to inadequate equipment and being lack of professional manpower (Erkmen et al, 2002; Casper and Murphy, 2003). Timely searching for victims and subsequent rescue operations from the rubble of collapsed buildings are highly required. Since these operations are very dangerous for human workers and even for trained dogs, autonomous mobile systems are highly needed to help in finding trapped victims. It is a great challenge to develop search and rescue robot that can actually work in the disaster site. The search and rescue robot research includes not only the robotics technology but also the rescue technology and the disaster science. Researches sponsored by the governments and the companies have resulted in the emergence of various kinds of search and rescue robots. Since earthquake happens in Japan frequently, intelligent rescue systems and robotic technology have been expected to mitigate disaster damages, especially after the 1995 Hanshin-Awaji Earthquake. From 2002, "Special Project for Earthquake Disaster Mitigation in Urban Areas" (a 5 years project which also called DDT project) was launched by Ministry of Education, Culture, Sports, Science and Technology, in Japan. Many kinds of search and rescue robots have been developed (Matsuno and Tadokoro, 2004), such as Souryu (Hirose and Fukushima, 2002), Moira (Osuka and Kitajima, 2003), and KOHGA (Tetsushi et al, 2004). In America the rescue robot research has attracted a lot attention too. Several robots were used for the search and detection operation in the collapsed World Trade Center building in September 2001 (Casper and Murphy, 2003). In the University of South Florida, Professor Murphy and her fellow have developed "Bujold", a kind of search and rescue robot that has the ability of shape shifting and has been equipped with many sensors (Murphy, 2002). In Carnegie Mellon Robotics Institute, researchers have developed multi-joint robot for inspection (Wolf et al, 2003). Foster-Miller Company also carries out TALON Robot series for search and rescue mission. In China, a series of search and rescue robots have been developed in recent years (Liu et al, 2007a; Zhang et al, 2006; Zou et al,

2006; Huang et al, 2006). Among them a shape-shifting robot named "AMOEBA-I" has been developed in the Shenyang Institute of Automation, CAS for search and rescue operation (Liu et al, 2004; Li et al, 2006). Since search and rescue robots are usually unmanned vehicles, autonomy or autonomous behaviours play key roles in their performances. In this chapter, we first provide a brief description of AMOEBA-I. Then, A-B Autonomy of "AMOEBA-I" has been discussed. Rules of autonomous shape-shifting for AMOEBA-I in urban terrain have been proposed according to the features of AMOEBA-I. Finally, experimental results from its autonomous motions (autonomous shape-shifting, autonomous navigation, autonomous climbing up obstacle and autonomous avoiding obstacle) have demonstrated that the AMOEBA-I shows great autonomy in urban terrain.

2. Mechanism of shape-shifting

In unstructured environment operations such as search and rescue, flexibility and adaptability are urgently required in such situations for mobile robots since the terrain is usually uneven and unpredictable. In nature, there are many kinds of creatures with high motilities. For instance, the link-type structure animals, such as snakes and centipedes, behave perfect mobility and flexibility in uneven terrain for their redundancy. The link-type structures also widely exist in snake-like robots, serpentine robots, and multi-joint robots. Inspired by the nature snake that has perfect mobility under various environments, link-type robots have been widely studied. Link-type structures are connected sequentially with active or passive joints between the adjacent units or modules. They have a large number of degrees of freedom. Powered links between the units allow the chain to adapt to the terrain, to adopt the desired shape for a particular task, and to lift the front modules to surmount obstacles. Generally, the link-type structures, which have been widely applied in snake-like robot, train-like robot, trailer-like robot, and multi-joint robot, have turned out to be effective mobile mechanisms for reasons listed as follows:

(1) A link-type robot with active body has excellent mobility. It can pass through narrow spaces in line as shown in Figure 1 (a), cross a ditch by stiffening the body to bridge the ditch as shown in Figure 1 (b), and lift the front part of the body to overcome the obstacles as shown in Figure 1 (d).

(2) With a low barycentre and sufficient contact to the ground, it has high longitudinal stability over uneven terrain and soft ground. As shown in Figure 1 (c), it can steadily cross a marsh by loosing joints to distribute its weight to the whole body.

(3) For a link-type structure with active joints, it can adopt manifold gaits of snakes such as serpentine motion, concertina motion, side-winding motion, rectilinear motion, thrusting motion, pushing motion, and jumping under different environments.

(4) The link-type structure can be an excellent manipulator by transforming the hyper redundant degrees of freedom just like the trunk of an elephant.

(5) The link-type structure is fault-tolerant to some extent. The reliability and maintainability can be high since the rest can still finish the mission even if a link cannot work. The malfunctioning segment can also easily be detached and replaced.

(6) The link-type structure can be easily modularized and reconfigurable. A single modular unit only has a limited mobility but a swarm of them generates better performances in many areas.

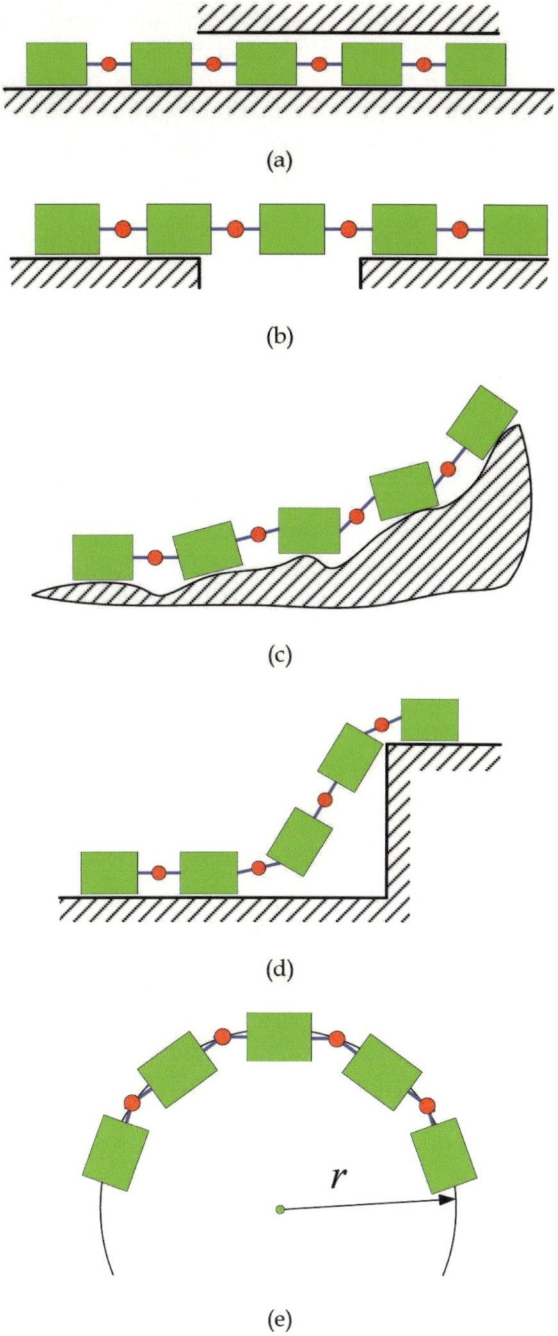

Fig. 1. Typical motilities of the link-type structure. (a) Pass through narrow space. (b) Cross a ditch. (c) Move over uneven terrain. (d) Overcome the obstacle. (e) Turn around.

However, being long and slim, the link-type structure also has its disadvantages in some situations such as in the debris which mainly include:

(1) Lateral stability: When moving over uneven terrain, it easily gets lateral tip over as shown in Figure 1 (c). If a tip over incident happens, it may result in a series of problems: loss of traction, entrapment, system damage, loss of control, difficulty in overturning back, and even mission failure.

(2) Turning mobility: For example, when turning, it needs a large radius as shown in Figure 1 (e). When in a narrow space, it is hard for the link-type structure to turn around effectively.

How to take the advantages and surmount the disadvantages of link-type structures is both interesting and challenging tasks. For a traditional link-type structure, it can only pose the straight or curve configuration since the degree of freedom between adjacent units is usually limited by joints' movable range. Therefore, we intend to develop a self-reconfigurable link-type robot to improve the tip over stability and the turning mobility of the link-type structure. The self-reconfigurable or shape-shifting robots have attracted great attentions since they can autonomously change their physical configurations or their structure to meet the requirement of the mission and the environment. And the tracked type robots have been widely developed for search and research operation. There are usually two kinds of reconfigurable track type mobile robot in the literatures. One kind is the track's shape being changeable. Reconfigurable tracks can change from one shape to another without modification of the vehicle. This approach effectively provides various dimensions of the track contacting the ground and the track above the ground. Bujold is a typical example of this category. It is a commercially available tracked chemical inspection microrobot which can change shape among three canonical configurations: sitting up and facing forward, sitting up and facing backward, and lying flat (Murphy, 2002). A new type of variable geometry single-track driving mechanism for a rescue robot has been proposed (Lim et al, 2005). This mechanism has a symmetrical configuration so that the robot advances in dual directions and prepares against overturning. Using transformation, it can reduce the energy consumption in steering and rotating while maximizing the capacity to overcome stairs. The other kind of the reconfigurable track type robot is similar to the self-reconfigurable modular robot in principle. For instance, NUGV is a novel multi-degree-of-freedom tracked robot that can change its conformation and dimensions, and negotiate a great range of environmental dimensionality (Blackburn et al, 2004).The tracked robot CUBIC-R's shape is a regular hexahedron and each surface has a crawler unit with a couple of crawlers. As each surface is united by transformational mechanism which has 1-DOF, it can overcome rough terrains (step, stairs, gap and so on) by using transformation (Tabata et al, 2005). The reconfigurable modular type tracked robots have attracted more and more attentions for their high mobility. And they will be the ideal selections for urban or field environments. To improve the environmental adaptability of the link-type tracked robots, we have proposed a novel link-type reconfigurable structure as shown in Figure 2. This link-type structure, with Yaw and Pitch offset joints at each module's both lateral sides, has enough flexibility for its shape change. The module body is a mobile unit with active tracks. The arms and joints play important roles in shape changing process. Figure 2 demonstrates a sequence of two-module's reconfiguration. In Figure 2 (a) the link-type is in a line type while in Figure 2 (e) it is in a row type. From line-type to row type, the Yaw joint rotates 180° and then so does the Pitch joint. It behaves the advantages of the link-type structure while in line, moreover, it can avoid the line-type's disadvantages while in row. For instance, the robot in row can turn easily with a zero radius by driving the lateral modules differentially. This link changes its configurations through a sequence of joints' rotation. Compared to reconfiguration by connection and disconnection, such kind of reconfiguration has higher reliability, less power consumption, and less time consumption.

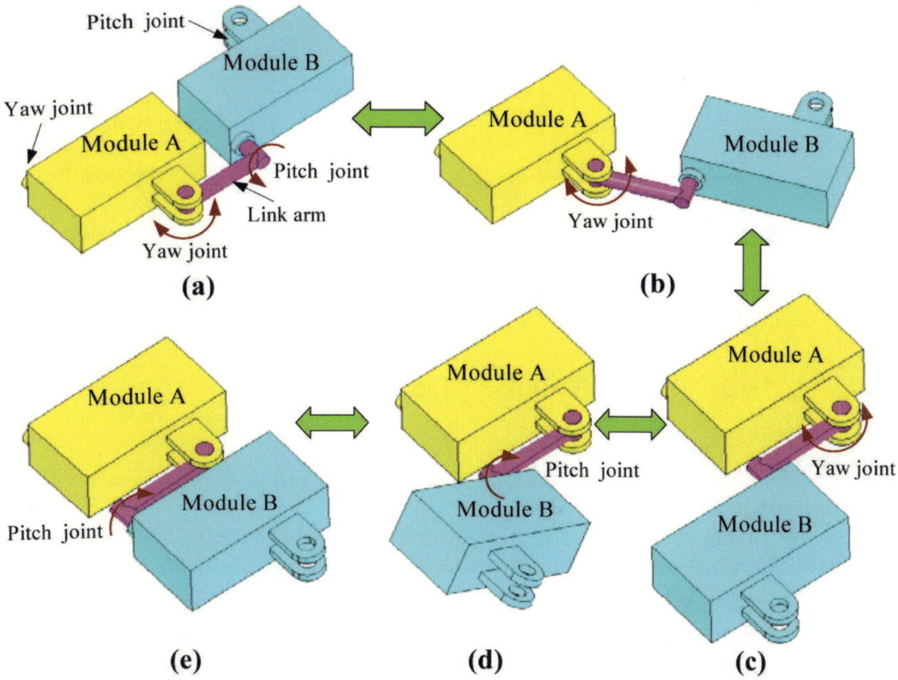

Fig. 2. Reconfiguration process of the proposed link-type structure

For two modules, there are three trim configurations shown in Figure 2 (a), Figure 2 (c) and Figure 2 (e), which are also called mobile configuration. The mobile configurations in this paper are defined to be the non-isomorphic configurations with parallel driving modules in the same plane. That is, in the mobile configuration, all the modules should contact the ground in parallel. In following sections, we merely take the mobile configurations into consideration in configuration analysis. Figure 3 (a) provides a top-view of these three mobile configurations. When designing the module and the linkage arm, the length of the module and the arm are strictly constrained to realize a trim figuration. The longitudinal dimension is designed in scale while the lateral dimension is free to some extent. The geometrical relationships in longitudinal direction shown in Figure 3 are given by

$$l_a = l_b + l_c \tag{1}$$

where l_a is the length of the module body, l_b is the length of the link arm, and l_c is the distance between Yaw joint to the front or the back side of the module body as shown in Figure 3(a).

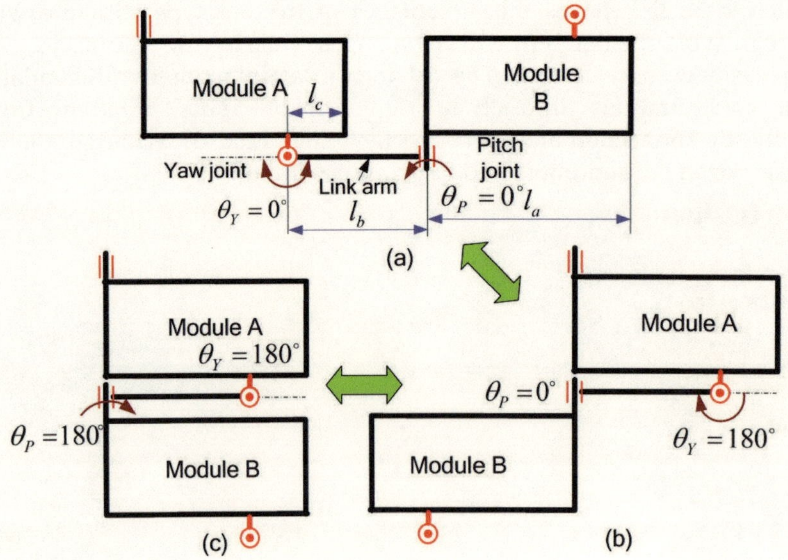

Fig. 3. Geometrical relationship of the link-type structures

3. The prototype of AMOEBA-I

3.1 Mechanical structure of AMOEBA-I

According to the principle mentioned above, a three-module robot AMOEBA-I has been developed. A single-module is mainly composed of a link arm, a track driving system, an offset Yaw joint driving system, a Pitch joint driving system. The module in Figure 4 is a standard one. There are often two or more wheels to support the track in the traditional track device. One of them is an active wheel and the others are passive. A modification is made on the wheel system as shown in Figure 4. First, we only use two wheels in the driving system. Second, the wheels are hollow inside. Such kind of wheel is light in mass. Moreover, it is possible for application of the system in water for its hollow body and lager volume. It has been taken into consideration that if the system is water proof, the robot can move both on land and in water like an amphibian vehicle. It has three DC motors which are for the track-driving, for pitch joint, and for yaw joint respectively. The motors are packed in the center box. Chain transmission device has been used in pitch joint and track's driving, while the yaw joint uses bevel gear pairs. Timing pulleys are used to driving the wheels forwards and backwards. The link arm and the link handle are used to connect and disconnect the adjacent modules. As shown in Figure 5, AMOEBA-I is repetitively composed of such kind module. It is a tracked robot which can overpass various terrains by transforming its configurations. The key advantage of this type over other link-type vehicle is its adaptability to environments through various configurations as shown in Figure 6.

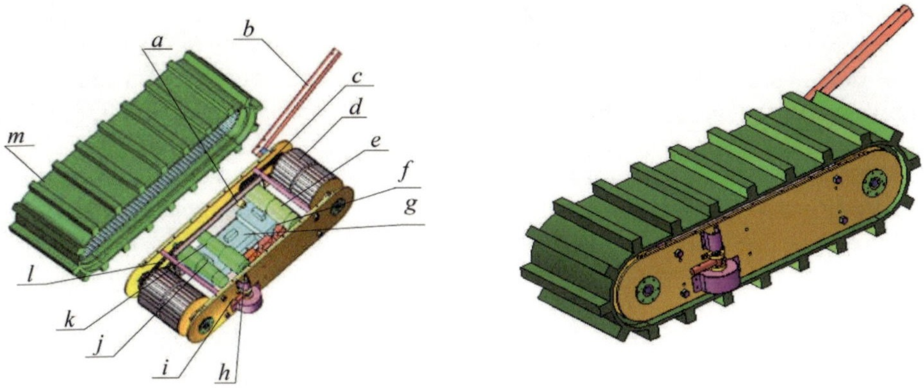

a. Pitch joint potentiometer; b. Link arm; c. Pitch joint chain driving; d. track wheel; e. Pitch joint motor; f. Motor controller; g. Motor drive; h. Yaw joint potentiometer; i. Yaw joint bevel gear driving; j. Track driving motor; k. Yaw joint motor; l. Track wheel chain driving; m. Track

Fig. 4. Structure of a single standard module of the prototype

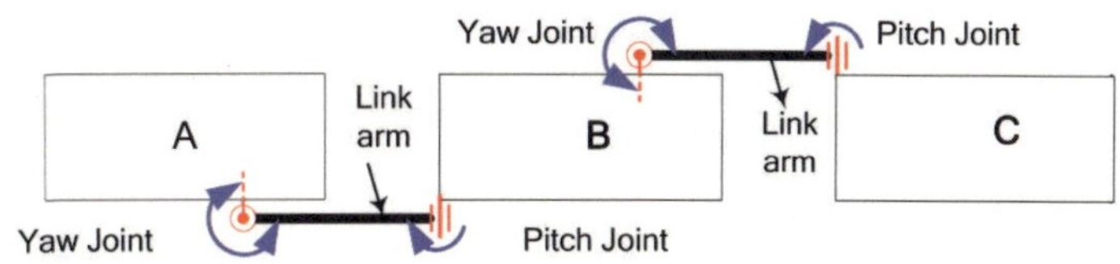

(a)

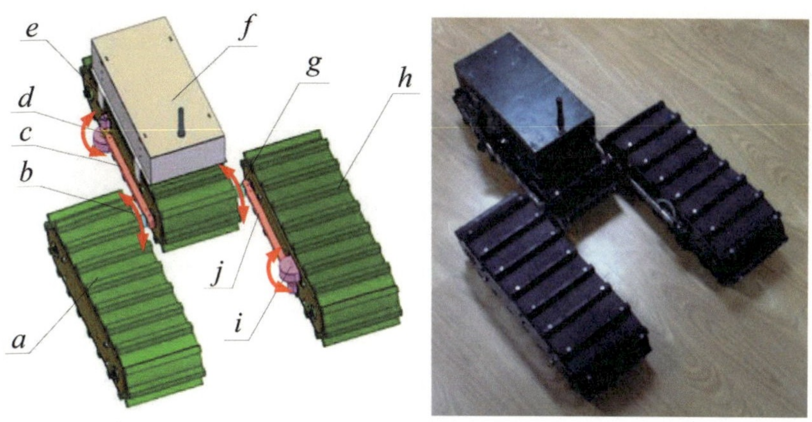

a. Module C; b. Pitch joint of Module C; c. Link arm between Module B and Module C; d. Yaw joint of Module B; e. Module B; f. Autonomous control system; g. Pitch joint of Module B; h. Module C; i. Yaw joint of Module A; j. Link arm between Module A and Module B. (b)

Fig. 5. Structure of AMOEBA-I

AMOEBA-I has 9 kinds of available mobile configurations (Liu et al, 2007b). And it has three kinds of symmetry configurations: the line type (a), the triangle type (e) and the row type (i). The robot may change its configuration to adapt the various environment and tasks. Each configuration has various gaits and dimensions. The characteristics of AMOEBA-I can be generalized as:

(1) They have many non-isomorphic configurations. Their configuration can change automatically to adapt to the environment. With two or more modules, they can pose line type and row type easily. For instance they can pass the narrow space and the hole in line and they can move on uneven terrain safely or steering easily in row.
(2) It is small, lightweight and easy to carry. And it can resist against dust, gas and other hostile environment.
(3) With hollow wheels and sealed body, it can move in shallow water. The wheel has been designed to be hollow inside for underwater case.
(4) It is modularized and reconfigurable. The modular robot has satisfied maintenance and interchangeableness. It is reconfigurable in urgent need. The connection and disconnection of the modules can be finished through the link arm and the link handle.

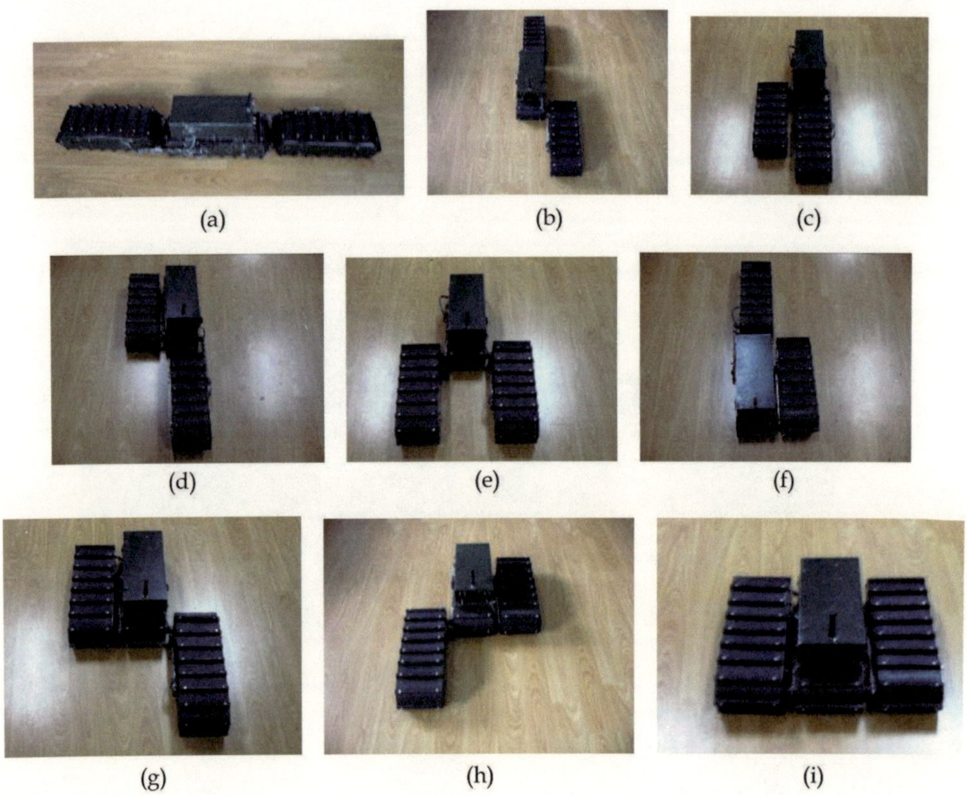

Fig. 6. Available configurations of AMOEBA-I

3.2 Structure of the control system

The control system is composed of a wireless module, a main control module, motor control modules and sensor-based feedback control modules. It adopts PC-and-MCU structure as shown in Figure 7. Modularized structure has been widely used in the control system of mobile robots to enhance the error tolerance. As shown in Figure 7, the hardware of control system for a single module is modularized for improving the error tolerance and exchangeable ability of the system. The control system is composed of a supervisor system, an autonomous control system, and an actuator system. Each motor control unit of the actuator system has a microprocessor to deal with the information and control command of the motors and sensors. To exchange data among micro-controllers, the robot needs a simple and high efficient bus to serve as the system bus. CAN bus is a kind of shared broadcast field bus, in which all the nodes can send and receive message. Error detection, correction, and exchange have been implemented by the CAN controller. Thus CAN bus is selected as the suitable information bus in control system of AMOEBA-I. The CAN bus bridges the information among the three independent modules.

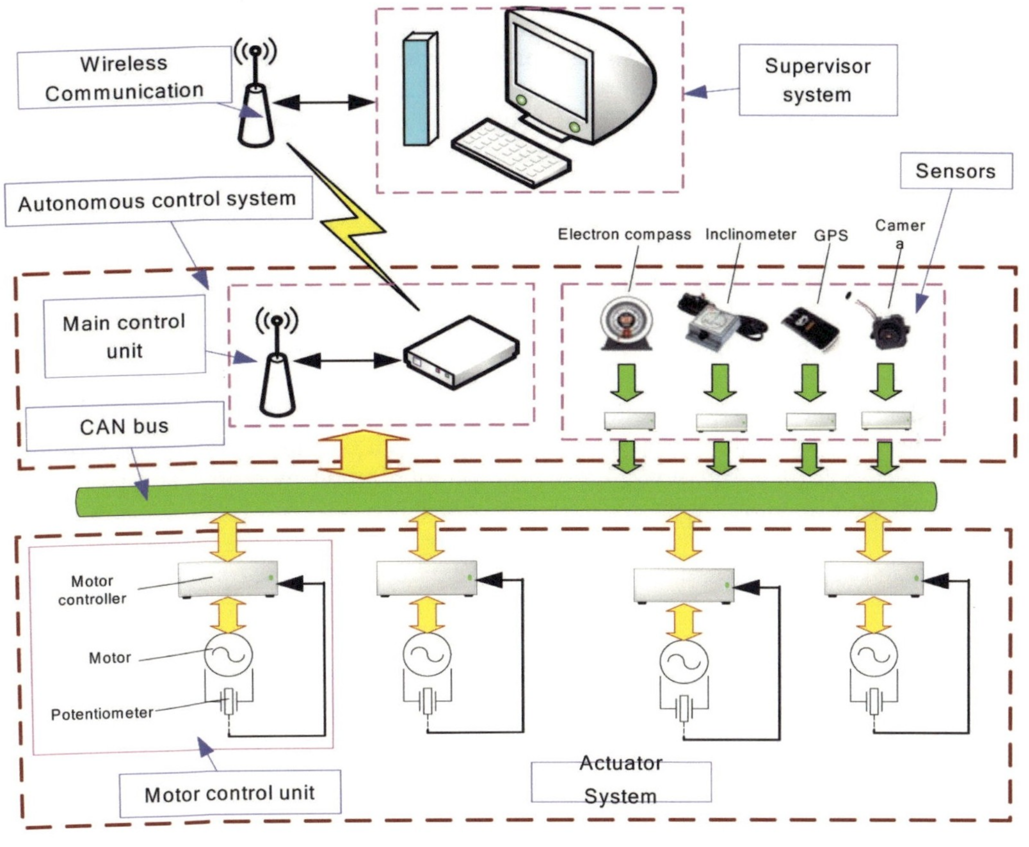

(a) Control system structure

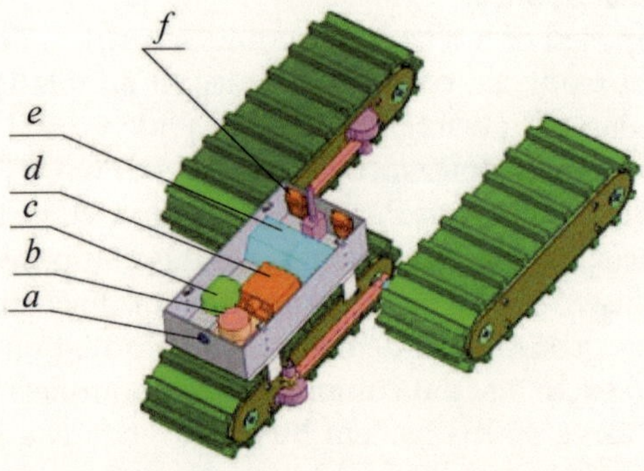

a. Camera; b. Electron compass; c. GPS; d. inclinometer; e. Power; f. Main control unit

(b) Details of the autonomous control system

Fig. 7. Control system of AMOEBA-I

Remote control and teleoperation control are required in a search and rescue robot for the real tasks since the disaster site is unreachable for the men. The supervisor system and main control unit in the control system communicate with each other through the wireless communication module. The supervisor system provides enough perception and command capabilities to the operator so that one can control the remote robot. The supervisor system not only sends the commands to the robot system, but also timely receives the status information returned by the robot, which include the configuration, joints angle, and the current consumed by the motor. The main unit is the kernel of the autonomous robot system. The main control unit is the decision-making unit of the robot, which estimates its own position, and status of the environment provided by the sensor-based feedback control unit and plans the movement of robot by using corresponding algorithms. Moreover, it is also used to globally program the movement of the robot and the change between the configurations. The main control unit has been designed and equipped in each mechanical module to implement its global modularization. It is also a communication interface between the robot and human-supervision platform by transmitting data from sensors to human-supervision platform through wireless communication module. The main control unit links all the dispersive motor control units, which are designed in the same method in order to implement the modularization of control system. It makes the whole system more extendable and stable. The main control unit and pinhole camera are placed in the control platform of middle module of AMOEBA-I. The motor control units implement the control for mechanical components in AMOEBA-I. All motor control units have the same structure, which receive the commands from the main control unit. After analysis and computation, actuators are set to a predefined position and rotational speed by using serial ports. When one module fails, it would be selected out of the whole control system to keep the system robust. The motor control unit is composed of the Fujitsu MCU, the CAN bus driver, the potentiometer, the DC motor and the DC controller. The motor control units collect each joint's angle form the A/D transformer in the MCU and then transfer it to main controller unit, which will help the main control unit plan the system globally.

The sensor system contains a camera, an inclinometer, an electron compass, and a GPS. It receives information from the environment and transmits the data to respective controllers in which pre-settled algorithms are used to deal with them, and then results are sent to the MCU.

4. A-B autonomy of AMOEBA-I

4.1 A-B autonomy with various configurations

A-B autonomy or A-B mobility is the key autonomous capability for unmanned system (Army Science Board, 2002; Research Council of the National Academies Technology, 2002; Quek, 2005). It is also absolutely important for urban search and rescue robots. When in rescue operations, the robots are usually arranged to arrive a designated place to search the survivals and/or reconnoitre the environments. When in AMOEBA-I's A-B autonomy consideration, AMOEBA-I can overpass various terrains by transforming its configurations. Figure 8 provide a conceptual layout of AMOEBA-I's A-B autonomy with comparison of various configurations. *Path d*, *Path c*, and *Path a* denote the robot's best available route in Row configuration, Line configuration or Triangle configurations respectively. However, AMOEBA-I maybe have a most-best route *Path b* with compound configurations which considered being shorter and/or safer.

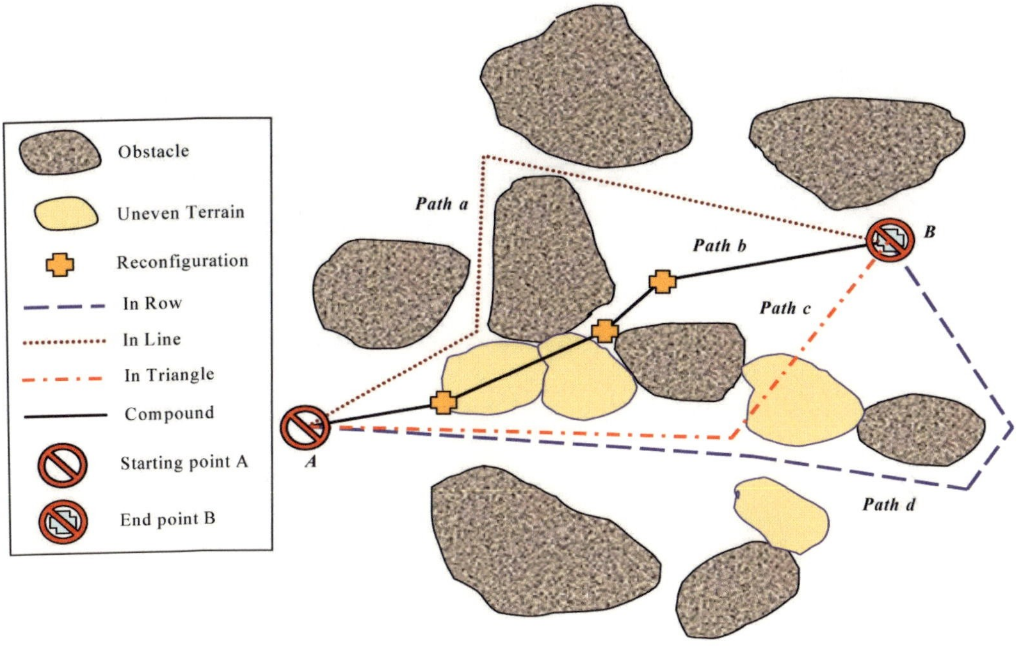

Fig. 8. A-B autonomy of AMOEBA-I with various configurations

When AMOEBA-I moves automatically in the unstructured environment, it must use the sensors to know the environment, and then select the optimal configuration according to the environment. So knowledge about the performance differences for various configurations and the rules of autonomous shape-shifting are very important for AMOEBA-I in its A-B autonomy.

4.2 Comparison of motion performances of AMOEBA-I

In order to constitute the rules of autonomous shape-shifting for AMOEBA-I, we test the adaptability of AMOEBA-I in symmetrical configurations as shown in Figure 9.

(a) Experiments under indoor environments

(b) Experiments under outdoor environments

Fig. 9. Experiments for the adaptability of AMOEBA-I

We have gotten the performances of AMOEBA-I by the experiment results as shown in Table 1.

Environment type	Line configuration	Triangle configuration	Row configuration
Uneven outdoors ground	II	I	III
Obstacles	I	II	III
Slopes	I	II	III
Stairs	I	II	III
Flat ground	III	II	I
Turning	III	II	I
Long and narrow spaces	I	III	II

Table 1. Motion performances of three symmetrical configurations

In Table 1, "I", "II" and "III" stand for three levels as "good", "common" and "poor", respectively. From Table II we can get conclusions as follows: The robot can climb up obstacles with satisfying performance as well as get across cabined spaces under Line configuration, which is a remarkable improvement for environment-adaptation. In Triangle configuration, stability and the mobility of climbing slopes are very notable. Zero-radius turning can be realized easily under Row configuration which makes it be more agilely and be minimum energy consumed.

4.3 Rules of autonomous shape-shifting in A-B autonomy

Since the robot mostly moves on flat ground, we choose Row configuration as the initial configuration. The robot takes the following methods to recognize different environments and adopts the most appropriate configuration. Here we will take the field, obstacle and slopes as examples respectively to show our rules of how to guide the robot to change its configuration.

(1) Environments of uneven ground in the field. In this situation, data from the inclinometer are taken as the most important into consideration. If the obliquity does not exceed the threshold of 10 degrees, we define the ground as general uneven ground. The states fall into 2 kinds according to the specific feedback from inclinometer.

   a. The slope's gradient changes not fast (if mean square errors are smaller than 20 with 10 times' continuously samplings), then the ground is thought to be flat and Row configuration is taken.

   b. The slope's gradient changes rapidly (if mean square errors exceed 20 with 10 times' continuously samplings), then the ground is thought to be uneven, and the robot takes the Triangle configuration.

(2) Environments of obstacles. When obstacles are in its way, the robot is in one of the three following configurations. Configuration changing is decided upon the feedback information from encoder and the motor currents.

   a. Row configuration. If data from motor encoder does not change in 10 seconds, and electric current value steps to a high level, we can conclude that the robot has some obstacles in its way and it cannot get over, so the robot changes into Triangle configuration. Otherwise, keep in current configuration and move on.

   b. Triangle configuration. If data from encoder does not change in 10 seconds, and current value steps to a high level, we can conclude that the robot meets some large obstacles hard to pass. In such case, the configuration alters into Line form. Otherwise, keep in current configuration and move on.

   c. Line configuration. If data from encoder does not change in 10 seconds and current value steps to a high level, which means large obstacles are in the way and the robot has to avoid them with utilizing information from electron compass and encoder. Otherwise, keep in current configuration and move on.

(3) Environments of slopes. Information from the inclinometer is at the first place to be considered when the robot gets over obstacles with certain gradients. The strategy of its locomotion is illuminated as:

   a. The gradients range in [0, 10°]: In this situation, there is no need to alter the robot's configuration.

b. The gradients range in [10°, 20°]: Triangle configuration which is prone to get over obstacles and has good stability is available.
c. The gradients range in [20°, 30°]: Take Line configuration for the robot may overcome tough obstacles in this kind.
d. The gradients exceed 30°: Steer clear of the slopes using data from electron compass and encoder.

In a word, the robot can judge its surroundings accurately by analyzing data from sensors. Moreover, it can take different configurations to adapt to the environment and complete the designated mission.

4.4 GPS based autonomous navigation

We used GPS to provide geographic coordinates. Some calculation should be done to get the robot's present position and the difference between its present and target position. Set these differences as the norm angles of electron compass, the robot will move towards target position. As is shown in Figure 10, O is set to be the mobile robot's initial position, whose geodetic coordinates are (L_O, B_O), and P is the target position with its geodetic coordinates at (L_P, B_P). The dashed circle is the confines of the range in which the robot may stop at the range of allowable error. After the robot covers a certain distance, it stops at A (L_A, B_A) to adjust its orientation. As it passed B (L_B, B_B), it enters GPS blind zone. At C, the robot makes its orientation adjustment utilizing geodetic coordinates according to information of inertial components. The robot leaves GPS blind zone from the moment it gets to D (L_D, B_D). Suppose it meets an obstacle at E, the navigation program will stop and a program of obstacle-get-over or obstacle-avoidance will take charge until it gets over or passes by the obstacle. The robot changes its configuration at F and alters into its former shape at G (L_G, B_G). In this graph, H (L_H, B_H) is its end point.

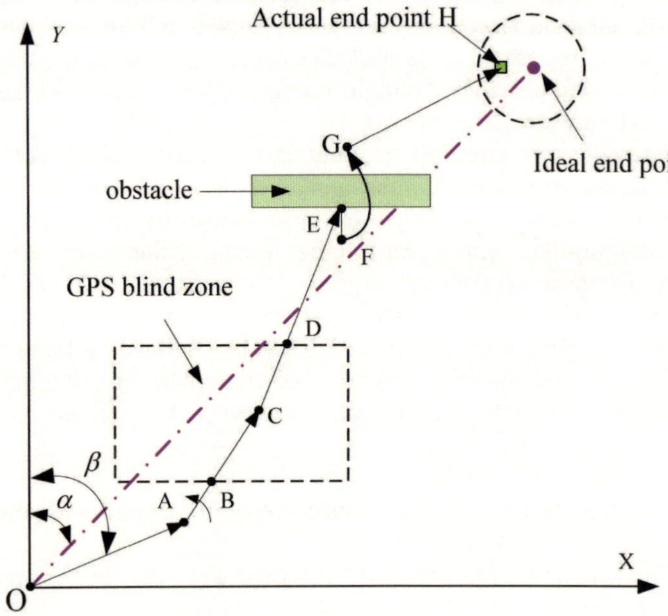

Fig. 10. Autonomous navigation of AMOEBA-I

After we select an end point, OP's direction represented by α can be ascertained.

$$\alpha = \frac{L_P - L_0}{B_P - B_0} \cos B_0 \tag{2}$$

The electron compass gives the robot's deflexion angle γ (the angle to Y axis) at its initial position. Therefore, at the beginning, the robot has to turn θ, which can be calculated by

$$\theta = \alpha - \gamma \tag{3}$$

If θ >0, it turns left, while it turns right if θ<0, otherwise it goes straightforward. Suppose the robot reaches a where GPS fails after certain periods (the interval between the robot samples the GPS values), it starts to calculate its position and orientation using the record of inertia components.
After one period, the robot moves to B, whose geodetic coordinates can be calculated by

$$B_B = B_A + \frac{vt \cos \gamma}{K_1} \tag{4}$$

$$L_B = L_A + \frac{vt \sin \gamma}{K_2} \tag{5}$$

In Eqs. (3) and (4), v is the velocity of robot, γ represents its including angle to Y axis and t is its motion period. K_1 is chosen to be 30.8m/s, which is the distance between the adjacent latitude divided by minutes along the same longitude.

$$K_2 = K_1 \cos B_A \tag{6}$$

where K_2 in Eq.(5) denotes the distance along BA latitude between every one minute longitudes. Whether it turns left or right can be concluded by Eqs. (2) and (3). Therefore, even if GPS system cannot receive the satellites' signals, orientation program by inertia components is automatically performed, in which data from the encoder, electron compass and inclinometer together with the previous GPS's information are used to conclude the robot's present geodetic coordinates to make sure its navigation works correctly. When it gets out of the blind zone, GPS system starts to work for the robot's orientation. E-F-G is the path followed by the robot when it gets over the obstacles after shape-shifting or round the obstacle directly. In this process, the robot checks the obstacles first. If the obstacle is too tough to get over, obstacle-avoidance program runs to make the robot pass by the obstacle. If it founds that the obstacle can be got over after shape-shifting, it will withdraw a certain distance and change into a more appropriate configuration. After getting over the obstacle, the robot will take a more efficient configuration, reorient itself and move towards the target.

In addition, the robot checks whether it reaches the target position every period by calculating the distance between its position and the target. Suppose that the robot's present position is C (L_C, B_C), then the distance to the target position P can be gotten by

$$L = \sqrt{(K_1(B_P - B_C))^2 + (K_2(L_P - L_C))^2} \tag{7}$$

As a measurement error lies in GPS system, the distance from P to C signed as L cannot be simply used to decide whether the robot has reached its target. If L's error lies at a tolerable range, the robot is thought to be get to the aim point; If not, then we let C replace B, recalculate α and do the conclusion and analysis that has been illuminated in the above paragraphs until we get an L that meets the error tolerance.

5. Autonomous motion experiments

Autonomous navigation experiments free of obstacles as well as experiments in the environment where the robot has to reshape itself to get over the small obstacles have been made with AMOEBA-I.

5.1 Experiment system and experimental field

The experiment system include one console and one AMOEBA-I. The experiment system and Experimental field to validate the autonomous navigation and shape-shifting algorithm are shown in Figure 11.

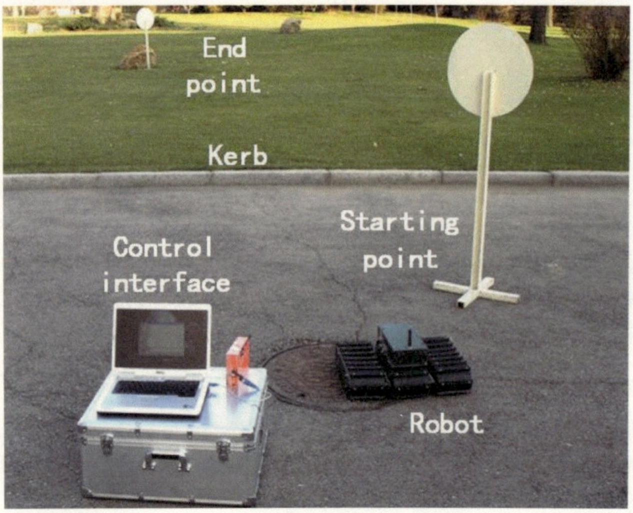

Fig. 11. Experiment system

5.2 Autonomous motion in the no-obstacle situation

AMOEBA-I has moved automatically from starting point to end point on the lawn. There is no-obstacle situation in the way that AMOEBA-I move on. Process of autonomous motion is shown in Figure 12.

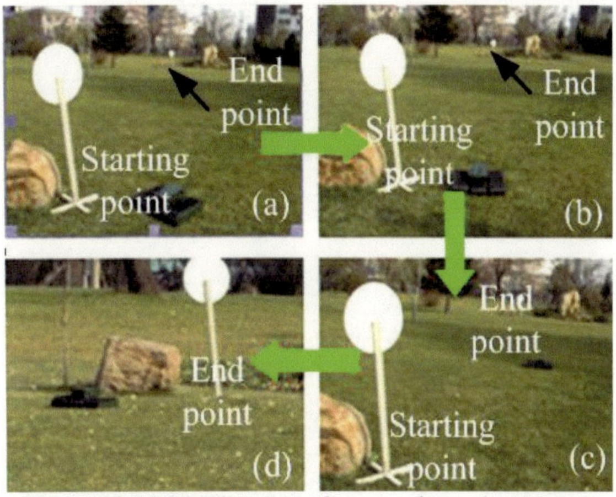

Fig. 12. Autonomous motion of AMOEBA-I in the no-obstacle situation

In the experiment, we set the target position at 41°45.6950'N/123°26.5642'E.
Figure 12 (a) shows the robot's orientation before the aim point was set, with its position at 41°45.6945'N/3°26.5345'E measured by GPS devices and the electron compass read 135 degrees.
Figure 12 (b) shows that after the aim point were set, the robot adjusted itself to face the target.
Figure 12 (c) shows one of the several adjustments on the way caused by the errors of GPS (<15m) and electron compass (-30~30).
Figure 12 (d) shows the robot stopped near the target at 41°45.6948'N/123°26.5340'E.

5.3 Autonomous motion in the small obstacle situation

The process of robot's autonomous navigation and autonomous shape-shifting is shown in Figure 13. Each step's motion is introduced as follows.
Figure 13(a): AMOEBA-I is located at the starting point.
Figure 13(b): After climbing obstacles, AMOEBA-I detected that Row configuration cannot cross.
Figure 13(c) and Figure 13(d): it returns and changes into the Triangle configuration with stronger ability to cross the obstacle.
Figure 13(e): The process of AMOEBA-I with Triangle configuration cross the obstacle.
Figure 13(f), Figure 13(g) and Figure 13(h): AMOEBA-I detected that the ground slope<10°, it turn back into Row configuration with good turning ability.
Figure 13(i): The point at which AMOEBA-I adjust its orientation.
Figure 13(j): AMOEBA-I arrives at target point and completes the process of autonomous navigation.
As the experimental results show, in the process of autonomous motion, when AMOEBA-I encountered a vertical obstruction or obstacle with small slope and angle, it can make a reaction in time and automatically change into other configuration with strong ability to climb up obstacle directly and approximate the target point.

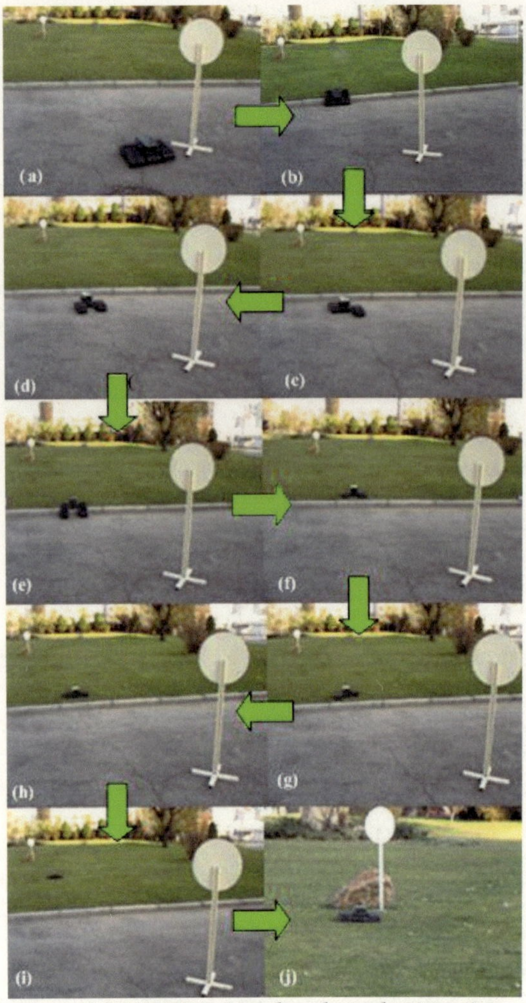

Fig. 13. Autonomous motion of AMOEBA-I in the obstacle situation

6. Conclusion

The frequent nature disasters and man-made catastrophes during the last decades such as earthquake, typhoon, hurricane, radiation and terrorist attack have greatly aroused people's attention on the importance of Urban Search and Rescue (USAR). In some extremely dangerous catastrophes, the rescue staffs are eager to know the inner situation while they cannot perform an in-depth reconnaissance. Robots are emergently needed for tactical search and rescue, which covers how the field teams actually find, support, and extract survivors. AMOEBA-I, a three-module shape-shifting robot with nine configurations from Shenyang Institute of Automation (SIA), Chinese Academy of Sciences (CAS), has been developed for such cases. In this research we mainly focused on the A-B autonomy of AMOEBA-I. With global considerations of the shape-shifting robot and its locomotion environments, this chapter presented a multi-sensor-based autonomous control system for the robot "AMOEBA-I" to move outdoors. With the advantages of various configurations,

AMOEBA-I has high flexibility under unstructured environments. When AMOEBA-I moves in autonomous mode, it can use its sensors to know about the environment, and then select the optimal configuration to achieve an A-B autonomy mission. Experiments have been mainly executed to validate the effectiveness of the system. Since the urban search and rescue environments are usually unstructured and unpredictable, we believe that the shape changing should be an ideal solution to improve the rescue robot's mobility, flexibility, and adaptability. This research result provides a fundamental approach for robot's possible operation in more complex environments.

7. Acknowledgement

This research is supported partly by the National Natural Science Foundation of China (Grant No. 60705029), the National High-Technology 863 Program (Grant Nos. 2006AA04Z254, 2007AA041502-5), Doctoral Startup Foundation of Liaoning Province and State Key Laboratory of Robotics foundation (Grant No. RLO200813).

8. References

Army Science Board (2002). Ad Hoc Study on Human Robot Interface Issues: Final Report, Virginia Blackburn, M.; Bailey, R. & Lytle, B. (2004). Improved mobility in a multi-degree-of-freedom unmanned ground vehicle. *Proceedings of SPIE, Vol. 5422,* pp. 124-134

Casper, J. & Murphy, R. (2003). Human-robot interaction during the robot-assisted urban search and rescue response at the world trade center. *IEEE Transaction on System, Man, and Cybernetics, Part B: Cybernetics,* Vol. 33, No. 3, pp. 367-383

Erkmen, I.; Erkmen, A. M.; Matsuno, F.; Chatterjee, R. & Kamegawa T. (2002). Snake robots to the rescue. *IEEE Robotics & Automation Magazine,* Vol. 9, No. 3, pp. 17-25

Hirose, S & Fukushima, E. (2002). Development of mobile robots for rescue operations. *Advanced Robotics,* Vol. 16, No. 6, pp. 509-112

Huang, B.; Li, M. & Sun, L.(2006) The research of a parent-children type robot system. *Proceedings of IEEE International Conference on Robotics and Biomimetics,* pp. 977-981

Li, B.; Ma S.; Liu, J. & Wang, Y. (2006). AMOEBA-I: an improved link-type shape shifting modular robot for search and rescue operation. *Proceedings of IEEE International Workshop on Safety, Security, and Rescue Robotics*

Lim, S.; Park, D.; Kwak, Y.; Kim, B. & Jeon, S. (2005). Variable geometry single-tracked mechanism for a rescue robot. *Proceedings of IEEE International Workshop on Safety, Security, and Rescue Robotics.* pp. 111-115

Liu, J.; Wang, Y.; Li, B. & Ma S. (2004). Link-type shape shifting modular robot for search and rescue. *High Technology Letters,* Vol. 10, sup, pp. 179-183

Liu, J.; Wang, Y.; Li, B. & Ma S. (2007a). Current research, key performance and future development of search and rescue robot. *Frontiers of Mechanical Engineering in China,* Vol. 2, No. 4, pp. 404-416

Liu, J.; Wang, Y.; Li, B.; Ma S. & Tan, D. (2007b). Center-configuration selection technique for the reconfigurable modular robot. *Science in China Series F-Information Sciences,* Vol. 50, No. 5, pp. 697-710

Matsuno, F. & Tadokoro, S. (2004). Rescue Robots and Systems in Japan. *Proceedings of IEEE International Conference on Robotics and Biomimetics*, pp. 12-20

Murphy, R. (2002). Marsupial and shape-shifting robots for urban search and rescue. *IEEE Intelligent Systems*, Vol. 20, No. 3/4, pp. 14-20

Murphy, R.; Tadokoro, S.; Nardi, D.; Jacoff, A.; Fiorini, P.; Choset, H. & Erkmen, A. (2007). Search and Rescue Robotics, in: *Springer Handbook of Robotics*, B. Siciliano et al. (Ed.), pp. 1151-1173, Springer Berlin Heidelberg, Berlin

Osuka, K. & Kitajima, H. (2003). Development of Mobile Inspection Robot for Rescue Activities: MOIRA. *Proceedings of IEEE/RSJ International Conference on Intelligent Robots and Systems*, pp. 3373-3377

Quek, B.K.; Ibanez-Guzman, J. & Lim, K.W. (2005). Feature-based perception for autonomous unmanned navigation. *Proceedings of* 31st Annual Conference of IEEE Industrial Electronics IECON, pp. 1791-1796

Research Council of the National Academies Technology (2002). Development for Army Unmanned Ground Vehicles, The National Academic Press, Washington D.C.

Tabata, K.; Inaba, A. & Amano, H. (2005). Development of a transformational mobile robot to search victims under debris and rubble -2nd report: improvement of mechanism and interface. *Proceedings of IEEE International Workshop on Safety, Security, and Rescue Robotics*. pp. 19-24

Tetsushi, K.; Yamasaki, T.; Igarashi, H. & Matsuno, F. (2004). Development of the snake-like rescue robot "KOHGA". *Proceedings of IEEE Conference on Robotics and Automation*, pp. 5081-5086

Wolf, A.; Brown, H.; Casciola, R.; Costa, A.; Schwerin, M.; Shamas, E. & Choset, H. (2003). A mobile hyper redundant mechanism for search and rescue tasks. *Proceedings of IEEE/RSJ International Conference on Intelligent Robots and Systems*, pp. 2889-2895

Zhang, H.; Wang, W.; Den, Z.; Zong, G & Zhang, J. (2006). A novel reconfigurable robot for urban search and rescue. *International Journal of Advanced Robotic Systems*, Vol. 3, No. 4, pp. 359-366

Zou, D.; Wang, T. & Liang, J. (2006). Reconfiguration research on modular mobile robot. *Proceedings of IEEE/RSJ International Conference on Intelligent Robots and Systems*, pp. 1082-1085

26

The Rh-1 full-size humanoid robot: Control system design and Walking pattern generation

Mario Arbulú, Dmitry Kaynov and Carlos Balaguer
Robotics lab, Department of Systems and Automation Engineering
University Carlos III of Madrid
Spain

1. Introduction

Since industrial robots cannot be easily adapted to assist human activities in everyday environments such as in hospitals, homes, offices, there is a growing need for robots that can interact with a person in a human-like manner. Wheel robots sometimes cannot be used in such kinds of environments because of the obvious restrictions posed by the use of wheels. For example, it is impossible for this kind of robot to go downstairs and upstairs or to clear some obstacles on the floor. What is more, humanoid robots are expected to play a more important role in the future.

One of the most exciting challenges that has faced the engineering community in recent decades was obtaining a machine of a similar form, a humanoid robot, that could do the same activities as a human being in addition to walking in the same manner (such as HONDA robots, Hirai et al. 1998; HRP robots, Kaneko et al. 2002, 2008; Johnnie, Loeffler et al.; LOLA, Lohmeier et al. 2006.)

There are several reasons to construct a robot with these characteristics. Humanoid robots will work in a human atmosphere with greater effectiveness than any other type of robots because the great majority of environments where they would interact reciprocally with a human are constructed, taking into account the dimensions of the latter. If it is supposed that a machine should complete dangerous tasks or work in extreme conditions, in the ideal case its anthropometric measures must be as close as possible to the ones of its prototype. Inclusively, there are professionals who adduce that for a human being to interact naturally with a machine, it must look like him.

The main goal of this project is the development of a reduced weight human size robot which can be a reliable humanoid platform for implementing different control algorithms, human interaction, etc.

The main assumption for the mechanical design started with the weight of a 1.20 m person and the desired walking motion of the humanoid robot. With these, the requirements for each joint's torque were calculated and then, by dynamical analysis, the structure was designed and the dimensions of the motors were determined. It was an iterative process for

obtaining the optimal torques which allows the anthropometrical walking of a 1.45m human.
Nowadays, the development of humanoid robots has become a very active area. However, it is still limited by the very high cost of maintenance and development.
The main parts of the hardware of the humanoid robot are the "custom-built" components. The software also does not have any standardization or common rules for the humanoid robot's programming. It implies the growth of usage of some technologies from the industrial automation field in humanoid robotics because of their low cost and reliability.
The control system of the Rh-1 robot was designed using the conventional electronic components of the automation industry in order to reduce the development time and cost and to have a flexible and easily upgradeable hardware system.
While generating walking patterns we can compute joint angular speed, acceleration and torque ranges (Arbulu et al. 2005 and Stramiglioli et al. 2002). There are two methods for designing gait patterns: The distributed mass model and the concentrated mass model. In our case, the concentrated mass model is used because the humanoid dynamics is simplified significantly (Kajita et al. 2003, Gienger et al. 2001). In order to obtain a natural and stable gait, the 3D-LIPM method is used, where the pendulum mass motion is constrained to move along an arbitrarily defined plane. This analysis leads us to a simple linear dynamics: the Three dimensional Linear Inverted Pendulum Mode (3D-LIPM, Kajita et al. 2003). Furthermore, the sagital and frontal motion can be studied in separate planes (Raibert 1986, Arbulu et al. 2006). The 3D-LIPM takes into account that the pendulum ball moves like a free ball in a plane following the inverted pendulum laws in the gravity field, so the ball motion has only three parameters: gravity, the plane position and the ball position. This model is applicable only during the single support phase. Another mass concentrated model is used, which is the Cart-table model (Kajita et. al. 2003), implemented in order to improve the walking patterns because the ZMP position can be predicted and a closer relationship with the COG is obtained. Smooth patterns by optimizing jerk are obtained; this will be seen in successful experiments. In order to apply the obtained trajectory to the humanoid robot Rh-1, the ball or cart motion drives the middle of the hip link. The foot trajectories are computed by single splines, taking into account foot position and orientation and the landing speed of each foot to keep the humanoid from falling down. Several direction patterns will be computed. Some simulation experiments have been done using a 21 DOF VRML robot model.
Then, using the Lie-logic method, the inverse kinematics problem for the entire robot body was solved and the trajectory vectors for each joint were obtained (Paden et. al. 1986). These trajectories are used as reference calculated motion patterns and denote feet, arms and the entire body trajectory of the robot.
An industrial robot usually operates in a well defined environment executing pre-programmed tasks or movement patterns. In the same way, the first approaches to make a humanoid robot walk (Shin et al. 1990) were based on the generation of stable offline patterns according to the ZMP (Zero Moment Point) concept (Vukobratovic et al. 1969). In contrast to industrial robots, a humanoid robot will interact with a person in a continuously changing workspace. Therefore, the use of only static motion patterns for the humanoid robot interaction is insufficient.
The other method is a real-time control, based on sensor information (Furusho et al. 1990 and Fujimoto et al. 1998). This approach requires a large amount of computing and

communication resources and sometimes it is not suitable for a humanoid robot with a high number of joints.
A humanoid robot can walk smoothly if it has previously defined walking patterns and an ability to react adequately to the disturbances caused by imperfections in its mechanical structure and irregular terrain properties. Previous works (Park et al. 2000 and Quang et al. 2000), rarely considered and published the detailed architecture for the online modification of dynamic motion patterns.
And finally, a new humanoid platform Rh-1 (Fig. 1) was constructed and successfully tested in a series of walking experiments.

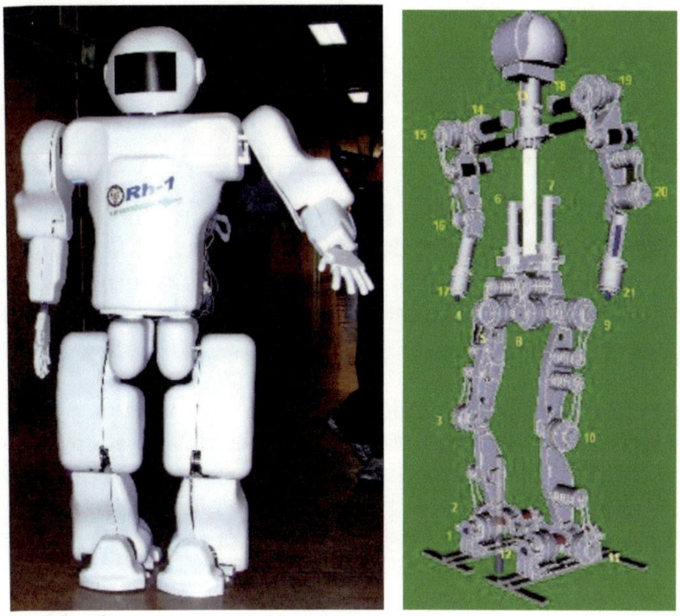

Fig. 1. Rh-1 humanoid robot

This paper presents the control architecture which combines use of the previously offline calculated motion patterns with online modifications for dynamic humanoid robot walking.

The main contributions of this work are:

- Validation of the novel approach for walking pattern generation: Local Axis Gait Algorithm (LAG). It permits walking in any direction and on an uneven surface (i.e. ramp, stairs).
- Implementation and validation of the use of screws provide a very geometric description of rigid motion, so the analysis of the mechanism is greatly simplified; furthermore, it is possible to carry out the same mathematical treatment for the different robot joints: revolute and prismatic.
- Development of hardware and software control architecture for the humanoid robot Rh-1. This allows us to obtain a more flexible and adaptable system capable of changing its properties according to user needs. Proposed hardware architecture is a novel solution for the area of humanoid robots that complies with modern tendencies in robotics. Software

architecture providing the robot with a standard functionality is easily upgraded and can use new one.
- Definition of purpose and validation of kinematics modelling of humanoids robots using screw theory and Paden-Kahan sub-problems, which have the following advantages:

> They avoid singularities because they offer a global description of rigid body motion; we only need to define two frames (base and tool) and the rotation axis of each DOF, to analyze the kinematics in a closed way.
>
> The Paden-Kahan sub problems allows for computing the inverse kinematics at position level.
>
> There is a faster computation time of the inverse kinematics with respect to the inverse Jacobean method, Euler angles or D-H parameters, so it contributes to real-time applications.

- Implementation and validation of the inverted pendulum and cart-table-based walking patterns for any humanoid robot, under the LAG algorithm.
- Development of new efficient algorithms for joint motion control and stabilization of the humanoid robot gait. These algorithms provide simple solutions allowing for fast and reliable integral control of a robot.

The paper is organized as follows. Section 2 deals with the human biomechanics study. Sections 3, 4 and 5 consider hardware, software and communication infrastructures of the humanoid robot Rh-1. Then, the section 6 presents walking pattern generation and some simulations results. Section 7 considers the control architecture implemented in order to control the robot's motion and stability. Experimental results will be shown in section 8. And finally, Section 9 presents conclusions of this work.

2. Biomechanics

2.1 Outline

The humanoid design starts from its motion requirements, so dimensions, joint range motion, joint velocities, forces and wrench should be studied. After that, the link design can start. This first humanoid robot prototype deals with the study of locomotion, so human locomotion will be analyzed. First, human biomechanics anthropometry is studied; next, human walking motion is analyzed.

2.2 Kinematics

The term used for these descriptions of human movement is kinematics. Kinematics is not concerned with the forces, either internal or external, which cause the movement, but rather with the details of the movement itself. In order to keep track of all the kinematic variables, it is important to establish a convention system. Thus if we wish to analyze movement relative to the ground or the direction of gravity, we must establish a spatial reference system (Fig. 2).

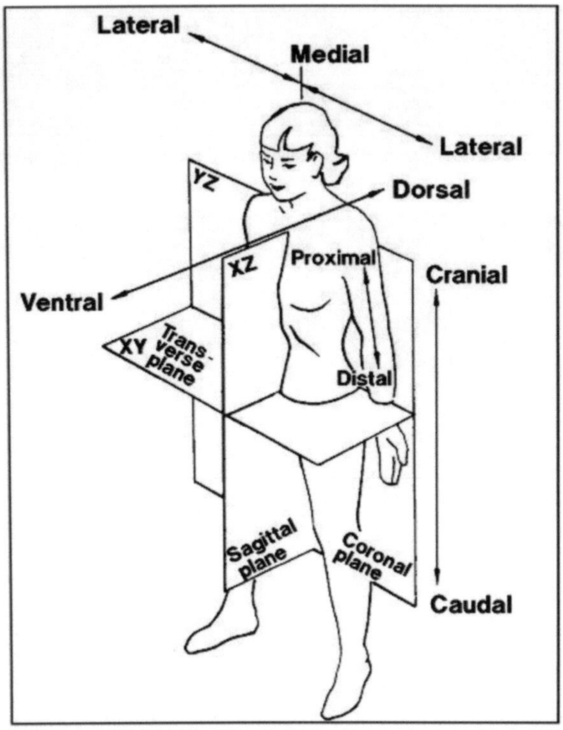

Fig. 2. Human motion planes, © NASA.

2.3 Human locomotion

For dividing the gait cycle in many stages or events, some considerations are taken into account such as the fact that the gait cycle is the period of time between any two identical events in the walking cycle (Ayyappa, E. 1997). As the gait cycle could be divided into events and the continuity between each of them must be maintained, any event could be selected as the starting of the gait cycle (that is in the ideal case because the terrain imperfections and human postures make gait cycle not periodic, see Fig. 3). So, the starting and finishing event are called the initial contact respectively. Otherwise, the gait stride is defined as the distance between two initial contacts of one foot.

The stance and swing are the events of the gait cycle. Stance is the event when the foot is in contact with the ground, (around 60 percent of the gait cycle). Swing is the event when the foot is in the air, (around 40 percent of the gait cycle).

2.4 Anthropomorphic human dimensions, volume and weight distribution

Human dimensions are taken into account as a reference because their proportions allow for stable walking and optimal distribution of the forces actuating while a human is walking. Biomechanics give us the relationship between human height and each link (Fig. 4, Winter, D. 1990), as well as in the same way as the mass.

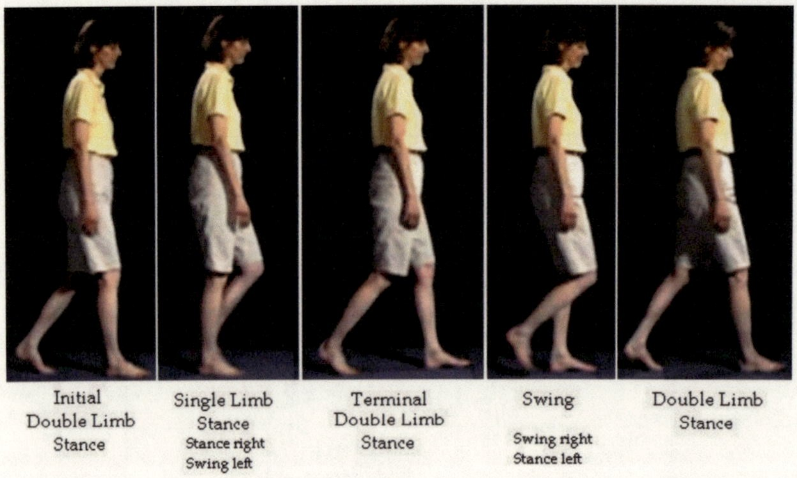

Fig. 3. The gait cycle has two phases: about 60-percent stance phase and about 40-percent swing phase with two periods of double support which occupy a total of 25 to 30 percent of the gait cycle.

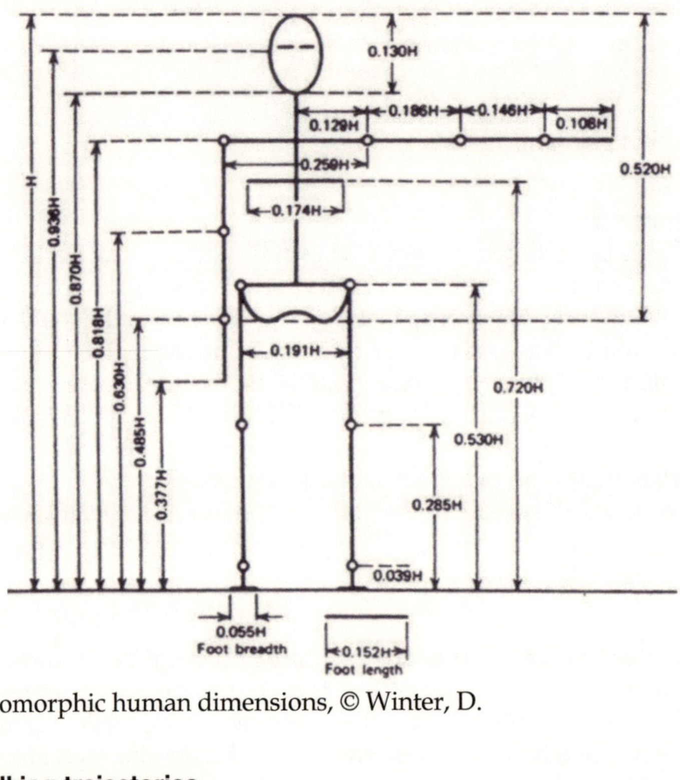

Fig. 4. Anthropomorphic human dimensions, © Winter, D.

2.5 Human walking trajectories

Human walking motion is studied in order to analyze the right motion of each link and joint during the step. The swing leg and hip motions must assure stable walking in any direction and speed.

The joint angular evolution during a walk should be measured with the appropriate devices, or by introducing the swing leg and hip trajectories as inputs of a kinematic model. For the humanoid robot, the joint angular evolution is the input for walking. The human swing foot normally falls to the ground when walking, while for a humanoid robot this must be avoided in order to protect the robot structure and force sensors of the soles. Thus, the adequate walking pattern should be generated for the COG and the swing foot. Normally, the human COG follows the laws of the inverted pendulum in the field of gravity during the walking motion, which is a hyperbolic orbit. It is suitable for making a smooth walking motion at the jerk level. However, the humanoid robot's swing foot motion should be faster than the human one in order not to fall while walking.

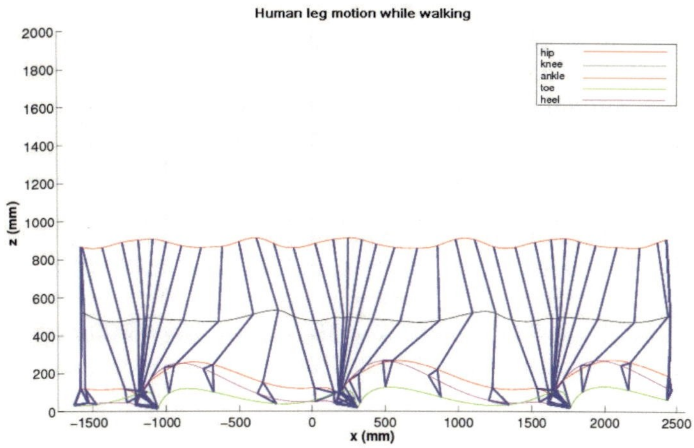

Fig. 5. Human leg motion, Sagital view.

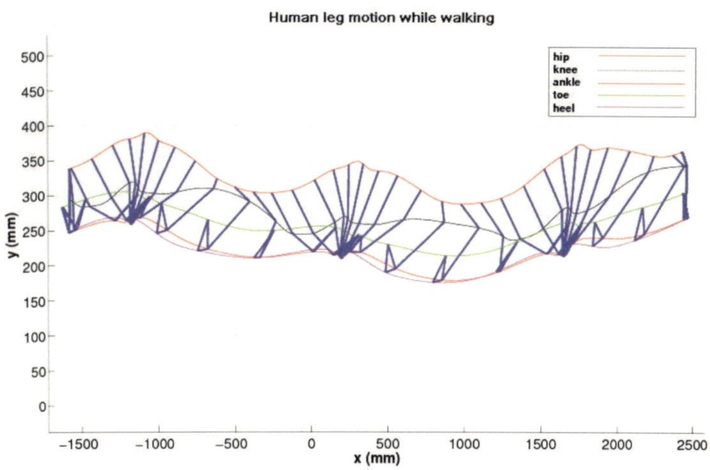

Fig. 6. Human leg motion, Top view.

Figures 5 and 6 show the leg motion and the hip, knee and feet trajectories (including the ankle, toe and heel). The hip trajectory is quite similar to the COG trajectory. In the sagital view, that trajectory climbs and descends cyclically. The falling motion increases the sole reaction force, so in the humanoid robot, it is better to have a motion on a horizontal plane; furthermore, the trajectory shape looks like the inverted pendulum motion (top view, Fig. 6), so we could approximate the humanoid robot in this way.

3. Hardware Architecture

The hardware architecture for the humanoid robot has some important restrictions imposed by the limited availability of space. In general, the basic requirements for hardware architecture of a humanoid robot are: scalability, modularity and standardized interfaces (Regenstein et. al. 2003). In the case of the Rh-1 robot with 21 DOF, which supposes the use of 21 DC motors in synchronized high-performance multi-axis application, it is first necessary to choose an appropriate control approach. The trend of modern control automation is toward distributed control. It is driven by one basic concept: by reducing wiring, costs can be lowered and reliability increased. Therefore, an electrical design of Rh-1 robot is based on a distributed motion control philosophy where each control node is an independent agent in the network. Figure 7 shows the physical distribution of the hardware inside the humanoid robot.

The architecture presented is provided with a large level of scalability and modularity by dividing the control task into Control, Device and Sensory levels (Fig. 8), (Kaynov et al. 2005).

The Control level is divided into 3 layers represented as a controller centered on its own tasks such as external communications, motion controller's network supervision or general control.

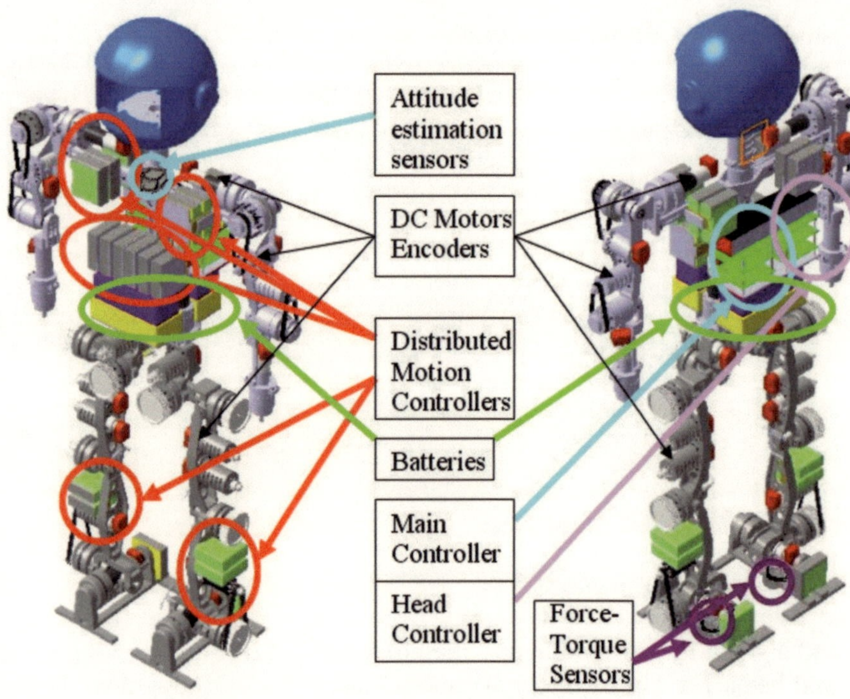

Fig. 7. Hardware distribution inside the humanoid robot.

In the Device level each servo drive not only closes the servo loop but calculates and performs trajectory online, synchronizes with other devices and can execute different movement programs located in its memory. These kinds of devices are located near the motors, thus benefiting from less wiring, which is one of the requirements for energy efficiency; they are lightweight and require less effort in cabling. Advanced and commercially available motion controllers were implemented in order to reduce development time and cost. Continuous evolution and improvements in electronics and computing have already made it possible to reduce the industrial controller's size to use it in the humanoid development project. Furthermore, it has the advantage of applying well supported and widely used devices from the industrial control field, and brings the commonly used and well supported standards into the humanoid robots development area.
On the Control level, the Main controller is a commercial PC/104+ single board computer because of its small size and low energy consumption. It was used instead of a DSP controller because it has a different peripheral interface as is the Ethernet and RS-232, and an easy programming environment. In addition, there is a great variety of additional extension modules as the PC/104+ bus like CAN-bus, digital and analog input-output, and PCMCIA cards. Selection criteria were fast CPU speed, low consumption and availability of expansion interfaces. The Main Controller provides general synchronization, updates sensory data, calculates the trajectory and sends it to the servo controllers of each joint. It also supervises data transmission for extension boards such as Supervisory Controller and ZMP Estimation Controller via PC/104+ bus.

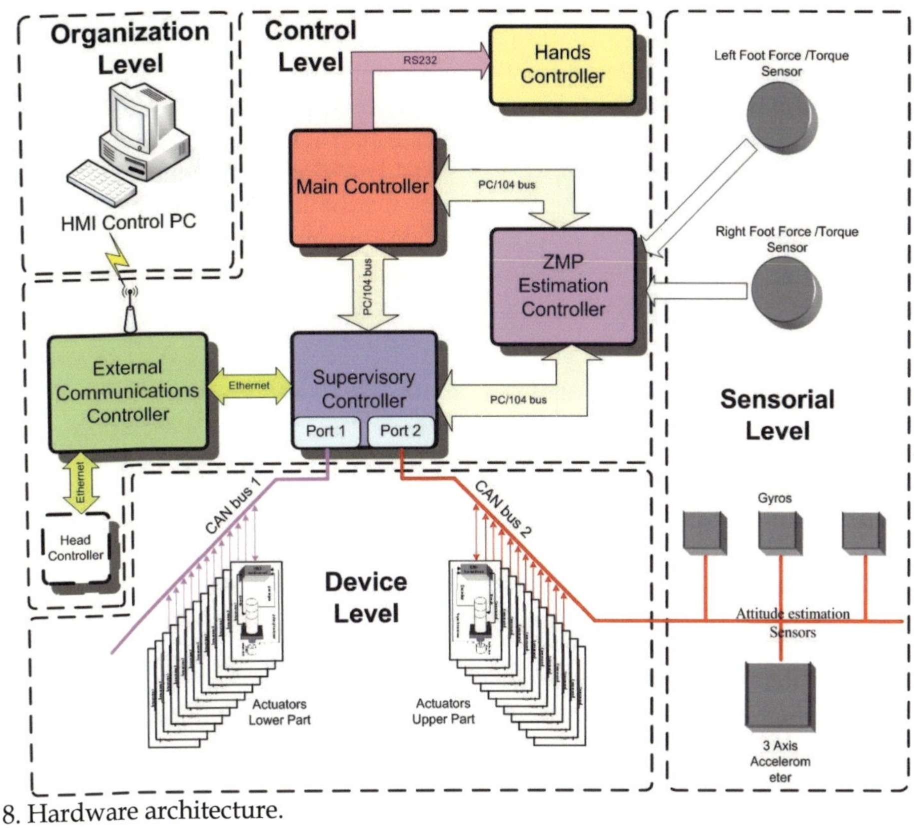

Fig. 8. Hardware architecture.

The Communication Supervisory Controller uses a network bus to reliably connect distributed intelligent motion controllers with the Main Controller.
The motion control domain is rather broad. As a consequence, communications standards to integrate motion control systems have proliferated. The available communication standards cover a wide range of capability and cost ranging from high-speed networked IO subsystems standards to distributed communications standards for integrating all machines on the shop floor into the wider company. The most appropriate solution to be implemented in the humanoid robot motion control system design seems to be the use of CAN-based standards. The CAN bus communication is used for the Sensory level and the CANOpen protocol on top of the CAN bus is used for the Device level of communications.
Thus, the control system adopted in the Rh-1 robot is a distributed architecture based on CAN bus. The CAN bus has also been chosen because of various characteristics, such as bandwidth up to 1 MBit/s that is of sufficient speed to control the axes of a humanoid robot, a large number of nodes (Rh-1 has 21 controllable DOF), differential data transmission, which is important for reducing the Electromagnetic Interference (EMI) effects caused by electric motors, and finally, the possibility for other devices such as sensors to reside in the same control network.
At the Device level, the Controller's network of the Rh-1 is divided into 2 independent CAN buses in order to reduce the load of the communication infrastructure. The Lower part bus controls 12 nodes of two legs and the Upper part bus controls 10 nodes of two arms and the trunk. To unify the data exchange inside the robot, the attitude estimation sensory system is also connected to the Upper part CAN bus. In this way, the communication speed of CAN bus used in Rh-1 is 1MBit/s. The synchronization of both parts is realized by the Supervisory Controller at the Control level of automation.
The External Communications module provides the Ethernet communication on the upper (Control) level of the automation with head electronics which comprise an independent vision and sound processing system. It also provides wireless communications with the Remote Client which sends operating commands for the humanoid robot. The proposed architecture complies with the industrial automation standards for the design of the motion control system.

4. Software Architecture

As mentioned above, a humanoid robot can be considered as a plant where the shop floor consists of a series of cells (intelligent motion controllers and sensors) managed by controllers (the Main Controller, Communication Supervisory Controller, etc.) In general, there are two basic control tasks for the control system of a humanoid robot. The first goal is to control all automation and supervise data transmission. Meanwhile, the second goal is to control and monitor the entire floor in order to detect failures as early as possible, and to report on performance indicators. In this context, the humanoid robot Rh-1 is provided with a software system allowing the implementation of the industry automation concepts (Kaynov et. al. 2007). The software architecture is based on the Server-Client model (Fig. 9).
For security reasons, the Control Server accepts the connections of other clients, such as the Head Client, responsible for the human-robot interaction, only if the Master Client allows it. If the connection is accepted, the Master Client only supervises the humanoid robot state and data transmission between the robot and other Client, but in the case of any conflict it always has top priority.

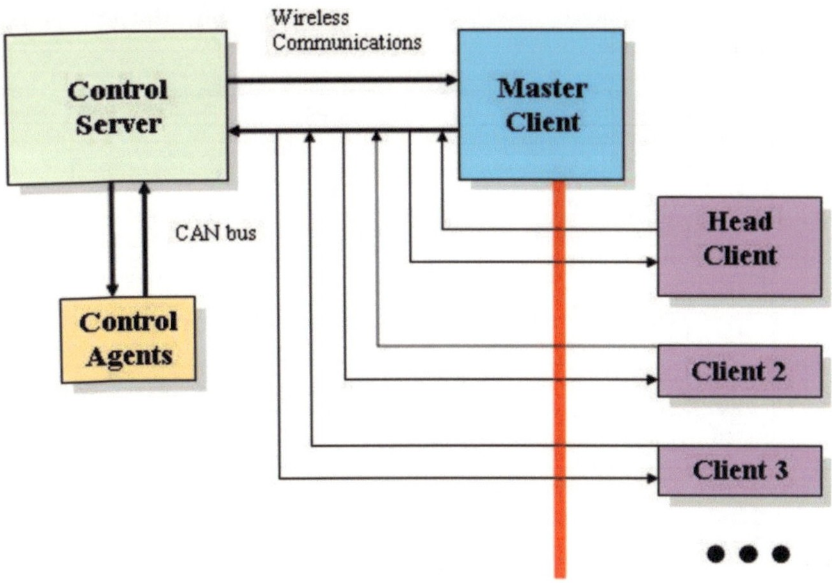

Fig. 9. System Architecture.

According to the Server-Client model, the humanoid robot is controlled by the passive Server, which waits for requests, and upon their receipt, processes them and then serves replies for the Client. On the other hand, the Server controls all Control Agents which reside in the CAN bus network. In that case, the Control Server is no longer a slave; it is a network master for Control Agents which performs their operations (motion control or sensing) and replies for the Server.

As a PLC in the automation industry, the Control Server is designed and programmed as finite state automata. Figure 10 shows the state diagram and Table II shows the state transition events of the humanoid robot Control Server functioning

Two basic types of incoming data are processed. A command is simple data, which can be executed by one Control Agent. The order is a complex command which needs the simultaneous action of many Control Agents and sensors which the humanoid robot possesses. After the connection of the Master Client, the humanoid robot stays in the Client Handling state waiting for an order or a simple command. The arrival of an order launches the User Program. The User Program is executed in the control area, the core of the humanoid robot Control Server software. It performs the data transmission between all Control Agents, sensory system and the Server. It performs trajectory execution at the synchronized multi-axis walking applications, controls the posture and ZMP errors at the dynamic walking mode, and reads the sensors' state, etc. The control area consists of different modules which provide the execution of motion control for stable biped locomotion of the humanoid robot. All tasks can be grouped by their time requirements.

The developed software provides the set of the C-based function to work with the robot and to generate the user's motions and control procedures that are not only for walking, but also for implementing different human-robot cooperation tasks. The code below shows the simple user program. The example in Figure 12 shows how the simple humanoid robot motion can be programmed. At the beginning, the synchronization procedure for every joint is performed, and then the motion is started. The robot will change the gait (walking mode) according to user request.

Event	Event Description
E1	The Client is connected
E2	An Order has arrived
E3	A Command has arrived
E4	A Command is sent to the Control Agent
E5	Agent's reply has arrived
E6	An Answer is sent to the Client
E7	The User Program has successfully terminated or an Error Event has occurred
E8	Connection with the Client is lost
E9	The Robot is staying in the secure position
E10	All processes are terminated

Table 1. State Transition Events

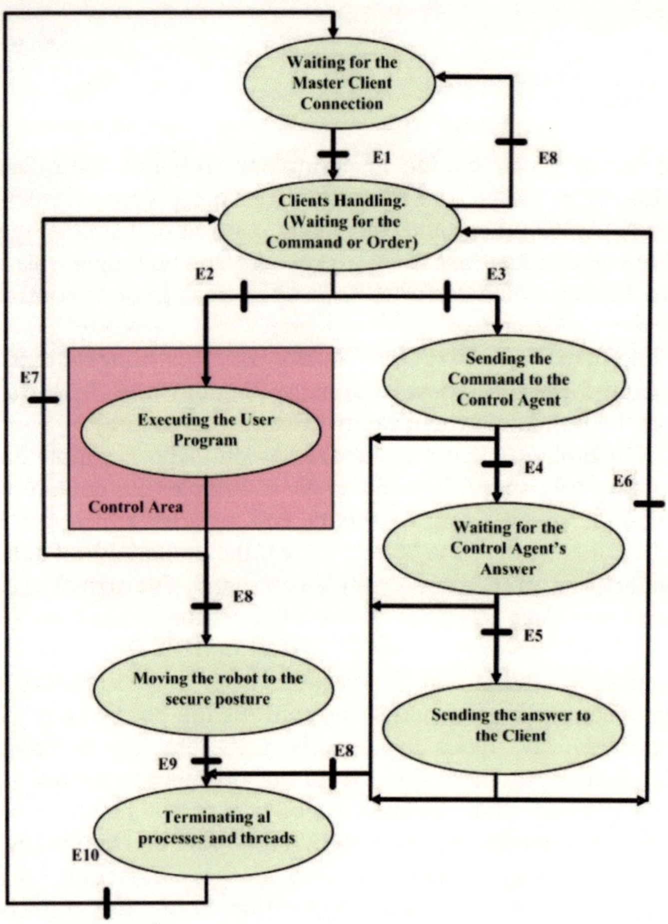

Fig. 10. Server functioning state diagram.

In the proposed software architecture, the Control Server is capable of accepting a large amount of clients' connections at the same time. It is evident that the Master Client, as the basic HMI of the humanoid robot, should provide and supervise the execution of the upper-level control tasks related with global motion planning, collision avoidance, and human-robot interaction. In general, these tasks are common for all mobile and walking robots and the design of these kinds of software systems is not considered in this paper. On the other hand, there are some bottom-level tasks that should be supervised such as sensory data acquisition, joint synchronization and walking stability control. In order to not overload the Master Client, which is more oriented to automation supervisory, these control tasks are processed with another client application. To provide the robot Rh-1 with bottom-level control, a SCADA system for the humanoid robot, called HRoSCoS (Humanoid Robot Supervisory Control System) Client was developed.

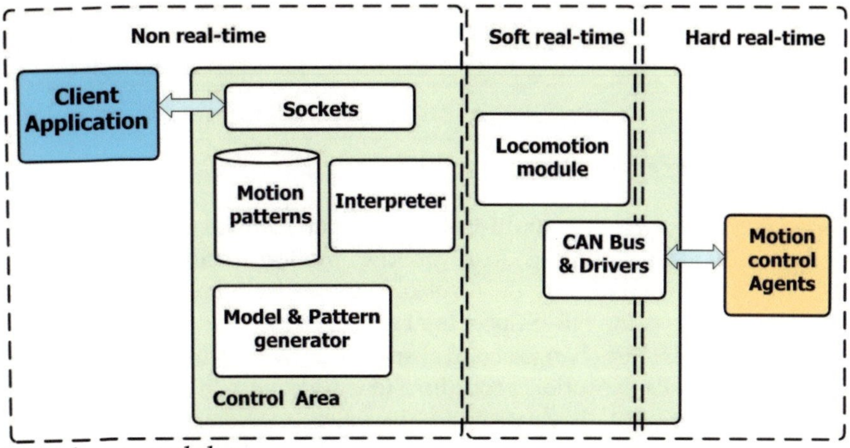

Fig. 11. Control area modules.

```
// Motion program example
#include "control.h"
#include "sensors.h"
#include "dynamic_gate.h"

main (void)
{
        EnableRobot();
        ReadJointsPosition();
        Sincronize();
        if ( Start() )
                do
                {
                        PostureControl();
                        DynamicGate(Gate1);
                        if (ChangeGate)

        DynamicGate(Gate2);
                }
                while (!Stop())
        DisableRobot();
}
```

Fig. 12. Motion program example.

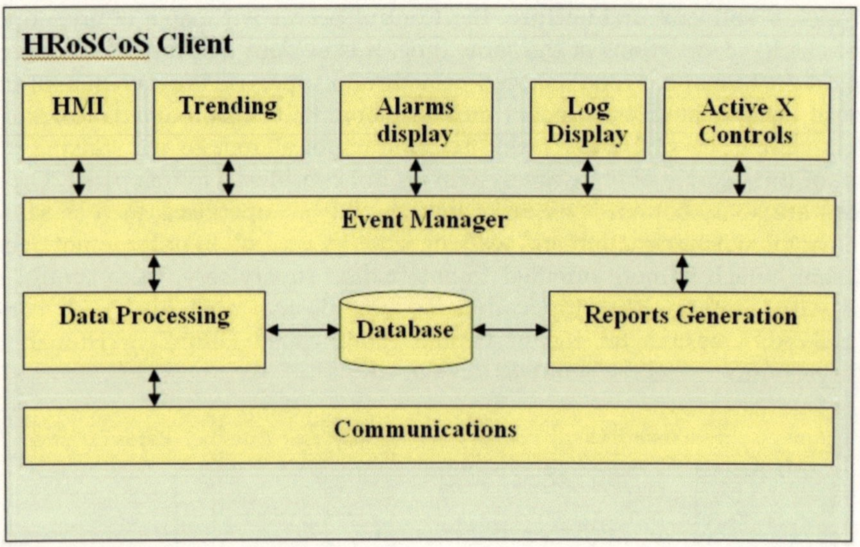

Fig. 13. HRoSCoS Client Architecture.

The developed software system is multi-tasking and the Control Server is also responsible for data acquisition and handling (e.g. polling motion controllers, alarm checking, calculations, logging and archiving) on a set of parameters when the HRoSCoS Client is connected. Figure 13 shows the HRoSCoS Client architecture.

The Client requires the data or changes control set points by sending commands. The arrival of a command launches its execution procedure (the right branch of the Server functioning State Diagram in Figure 10). It consists of the interpretation and transmission of the Command to a Control Agent. When the answer is received, it is converted and transmitted to the HRoSCoS Client to be processed and visualized.

The HRoSCoS Client provides the trending of different parameters of the robot, such as the joint velocities, accelerations, currents, body inclinations, forces and torques which appear during humanoid robot walking. Real-time and historical trending is possible, although generally not in the same chart. Alarm handling is based on limit and status checking and is performed in the Control Server (for example current limit or physical limit of the joint) and then the alarm reports are generated into the HRoSCoS Client application. More complicated expressions (using arithmetic or logical expressions) are developed by creating derived parameters on which status or limit checking is then performed. Logging of data is performed only if some value changes. Logged data can be transferred to an archive once the log is full. The logged data is time-stamped and can be filtered when viewed by a user. In addition, it is possible to generate different reports on the humanoid robot state at any time.

The HRoSCoS Client system presents the information to the operating personnel graphically. This means that the operator can observe a representation of the humanoid robot being controlled (Fig. 11).

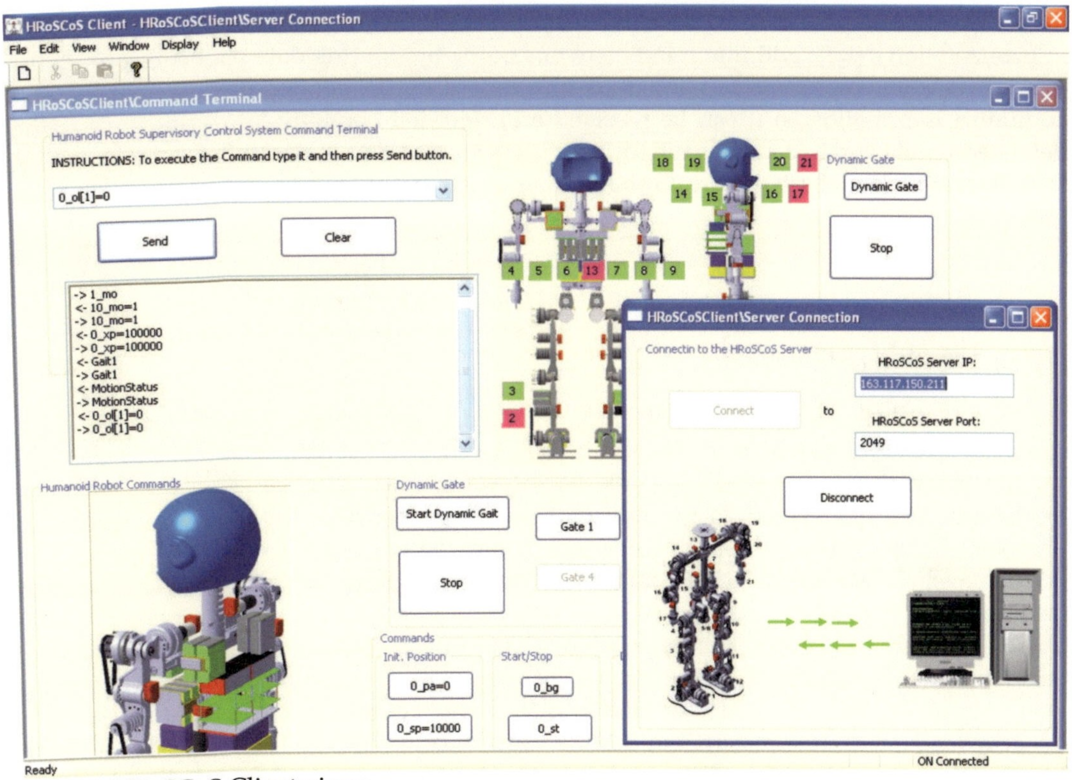

Fig. 14. HRoSCoS Client views.

The HMI supports multiple screens, which can contain combinations of synoptic diagrams and text. The whole humanoid robot is decomposed in "atomic" parameters (e.g. a battery current, its maximum value, it's on/off status, etc.) to which a Tag-name is associated. The Tag-names are used to link graphical objects to devices. Standard windows editing facilities are provided: zooming, re-sizing, scrolling, etc. On-line configuration and customization of the HMI is possible for users with the appropriate privileges. Links are created between display pages to navigate from one view to another

5. Communication Infrastructure and Methods

When building automation applications, communication with the host is often a crucial part of the project. Nodes of the network always function as data servers because their primary role is to report information (status, acquired data, analyzed data, etc.) to the host at constant rates.

As shown in Figure 9, Hardware Architecture consists of three basic levels of automation which uses its own communication systems. The upper (Control) level uses a TCP/IP based communication protocol. Ethernet communication is one of the most common methods for sending data between computers. The TCP/IP protocol provides the technology for data sharing, but only the specific application implements the logic that optimizes performance and makes sense of the data exchange process. When data transmission begins, the sender should packetize each piece of data with an ID code that the receiver can use to look up the decoding information. In this way, developed communication protocol hides the TCP implementation details and minimizes network traffic by sending data packages only when they are needed. When a data variable is transmitted by the sender, it is packetized with additional information so it can be received and decoded correctly on the receiving side. Before each data variable is transmitted, a packet is created that includes fields for Data Size, Data ID and the data itself. Figure 15 shows the packet format.

Data Size (8 bits)	Data ID (16 bits)	Data (32 bytes)

Fig. 15. The package format.

The Data ID field is populated with the index of the data array element corresponding to the specified variable. Since the receiving side also has a copy of the data array, it can index it to get the properties (name and type) of the incoming data package. This very effective mechanism is implemented to provide data exchange between the Control Server and different Clients on the Control level of automation of the humanoid robot.

Bottom-level (Sensory and Field) communications are realized using CAN and CanOpen protocols (Fig. 16).

These communication protocols provide data transmission in broadcast type of communication. A sender of information transmits to all devices on the bus. All receiving devices read the message and then decide if it is relevant to them. This guarantees data integrity as all devices in the system use the same information. The sensory system of the humanoid robot makes data exchange under lower CAN protocol and the intelligent motion controllers uses upper-level CANOpen protocol. The same physical layer of these protocols allows them to reside in the same physical network.

The communication implemented on the bottom level involves the integration of CANOpen (Drives and Motion Control Device Profile) and the introduction of new functionality which is not contained within the relevant device profiles for the sensory data processing.

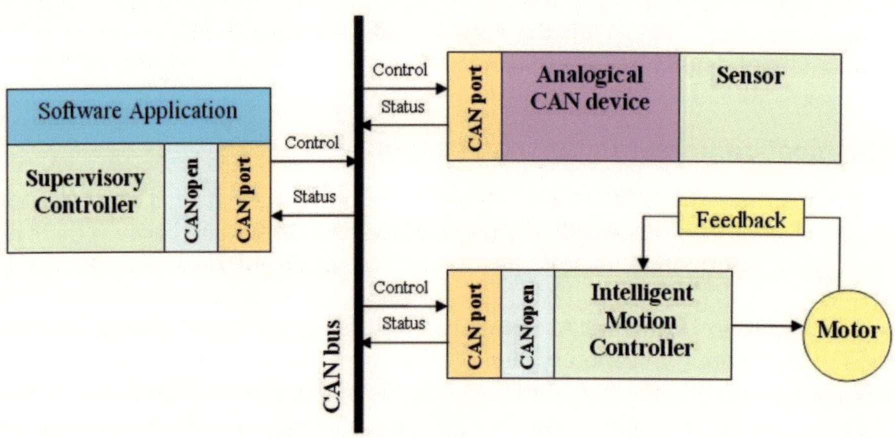

Fig. 16. CAN bus-based communication system.

6. Walking Pattern Generation

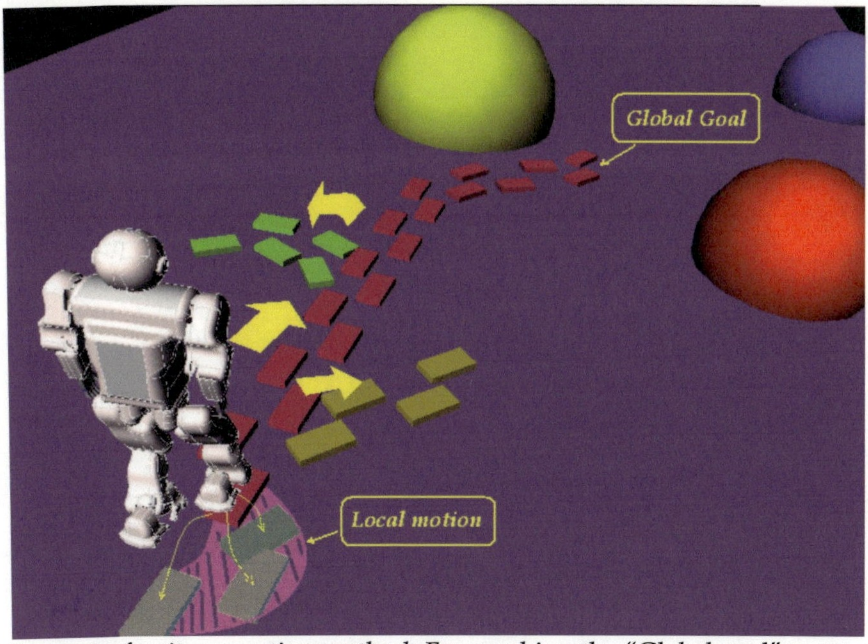

Fig. 17. Concept of gait generation method. For reaching the "Global goal", a set of "Local motions" must be generated. Thus the local motion decide the better foot location for going ahead, go back, turn left, turn right, doing lateral step, climbing a stair or ramp.

There are many propositions for generating the walking patterns of humanoid robots, some of them a mass distributed based model, (Hirukawa et. al. 2007) and other ones a mass concentrated based model, (Kajita et. al. 2004 and Gienger et. al. 2001). The first approach describes the motion accurately, but it has a high computation cost, which is not suitable for real-time applications. On the other hand, the second approach saves computation time and performs the walking motion suitably. In this section, two kinds of mass concentrated models will be explained and discussed, that is: the inverted pendulum model and the cart-table model. Both models have been tested on the Rh-1 humanoid robot platform in order to generate stable walking patterns. At first, the 2D inverted pendulum model will be detailed, for introducing pendulum laws; next the 3D version is developed; after that, the Cart-table model will be introduced and its advantages with respect to the inverted pendulum are explained; next, the walking pattern strategy is proposed with the "Local Axis Gait" (Fig. 18) algorithm, (Arbulu et. al. 2007, 2008). Finally, in order to compute joint patterns the inverse kinematics model is proposed, by using the screw theory and Lie groups.

6.1 2D Inverted pendulum model

The gait pattern generation for a humanoid robot could be simplified as: studying the motion in the sagital plane and concentrating all the body mass in the COG. In this way, it is possible to use the 2D Inverted pendulum model to obtain stable and smooth walking motion.

The 2D inverted pendulum model is composed of a mass and a telescopic leg without mass (Fig. 18).

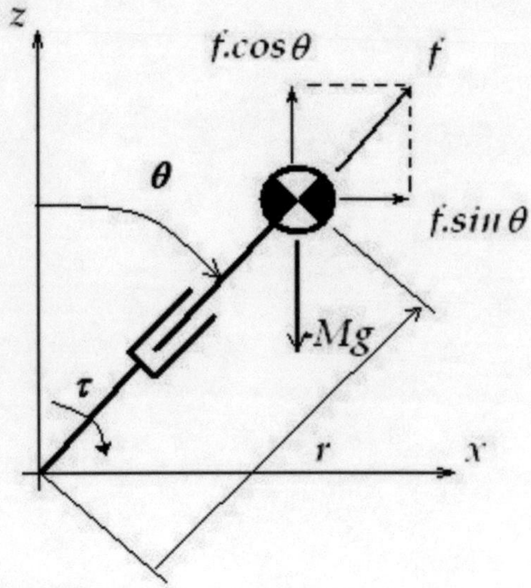

Fig. 18. The 2D Inverted Pendulum Model with motion in the x-z plane.

So, the model is described in the next state variables:

r : *Radius of Position vector (massless and telescopic leg)*

θ : *Pitch angle*

f :*Reaction Force on pendulum*

τ : *External Pitch torque*

From the free body diagram of the pendulum ball, the dynamics equations should be written as follows:

$$F_x = f \sin\theta + \frac{\tau}{r}\cos\theta \quad (1)$$

$$F_z = f \cos\theta - Mg - \frac{\tau}{r}\sin\theta \quad (2)$$

It is known that ***p=(x, z)***, so the dynamics equations of the pendulum ball motion are:

$$F_x = M\ddot{x} = f\sin\theta + \frac{\tau}{r}\cos\theta \tag{3}$$

$$F_z = M\ddot{z} = f\cos\theta - Mg - \frac{\tau}{r}\sin\theta \tag{4}$$

There are several solutions for the ball pendulum motion from this complex dynamic model. In order to simplify the dynamic problem, some constraints could be taken:

1) Motion at constraint height:

$$z = z_c \tag{5}$$

$$\ddot{z} = 0 \tag{6}$$

2) It is possible to consider natural pendulum ball motion, so the input torque turns to zero:

$$\tau = 0 \tag{7}$$

From these constraints the dynamic equations (3) and (4) reduce the dynamic motion to a linear one:

$$F_x = M\ddot{x} = f\sin\theta \tag{8}$$

$$F_z = 0 = f\cos\theta - Mg \tag{9}$$

By combining equations (8) and (9), the dynamic pendulum ball motion is obtained as:

$$\ddot{x} = g\frac{x}{z_c} \tag{10}$$

The natural motion of the pendulum ball depends on the potential gravity field *(g)*, position *(x, z)* and distance from the pendulum base *(z_c)*. Thus, no linear dynamic motion equations are conversed to linear ones; this way a single solution could be found and this kind of trajectory is applicable in real-time applications of walking locomotion.

In order to design walking patterns and to determine the spatial geometry of trajectories, the concept of orbital energy is introduced. Orbital energy evaluates the pendulum ball energy at the level of the motion plane. It is composed of the potential and kinetic energy of the pendulum ball. In this way, it is possible to determine if the pendulum motion is in a state of equilibrium, going forward, or if it never passes the zero position.

The mathematical expression of orbital energy could be developed by multiplying equation (10) by and integrating it.

$$\dot{x}\left(\ddot{x} - g\frac{x}{z_c}\right) = 0 \quad (11)$$

$$\int\left(\ddot{x}\dot{x} - g\frac{x}{z_c}\dot{x}\right)dt = const \quad (12)$$

$$\frac{1}{2}\dot{x}^2 - \frac{1}{2}\frac{g}{z_c}x^2 = E \quad (13)$$

Equation (13) shows that a kind of energy is conserved, which is called orbital energy. The first term represents the kinetic energy per unit mass of the body, while the second one is the virtual energy caused by a force field that generates a force ***(g/z$_c$).x*** on the unit mass located at x. Furthermore (Fig. 19), ***E>0*** means that the pendulum mass swings forward; ***E=0*** represents: the equilibrium state, the pendulum mass swinging toward the equilibrium point or the pendulum mass swinging out from the equilibrium point; finally ***E<0*** means that the body never passes point ***x=0***.

At this point, it is possible to generate two-dimensional stable natural walking patterns. This study is the basis for obtaining the solution of three-dimensional walking patterns, suitable for any humanoid robot. Human-like walking motion can be obtained because biomechanical studies demonstrate that ***COG*** human motion on the walking cycle could be approached by an inverted pendulum motion. The next section focuses on three-dimensional pendulum motion.

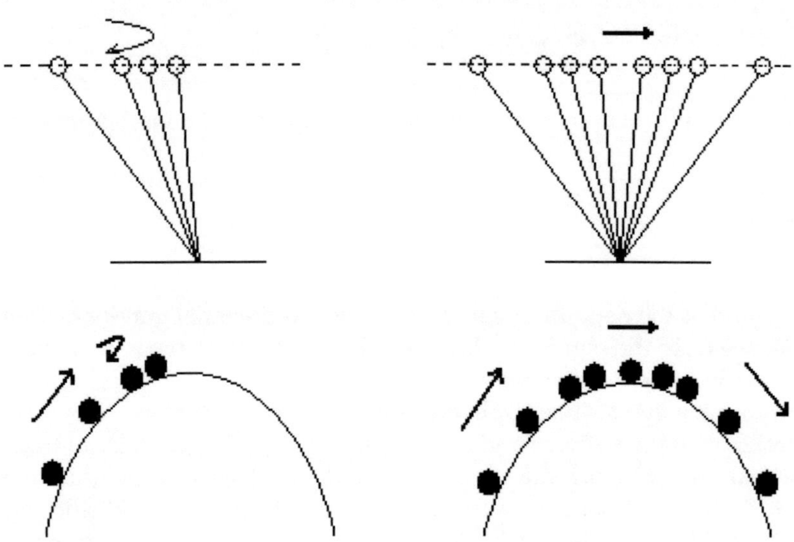

Fig. 19. Pendulum ball rolling in potentials.

6.2 Motion Laws of 3D-LIPM. Inverted Pendulum Model

In Figure 20, the three dimensional Inverted pendulum model is shown consisting of a point mass (*p*) and a mass less telescopic leg, where ***p=(x, y, z)*** is the position of the mass ***M***, which is uniquely specified by variables ***q= (θr,θp,r)***, so:

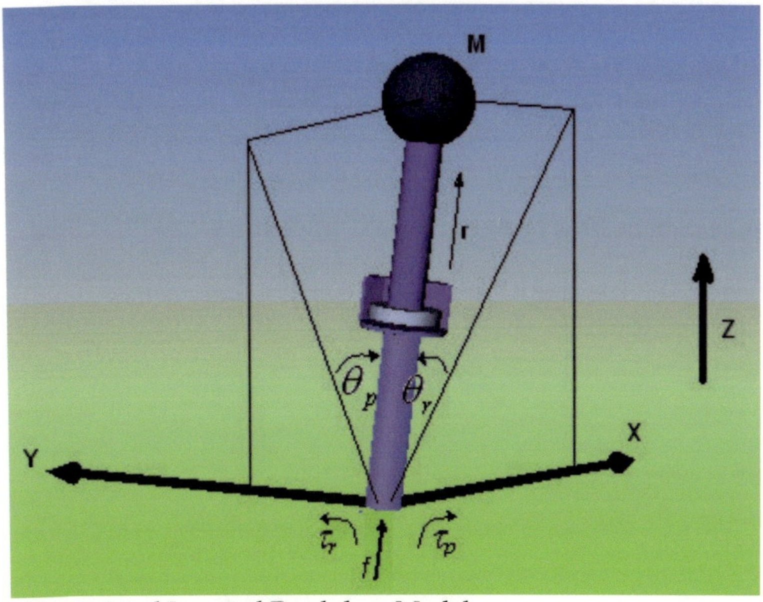

Fig. 20. Three dimensional Inverted Pendulum Model.

Following the right hand rule:

$$x = r\sin(\theta_p) \tag{14}$$

$$y = -r\sin(\theta_r) \tag{15}$$

$$z = r\sqrt{1-\sin(\theta_r)^2 - \sin(\theta_p)^2} \tag{16}$$

The motion equation of the inverted pendulum in Cartesian coordinates is:

$$\begin{pmatrix}\tau_r \\ \tau_p \\ f\end{pmatrix} = m\begin{pmatrix} 0 & -r.\cos(\theta_r) & -\dfrac{r.\cos(\theta_r).\sin(\theta_r)}{\sqrt{1-\sin(\theta_r)^2-\sin(\theta_p)^2}} \\ r.\cos(\theta_p) & 0 & -\dfrac{r.\cos(\theta_p).\sin(\theta_p)}{\sqrt{1-\sin(\theta_r)^2-\sin(\theta_p)^2}} \\ \sin(\theta_p) & -\sin(\theta_r) & \sqrt{1-\sin(\theta_r)^2-\sin(\theta_p)^2}\end{pmatrix}\begin{pmatrix}\ddot{x} \\ \ddot{y} \\ \ddot{z}\end{pmatrix} + mg\begin{pmatrix} -\dfrac{r.\cos(\theta_r).\sin(\theta_r)}{\sqrt{1-\sin(\theta_r)^2-\sin(\theta_p)^2}} \\ -\dfrac{r.\cos(\theta_p).\sin(\theta_p)}{\sqrt{1-\sin(\theta_r)^2-\sin(\theta_p)^2}} \\ \sqrt{1-\sin(\theta_r)^2-\sin(\theta_p)^2}\end{pmatrix} \tag{17}$$

So the dynamics along the ***x-axis*** is given by:

$$m(z\ddot{x}-x\ddot{z})=\frac{\sqrt{1-\sin(\theta_r)^2-\sin(\theta_p)^2}}{\cos(\theta_p)}\tau_p+mgx \tag{18}$$

And the equation for the dynamics along the y-axis is given by:

$$m(-z\ddot{y}+y\ddot{z})=\frac{\sqrt{1-\sin(\theta_r)^2-\sin(\theta_p)^2}}{\cos(\theta_r)}\tau_r-mgy \tag{19}$$

6.3 Natural three dimensional Linear Inverted Pendulum Mode (3D-LIPM)

In order to reduce the motion possibilities of the pendulum, we introduce some constraints to limit this motion. One constraint limits the motion in a plane, so:

$$z=k_x x+k_y y+z_c \tag{20}$$

Where zc is the distance from the xy-plane to the pendulum mass. Replacing (20) and its second derivative into (18) and (19), we get:

$$\ddot{x}=\frac{g}{z_c}x+\frac{k_y}{z_c}(x\ddot{y}-\ddot{x}y)+\frac{1}{mz_c}\frac{\sqrt{1-\sin(\theta_r)^2-\sin(\theta_p)^2}}{\cos(\theta_p)}\tau_p \tag{21}$$

$$\ddot{y}=\frac{g}{z_c}y-\frac{k_x}{z_c}(x\ddot{y}-\ddot{x}y)-\frac{1}{mz_c}\frac{\sqrt{1-\sin(\theta_r)^2-\sin(\theta_p)^2}}{\cos(\theta_r)}\tau_r \tag{22}$$

The above equations allow pendulum motion in any plane and slope, if the motion is constrained to a flat plane (k_x=0 and k_y=0), so:

$$\ddot{y}=\frac{g}{z_c}y-\frac{1}{mz_c}\frac{\sqrt{1-\sin(\theta_r)^2-\sin(\theta_p)^2}}{\cos(\theta_r)}\tau_r \tag{23}$$

$$\ddot{x}=\frac{g}{z_c}x+\frac{1}{mz_c}\frac{\sqrt{1-\sin(\theta_r)^2-\sin(\theta_p)^2}}{\cos(\theta_p)}\tau_p \tag{24}$$

Note that (23) and (24) are independent equations and no linear dynamics is simplified to a linear one.

The natural 3D-LIPM takes into account the trajectories of the Inverted Pendulum Model without input torques. Hence, the equations (23) and (24) are simplified to:

$$\ddot{x}=\frac{g}{z_c}x \tag{25}$$

$$\ddot{y} = \frac{g}{z_c} y \tag{26}$$

Solving those equations (25) and (26), a three dimensional pendulum ball motion is obtained in the gravity field. Figure 21 shows an example of the pendulum motion for a different support foot, i.e. blue pendulum motion for the left support foot at its local frame and red pendulum motion for right support foot at its local frame.

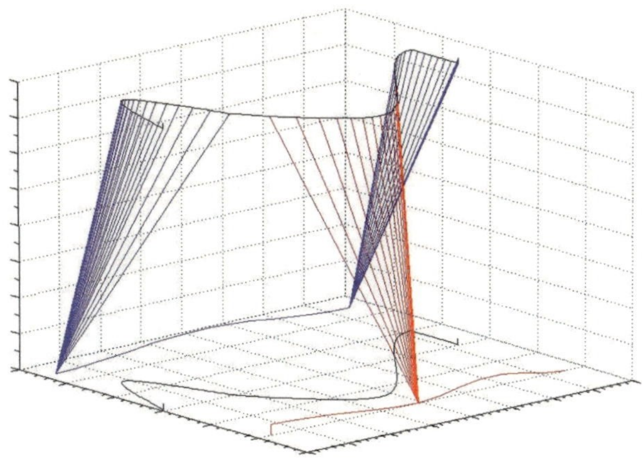

Fig. 21. Inverted Pendulum motion under natural 3D-LIPM.

6.4 Geometry of trajectory

The pendulum spatial motion in the gravity field should be studied and analyzed in order to predict the stability and suitable three-dimensional local motion. So, describing the local motion (Fig. 22) in any rotational axis, it is possible to study the gravitational effects on natural pendulum motion, which is like potential energy acting on a space shuttle.
Orbital energy on each axis being:

$$E_x = \frac{1}{2}\dot{x}^2 - \frac{1}{2}\frac{g}{z_c}x^2 \tag{27}$$

$$E_y = \frac{1}{2}\dot{y}^2 - \frac{1}{2}\frac{g}{z_c}y^2 \tag{28}$$

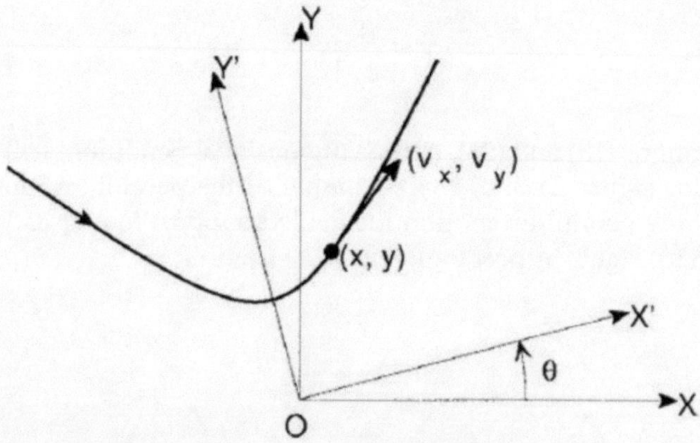

Fig. 22. 3D-LIPM projected onto the XY plane at the local axis.

The orbital energy on X'Y' axis is obtained as follows:

$$E'_x = \frac{1}{2}(\dot{x}\cos\theta + \dot{y}\sin\theta)^2 - \frac{1}{2}\frac{g}{z_c}(x\cos\theta + y\sin\theta)^2 \quad (29)$$

$$E'_y = \frac{1}{2}(-\dot{x}\sin\theta + \dot{y}\cos\theta)^2 - \frac{1}{2}\frac{g}{z_c}(-x\sin\theta + y\cos\theta)^2 \quad (30)$$

We can calculate the axis of symmetry by solving the variation of orbital energy with respect to the rotation angle, and in this way the maximum is found; the mathematical expression is developed as follows:

$$\frac{\partial E'_x}{\partial\theta} = A[(\sin\theta)^2 - (\cos\theta)^2] + B\sin\theta\cos\theta = 0 \quad (31)$$

$$A = \left(\frac{g}{z_c}\right)xy - \dot{x}\dot{y} \quad (32)$$

$$B = \left(\frac{g}{z_c}\right)(x^2 - y^2) - (\dot{x}^2 - \dot{y}^2) \quad (33)$$

Finding the symmetry axis from equation (31), by trigonometric identities:

$$\frac{A}{B} = -\frac{\sin\theta.\cos\theta}{(\sin\theta)^2 - (\cos\theta)^2} \quad (34)$$

$$\theta = \frac{1}{2}\tan^{-1}\left(\frac{2A}{B}\right) \quad (35)$$

It is well known that the ***y-axis*** is the axis of symmetry for $\boldsymbol{\theta = 0}$, so:

$$A = \left(\frac{g}{z_c} \right) xy - \dot{x}\dot{y} = 0 \tag{36}$$

$$\left(\frac{g}{z_c} \right) xy = \dot{x}\dot{y} \tag{37}$$

Eq. (37) could be used for computing the 3D-LIPM geometric shape with the orbital energy mathematical expressions from eq. (27) and (28):

$$\left(\frac{g}{z_c} \right)^2 x^2 y^2 = (2E_x + (g/z_c)x^2)(2E_y + (g/z_c)y^2) \tag{38}$$

By simplifying the last equation an interesting expression is found which describes the shape of the pendulum mass trajectory in the gravity field (eq. 39):

$$\frac{g}{2z_c E_x} x^2 + \frac{g}{2z_c E_y} y^2 = -1 \tag{39}$$

It is possible to deduce that $E_x > 0$, because the *x-axis* pendulum passes ***0*** of the local frame and $E_y < 0$, due to the fact that the *y-axis* pendulum does not pass ***0*** of the local frame (Fig. 19). These facts show us that the pendulum mass trajectory shape is a hyperbolic curve described by eq. (39). Furthermore, the natural pendulum mass motion in three dimensions gives us information about the motion range for several initial conditions, which could be applied to the single support phase of the humanoid body motion.

6.5 Temporal equations

With initial conditions $(x_i, \dot{x}_i)$ and $(y_i, \dot{y}_i)$ at time t_i, the mass trajectory is calculated by solving differential equations (25) and (26):

$$x(t) = x_i \cosh(\frac{t - t_i}{T_c}) + T_c \dot{x}_i \sinh(\frac{t - t_i}{T_c}) \tag{40}$$

$$\dot{x}(t) = \frac{x_i}{T_c} \sinh(\frac{t - t_i}{T_c}) + \dot{x}_i \cosh(\frac{t - t_i}{T_c}) \tag{41}$$

$$y(t) = y_i \cosh(\frac{t - t_i}{T_c}) + T_c \dot{y}_i \sinh(\frac{t - t_i}{T_c}) \tag{42}$$

$$\dot{y}(t) = \frac{y_i}{T_c} \sinh(\frac{t - t_i}{T_c}) + \dot{y}_i \cosh(\frac{t - t_i}{T_c}) \tag{43}$$

6.6 Cart-table model

In order to establish a relationship between the COG and ZMP motion, the cart-table model is proposed. This model, by controlling cart acceleration, gives us an interesting relationship between the ZMP and the COG.

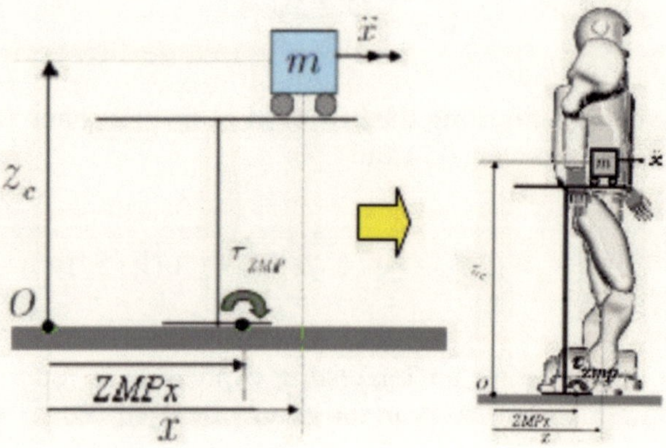

Fig. 23. Cart-table model.

By evaluating the torque in the ZMP (Fig. 23):

$$\tau_{zmp} = mg(x - ZMP_x) - m\ddot{x}Z_c \tag{44}$$

As we know the torque in the ZMP is zero, thus from the eq. (44), ZMPx is being by:

$$ZMP_x = x - \frac{Z_c}{g}\ddot{x} \tag{45}$$

Note that eq. (45) is similar to the inverted pendulum (24), with the main difference being that ZMP_x is constrained to **0**, while if we knew it, it would be fixed to any point on the local axis. In the y direction, a similar equation could be obtained. In order to get the COG motion as an inverse problem from the ZMP one, the solution of eq. (45) should be treated as a servo control problem.

6.7 Comparing pendulum and Cart-table models

In the inverted pendulum model the input reference is the ZMP and the output is the COG pattern. Note that the ZMP is always at the base of the pendulum (i.e. Fig. 24b). In the cart-table model, ZMP motion is instead generated by the COG as reference (i.e. Fig. 24a).

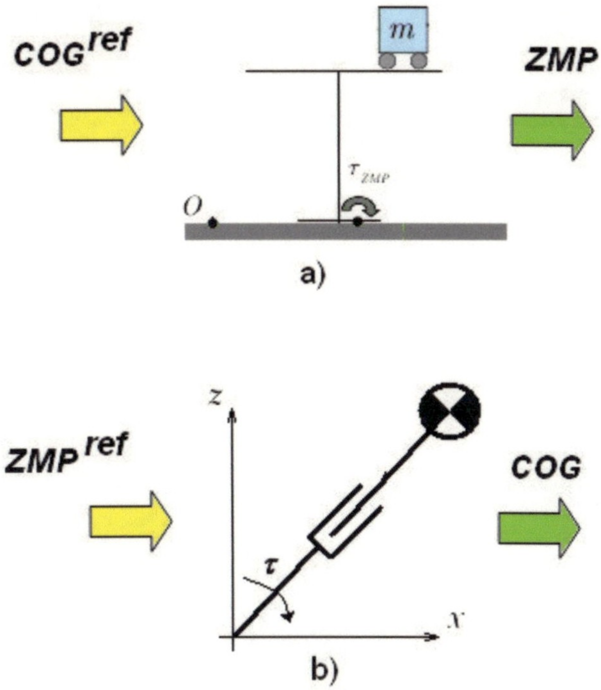

Fig. 24. Comparison between the: a) Cart-table, and b) Inverted pendulum.

Other facts are that there is a discontinuity in the change from a single support phase to a double one, so at high walking velocity, jerk is an important fact; it could be improved by using high-order splines. Thus, the cart-table model optimizes jerk and continuity is maintained at all times, no matter the change of phase, and in this way a high walking speed is possible.

6.8 Walking pattern strategy

Figure 25 show us the steps of the walking strategy, in the single support phase the pendulum ball follows 3D-LIPM laws (A to B, C to D and E to F); in the double support phase, the pendulum ball moves at a constant speed (B to C and D to E). This motion drives the COG of the humanoid robot. We could assume that the COG is in the middle of the hip joint. Foot trajectories are computed by single splines taking into account some constraints such as step length, maximum height of the foot, lateral foot motion, foot orientation and speed in order to avoid falling down and to reduce the impact force on the landing foot (Fig. 26).

6.9 Local Axis Gait algorithm (LAG)

In order to generalize the walking patterns of any direction and surface such as stairs or slopes; the "Local Axis Algorithm" (LAG) is proposed, (Arbulu et al. 2007) so as to plan the stable local walking motion. The LAG is divided into several stages: computation of the footprints; the decision of the ZMP limits around the footprints; the dynamic humanoid COG motion generation based on the mass concentrated model; and finally joining the footprints of the swing foot by splines. In this way, it is possible to generate each step online, using the desired footprints as input.

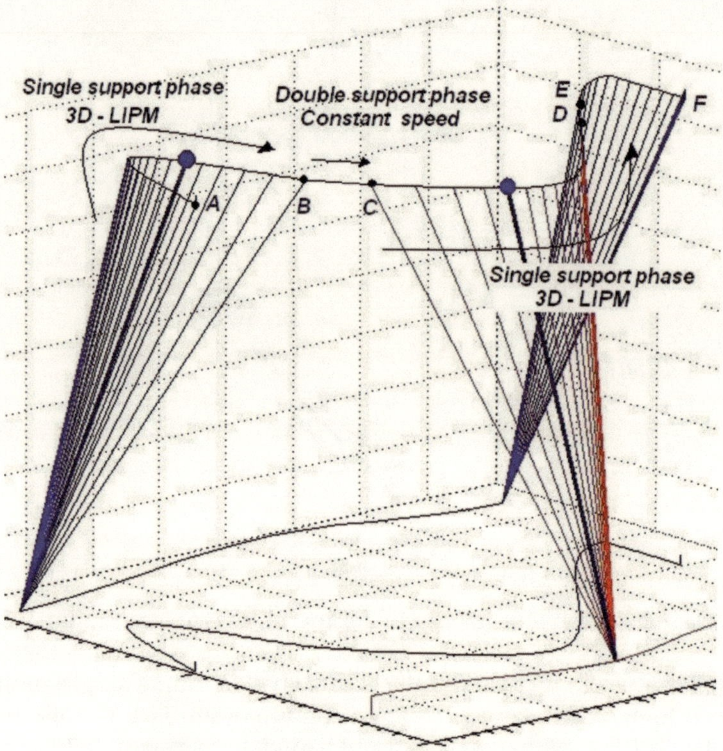

Fig. 25. Walking pattern strategy

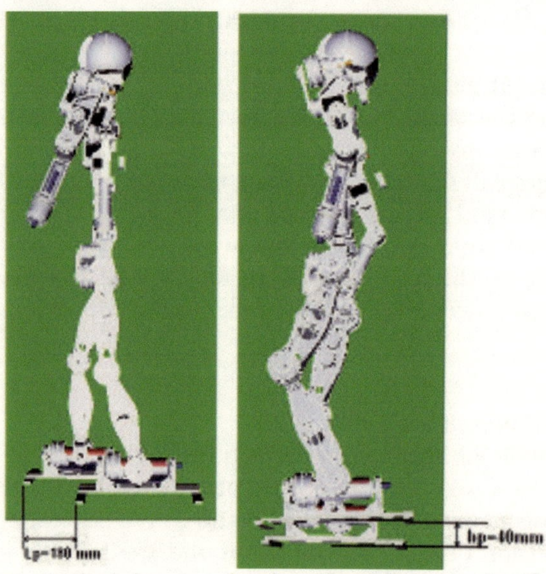

Fig. 26. Some foot trajectory constraints: max step length and max swing foot height.

The footprints (Fig. 27) for doing an n-th step can be computed as follows:

$$P^n = P^n + R\left(\theta_z^n\right)^T . L^n \tag{46}$$

Where:

$$P^n = \left(p_x^n \quad p_y^n \quad p_z^n\right)^T,$$

$$L^n = \left(L_x^n \quad -(-1)^n L_y^n \quad L_z^n\right)^T$$

$\sum, \sum^n, \sum^{n-1}, \sum^{n+1}$: World and feet frames

P^n, P^{n-1}, P^{n+1} : feet position

$L_x^{n+1}, L_y^{n+1}, L_z^{n+1}$: swing foot displacements

$\theta_x^{n+1}, \theta_y^{n+1}, \theta_z^{n+1}$: rotations about world frame

The walking patterns developed are introduced into the inverse kinematics algorithm (Arbulu et al. 2005) to obtain the angular evolution of each joint; those are the reference patterns of the humanoid robot.

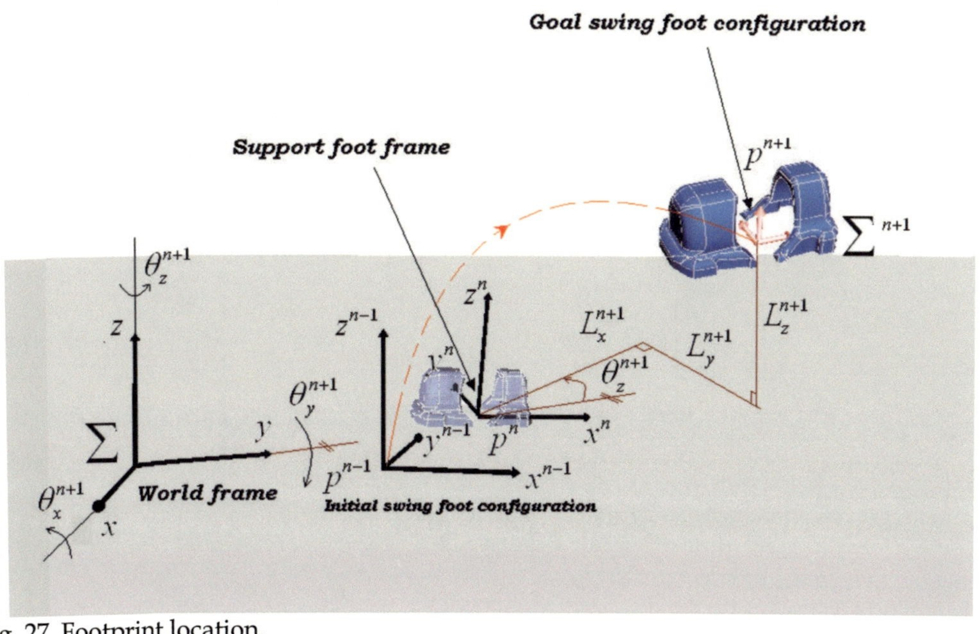

Fig. 27. Footprint location.

6.10 Inverse Kinematics model

In order to compute the robot's joint motion patterns some kinematics considerations must be made. Due to the fact that the kinematics control is based on screw theory and Lie logic techniques, it is also necessary to present a basic explanation.

Lie logic background

Lie groups are very important for mathematical analysis and geometry because they serve to describe the symmetry of analytical structures (Park et. al. 1985). A Lie group is an analytical manifold that is also a group. A Lie algebra is a vectorial space over a field that completely captures the structure of the corresponding Lie group. The homogeneous representation of a rigid motion belongs to the special Euclidean Lie group (SE(3)) (Abraham et. al. 1999). The Lie algebra of ***SE(3)***, denoted ***se(3)***, can be identified with the matrices called twists "$\xi^\wedge$", (eq. 47), where the skew symmetric matrix "$\varpi^\wedge$", (eq. 48) is the Lie algebra ***so(3)*** of the orthogonal special Lie group (***SO(3)***), which represents all rotations in the three-dimensional space. A twist can be geometrically interpreted using screw theory (Paden 1986), as Charles's theorem proved that any rigid body motion could be produced by a translation along a line followed by a rotation around the same line;, this is a screw motion, and the infinitesimal version of a screw motion is a twist.

$$\xi^\wedge = \begin{bmatrix} \varpi^\wedge & \upsilon \\ 0 & 0 \end{bmatrix} \in se(3) / se(3) = \{(\upsilon, \varpi^\wedge) : \upsilon \in \Re^3, \varpi^\wedge \in so(3)\} \in \Re^{4x4} \quad (47)$$

$$\varpi^\wedge = \begin{bmatrix} 0 & -\varpi_3 & \varpi_2 \\ \varpi_3 & 0 & -\varpi_1 \\ -\varpi_3 & \varpi_1 & 0 \end{bmatrix} / \forall \varpi = \begin{bmatrix} \varpi_1 \\ \varpi_2 \\ \varpi_3 \end{bmatrix} \wedge \upsilon = \begin{bmatrix} \upsilon_1 \\ \upsilon_2 \\ \upsilon_3 \end{bmatrix} \Rightarrow \varpi \times \upsilon = \varpi^\wedge . \upsilon \quad (48)$$

The main connection between ***SE(3)*** and ***se(3)*** is the exponential transformation (eq. 49). It is possible to generalize the forward kinematics map for an arbitrary "open-chain" manipulator with ***n*** DOF of magnitude ***g(θ)***, through the product of those exponentials, expressed as **POE** (eq. 50), where ***g(0)*** is the reference position for the coordinate system.

$$\left.\begin{aligned} e^{\xi^\wedge \theta} &= \begin{bmatrix} e^{\varpi^\wedge \theta} & (I - e^{\varpi^\wedge \theta})(\varpi \times \upsilon) + \varpi \varpi^T \upsilon \theta \\ 0 & 1 \end{bmatrix} \in SE(3); \varpi \neq 0 \\ e^{\xi^\wedge \theta} &= \begin{bmatrix} I & \upsilon \theta \\ 0 & 1 \end{bmatrix} \in SE(3); \varpi = 0 \\ e^{\xi^\wedge \theta} &= I + \varpi^\wedge \sin\theta + \varpi^{\wedge 2}(1 - \cos\theta) \end{aligned}\right\} \quad (49)$$

$$g(\theta) = \prod_{i=1}^{n} e^{\xi_i^\wedge \theta_i} . g(0) \quad (50)$$

A very important payoff for the POE formalism is that it provides an elegant formulation of a set of canonical problems, the Paden and Kahan sub-problems, (Pardos et. al. 2005, Arbulu et. al. 2005) among others, which have a geometric solution for their inverse kinematics. It is possible to obtain a close-form solution for the inverse kinematics problem of complex mechanical systems by reducing them into the appropriate canonical sub-problems.
The Paden and Kahan sub problems are introduced as following (Murray et al. 1994):

Paden-Kahan 1: Rotation about a single axis

Finding the rotation angle using "screw theory" and Lie groups, at first, the point rotation expression from "*p*" to "k" is expressed by (Fig. 28):

$$e^{\hat{\xi}\theta} p = k \qquad (51)$$

The twist and projection vectors on the rotation plane are as follows:

$$\xi = \begin{bmatrix} v \\ \varpi \end{bmatrix} = \begin{bmatrix} -\varpi \times r \\ \omega \end{bmatrix} \qquad (52)$$

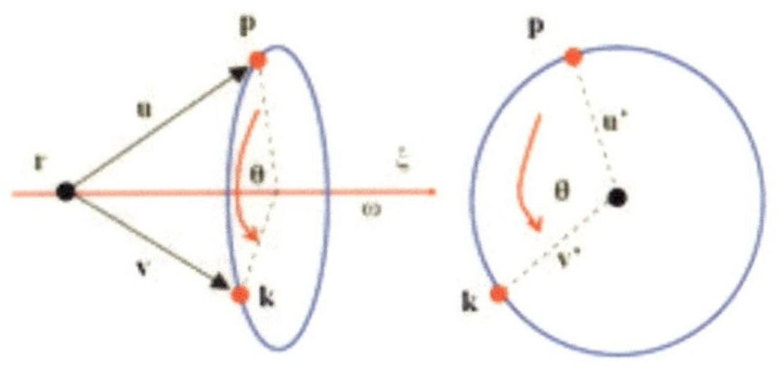

Fig. 28. Rotation on single axis "ω" from point "p" to point "k".

$$u' = u - \omega\omega^T u \qquad (53)$$

$$\upsilon' = \upsilon - \omega\omega^T \upsilon \qquad (54)$$

Finally, the rotation angle is calculated with the following expression:

$$\theta = a\tan 2\left[\omega^T (u' \times \upsilon'), u'^T . \upsilon'\right] \qquad (55)$$

Paden-Kahan 2: Rotation about two subsequent axes

The rotation expression is the following (Fig. 29):

$$e^{\hat{\xi}_1 \theta_1} e^{\hat{\xi}_2 \theta_2} p = e^{\hat{\xi}_1 \theta_1} c = k \tag{56}$$

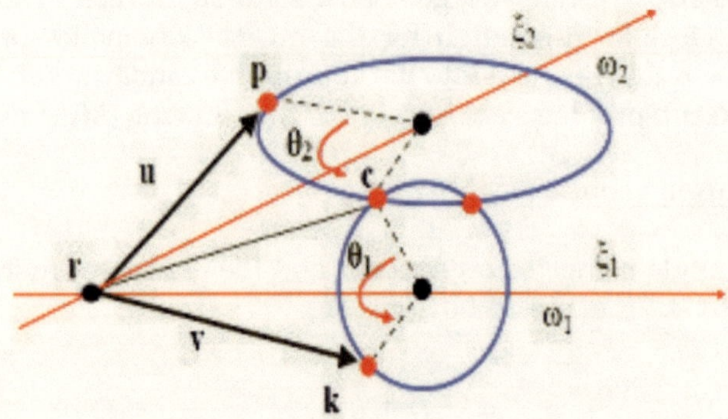

Fig. 29. Rotation on two subsequent axes "ω_1" and "ω_2" from "p" to "c" and from "c" to "k".

The respective twists are described as follows:

$$\xi_1 = \begin{bmatrix} -\omega_1 \times r \\ \omega_1 \end{bmatrix} \wedge \xi_2 = \begin{bmatrix} -\omega_2 \times r \\ \omega_2 \end{bmatrix} \tag{57}$$

Some values are computed in order to obtain the point "c" with the following expressions:

$$\alpha = \frac{\left(\omega_1^T \omega_2\right)\omega_2^T u - \omega_1^T \upsilon}{\left(\omega_1^T \omega_2\right)^2 - 1} \tag{58}$$

$$\beta = \frac{\left(\omega_1^T \omega_2\right)\omega_1^T \upsilon - \omega_2^T u}{\left(\omega_1^T \omega_2\right)^2 - 1} \tag{59}$$

$$\gamma^2 = \frac{\|u\|^2 - \alpha^2 - \beta^2 - 2\alpha\beta\omega_1^T \omega_2}{\|\omega_1 \times \omega_2\|^2} \tag{60}$$

Obtaining the point "c":

$$c = r + \alpha\omega_1 + \beta\omega_2 \pm \gamma\left(\omega_1 \times \omega_2\right) \tag{61}$$

Once we get "c" for the second sub-problem, we can apply the first Paden-Kahan sub-problem to obtain the solutions for $\boldsymbol{\theta_1}$ and $\boldsymbol{\theta_2}$. Note that there might be two solutions for "c", each of them giving a different solution for $\boldsymbol{\theta_1}$ and $\boldsymbol{\theta_2}$.

Paden-Kahan 3: Rotation to a given distance

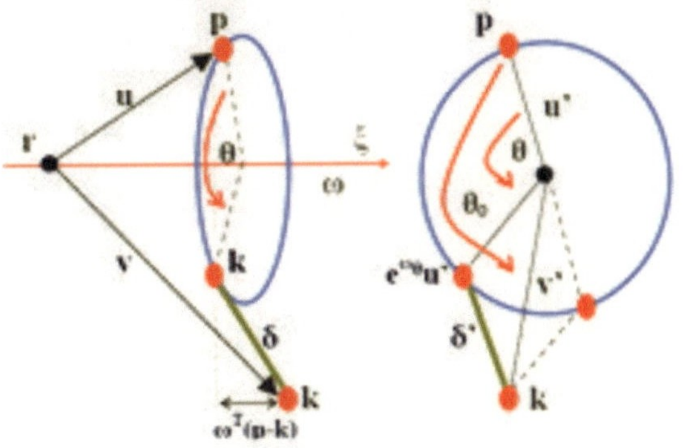

Fig. 30. Rotation of point "p" to "k" which is a distance "δ" from "q".

The distance "δ" is shown as follows:

$$\left\| e^{\xi^{\wedge}\theta} p - q \right\| = \delta \tag{62}$$

The associate twist and vectors projection in the perpendicular plane of rotation axis could be computed as:

$$\xi = \begin{bmatrix} \upsilon \\ \omega \end{bmatrix} = \begin{bmatrix} -\omega \times r \\ \omega \end{bmatrix} \tag{63}$$

$$u' = u - \omega\omega^{T} u \tag{64}$$

$$\upsilon' = \upsilon - \omega\omega^{T} \upsilon \tag{65}$$

Projecting "δ" in "ω" direction:

$$\delta'^{2} = \delta^{2} - \left|\omega^{T}(p-q)\right|^{2} \tag{66}$$

If we let "θ_0" be the angle between the vectors "u" and "υ", we have:

$$\theta_0 = a\tan 2\left[\omega^{T}(u' \times \upsilon'), u'^{T}.\upsilon'\right] \tag{67}$$

Finally, we obtain the rotation angle by:

$$\theta = \theta_0 \pm \cos^{-1}\left(\frac{\|u'\|^2 + \|\upsilon'\|^2 - \delta'^2}{2\|u'\|\|\upsilon'\|}\right) \tag{68}$$

The algorithm developed is called Sagital Kinematics Division (SKD). It divides the robot body into two independent manipulators, one for the left and one for the right part of body (Fig. 31), subject to the following constraints at any time: keeping the balance of the humanoid ZMP and imposing the same position and orientation for the common parts (pelvis, thoracic, cervical) of the four humanoid manipulators.

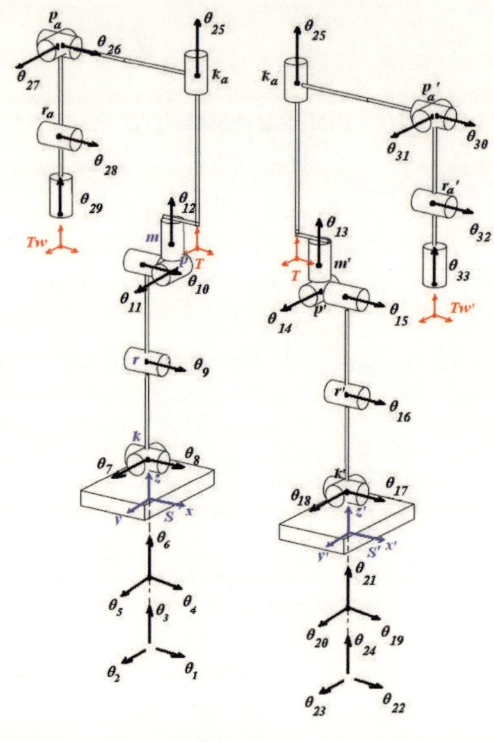

Fig. 31. Rh-1 Sagital Kinematics Division (SKD).

<u>Solving the kinematics problem</u>

It is possible to generalize the leg forward kinematics map with 12 DOF **($\theta_1 ... \theta_{12}$)**. The first six DOF correspond to the position **(θ_1, θ_2, θ_3)** and orientation **(θ_4, θ_5, θ_6)** of the foot. Note that these DOF do not correspond to any real joint and for that reason we call them "non-physical" DOF.
The other DOF are called "physical DOF" because they correspond to real motorized joints. These are: **θ_7** for the hindfoot, **θ_8** for the ankle, **θ_9** for the knee, **θ_{10}** for the hip on the ***x axis***, **θ_{11}** for the hip on the ***y axis*** and **θ_{12}** for the hip on the ***z axis***. Let ***S*** be a frame attached to the base system (support foot) and ***T*** be a frame attached to the humanoid hip.

The reference configuration of the manipulator is the one corresponding to $\theta_i = 0$, and $g_{st}(0)$ that represents the rigid body transformation between T and S when the manipulator is at its reference configuration.
Then, the product of exponentials formula for the right and left legs forward kinematics is $g_{st}(\theta)$ and $g_{s't'}(\theta)$, being ξ^ the 4x4 matrices called "*twists*".

$$g_{st}(\theta) = e^{\xi_1^\wedge \theta_1} \cdot e^{\xi_2^\wedge \theta_2} \cdots e^{\xi_{29}^\wedge \theta_{29}} \cdot g_{st}(0) \tag{69}$$

$$g_{s't'}(\theta) = e^{\xi_{24}^\wedge \theta_{24}} \cdot e^{\xi_{23}^\wedge \theta_{23}} \cdots e^{\xi_{33}^\wedge \theta_{33}} \cdot g_{s't'}(0) \tag{70}$$

The inverse kinematics problem i.e. for the right leg (see Fig. 31) consists of finding the joint angles, that is, the six physical DOF $(\theta_7 \ldots \theta_{12})$, given the non-physical DOF $(\theta_1 \ldots \theta_6)$ from the humanoid footstep planning, the hip orientation and position $g_{st}(\theta)$, which achieve the ZMP humanoid desired configuration. Using the PoE formula for the forward kinematics it is possible to develop a numerically stable geometric algorithm, to solve this problem, by using the Paden-Kahan (P-K) geometric sub-problems. It is straightforward to solve the inverse kinematics problem in an analytic, closed-form and geometrically meaningful way, with the following formulation.

At first, twist and reference configurations are computed:

$$g_{st}(0) = \begin{bmatrix} 1 & 0 & 0 & T_x - S_x \\ 0 & 1 & 0 & T_y - S_y \\ 0 & 0 & 1 & T_z - S_z \\ 0 & 0 & 0 & 1 \end{bmatrix} \tag{71}$$

$$\upsilon_1 = \begin{bmatrix} 1 \\ 0 \\ 0 \end{bmatrix}; \upsilon_2 = \begin{bmatrix} 0 \\ 1 \\ 0 \end{bmatrix}; \upsilon_3 = \begin{bmatrix} 0 \\ 0 \\ 1 \end{bmatrix}; \omega_4 = \begin{bmatrix} 1 \\ 0 \\ 0 \end{bmatrix}; \omega_5 = \begin{bmatrix} 0 \\ 1 \\ 0 \end{bmatrix}; \omega_6 = \begin{bmatrix} 0 \\ 0 \\ 1 \end{bmatrix} \tag{72}$$

$$\omega_7 = \begin{bmatrix} 0 \\ 1 \\ 0 \end{bmatrix}; \omega_8 = \begin{bmatrix} 1 \\ 0 \\ 0 \end{bmatrix}; \omega_9 = \begin{bmatrix} 1 \\ 0 \\ 0 \end{bmatrix}; \omega_{10} = \begin{bmatrix} 1 \\ 0 \\ 0 \end{bmatrix}; \omega_{11} = \begin{bmatrix} 0 \\ 1 \\ 0 \end{bmatrix}; \omega_{12} = \begin{bmatrix} 0 \\ 0 \\ 1 \end{bmatrix} \tag{73}$$

$$\xi_1 = \begin{bmatrix} \upsilon_1 \\ 0 \end{bmatrix}; \quad \xi_2 = \begin{bmatrix} \upsilon_2 \\ 0 \end{bmatrix}; \quad \xi_3 = \begin{bmatrix} \upsilon_3 \\ 0 \end{bmatrix} \tag{74}$$

$$\xi_4 = \begin{bmatrix} -\omega_4 \times S \\ \omega_4 \end{bmatrix}; \xi_5 = \begin{bmatrix} -\omega_5 \times S \\ \omega_5 \end{bmatrix}; \xi_6 = \begin{bmatrix} -\omega_6 \times S \\ \omega_6 \end{bmatrix} \tag{75}$$

$$\xi_7 = \begin{bmatrix} -\omega_7 \times k \\ \omega_7 \end{bmatrix}; \xi_8 = \begin{bmatrix} -\omega_8 \times k \\ \omega_8 \end{bmatrix}; \xi_9 = \begin{bmatrix} -\omega_9 \times r \\ \omega_9 \end{bmatrix} \tag{76}$$

$$\xi_{10} = \begin{bmatrix} -\omega_{10} \times p \\ \omega_{10} \end{bmatrix}; \xi_{11} = \begin{bmatrix} -\omega_{11} \times p \\ \omega_{11} \end{bmatrix}; \xi_{12} = \begin{bmatrix} -\omega_{12} \times p \\ \omega_{12} \end{bmatrix} \tag{77}$$

Next, it is possible to compute the inverse kinematics as follows: angle θ_9 is obtained using the third ***P-K*** sub problem. We pass all known terms to the left side of the equation (69), apply both sides to point ***p***, subtract point ***k***, and apply the norm. We operate in such a way because the resulting equation (78) is only affected by θ_9, and therefore we can rewrite the equation as (79), which is exactly the formulation of the Paden-Kahan canonical problem for a rotation to a given distance. Thus, we can geometrically obtain the two possible values for the variable θ_9.

$$\left\| e^{-\hat{\xi}_6 \theta_6} \cdots e^{-\hat{\xi}_1 \theta_1} g_{st}(\theta) g_{st}(0)^{-1} p - k \right\| = \left\| e^{\hat{\xi}_7 \theta_7} \cdots e^{\hat{\xi}_{12} \theta_{12}} p - k \right\| \tag{78}$$

$$\delta = \left\| e^{\hat{\xi}_9 \theta_9} p - k \right\| \xrightarrow{P-K-3} \theta_9 \tag{79}$$

Next, θ_7 and θ_8 are obtained using the second ***P-K*** sub problem. We pass all possible terms to the left side of the equation (51) and apply both sides to point ***p***. In so doing, the resulting equation (80) is only affected by θ_7, θ_8 and θ_9, and therefore, we can rewrite the equation as (81), which is exactly the formulation of the Paden-Kahan canonical problem for two successive rotations.

Therefore, we can geometrically obtain the two possible values, for the pair of variables θ_7 and θ_8.

$$e^{-\hat{\xi}_6 \theta_6} \cdots e^{-\hat{\xi}_1 \theta_1} g_{st}(\theta) g_{st}(0)^{-1} p = e^{\hat{\xi}_7 \theta_7} \cdots e^{\hat{\xi}_{12} \theta_{12}} p \tag{80}$$

$$q' = e^{\hat{\xi}_7 \theta_7} e^{\hat{\xi}_8 \theta_8} p' \xrightarrow{P-K-2} \theta_7, \theta_8 \tag{81}$$

After that, θ_{10} and θ_{11} are obtained using the second ***P-K*** sub problem. We pass all known terms to the left side of the equation (69) and apply both sides to point ***m***. As a result of these operations, the transformed equation (82) is only affected by θ_{10} and θ_{11}, and we can rewrite the equation as (83), which is again the formulation of the Paden-Kahan canonical problem for two successive rotations around crossing axes. Hence, we can geometrically solve the two possible values for the pair of variables θ_{10} and θ_{11}.

$$e^{-\hat{\xi}_9\theta_9} \cdots e^{-\hat{\xi}_1\theta_1} g_{st}(\theta) g_{st}(0)^{-1} m = e^{\hat{\xi}_{10}\theta_{10}} e^{\hat{\xi}_{11}\theta_{11}} e^{\hat{\xi}_{12}\theta_{12}} m \tag{82}$$

$$q'' = e^{\hat{\xi}_{10}\theta_{10}} \cdot e^{\hat{\xi}_{11}\theta_{11}} m \xrightarrow{P-K-2} \theta_{10}, \theta_{11} \tag{83}$$

Finally, $\boldsymbol{\theta_{12}}$ is obtained using the first ***P-K*** sub-problem. We pass all known terms to the left side of the equation (69) and apply both sides to point *S*. As a result, the equation is transformed into (84), which is obviously only affected by $\boldsymbol{\theta_{12}}$, and we can rewrite it as (85), which is the formulation of the Paden-Kahan canonical problem for a rotation around an axis. Thus, we can geometrically obtain the single possible value for variable $\boldsymbol{\theta_{12}}$.

$$e^{-\hat{\xi}_{11}\theta_{11}} \cdots e^{-\hat{\xi}_1\theta_1} g_{st}(\theta) g_{st}(0)^{-1} S = e^{\hat{\xi}_{12}\theta_{12}} S \tag{84}$$

$$q''' = e^{\hat{\xi}_{12}\theta_{12}} S \xrightarrow{P-K-1} \theta_{12} \tag{85}$$

The arm motion could be implemented as follows: i.e. the $\boldsymbol{\theta_{25}}$ to $\boldsymbol{\theta_{29}}$ solutions. The manipulator shoulder and wrist do not intervene in locomotion and therefore $\boldsymbol{\theta_{25}}$ and $\boldsymbol{\theta_{29}}$ are zero for the analyzed movement. The other arm DOF may or may not contribute to the locomotion, helping the balance control to keep the ***COG*** as close as possible to its initial geometric position; but to achieve this behavior, we must solve the arm inverse dynamics problem, which is beyond the scope of this paper. A very simple but effective practical arm kinematics solution takes advantage of the necessary body sagital coordination (see the SKD model in Figure 31), and the right arm DOF is made equal or proportional to its complementary left leg DOF. Therefore, the values for the variables $\boldsymbol{\theta_{25}}$ to $\boldsymbol{\theta_{29}}$ are defined as (86).

$$\theta_{25} = 0; \ \theta_{26} = \theta_{15}; \ \theta_{27} = \theta_{14}; \ \theta_{28} = \theta_{16}; \ \theta_{29} = 0 \tag{86}$$

With these computations, the right manipulator inverse kinematics problem is solved in a geometric way, and what is more, we have not only one solution but the set of all possible solutions. For instance, the right leg has eight theoretical solutions, which are captured with the approach shown in this paper (87), if they exist.

$$Solutions = \theta_9 Double \times \theta_8\theta_7 Double \times \theta_{11}\theta_{10} Double \times \theta_{12} Single = 8 \tag{87}$$

After repeating exactly the same technique for the left manipulator, the complete Rh-1 humanoid inverse kinematics problem is, in fact, totally resolved.

6.11 Simulation results

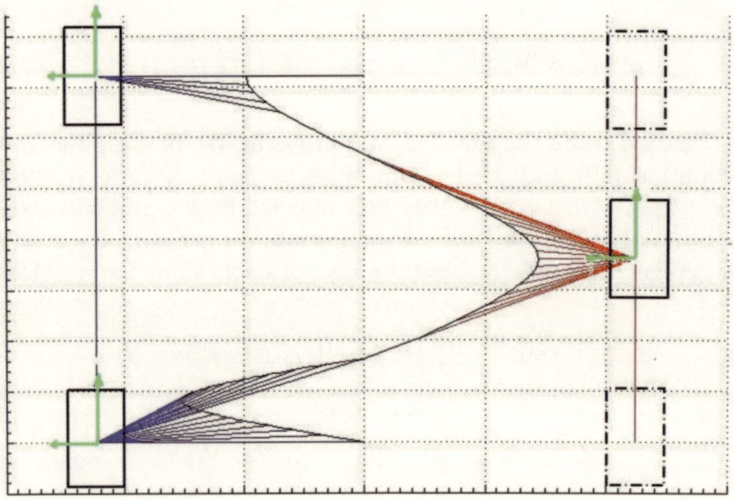

Fig. 32. *COG* hyperbolic trajectories in local axes (green)

For three steps, Figure 32 shows the spatial motion of the pendulum mass, and the local frame (green frames) of hyperbolic trajectories obtained in the single support phase; the trajectories shape looks like a hyperbolic curve as deduced above. It is a hyperbolic trajectory because the orbital energy in "*y-direction*" is negative (this is due to the fact that the pendulum frontal motion accelerates and decelerates without passing the equilibrium point, as shown in Figure 19), so the eq. 39 describes a hyperbole. The passive walkers have another walking principle, which is based on a limit cycle, when the gravity fields act on the device to achieve motion. In our case, we introduce the reference *COG* motion to make the robot walk, so we can pre-plane the stable walking pattern and introduce it to the humanoid robot. It is noted that the pendulum base is centered in the middle of the support foot and the natural ball pendulum motion follows a hyperbolic trajectory; the smooth pattern found drives the *COG* of the humanoid robot; natural and stable walking motion is obtained as will be demonstrated in several simulations and experimental results explained in the next paragraphs and sections. Figure 33 shows the temporal pendulum mass trajectories, in this walking pattern the single support phase takes 1.5 seconds and the double one 0.2 seconds. After computing the inverse kinematics at each local axis (Fig. 32), the joint patterns of the right humanoid leg and angular velocities obtained are shown in Figure 34. Those allow for checking the joint constraints in order to satisfy the actuator's performances.

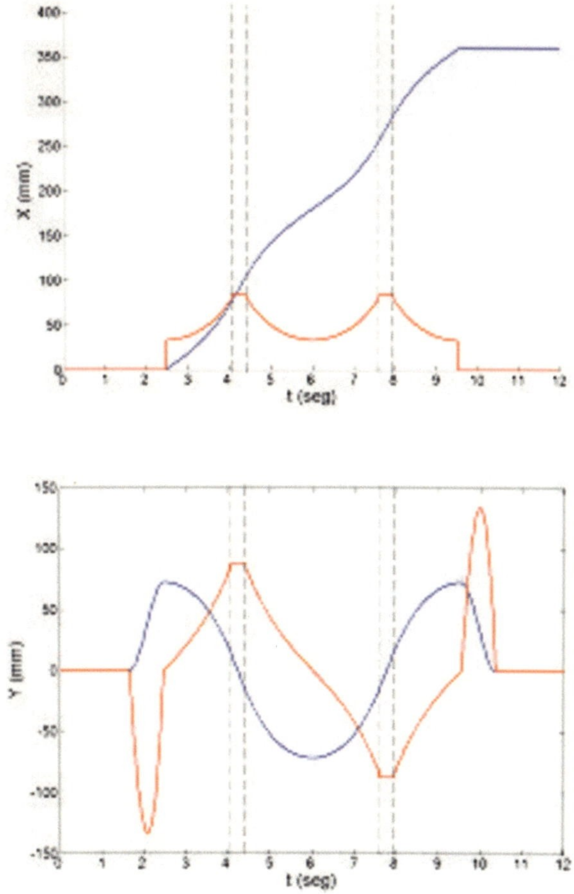

Fig. 33. ***COG*** temporal position (blue) and velocity (red) patterns for doing three steps. In the double support phase (between vertical dashed lines) constant speed maintains the trajectory's continuity.

Rh-1 simulator results are shown in Figure 35, from the VRML environment developed, which let us test the walking pattern previously so as to test it in the real humanoid robot. This environment lets us evaluate the angle motion range of each joint, avoid self-collision and obstacle collision, in order to obtain adequate walking patterns considering the robot's dimensions and mechanical limitations. It is verified by several simulation tests that smooth, fast and natural walking motion is obtained using the 3D-LIPM and foot motion patterns.

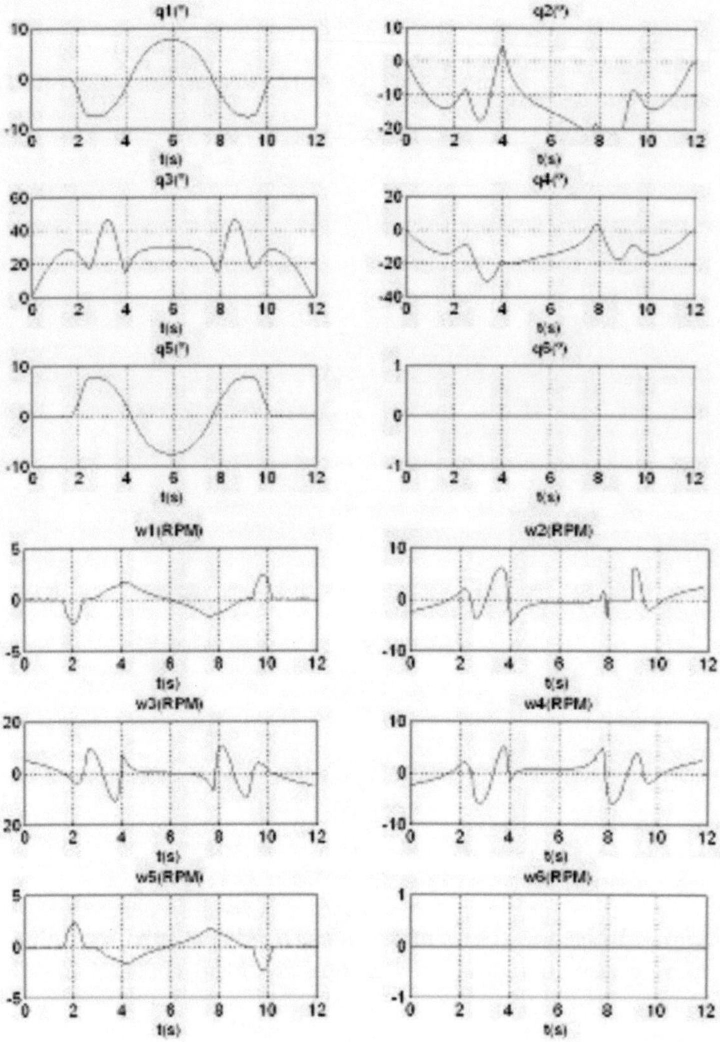

Fig. 34. Simulation of joint angle evolution and joint velocity evolution of right leg for three steps walking in a straight line (note that θ_7, to θ_{12} correspond to q_1 to q_6)

In order to obtain global humanoid motion for avoiding an external obstacle (Yoshida et al. 2005), walking patterns in any direction could be developed using a rotation matrix around the z-axis of the local frame and the proper boundary conditions such as position and velocity at the local frame of the ***COG***, and foot trajectories while changing the support foot in order to obtain smooth walking patterns. Figure 36 and Figure 37 shows us an example of planning walking motion with a change in direction. The walking pattern generated in each local frame maintains continuity with the previous and the next walking pattern. In addition, real-time walking pattern generation is possible, which changes direction, length and step width at any time using the information from sensors or by external command of the humanoid robot, according to the LAG algorithm.

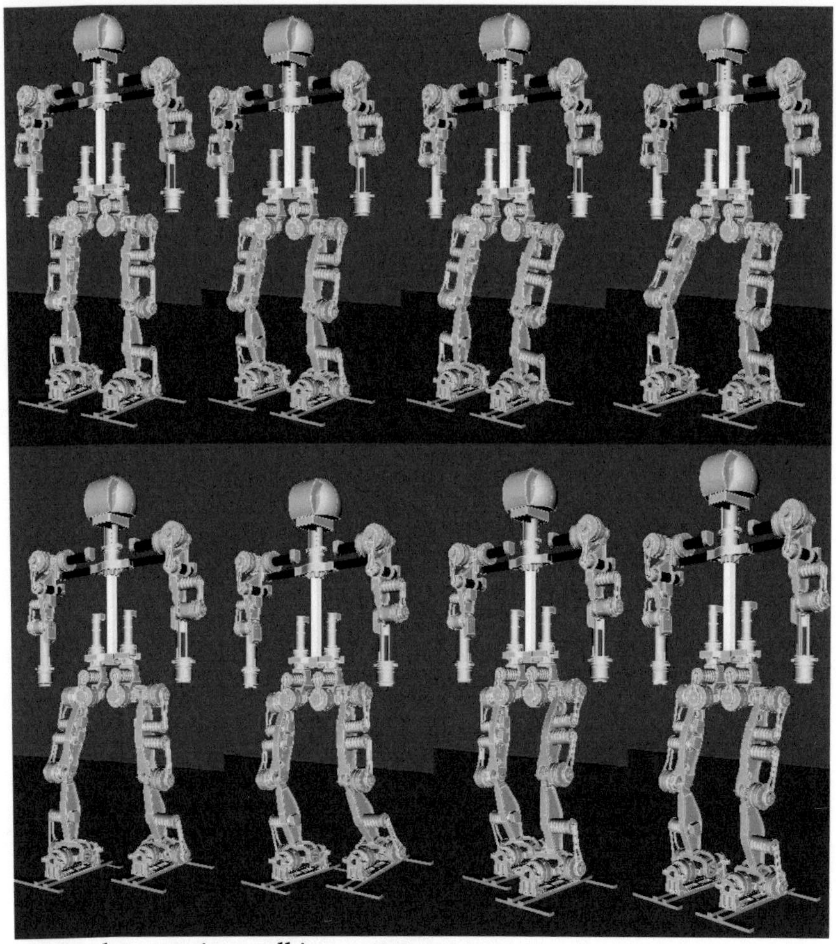

Fig. 35. Rh-1 simulator testing walking patterns.

In order to correct mechanical flexion and terrain irregularities, some joint patterns should be modified (i.e. ankle and hip joints), by offline and online control gait. The offline control reduces high joint acceleration at the beginning of the single support phase, because correction starts at this time. Online correction compensates for the actual environment changes and the mechanical imperfections. In this way, stable walking is obtained, by maintaining the body's orientation and ZMP in the right position.

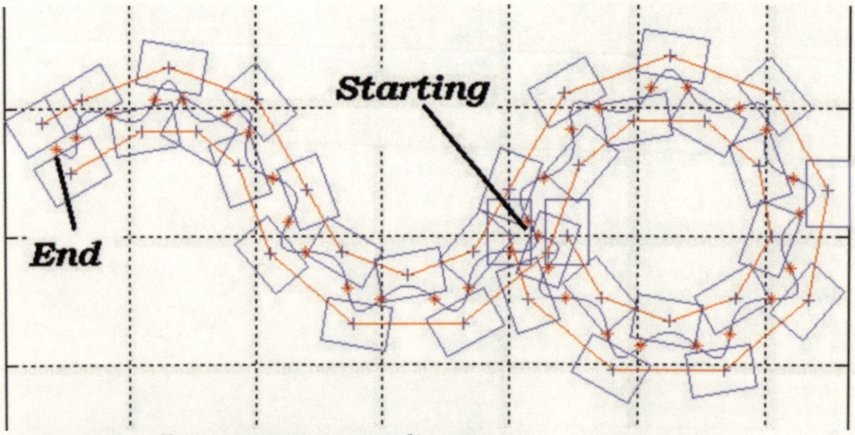

Fig. 36. Generating walking patterns in any direction.

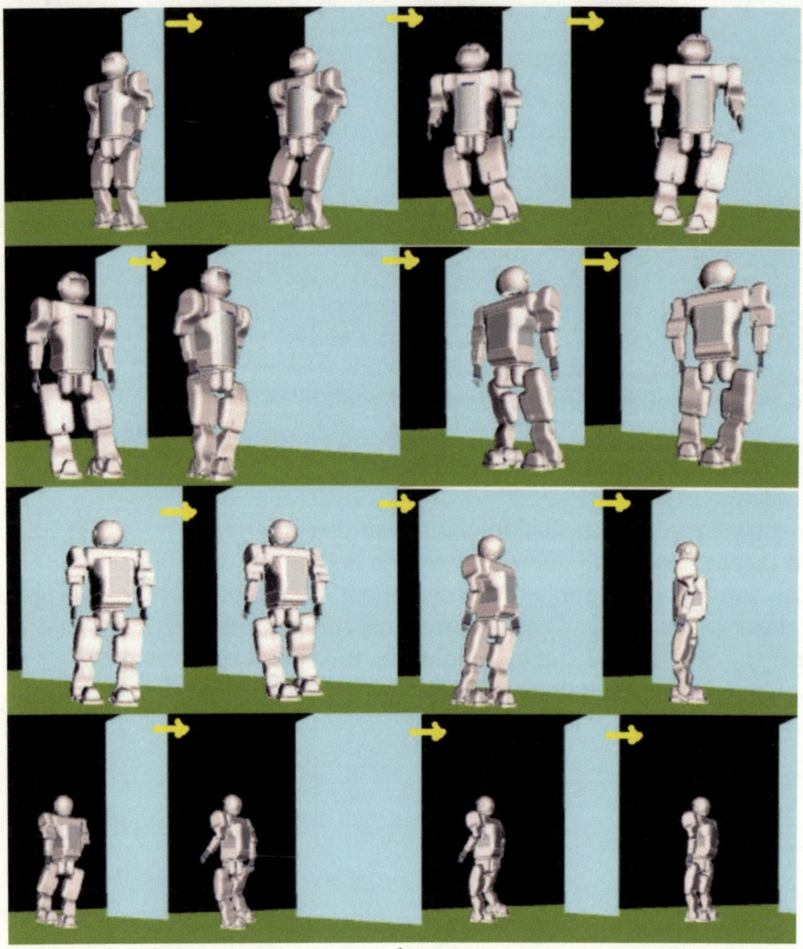

Fig. 37. Snapshots of walking patterns in any direction.

7. Control Architecture

7.1 Control overview

Humans usually walk using a specific walking pattern in normal environments. In cases of unexpected disturbances, this pattern can be changed immediately to another one in order to adapt to terrain requirements. The humanoid robot operates in the same space as the human and needs the same mechanism to adapt its walking to changing conditions.

Although planned motion patterns satisfy the stability constraints, some errors caused by the irregularity of the terrain or some external forces can cause the humanoid robot to fall down. Moreover, the flexion of the mechanical structure of the humanoid robot is the largest source of errors affecting its walking. To reduce the influence of these errors on walking stability, the control architecture (Fig. 38) for online motion patterns modification has been implemented.

The control architecture consists of two basic parts: Joint Control and Stabilization Control. The Joint Control is the core of the control scheme. When the offline calculated motion pattern is received and the motion is started, the Adaptive Control Algorithm adjusts the motion controller of each articulation to the diverse postures of the humanoid robot. If the sensorial system of the humanoid robot detects an error in the body position, Stabilization Control corrects it and then tries to recover each joint's trajectory in order to execute the previously calculated motion pattern.

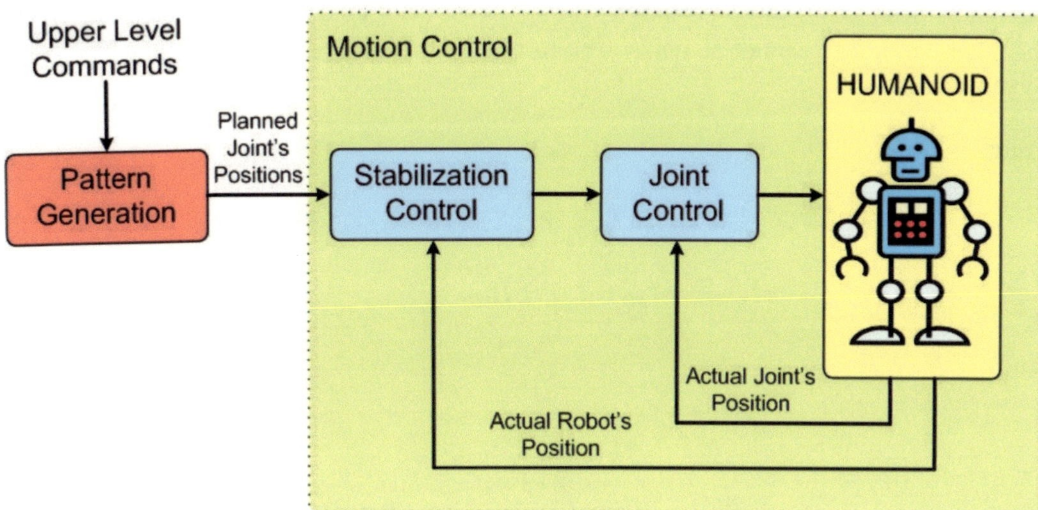

Fig. 38. Control architecture.

When the robot is working in the operational interaction mode (walking, object manipulation, etc.), there are several computing and communication tasks that need to be performed in a cyclical mode and be fast enough to avoid any possible loss of control. The periodic (with period Ts) chain (Fig. 39) begins with the sensing task, taking the time interval tatt for attitude estimation gyros and accelerometers readings and tzmp for ZMP force-torque sensor readings.

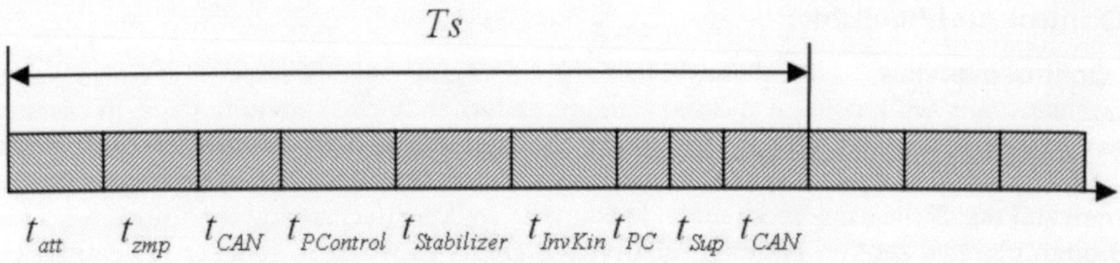

Fig. 39. Main computing and communication tasks for the Rh-1.

These tasks are followed by the tasks performing CAN bus communications, Posture Control and then Stabilizer and Inverse Kinematics Calculator computing, internal PC bus communications, Supervisory controller, and CAN bus transmission of new reference for each joint of the humanoid robot.

The period Ts should be small and compatible with the dynamics of the humanoid robot movement. On the one hand, Ts cannot be made arbitrarily small because the various computing and communication tasks with execution times tatt, tzmp, t_{CAN}, t_{PC}, $t_{Stabilizer}$, etc., cannot themselves be made arbitrarily small. Also, a small value for Ts would generate too many messages in communication lines (PC bus, CAN bus) that would overload it. On the other hand, Ts cannot be made arbitrarily large because of the dynamics of the robot (Nyquist criterion).

Thus,

$$Ts > t_{att} + t_{zmp} + t_{CAN} + t_{PControl} + t_{Stabilizer} + \\ + t_{InvKin} + t_{PC} + t_{Sup} + t_{CAN} \tag{88}$$

And

$$Ts < \frac{1}{2 \cdot Fr} \tag{89}$$

Where Fr is the highest movement frequency on any robot link.
Assuming that the robot walks at about the same speed as a human walking normally, $Fr = 2$ Hz.

Thus,

$$Ts < 250 \text{ ms} \tag{90}$$

This upper limit of the sample time is a strict real-time requirement considering the complexity of the computing and communication tasks to be performed within this time limit.
The following sections present the detailed design of Joint and Stabilization Controls.

7.2 Joint Control

Different postures of the humanoid robot radically change the motors' dynamics and require adaptive algorithms to improve the control of each joint. The main goal is to achieve the adequate systems response for the desired humanoid robot movements.
In the humanoid robot Rh-1, each motor is controlled by the motion controller. This motion controller includes the bottom part - three cascade control loops (Fig. 40): position loop (proportional), velocity loop (proportional - integral) and current loop (proportional - integral).

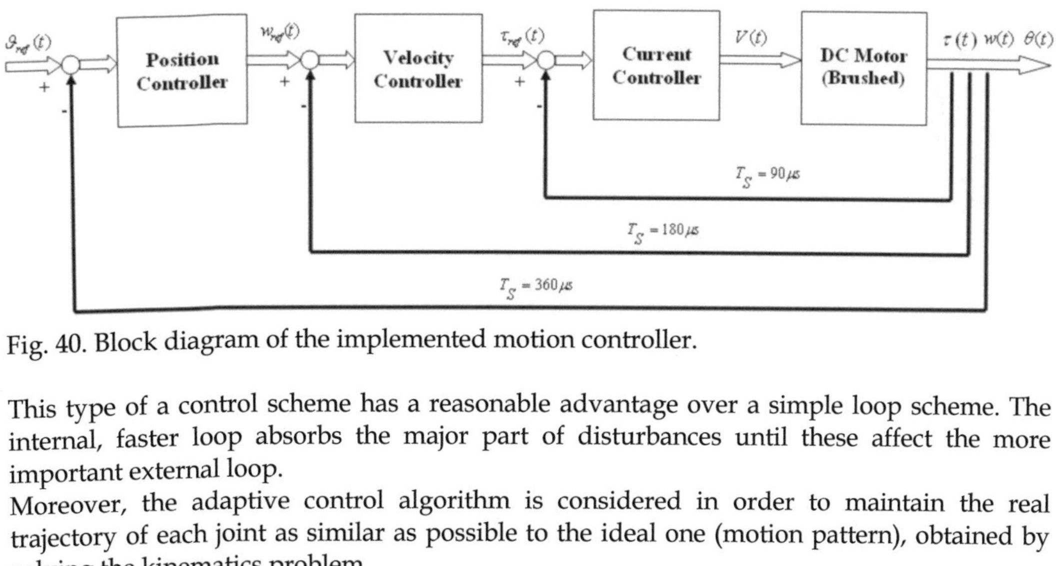

Fig. 40. Block diagram of the implemented motion controller.

This type of a control scheme has a reasonable advantage over a simple loop scheme. The internal, faster loop absorbs the major part of disturbances until these affect the more important external loop.
Moreover, the adaptive control algorithm is considered in order to maintain the real trajectory of each joint as similar as possible to the ideal one (motion pattern), obtained by solving the kinematics problem.
Equation (73) presents a classical dynamic model of a loaded DC motor which is usually used to move the joints of a humanoid robot,

$$\Omega(s) = \frac{Ks}{(D + s \cdot J)} \cdot I_a(s) - \frac{1}{(D + s \cdot J)} \cdot T_L(s) \tag{91}$$

where $\Omega(s)$ is the velocity of rotor of the DC motor, $I_a(s)$ - current, Ks - velocity constant, D - viscous friction or viscous damping inherent in the system, J - total moment of inertia of a joint, and $T_L(s)$ - load torque. Evidently, when a humanoid robot moves, it assumes different postures characterized by different configurations of joints. This leads to continuous changing in the model of a DC motor. Thus, the adaptive algorithm, tuning the controller of each joint when the model of a motor changes, is required. It carries out online parameter modification of the motion controller in compliance with the system's behaviors. The most suitable way to implement an adaptive control scheme is the gain-scheduling control. It does not need online identification. The offline software integrating identification and controller design algorithms estimate position and velocity controllers' gains for every possible configuration of the humanoid robot. The identification is made for real gaits (forward walking, backward walking and turning) in order to consider the influence of the posture of the robot on the system dynamics at every moment of the motion. In other words, every offline trajectory (gait) is divided into small parts. For each part, its own values of controller's gains are estimated and a multidimensional table of gains is created. Figure 41 shows the block diagram of the designed gain-scheduling motion controller.

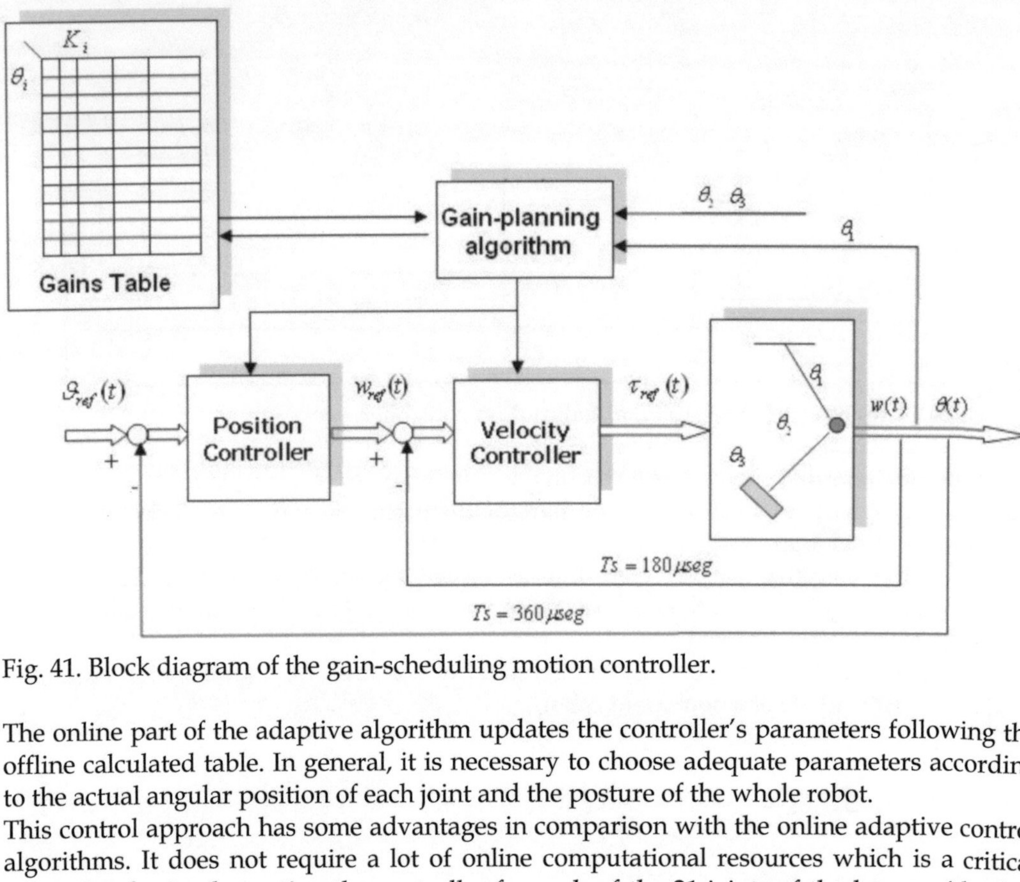

Fig. 41. Block diagram of the gain-scheduling motion controller.

The online part of the adaptive algorithm updates the controller's parameters following the offline calculated table. In general, it is necessary to choose adequate parameters according to the actual angular position of each joint and the posture of the whole robot.

This control approach has some advantages in comparison with the online adaptive control algorithms. It does not require a lot of online computational resources which is a critical restriction for implementing the controller for each of the 21 joints of the humanoid robot Rh-1.

7.3 Stabilization Control Joint Control

As was shown in the previous section, the humanoid robot's dynamics is governed by the ZMP equation 45. From these equations it can be concluded that there are two main variables to control:

$$e_{ZMP} = ZMP^d - ZMP^a \quad (92)$$

$$e_{COG} = COG^d - COG^a \quad (92)$$

where ZMP error is denoted as e_{ZMP} and **COG** position error is denoted as e_{COG} . ZMP^d , COG^d are desired and ZMP^a , COG^a are actual ZMP and COG positions. Equations (74) and (75) state that two different controls should be implemented in order to stabilize humanoid robot walking. Figure 42 presents the main structure of the stabilization controller for the humanoid robot.

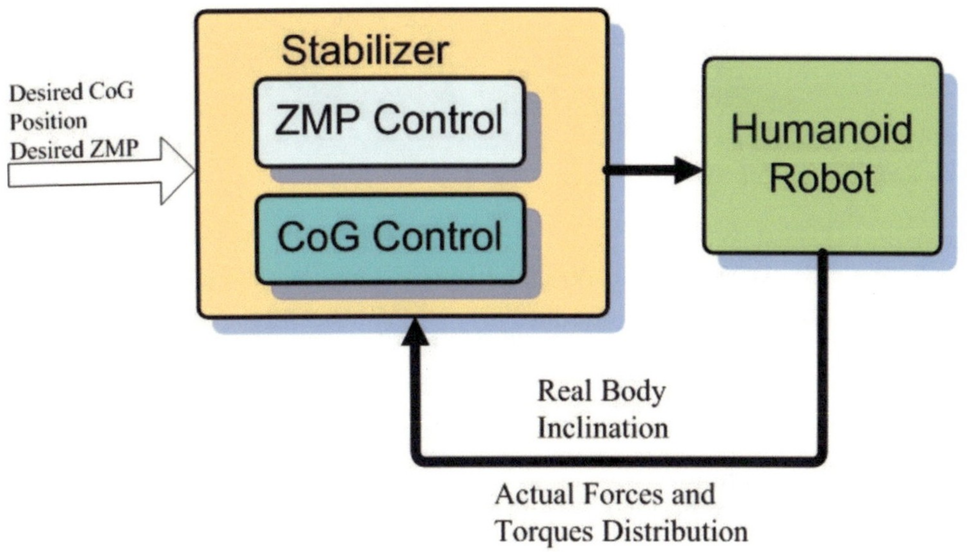

Fig. 42. Stabilizer structure.

Previously computed desired COG and ZMP positions generate joint trajectories which are sent for the humanoid motion. During the execution of motion patterns, the real ZMP and COG position should be measured. The Stabilizer compares it with the ideal ones and generates a new corrected motion pattern for the next moment. This control scheme provides a simple and effective way of controlling humanoid walking stability.

7.4 ZMP control

The implementation of the ZMP control should provide an easy and robust method to move the actual ZMP to the point where the target ZMP is located. The easiest way here is to neglect the influence of the part of the humanoid robot above its ankle joint. It can be made

by rotation of the upper part of the mechanism with respect to the ankle joint as shown in Figure 43.
Although the proposed compensational mechanism allows the actual ZMP to be moved to the point where the target ZMP is located, some problems appear. When the upper body is rotated with respect to the ankle joint, body angle errors in frontal $\Delta q_{Bf}(t)$ and sagital $\Delta q_{Bs}(t)$ planes (Fig. 43) appear.

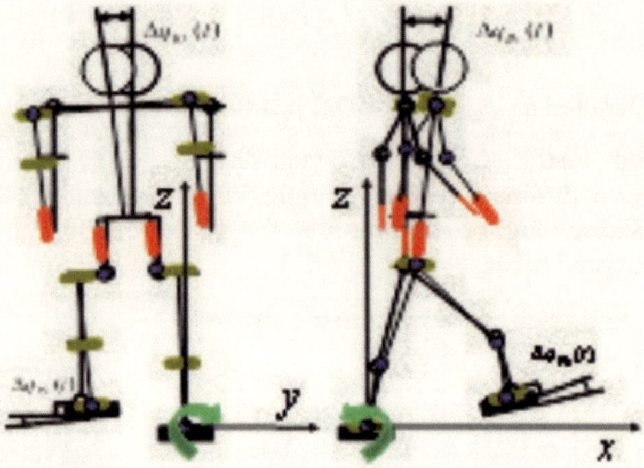

Fig. 43. ZMP error compensation.

These errors in frontal and sagital planes cause the appearance of tilting torques and errors in the frontal $\Delta q_{Ff}(t)$ and sagital $\Delta q_{Fs}(t)$ planes during the positioning of the hanging foot. It can overturn the robot or introduce very strong instability and vibrations when the foot is landing. Therefore, the mechanism for compensating for the body inclination errors in the frontal $\Delta q_{Bf}(t)$ and sagital $\Delta q_{Bs}(t)$ planes should complement the ZMP control.

7.5 COG - Attitude control

Using a 3D-LIPM (or other method) for trajectory generation we assume that the humanoid's upper body does not have any influence on the dynamics of a model. In the real walking case it leads to unbounded errors in the COG position. It differs radically from the ideal position computed for a motion pattern. The tilting moment M_{CoG} compounded by M_x and M_y components denotes the robot's upper body dynamics (Fig. 44) and should be compensated by a control algorithm. The Attitude control is a special sort of a COG control which tries to maintain the trunk of the robot strictly vertical in every stage of its motion thus eliminating the tilting moment and body inclination errors.

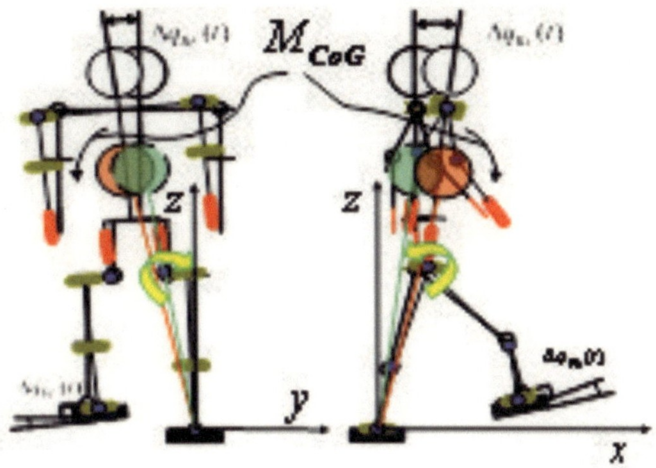

Fig. 44. COG error compensation - Attitude control.

The attitude of the robot is normally determined by a combination of gyroscopes and accelerometers placed in the waist, where the approximate center of the mass or the upper part of its backbone is located. As soon as the sensorial information is obtained, the most effective way to control the humanoid robot's body inclination is to maintain its backbone strictly vertical at every stage of the trajectory. In this case, it is enough to control only the hip joints in the frontal and sagital planes

7.6 Double Inverted Pendulum

As was shown in the previous sections, to maintain the stability of a humanoid robot it is necessary to implement simultaneously both the ZMP and the Attitude controls acting on the ankle and hip joints. The humanoid robot in that case should be modeled as an inverted double pendulum shown in Figure 45(a).

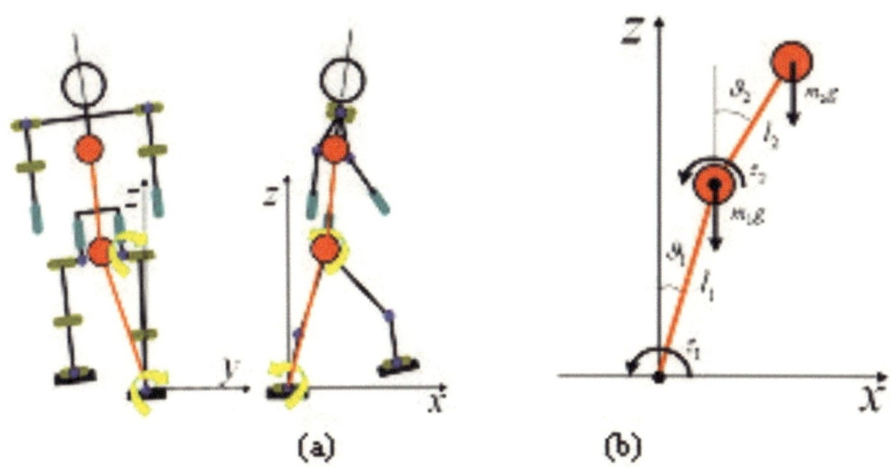

Fig. 45. a) Humanoid robot motion modelling. b)Double Inverted Pendulum.

A double pendulum consists of one pendulum attached to another. Consider a double bob pendulum with masses m_1 and m_2 ($m_1 + m_2 = M$ - the total mass of the humanoid robot) attached by rigid mass less wires of lengths l_1 and l_2. Furthermore, let the angles that the two wires make with the vertical be denoted ϑ_1 (ankle rotation) and ϑ_2 (hip rotation), as illustrated in Figure 45(b). The position of the center of mass of the two rods may be written in terms of these angles. The potential energy of the system is given by:

$$V = (m_1 + m_2) g l_1 \cos\vartheta_1 + m_2 g l_2 \cos\vartheta_2 \tag{94}$$

The kinetic energy of the system is given by:

$$T = \frac{1}{2} m_1 l_1^2 \dot{\vartheta}_1^2 + \frac{1}{2} m_2 \left[l_1^2 \dot{\vartheta}_1^2 + l_2^2 \dot{\vartheta}_2^2 + 2 l_1 l_2 \dot{\vartheta}_1 \dot{\vartheta}_2 \cos(\vartheta_1 - \vartheta_2) \right] \tag{95}$$

Writing the Lagrangian of a system using equations (94) and (95) for ϑ_1, ϑ_2 and simplifying we get:

$$\begin{aligned} &(m_1 + m_2) l_1^2 \ddot{\vartheta}_1 + m_2 l_1 l_2 \ddot{\vartheta}_2 \cos(\vartheta_1 - \vartheta_2) + \\ &+ \mathrm{m}_2 l_1 l_2 \dot{\vartheta}_2^2 \sin(\vartheta_1 - \vartheta_2) - l_1 g (\mathrm{m}_1 + \mathrm{m}_2) \sin\vartheta_1 = \tau_1 \end{aligned} \tag{96}$$

$$\begin{aligned} &m_2 l_2^2 \ddot{\vartheta}_2 + m_2 l_1 l_2 \ddot{\vartheta}_1 \cos(\vartheta_1 - \vartheta_2) - \\ &- \mathrm{m}_2 l_1 l_2 \dot{\vartheta}_1^2 \sin(\vartheta_1 - \vartheta_2) - l_2 \mathrm{m}_2 \mathrm{g} \sin\vartheta_2 = \tau_2 \end{aligned} \tag{97}$$

The aim of the present research is to develop a simple control strategy allowing for the use of traditional controllers such as the linear quadratic regulator (LQR) or a simple PID controller to regulate the inverted pendulum about the upright equilibrium point. As the name may suggest, the LQR controller requires a linear system for which it will generate constant gains for full state feedback to make the equilibrium point globally asymptotically stable. However, the dynamics of double inverted pendulum systems are inherently nonlinear.

The approach chosen was to linearize the equations of motion regarding the operating point and define a domain of attraction within which the constant gain controller results in local asymptotic stability.

Take the approximation that perturbations in the problem are very small, and those terms of second order and higher are negligible. In this limit $\cos\vartheta_i = 1$, $\sin\vartheta_i = \vartheta_i$, $\cos(\vartheta_1 - \vartheta_2) = 1$. Applying a small angle approximation we get the following linearization:

$$(m_1 + m_2) l_1^2 \ddot{\vartheta}_1 + m_2 l_1 l_2 \ddot{\vartheta}_2 - l_1 g (\mathrm{m}_1 + \mathrm{m}_2) \vartheta_1 = \tau_1 \tag{98}$$

$$m_2 l_2^2 \ddot{\vartheta}_2 + m_2 l_1 l_2 \ddot{\vartheta}_1 - l_2 \mathrm{m}_2 \mathrm{g} \vartheta_2 = \tau_2 \tag{99}$$

The state representation of the dynamical system in standard form is:

$$\begin{bmatrix} \dot{x}_1 \\ \dot{x}_2 \\ \dot{x}_3 \\ \dot{x}_4 \end{bmatrix} = \begin{bmatrix} 0 & 1 & 0 & 0 \\ \frac{g(m_1+m_2)}{m_1 l_1} & 0 & -\frac{m_2 g}{m_1 l_1} & 0 \\ 0 & 0 & 0 & 1 \\ -\frac{g(m_1+m_2)}{m_1 l_2} & 0 & \frac{g(m_1+m_2)}{m_1 l_2} & 0 \end{bmatrix} \begin{bmatrix} x_1 \\ x_2 \\ x_3 \\ x_4 \end{bmatrix} + \begin{bmatrix} 0 & 0 \\ \frac{1}{m_1 l_1^2} & -\frac{1}{m_1 l_1 l_2} \\ 0 & 0 \\ -\frac{1}{m_1 l_1 l_2} & \frac{(m_1+m_2)}{m_1 m_2 l_2^2} \end{bmatrix} \begin{bmatrix} u_1 \\ u_2 \end{bmatrix} \tag{100}$$

With the output equation:

$$\begin{bmatrix} y_1 \\ y_2 \end{bmatrix} = \begin{bmatrix} 1 & 0 & 0 & 0 \\ 0 & 0 & 1 & 0 \end{bmatrix} \begin{bmatrix} x_1 \\ x_2 \\ x_3 \\ x_4 \end{bmatrix} \tag{101}$$

where state variables x_1, x_2, x_3 and x_4 are defined as: $x_1 = \vartheta_1$, $x_2 = \dot{\vartheta}_1$, $x_3 = \vartheta_2$, $x_4 = \dot{\vartheta}_2$.

LQR controller design considers an optimal control problem that for given system equations (100) and (101) determines the matrix $\mathbf{K}$ of the optimal control vector:

$$\mathbf{u}(t) = -\mathbf{K}\mathbf{x}(t) \tag{102}$$

The block diagram showing the optimal controller configuration for the double inverted pendulum system is presented in Figure 46.

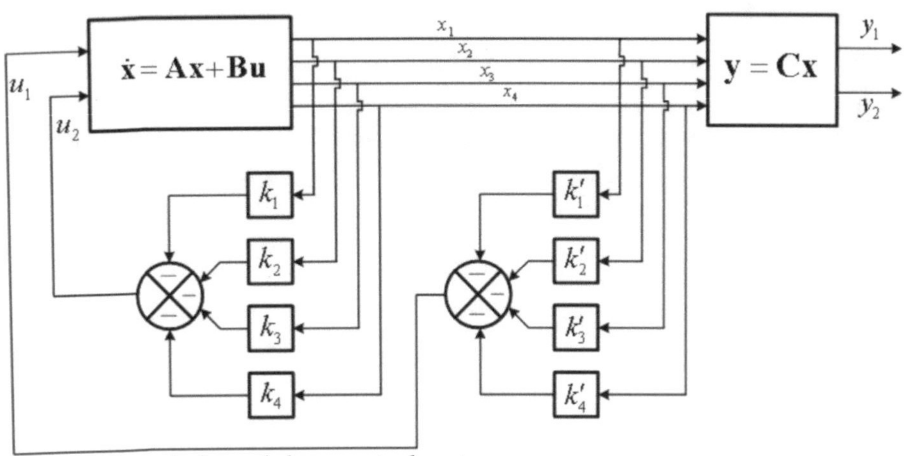

Fig. 46. Double inverted pendulum control system.

Note that the control system presented in Figure 46 is the regulator system. In this case the controller maintains desired angles ϑ_1 and ϑ_2 of the double pendulum close to zero.

7.7 Decoupled Control

Although the stabilization method for the humanoid robot based on the double inverted pendulum dynamics presented in the previous section seems to be the most appropriate for controlling the mechanism, in practice, it has many drawbacks. The main reason is that the double inverted pendulum considers control torques of both (hip and ankle) joints of the humanoid robot involved in the stabilization control. This means it is necessary use the torque controlled actuators to implement this control algorithm. Most contemporary humanoid robots are driven by DC motors with position control where torque control cannot be implemented. A decoupled approach considers that ZMP and Attitude can be implemented as totally independent and the influence of one on another will be negligibly small. To prove the appropriateness of this, let us examine the functioning of the already developed double inverted pendulum system in diverse initial positions of the pendulum (Figure 47), which correspond to different perturbations acting on the humanoid robot:

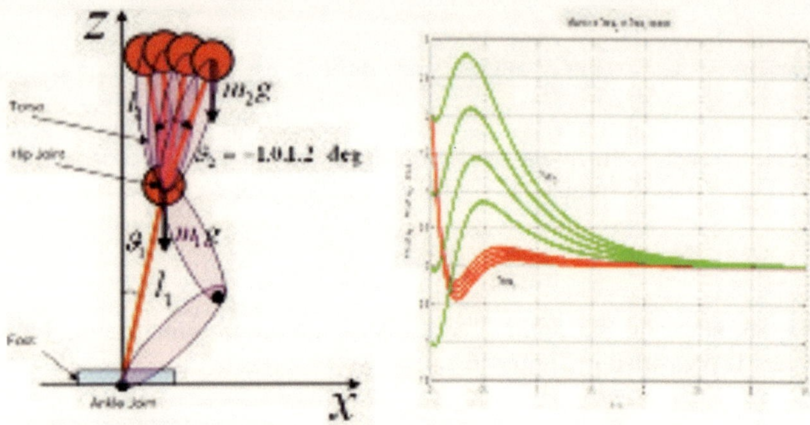

Fig. 47. a) Double Inverted Pendulum ϑ_2 variation b) Influence of the ϑ_2 on ϑ_1 variation.

From Figure 47(b) it can be observed that a large variation in ϑ_2 does not have much influence on ϑ_1. Its variations stay practically at the same level in every experiment. Therefore, we take into consideration that the control of ϑ_2 will have more priority (or gain) and thus, ϑ_2 variation will be practically zero. On the other hand, variation of ϑ_1 sensed by ϑ_2 can be considered as an additional perturbation and will be compensated for by the controller. Moreover, it should be noted that the dynamics of the real robot will be attenuated by the reduction between its links. Thus, stabilizer control of the humanoid robot can be considered as a sum of two decoupled components related to Posture control (ϑ_2) and ZMP control (ϑ_1). In practice, decoupled ZMP and Attitude controllers were designed as LQR controllers.

After the foregoing detailed development of all the parts, we now turn to the detailed control architecture (Figure 48).

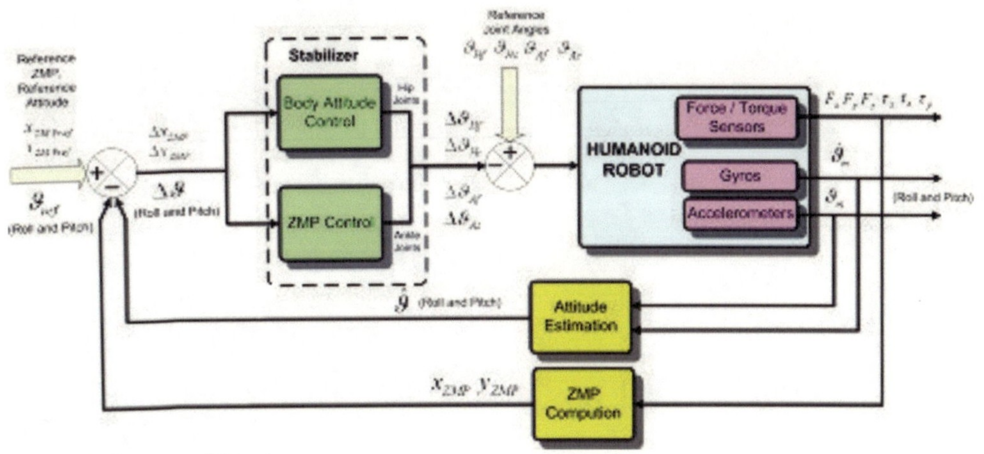

Fig. 48. Stabilizer architecture.

A sensorial system of the robot consisting of two six-axis force-torque should provide the controller with the real distribution of the forces and torques F_x, F_y, F_z, τ_x, τ_y, τ_z at the contact point of the foot with the ground. The 3-axis Gyro and Accelerometer provides the measurements of angular position ϑ_m and angular velocity $\dot{\vartheta}_m$ of the upper body (trunk) of the robot in the frontal and sagital planes (Roll and Pitch), (Löffler et. Al., 2003 and Baerveldt et. al., 1997). After the actual ZMP position x_{ZMP}, y_{ZMP} is computed by the ZMP Computational module and the real attitude is estimated in the Attitude Estimation module ZMP Δx_{ZMP}, Δx_{ZMP} and Attitude $\Delta\vartheta$, errors can be estimated. These errors are the input data for the Stabilizer. The Stabilizer is designed as a decoupled controller. It controls error in ZMP and Attitude positioning of the humanoid robot by the motion of the ankle and hip joints as was discussed above. Finally, the compensational motion of the ankle $\Delta\vartheta_{Af}$, $\Delta\vartheta_{As}$ and hip $\Delta\vartheta_{Hf}$, $\Delta\vartheta_{Hs}$ joints in the frontal and sagital planes should be superimposed with their reference trajectories ϑ_{Af}, ϑ_{As}, ϑ_{Hf}, ϑ_{Hs} which are responsible for the entire walking process. By doing this, new motion patterns for ankle and hip joints are generated. The implementation of the decoupled stabilizer provides fast and easy control of the walking stability of the humanoid. All changes are applied to the ankle and hip joints eliminating the need for inverse kinematics computation.

8. Experimental Results

Once the design, construction and assembly stages were carried out, the robot's experimental results followed.
Experimental results applying the natural 3D-LIPM and Cart-table models with Body Posture Control algorithms on the Rh-1 humanoid robot platform are discussed in this section. At the beginning, snapshots of dynamic walking of the Rh-1 humanoid robot are shown in Figure 49. As shown in the VRML (Figure 35) environment, the same walking motion pattern is followed.

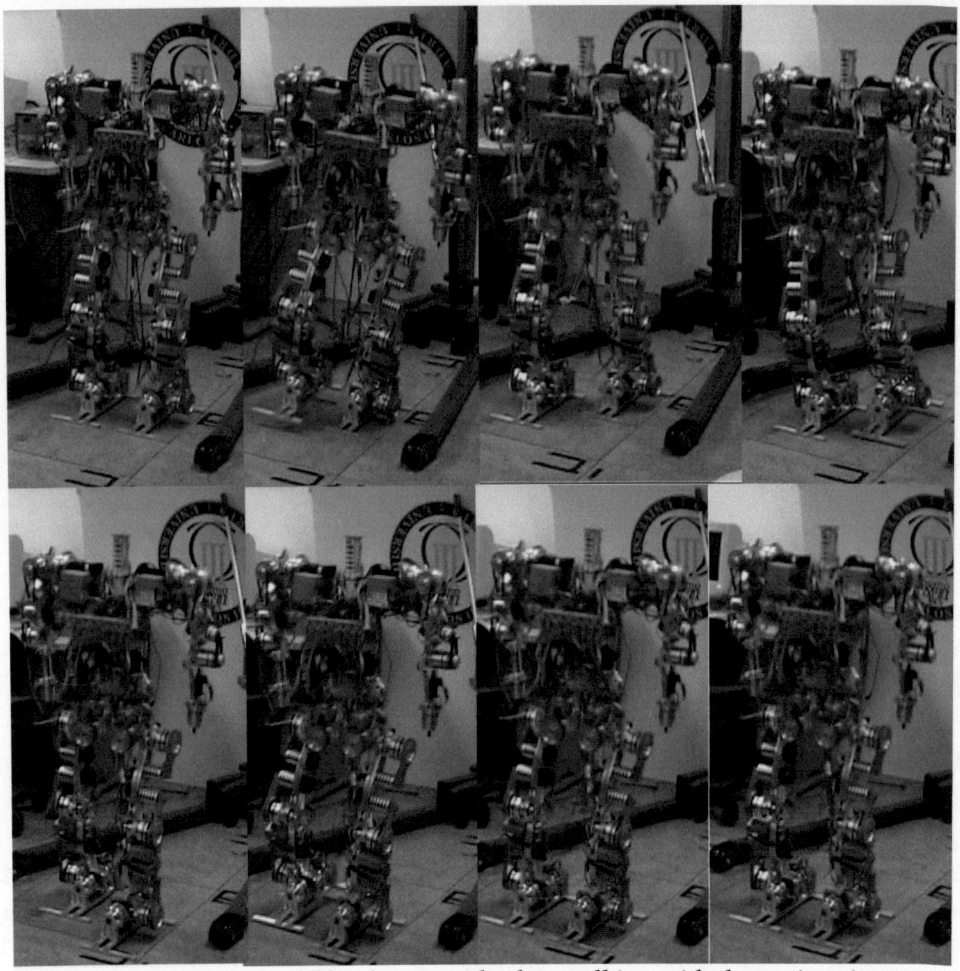

Fig. 49. Snapshots of the actual Rh-1 humanoid robot walking with dynamic gait.

The gait patterns proposed above allow for stable walking at 1 Km/h. Smooth and natural walking motions are obtained by introducing the adequate initial conditions on COG body motion and correct boundary conditions in foot planning. Some results are shown in Figure 50, where the frontal ankle joint's evolution can be seen in the chart, i.e. q_1 is the frontal right ankle joint, the blue line is the pattern obtained from the inverse kinematics model using the COG and foot patterns as inputs, for doing three steps; the red line is the on-line compensated reference pattern which allows for frontal humanoid body reorientation. This compensation is necessary because the mechanical elasticity tilts the humanoid into an unstable zone. On the other hand, w_1 is the frontal joint angular velocity of the right ankle, the blue line is the velocity pattern obtained without compensation, the red one is the compensated velocity; adequate interpolators have been developed in order to maintain the humanoid body in the stable zone. Compensated curves are delayed with respect to reference ones, because there exists a compensation time which starts in the single support phase. The q_{12} and w_{12} patterns correspond to the left ankle frontal joint. In the sagital motion, a similar compensation had been made.

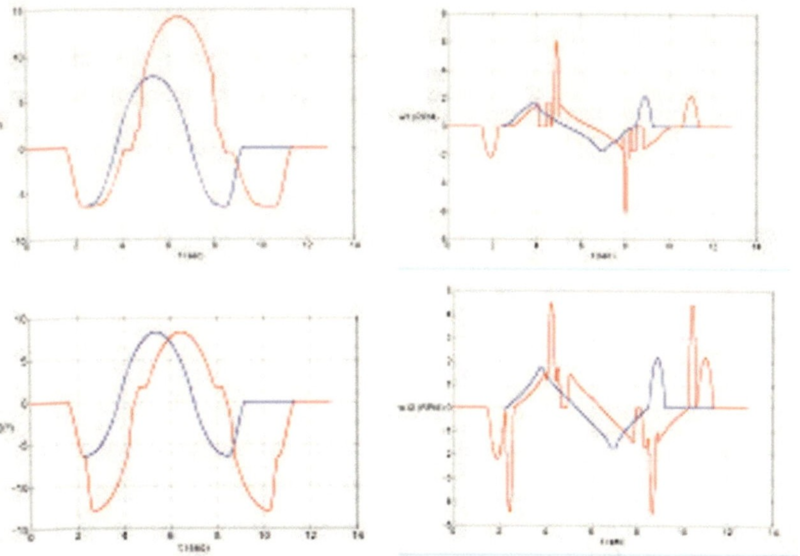

Fig. 50. The ankle joint's frontal motion (angular and velocity evolution), original evolution (blue line), compensated evolution (red line).

This compensation maintains the COG following pendulum laws, and thus, natural and stable walking is obtained. Posture compensation allows for a stable gait because the natural 3D-LIPM spatial trajectories are followed. Offline and online posture control compensates for several things: the gravitational effect on the humanoid structure in the sagital and frontal motion planes, which tilts the humanoid to an unstable position; terrain irregularities and errors in the concentrated mass model that we have used for generating walking patterns. The offline compensated and real measured joint pattern references for both legs are shown in Figure 51 and Figure 52; furthermore the joints' current consumptions are presented in Figure 53, in order to check the actuators current limits. Next, snapshots following the Cart-table model are shown in Figure 54. Successful results of straight walking motion in the laboratory and hall have been obtained. The step length is about 180mm, max feet height 40mm and step time about 1.25 sec. The joint patterns and angular velocities of the right leg results are shown in Figure 55 and Figure 56. The reference, offline corrected and real measure patterns of each joint for taking three steps forward are overlapped. The main differences are overlapped between the offline compensation and measured joint evolution is because of the variations, friction on the surface and mechanical imperfections of the robot.

We test that the control system allows for a stable walking motion in a straight line. Further improvements on the mechanical structure, walking pattern generation and control (i.e. foot landing with compliance control, for reducing the impact forces) in order to compensate for whole body moments, will induce a smooth and natural walking motion.

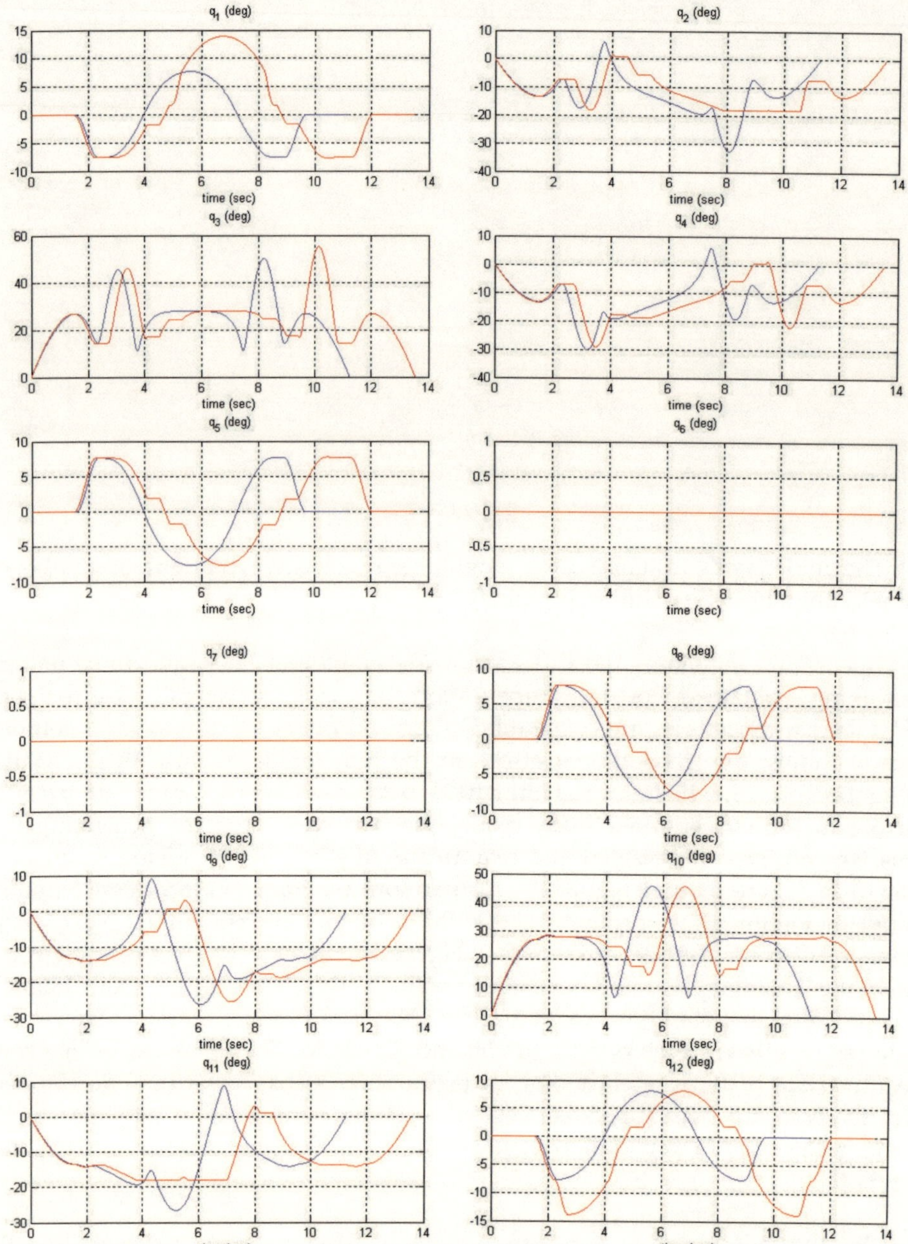

Fig. 51. Both leg joint patterns. Reference (blue) and offline corrected (red) overlapped.

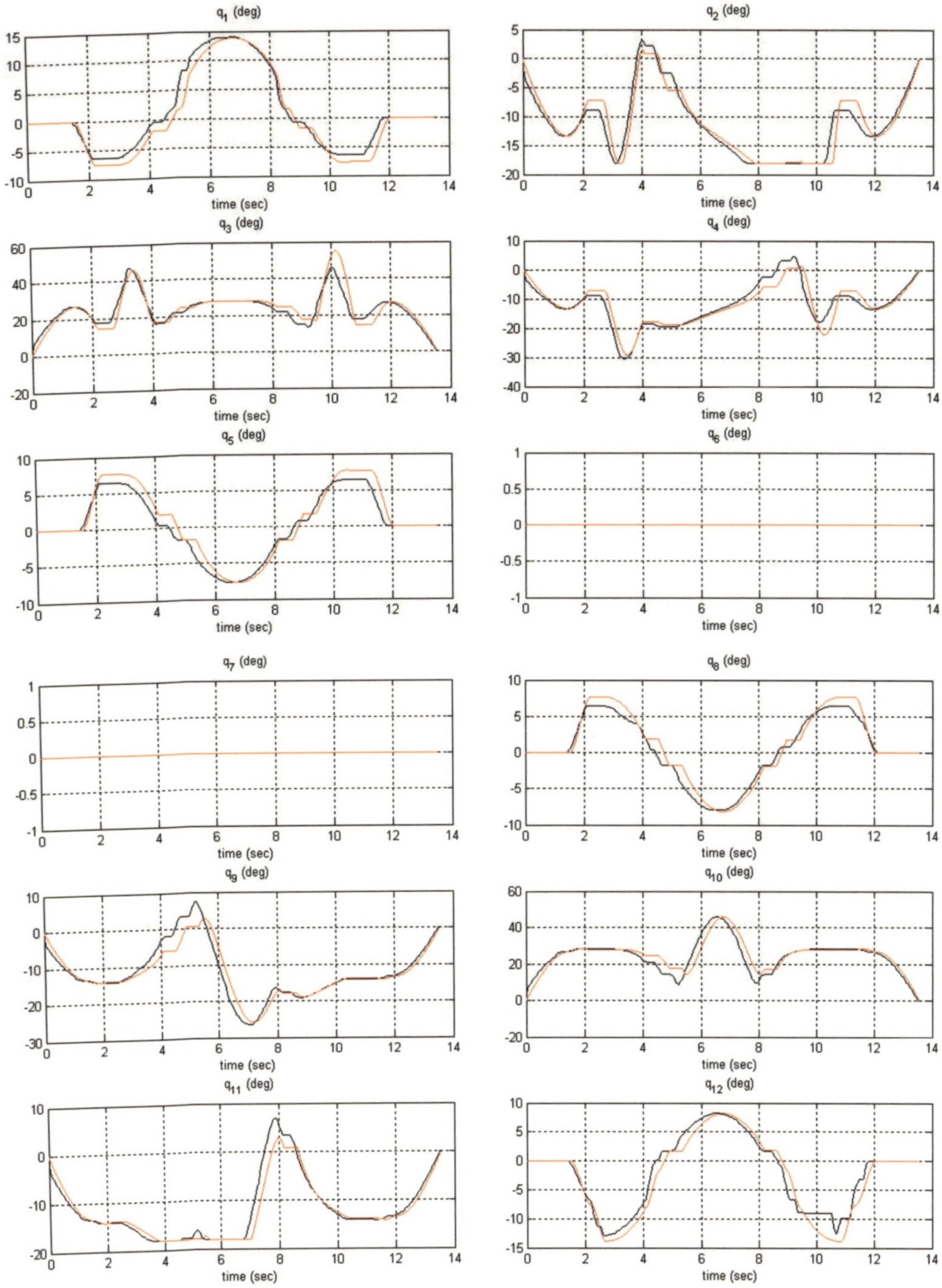

Fig. 52. Both leg joint patterns. a) Reference (blue) and offline corrected (red) overlapped; b) Offline corrected and measured (black) overlapped.

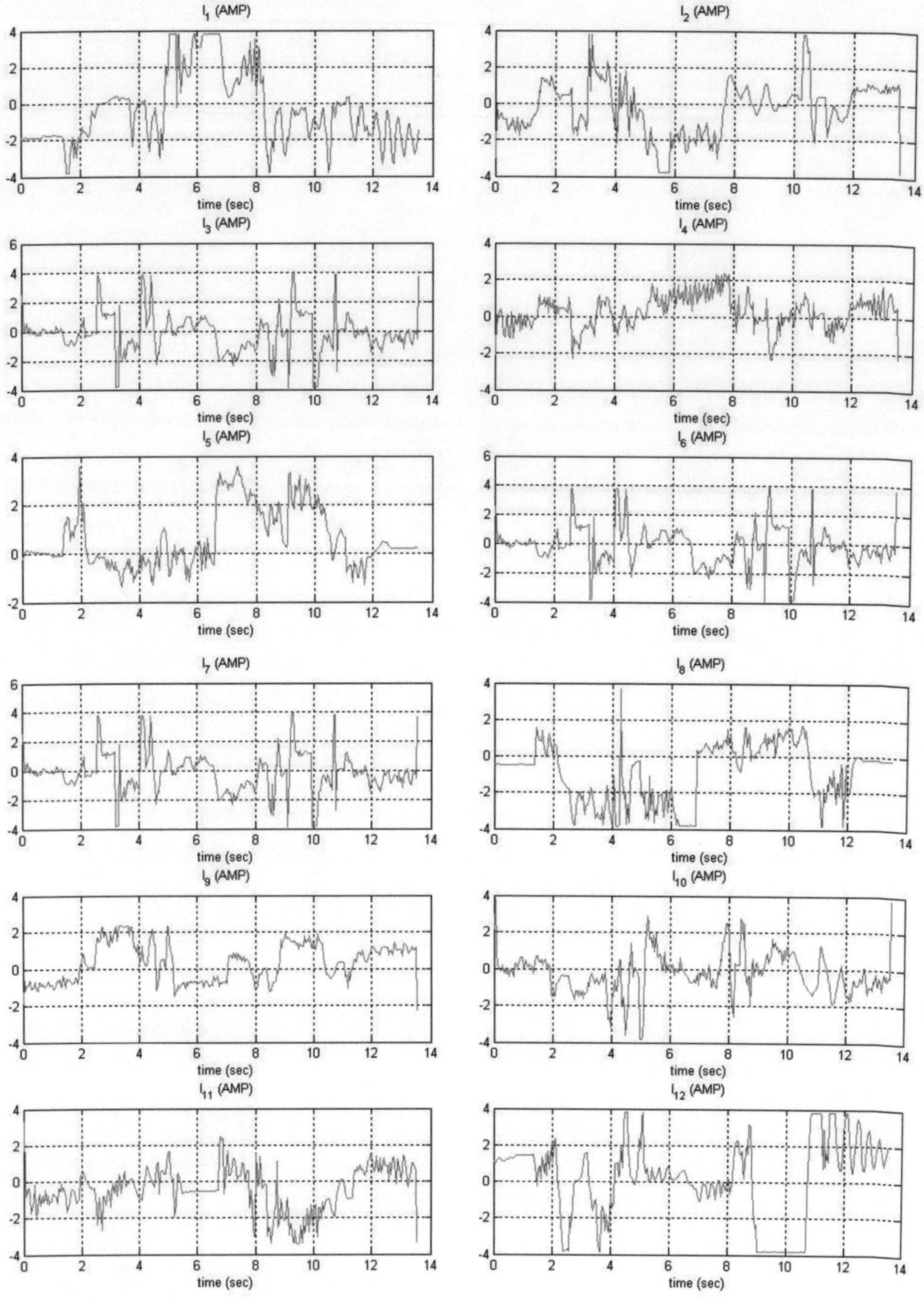

Fig. 53. Both legs' current consumption.

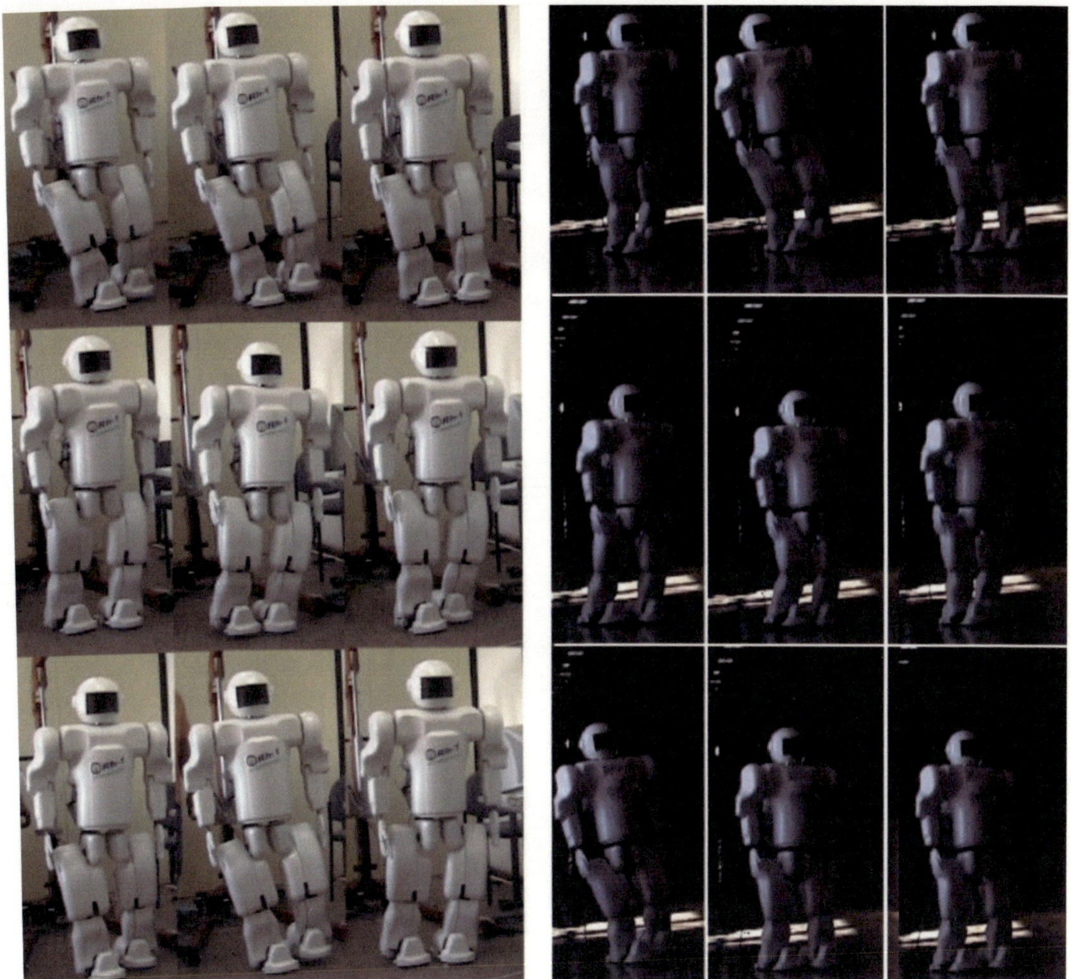

Fig. 54. Rh-1 snapshots walking forward.

9. Conclusions

Normal bipedal gait is achieved through a complex combination of automatic and volitional postural components. Normal walking requires stability to provide antigravity support of body weight, mobility of body segments and motor control to sequence multiple segments while transferring body weight from one limb to another. The result is energy-efficient forward progression. The human "gait cycle" has been analyzed in order to understand biped walking motion in its main phases, single support and double support phases and their properties: force reaction, cycle time, foot, knee, hip and body motion trajectories. In this way humanoid robot trajectories can be created on the order of human ones. It is demonstrated that the COG human motion follows the inverted pendulum laws at normal walking velocity, which is an important fact for maintaining stability while walking.

Concerning the facts previously explained, it is possible to state that very satisfactory results were obtained, thus being a starting point for innumerable investigations in the future.

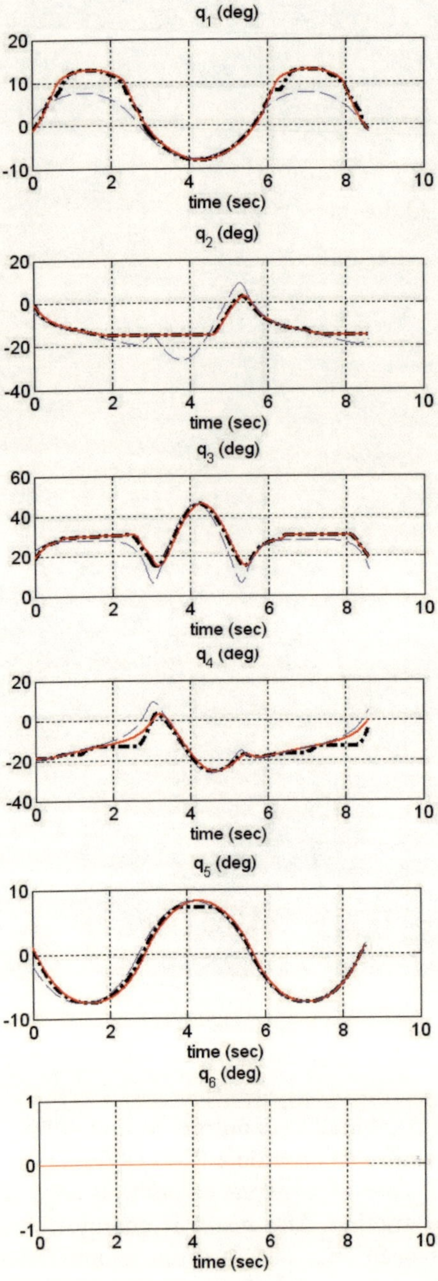

Fig. 55. Real joint angular evolution (dot dashed black line), offline compensated (continuous red line) and reference (blue dashed line). Right leg.

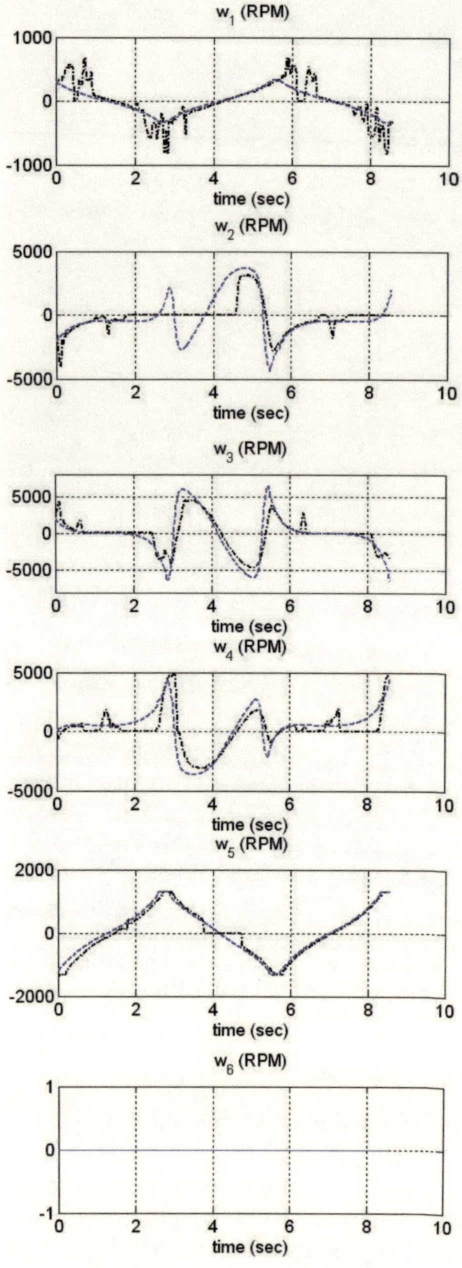

Fig. 56. Real (dot dashed black line) and reference (blue dashed line) motor angular velocity evolution. Right leg.

At the moment, many improvements and corrections are to be done to the mechanical parts. Due to the great amount of elements working together, some unwanted clearances and movements in the mechanical structure of the robot may appear. Furthermore, the robot is in its second evaluation stage and the number of elements that make up the robot must be decreased, either by redesigning the most critical ones or by fusing several of them into one. Compliance foot improvements will be implemented in order to reduce the efforts on each joint and overall structure.

Considering the hardware and software architecture of the Rh-1 robot, we must point out that this work makes an effort to show that there is a possibility of bringing some basic aspects of industrial automation and control to the other, more sophisticated fields of robotics, in order to extend further standardization and unification of the design processes. Moreover, the proposed approach allows for consideration of humanoid robot locomotion inside the global automation problem.

Dynamic walking was successfully implemented in the Rh-1 humanoid robot. It can walk smoothly and about twenty times faster than when using a static walking pattern, as was studied in previous works.

The SKD humanoid model makes it easy to solve the inverse kinematics problem using Lie groups math techniques, such as the POE. For bipedal locomotion, 3D-LIPM and Cart-table models of the COG motion can be computed in real time and be dynamically stable. The algorithms introduced have closed-form solutions with clear geometric meaning, and therefore can be useful for developing robust real-time applications.

It was demonstrated that offline compensation of the body orientation contributes to online control, reducing high joint accelerations. As a result, a stable motion was obtained.

The Stabilizer was designed as a decoupled controller. It controls the error in ZMP and Attitude positioning of the humanoid robot by the motion of the ankle and hip joints.

The humanoid robot Rh-1 provided with the developed control architecture is able to walk stably on a plain surface and to absorb some external disturbances. Future work will be focused on adding other elements to the proposed control architecture such as a foot landing control in order to correct for structural and walking surface imperfections, and to reduce the mechanical landing impact on the humanoid structure, which are the essential conditions for achieving stable humanoid robot walking on irregular terrain. Also further improvements on existing mechanical, hardware and software architecture will be continued.

10. References

K. Hirai, M. Hirose, Y. Hikawa and T. Takanaka, The development of Honda humanoid robot, IEEE International Conference on Robotics and Automation ICRA 1998) Leuven (Belgium)

K. Kaneko, F. Kanehiro, S. Kajita, K. Yokoyama, K. Akachi, T. Kawasaki, S. Ota and T. Isozumi, "Design of prototype humanoid robotics platform for HRP", Proc. of IEEE/RSJ Int. Conference on Intelligent Robots and Systems, pp. 2431-2436, 2002.

J.M. Pardos; C.Balaguer, Rh-0 Humanoid Robot Bipedal Locomotion and Navigation Using Lie Groups and Geometric Algorithm. International Conference on Intelligent Robots and Systems (IROS'2005). Edmonton. Canada. Aug, 2005

M. Arbulú, J.M. Pardos, L.M. Cabas, P. Staroverov, D. Kaynov, C. Pérez, M.A. Rodríguez; C. Balaguer, Rh-0 humanoid full size robot`s control strategy based on the Lie logic technique, IEEE-RAS International Conference on Humanoid Robots (Humanoids'2005). Tsukuba. Japan. Dec, 2005

S. Stramigioli, B. Mashke, C. Bidard, On the geometry of rigid body motions: the relation between Lie groups and screws, Journal of Mechanical Engineering Science, Vol. 216, n. C1, pp 13-23, 2002.

M. Arbulú, F. Prieto, L. Cabas, P. Staroverov, D. Kaynov, C. Balaguer, ZMP Human Measure System. 8th International Conference on Climbing and Walking Robots (Clawar'2005). London. United Kingdom. Sep, 2005

J. Yamaguchi, E. Soga, S. Inoue A. and Takanishi, Development of a bipedal humanoid robot control method of whole body cooperative dynamic bipedal walking, IEEE International Conference on Robotics and Automation (ICRA' 1999), Detroit, (USA)

S. Kajita, F. Kaneiro, K. Kaneko, K. Fujiwara, K. Yokoi. and H. Hirukawa, Biped walking pattern generation by a simple 3D inverted pendulum model, Autonomous Robots, vol 17, nª2, 2003

M.H. Raibert, Legged robots that balance, MIT Press:Cambridge, 1986

M. Arbulú; L.M. Cabas; P. Staroverov; D. Kaynov; C. Pérez; C. Balaguer. On-line walking patterns generation for Rh-1 Humanoid Robot using a simple three-dimensional inverted pendulum model. 9th Internacional Conference on Climbing and Walking Robots (Clawar 2006). Brussels. Belgium. Sep, 2006.

C.L. Shin, Y.Z.`Li, S.Churng, T.T. Lee and W.A. Cruver. Trajectory Synthesis and Physical Admissibility for a Biped Robot During the Single-Support Phase, Proc. of IEEE International Conference on Robotics and Automation, pp. 1646-1652, 1990

M. Vukobratovic, D. Juricic. Contribution to the Synthesis of Biped Gait. IEEE Tran. On Bio-Medical Engineering, Vol. 16, No. l, pp. 1-6, 1969

J. Furusho and A. Sano, Sensor-Based Control of a Nine-Link Biped, Int. J. on Robotics Research, Vol.. 9, No. 2, pp. 83-98, 1990

Y. Fujimoto, S. Obata and A. Kawamura. Robust Biped Walking with Active Interaction Control between Foot and Ground, Proc. of IEEE International Conference on Robotics and Automation, pp. 2030-2035, 1998

J. H. Park and H. C. Cho. An On-line Trajectory Modifier for the Base Link of Biped Robots to Enhance Locomotion Stability, Proc. of the IEEE ICRA2000, pp. 3353-3358, 2000.

Q. Huang; K. Kaneko; K. Yokoi; S. Kajita; T. Kotoku; N. Koyachi; H. Arai; N. Imamura; K. Komoriya; K. Tanie. Balance Control of a Biped Robot Combining Off-line Pattern with Real-time Modification, Proc. of IEEE International Conference on Robotics and Automation, 2000.

L. Cabas, S. de Torre, I. Prieto, M. Arbulu, C. Balaguer, Development of the lightweight human size humanoid robot RH-0. CLAWAR 2004, Madrid September 2004.

L.M. Cabas; R. Cabas; P. Staroverov; M. Arbulú; D. Kaynov; C. Pérez; C. Balaguer. Challenges in the design of the humanoid robot RH-1. 9th Internacional Conference on Climbing and Walking Robots (Clawar 2006). Brussels. Belgium. Sep, 2006.

A. Bicchi, G. Tonietti, and R. Schiavi. Safe and Fast Actuators for Machines Interacting with Humans. In Proc. of the 1st Technical Exhibition Based Conference on Robotics and Automation, TExCRA2004, November 18-19, TEPIA, Tokyo, Japan, 2004.

L.M. Cabas; R. Cabas; P. Staroverov; M. Arbulú; D. Kaynov; C. Pérez; C. Balaguer. Mechanical Calculations on a Humanoid Robot. 9th Internacional Conference on Climbing and Walking Robots (Clawar 2006). Brussels. Belgium. Sep, 2006.

K. Regenstein and Rudiger Dillmann, Design of an open hardware architecture for the humanoid robot ARMAR, Proc. of IEEE Int. Conference on Humanoid Robots, 2003.

D. Kaynov; M.A. Rodríguez; M. Arbulú; P. Staroverov; L.M. Cabas; C. Balaguer. Advanced motion control system for the humanoid robot Rh-0. 8th International Conference on Climbing and Walking Robots (Clawar 2005), 2005.

D. Kaynov, C.Balaguer. Industrial automation based approach to design control system of the humanoid robot. Application to the Rh-1 humanoid robot. Accepted for IEEE International Symposium on Industrial Electronics (ISIE2007)

E. Yoshida, I. Belousov, C. Esteves and J. P. Laumond. Humanoid Motion Planning for Dynamic Tasks, Proceedings of IEEE-RAS International Conference on Humanoid Robots (Humanoids 2005), pp. 1-6, 2005.

Löffler, M. Giender and F. Pfeifer. Sensors and Control Design of a Dynamically Stable Biped Robot, Proc. of IEEE Int. Conference on Robotics and Automation, pp. 484-490, 2003

M. Gienger, K. Löffler, and F. Pfeifer, "Towards the design of biped jogging robot", Proc. of IEEE Int. Conference on Robotics and Automation, pp. 4140-4145, 2001.

A.-J. Baerveldt, R. Klang. A low cost and Low-weight Attitude Estimation System for an Autonomous Helicopter. Proc. of IEEE International Conference on Intelligent Engineering Systems, pp. 391-391, 1997.

H. Hirukawa, S. Hattori, S. Kajita, K. Harada, K. Kaneko, F. Kanehiro, M. Morisawa, and S. Nakaoka, A pattern generator of humanoid robots walking on a rough terrain, in IEEE International Conference on Robotics and Automation, Roma and Italy, April 10-14 2007, pp. 2781- 2187.

S. Kajita, F. Kanehiro, K. Kaneko, K. Fujiwara, K. Harada, K. Yokoi, and H. Hirukawa, Biped walking pattern generation by using preview control of zero-moment point, in IEEE International Conference on Robotics Automation, Taipei and Taiwan, September 14-19 2003, pp. 162-1626.

M. Arbulu and C. Balaguer, Real-time gait planning for Rh-1 humanoid robot, using local axis gait algorithm, in 7th IEEE-RAS International Conference on Humanoid Robots, Pittsburgh, USA, Nov. 29-Dec. 2 2007.

F.C. Park, J.E. Bobrow, and S.R. Ploen, "A Lie group formulation of robot dynamics," Int. J. Robotics Research. Vol. 14, No. 6, pp. 609-618, 1995.

R.A. Abraham, and J.E. Marsden, Foundations of Mechanics. Perseus Publishing, 1999.

B. Paden. Kinematics and Control Robot Manipulators. PhD thesis, Department of Electrical Engineering and Computer Sciences, University of California, Berkeley, 1986.

S. Torre; L.M. Cabas; M. Arbulú; C. Balaguer. Inverse Dynamics of Humanoid Robot by Balanced Mass Distribution Method. IEEE/RSJ International Conference on Intelligent Robots and Systems (IROS'2004). Sendai. Japan. Sep, 2004.

M. Arbulu and C. Balaguer, Real-time gait planning for Rh-1 humanoid robot, using local axis gait algorithm, in International Journal of Humanoid Robotics. Print ISSN: 0219-8436. Online ISSN: 1793-6942. Vol. 6. No. 1. pp.71-91. 2009

R. M. Murray, Z. Li, and S. S. Sastry. Mathematical Introduction To Robotic Manipulation. CRC Press, 1994.

E. Ayyappa. Normal human locomotion, part 1: Basic concepts and terminology. Journal of Prosthetics and Orthotics, pages 10–17, 1997.

D. A.Winter. Biomechanics And Motor Control of Human Movement. A Wiley-Interscience Publication, 1990.

K. Loeffler, M. Gienger, F. Pfeiffer, and H. Ulbrich. Sensors and control concept of a biped robot. IEEE TRANSACTIONS ON INDUSTRIAL ELECTRONICS, 51:1–9, 2004.

S. Lohmeier, T. Buschmann, H. Ulbrich, F. Pfeiffer: Modular joint design for performance enhanced humanoid robot LOLA. In: Proc. IEEE Int. Conf. Rob. Aut. (ICRA), pp. 88–93 (2006)

K. Kaneko, K. Harada, F. Kanehiro, G. Miyamori, K. Akachi, Humanoid Robot HRP-3, IEEE-RAS International Conference on Humanoid Robots (Humanoids'2008). Nice. France.